Bioactive Compounds in Fermented Foods

Health Aspects

Books Published in *Food Biology* series

Food Biology Series

Bioactive Compounds in Fermented Foods

Health Aspects

Editors

Amit Kumar Rai

Scientist C, Institute of Bioresources and Sustainable
Development, Regional Centre, Sikkim, India

Anu Appaiah K. A.

Senior Principal Scientist (Rtd.), Central Food Technological
Research Institute, Mysore, Karnataka, India

CRC Press
Taylor & Francis Group
Boca Raton   London   New York

CRC Press is an imprint of the
Taylor & Francis Group, an **informa** business

A SCIENCE PUBLISHERS BOOK

First edition published 2022
by CRC Press
6000 Broken Sound Parkway NW, Suite 300, Boca Raton, FL 33487-2742

and by CRC Press
2 Park Square, Milton Park, Abingdon, Oxon, OX14 4RN

© 2022 Taylor & Francis Group, LLC

CRC Press is an imprint of Taylor & Francis Group, LLC

Library of Congress Cataloging-in-Publication Data

NNames: Rai, Amit Kumar, 1984- editor. | Appaiah, Anu, 1961- editor.
Title: Bioactive compounds in fermented foods : health aspects / editors,
  Amit Kumar Rai, Anu Appaiah.
Other titles: Food biology series.
Description: First edition. | Boca Raton : CRC Press, Taylor & Francis
  Group, 2021. | Series: Food biology series | "A science publishers
  book." | Includes bibliographical references and index. | Summary: "The
  book entitled "Bioactive compounds in fermented foods: health aspects"
  discusses the recent advances in health benefits of the different types
  of bioactive components produced during food fermentation. In recent
  years a lot of attention has been given on characterization of bioactive
  compounds and their impact on human health to develop functional foods against different
  metabolic disorders. The bioactive compounds are
  produced either during biotransformation of food metabolites or directly
  produced by microorganisms. The book discusses different types of
  bioactive compounds and concludes with the impact of gut fermentation on
  production of bioactive compounds with specific human health"-- Provided
  by publisher.
Identifiers: LCCN 2021017086 | ISBN 9780367136000 (hbk) | ISBN
  9781032025254 (pbk) | ISBN 9780429027413 (ebk)
Subjects: MESH: Fermented Foods and Beverages--microbiology | Fermented
  Foods and Beverages--analysis | Functional Food--microbiology |
  Phytochemicals
Classification: LCC TP371.44 | NLM QW 85 | DDC 664/.024--dc23
LC record available at https://lccn.loc.gov/2021017086

ISBN: 978-0-367-13600-0 (hbk)
ISBN: 978-1-032-02525-4 (pbk)
ISBN: 978-0-429-02741-3 (ebk)

DOI: 10.1201/9780429027413

Typeset in Times New Roman
by Innovative Processors

Preface to the Series

Food is the essential source of nutrients (such as carbohydrates, proteins, fats, vitamins, and minerals) for all living organisms to sustain life. A large part of daily human efforts is concentrated on food production, processing, packaging and marketing, product development, preservation, storage, and ensuring food safety and quality. It is obvious therefore, our food supply chain can contain microorganisms that interact with the food, thereby interfering in the ecology of food substrates. The microbe-food interaction can be mostly beneficial (as in the case of many fermented foods such as cheese, butter, sausage, etc.) or in some cases, it is detrimental (spoilage of food, mycotoxin, etc.). The *Food Biology* series aims at bringing all these aspects of microbe-food interactions in form of topical volumes, covering food microbiology, food mycology, biochemistry, microbial ecology, food biotechnology and bio-processing, new food product developments with microbial interventions, food nutrification with nutraceuticals, food authenticity, food origin traceability, and food science and technology. Special emphasis is laid on new molecular techniques relevant to food biology research or to monitoring and assessing food safety and quality, multiple hurdle food preservation techniques, as well as new interventions in biotechnological applications in food processing and development.

The series is broadly broken up into food fermentation, food safety and hygiene, food authenticity and traceability, microbial interventions in food bio-processing and food additive development, sensory science, molecular diagnostic methods in detecting food borne pathogens and food policy, etc. Leading international authorities with background in academia, research, industry and government have been drawn into the series either as authors or as editors. The series will be a useful reference resource base in food microbiology, biochemistry, biotechnology, food science and technology for researchers, teachers, students and food science and technology practitioners.

Ramesh C Ray
Series Editor

Preface

The book entitled "Bioactive compounds in fermented foods: Health aspects" consists of 14 chapters reviewing different types of bioactive components produced in fermented foods and their impact on human health. This book is a part of food biology series covering different aspects of advancement of food research (Series Editor – Ramesh C. Ray). The diversity of microorganism responsible for the production of different types of fermented foods and beverages includes Lactic acid bacteria, *Bacillus* species, yeasts, and filamentous fungi. Biotransformation of food constituent by microorganisms occurs during fermentation processes for the production of fermented food and in the gastrointestinal tract by gut microorganisms. These biotransformation results in production of specific bioactive compounds those are responsible for a wide range of health benefits. The bioactive compound discussed in this book includes bioactive peptides, polyphenols, fibrinolytic enzymes, exopolysaccharides, prebiotic, symbiotic and antinutritional factors. These bioactive compounds are responsible for health benefits such as antioxidant, antihypertension, antimicrobial, cholesterol lowering, anticancer, obesity and antithrombotic properties. The book has chapters focusing on different types of fermented foods (i) Fermented soybean products, (ii) Fermented milk products, (iii) Fermented meat and fish products, and (iv) Fermented alcoholic beverages. Advanced research in the field of food fermentation and their health benefits have resulted in commercialization of some of the fermented foods as functional foods. The traditional fermented foods consumed in different parts of the world and their health benefits are presented in detail. The book concludes with recent advances in microbial transformation during gut fermentation and their impact on human health. There has been increasing interest among researchers on the proposed title in last one decade and the book brings the updated information on research and advances in different types of health benefits exhibited by bioactive compounds in wide range of fermented foods.

We highly appreciate the sincere contribution of authors from different part of the globe in presenting the comprehensive scientific and relevant state-of-art information in different aspects of bioactive compounds in fermented foods. We strongly believe that the updated scientific content of the book will be useful to diverse sections of scientific and academic communities. We sincerely acknowledge the efforts of reviewers for their timely efforts in reviewing the chapters of the book, which led to scientific improvement. We thank CRC Press team comprising Mr. Vijay Primlani, Acquisitions Editor and the entire production team for their support in rapid publication of this book. We would like to thank Prof. Ramesh C. Ray, the series editor of the 'Food Biology' book series for giving us the opportunity to edit this book and guide us throughout the publication process for completing this assignment.

Amit Kumar Rai
Anu Appaiah K. A.

Contents

Part III: Traditional Fermented Products and Health Benefits

Part IV: Recent Advances in Food Fermentation

Part I
Introduction

Microorganisms Associated with Food Fermentation

Spiros Paramithiotis

Laboratory of Food Quality Control and Hygiene, Department of Food Science and Human Nutrition, Agricultural University of Athens, Iera Odos-75, 11855 Athens, Greece

1. Introduction

Food fermentation may be defined as a microbe-driven bio-transformation of substrates adopted for human consumption into another form with modified organoleptic properties, enhanced shelf-life and, in most of the cases, nutritional value. These characteristics set the borderline between fermentation and spoilage; the latter includes non-desirable modification of sensorial properties and possibly toxic metabolic end products. Fermentation has been applied for centuries as a way to preserve excess foodstuff (e.g. lactic acid fermentation) or as a step in the production procedure of a variety of products (e.g. acetic acid fermentation for vinegar production, alcoholic fermentation for beverage production, sourdough for rye breadmaking, etc).

There are three parameters that direct the onset and evolution of the microcommunity that drives spontaneous fermentation: the type and microbiological quality of the raw materials, the selective agents added or released during processing and the incubation temperature. Proper control of these parameters is a prerequisite for successful fermentation. Depending on the main metabolic end product, as a consequence of the microbiota that drives fermentation, food fermentation may be divided into: (i) acid fermentation, the main metabolic end product is an organic acid, which is in the vast majority of cases lactic, acetic or propionic acid, the accumulation of which results in the drop of pH value and accumulation of acidity; (ii) alkaline fermentation, in which the main metabolic end product is ammonia, the accumulation of which results in the increase of pH value; and (iii) alcoholic fermentation, in which the main metabolic end product is ethanol. In addition, another type of fermentation exists but may not fit into the above categories – it is driven by moulds and therefore termed mold fermentation.

Email: sdp@aua.gr

The aim of the present chapter is to provide an overview of the microbiota associated with each type of fermentation as well as the biotic and abiotic factors that ensure the presence and dominance of specific microorganisms.

2. Acid Fermentation

2.1 Lactic Acid Fermentation

Lactic acid fermentation is principally driven by lactic acid bacteria (LAB). Lactic acid is produced through homofermentative or heterofermentative catabolism (Fig. 1). Regarding the first, catabolism of carbohydrates to pyruvate takes place through the Embden-Meyerhof-Parnas pathway with pyruvate reduced to lactate through simultaneous NAD+ regeneration. Heterofermentative catabolism takes place through the pentose phosphoketolase pathway leading to the production of lactate, acetate and ethanol. The presence of molecules that may be used as electron acceptors may favour the production of one metabolite over another. A wide variety of substrates, including vegetables, fruits, cereals, meat and milk have been exploited

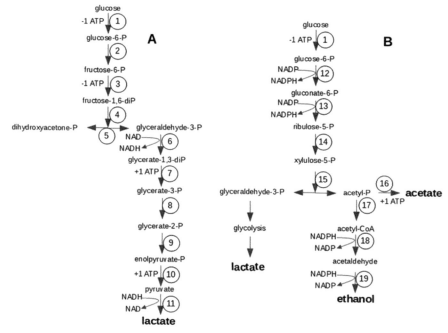

Fig. 1: Glucose catabolism through the homofermentative (A) and heterofermentative (B) pathways. The reactions are catalysed by the enzymes: 1: glucokinase; 2: glucose-phosphate isomerase; 3: phosphofructokinase; 4: fructose-bisphosphate aldolase; 5: triose-phosphate isomerase; 6: glyceraldehyde-phosphate dehydrogenase; 7: phosphoglycerate kinase; 8: phosphoglycerate mutase; 9: enolase; 10: pyruvate kinase; 11: lactate dehydrogenase; 12: glycose-6-P dehydrogenase; 13: gluconate-6-P dehydrogenase; 14: ribulose-5-P-3-epimerase; 15: D-xylulose-5-P phosphoketolase; 16: acetate kinase; 17: phosphotransacetylase; 18: acetaldehyde dehydrogenase; and 19: alcohol dehydrogenase

through lactic acid fermentation, resulting in a wealth of products. Research was primarily focused on documentation of the production procedures and the ecological factors that direct the development of the microecosystem that dominates these fermentations. In addition, the properties that affect the safety, the nutritional value as well as the physicochemical and sensorial qualities of the final products, attracted specific interest. Soon, it was realised that LAB through their metabolic activities may produce bioactive compounds and, therefore, improve the functionality of these products, exerting, thus, health benefits to the consumers. The latter is currently the epicentre of intensive research with quite promising results. Lactic acid fermentation of fruits and vegetables, meat and milk is the most widely studied and therefore will be considered in this chapter.

2.1.1 *Lactic Acid Fermentation of Fruits and Vegetables*

An extended range of fruits and vegetables, according to geographical location as well as seasonal availability, have been used as substrates for lactic acid fermentation. However, only four of them have been worldwide recognized and commercially successful, namely fermented olives, fermented cucumbers, sauerkraut and kimchi.

In general, the production procedure usually involves four steps:

1. *Collection and primary processing:* Depending on the final product, the respective raw materials are collected and accordingly processed, e.g. for sauerkraut production, cabbage is washed and chopped. Other processing steps may include blanching, peeling, cooking or more specific steps, such as lye treatment that is required for olive fermentation.
2. *Salting:* Salt can be applied in a dry or a brine form. There is also a group of products that lactic acid fermentation is initiated without the need of a salting step (Gashe, 1987; Tamang *et al.*, 2005). In dry-salting method, salt is applied directly on the chopped vegetables and brine is formed through the extraction of water and nutrients. In brine-salting method, an initial brine of 5-20 per cent, depending on the raw materials, is prepared and the respective fruits and vegetables are submerged. Extraction of water and nutrients takes place and finally equilibration is achieved after one day.
3. *Fermentation:* It usually takes place at ambient temperature. The micro-ecosystem development is accompanied by a series of physicochemical changes, according to the raw materials used.
4. *Secondary treatment:* This step includes any treatment that may extend the shelf-life and assist commercialisation of the final product, such as drying, pasteurisation, boiling, etc.

The microbiota that resides on the surface of intact fruits and vegetables consists mainly of members of the Enterobacteriaceae family and yeasts, though LAB are occasionally enumerated. Indeed, total aerobic mesophilic, Enterobacteriaceae and yeast/mold counts of cauliflower, green tomatoes and asparagus surface was reported to range between 3.5-5.7, 3.5-5.17 and 3.2-5.39 log CFU/g, respectively (Paramithiotis *et al.*, 2010a, 2014a, b). LAB were only enumerated in the latter two cases, ranging between 3.14-5.32 log CFU/g. Dominance of LAB in the microecosystem that is developed through time is ensured by salting, incubation temperature and the microbiological quality of the raw materials. Effective control of the above parameters will result in enhancement of LAB growth during the first hours of fermentation, which is crucial due to concomitant acidity accumulation. The latter will prohibit antagonistic microbiota from developing in large numbers.

In Table 1, the lactic acid microbiota involved in representative lactic acid fermented products is shown. Certain LAB species are characterised by certain properties that justify their existence in fruit and vegetable fermentation. More specifically, *Leuconostoc mesenteroides* is known for a relatively shorter generation time, the ability to tolerate elevated NaCl concentration but sensitivity to acidity. These properties justify the frequent occurrence of this species in the early stages of fermentation. *Ln. gasicomitatum* has exhibited greater acid tolerance than *Ln. mesenteroides* and *Ln. citreum* and therefore its presence in later stages of fermentation may be expected. *Lactobacillus plantarum* is characterised by tolerance to acidic conditions and a large metabolic capacity, enabling it to utilise a wide range of carbon sources. Therefore, this species very often dominate the final stages of fermentation. *Ln. gelidum* has been reported as capable of growing at temperatures as low as 10°C and *Weissella koreensis* at even lower. Therefore, these species may dominate fermentations taking place at such low temperatures. Finally, *Lb. sakei, Lb. brevis, Pediococcus pentosaceus* and *Weissella* spp. can tolerate acidic conditions and thus their occurrence as part of a secondary microbiota is very often reported (Stamer *et al.*, 1971; Daeschel *et al.*, 1987; Cho *et al.*, 2006; Jung *et al.*, 2013; Jeong *et al.*, 2013a).

Based on the above, the species succession that takes place very often in spontaneous fermentations may be explained. The most known example of such a succession is the one occurring during sauerkraut fermentation, in which *Ln. mesenteroides* initiates fermentation and as acidity accumulates, it is gradually replaced by *Lb. plantarum*. Similarly, cauliflower fermentation is also characterised by the same species succession (Paramithiotis *et al.*, 2010a). However, there are cases that may not be explained by the above-mentioned characteristics. This is the case of spontaneous fermentation of unripe tomatoes (Paramithiotis *et al.*, 2014a). In that fermentation, *Ln. mesenteroides* initiated fermentation, despite the very low pH value (3.8-4.8), which indicated the complexity of trophic relationships between abiotic and biotic factors that drive spontaneous fermentations and that many aspects of the physiology of lactic acid bacteria are yet to be discovered.

2.1.2 *Spontaneous Sausage Fermentation*

In general, production of spontaneously fermented sausages includes mixing of minced meat with fatty tissue, curing agents and spices, stuffing into casings, fermentation and ripening. The latter may take place in one, two or three steps. The three-step procedure consists of a drying stage at 24°C and 80 per cent relative humidity (RH) for 24 hours, that is followed by a fermentation step at 18-21°C and 75-89 per cent RH for five days and 15-20°C and 80-85 per cent RH for two months. Such a procedure is followed for the production of salciccia sotto sugna (Gardini *et al.*, 2001; Parente *et al.*, 2001). The second step may take place in dynamic conditions, such as in the case of dry fermented sausages that are produced

Table 1: Lactic Acid Bacteria Involved in Representative Spontaneously Fermented Fruits and Vegetables

Product	Lactic Acid Bacteria	References
Fermented olives	*E. casseliflavus, E. faecium, Lb. brevis, Lb. casei, Lb. paracasei, Lb. parafarraginis, Lb. pentosus, Lb. plantarum, Lb. sanfranciscensis, Ln. mesenteroides*	Hurtado *et al.* (2012), Benítez-Cabello *et al.* (2016)
Sauerkraut	*E. faecalis, Lb. brevis, Lb. confusus, Lb. curvatus, Lb. plantarum, Lb. sakei, Lc. lactis subsp. lactis, Ln. fallax, Ln. mesenteroides, Pd. pentosaceus*	Harris (1998), Barrangou *et al.* (2002), Beganovic *et al.* (2014)
Kimchi	*Lb. brevis, Lb. curvatus, Lb. parabrevis, Lb. pentosus, Lb. plantarum, Lb. sakei, Lb. spicheri, Lc. gelidum, Ln. carnosum, Ln. citreum, Ln. gasicomitatum, Ln. gelidum, Ln. holzapfelii, Ln. inhae, Ln. kimchii, Ln. lactis, Ln. mesenteroides, W. koreensis, W. cibaria, W. confusa, W. kandleri, W. soli*	Lee *et al.* (2005), Cho *et al.* (2006), Chang *et al.* (2008), Jung *et al.* (2013), Jeong *et al.* (2013a, b), Yeun *et al.* (2013), Chang, 2018
Fermented cucumber	*Lb. brevis, Lb. plantarum, Lactococcus* spp., *Leuconostoc* spp., *Pd. ethanolidurans, Pd. pentosaceus, Weissella* spp.	Fleming *et al.* (1995), Perez-Diaz *et al.*, 2017
Mustard	*E. durans, Lb. alimentarius, Lb. brevis, Lb. casei, Lb. coryniformis, Lb. farciminis, Lb. fermentum, Lb. pentosus, Lb. plantarum, Lb. versmoldensis, Ln. citreum, Ln. mesenteroides, Ln. pseudomesenteroides, Pd. pentosaceus, W. cibaria, W. paramesenteroides*	Tamang *et al.* (2005), Chen *et al.* (2006), Chao *et al.* (2009), Nguyen *et al.* (2013), Goswami *et al.* (2017)
Leek	*C. maltaromaticum, Lb. brevis, Lb. crustorum, Lb. hammessii/parabrevis, Lb. nodensis, Lb. parabrevis, Lb. plantarum, Lb. sakei, Lb. sakei/curvatus, Lc. lactis, Lc. raffinolactis, Ln. gasicomitatum, Ln. gelidum, Ln. kimchii, Ln. lactis/citreum, Ln. lactis/garlicum, Ln. mesenteroides, W. soli*	Wouters *et al.* (2013a)
Cauliflower	*E. faecalis* sensu lato, *E. faecium* sensu lato, *Lb. brevis, Lb. plantarum* sensu lato, *Lb. sakei/curvatus, Ln. mesenteroides* sensu lato, *Pd. pentosaceus, W. kimchii, W. viridescens*	Paramithiotis *et al.* (2010a), Wouters *et al.* (2013b)
Green tomatoes	*Lb. casei, Lb. curvatus, Ln. citreum, Ln. mesenteroides*	Paramithiotis *et al.* (2014a)
Asparagus	*E. faecium* sensu lato, *Lb. sakei, W. cibaria, W. viridescens*	Paramithiotis *et al.* (2014b)
Radish roots	*Lb. brevis, Lb. plantarum, Pd. pentosaceus*	Pardali *et al.* (2017)

(Contd.)

Table 1: *(Contd.)*

Product	Lactic Acid Bacteria	References
Shalgam	*Lb. brevis, Lb. fermentum, Lb. paracasei* subsp. *paracasei, Lb. plantarum, Ln. mesenteroides* subsp. *mesenteroides, Pd. pentosaceus, Lb. delbrueckii* subsp. *delbrueckii,*	Tanguler and Erten (2012)
Gilaburu	*Lb. brevis, Lb. buchneri, Lb. casei, Lb. cornyformis, Lb. harbinensis, Lb. hordei, Lb. pantheris, Lb. parabuchneri, Lb. paraplantarum, Lb. plantarum, Ln. mesenteroides, Ln. pseudomesenteroides*	Sagdic *et al.* (2014)
Hardaliye	*Lb. acetotolerans, Lb. brevis, Lb. casei* subsp. *pseudoplantarum, Lb. paracasei* subsp. *paracasei, Lb. pontis, Lb. sanfranciscensis, Lb. vaccinostercus*	Arici and Coskun (2001)

C.: Carnobacterium; E.: Enterococcus; Lb.: Lactobacillus; Lc.: Lactococcus; Ln.: Leuconostoc; Pd.: Pediococcus; W.: Weissella in Greece. In that case, the fermentation step takes place at a temperature and RH that decrease gradually from 18-24°C and 60-94 per cent RH to 11-19°C and 60-90 per cent RH within five to seven days (Coppola *et al.*, 1998; Metaxopoulos *et al.*, 2001; Papamanoli *et al.*, 2003; Comi *et al.*, 2005; Greco *et al.*, 2005; Drosinos *et al.*, 2005, 2007). The second step may as well be omitted and the whole procedure may be carried out in two steps, as in the case of chorizo, salchichon or the Naples-type salami (Lizaso *et al.*, 1999; Garcia-Varona *et al.*, 2000; Mauriello *et al.*, 2004). Finally, Sobrasada and Salame di Senise have been reported to occur in one step (Rossello *et al.*, 1995; Baruzzi *et al.*, 2006).

Physicochemical and microbiological characteristics of a wide range of spontaneously fermented sausages are currently available in literature. The variability of the raw materials employed as well as fermentation and ripening conditions are reflected in the extended diversity of products with unique organoleptic qualities. The microecosystem of fresh meat reflects the conditions of slaughter and primary processing. Usually it is dominated by *Pseudomonas* spp. and micrococci; members of the Enterobacteriaceae family as well as Gram-positive bacteria, yeasts and moulds constitute the secondary microbiota (von Holy *et al.*, 1992; Gill, 2005; Nychas *et al.*, 2008). Spontaneous meat fermentation is initially dominated by a microbial consortium consisting of lactic acid bacteria, members of the Micrococcaceae family, enterococci and yeasts. As fermentation proceeds, LAB dominate the microecosystem, usually after three to four days in the case of dry sausages and one to two days in the case of semi-dry ones – the latter due to the higher fermentation temperatures employed. The pH value of the final product depends upon the type of raw materials used as well as the fermentation and ripening conditions that may range from 4.4 to 6.5 (Vignolo *et al.*, 2010). The diversity of the recipes, fermentation and ripening conditions is reflected to the diversity of the LAB and staphylococci that contribute to fermentation.

LAB affect the development of the microecosystem through the production of lactic and acetic acids. This acidification, apart from inhibiting the growth of spoilage and pathogenic microorganisms, affects the technological characteristics of the product; it lowers meat water-holding capacity, leading to improvement in cohesiveness and firmness and enhancing colour development. Gram-positive catalase-positive cocci have only a marginal effect on the development of the microecosystem through the antagonism for the available resources. On the other hand, they hold a very important technological role as they contribute to colour stability through their catalase and nitrate reductase activities and flavour development through their lipolytic and proteolytic activities. The latter is also the main contribution of the yeasts that may be present as secondary microbiota.

An extensive variety of LAB and Gram-positive catalase-positive cocci has been isolated from spontaneously-fermented meat products. In Table 2, the bacteria that have been involved in the fermentation of a wide range of products, is presented. Regarding LAB, the most frequently isolated one, in population suggesting dominance over the lactic acid microbiota is *Lb. sakei* followed by *Lb. curvatus* and *Lb. plantarum*. Regarding the first two species, this frequency indicates an adaptation to the respective microenvironment, while the large metabolic capacity of the latter has been already mentioned. In the case of *Lb. sakei*, this has been verified

Table 2: Bacteria Involved in Spontaneous Sausage Fermentation

Origin	Product	Bacteria Species	References
Argentina	Dry fermented sausage	*Lb. plantarum, Lb. sakei, St. saprophyticus*	Fontana *et al.* (2005)
France	Dry fermented sausages	*E. faecalis, E. faecium, Lb. sakei, Lc. garvieae, Ln. mesenteroides, St. carnosus, St. epidermidis, St. equorum, St. saprophyticus, St. sciuri, St. succinus, St. warneri, V. carniphilus*	Ammor *et al.* (2005), Coton *et al.* (2010)
Greece	Dry fermented sausages	*Carnobacterium* spp., *D. nishinomiyaensis, E. faecium, Ko. varians, Lb. alimentarius, Lb. brevis, Lb. buchneri, Lb. casei, Lb. coryniformis, Lb. curvatus, Lb. farciminis, Lb. paracasei, Lb. paraplantarum, Lb. pentosus, Lb. plantarum, Lb. rhamnosus, Lb. sakei, Lb. salivarius, Lc. lactis, Ln. mesenteroides, Ln. pseudomesenteroides, Pediococcus* spp., *St. aureus/intermedius, St. auricularis, St. capitis, St. caprae, St. carnosus, St. cohnii, St. epidermidis, St. equorum, St. gallinarum, St. haemolyticus, St. hominis, St. hyicus, St. lentus, St. saprophyticus, St. sciuri, St. simulans, St. vitulinus, St. warneri, St. xylosus, W. hellenica, W. minor, W. paramesenteroides, W. viridescens*	Papamanoli *et al.* (2002, 2003), Drosinos *et al.* (2005, 2007), Samelis *et al.* (1994, 1998), Rantsiou *et al.* (2005, 2006), Paramithiotis *et al.* (2008)
Italy	Brianza	*Lb. curvatus, Lb. sakei, Pd. pentosaceus, St. xylosus*	Di Cagno *et al.* (2008)
	Ciauscolo	*Carnobacterium* spp., *E. faecalis, Lb. brevis, Lb. curvatus, Lb. johnsonii, Lb. paracasei, Lb. paraplantarum, Lb. plantarum, Lb. sakei, Lc. lactis, Ln. mesenteroides, Pd. acidilactici, Pd. pentosaceus, St. xylosus, W. hellenica*	Aquilanti *et al.* (2007), Federici *et al.* (2014), Silvestri *et al.* (2007)
	Cinta Senese	*Carnobacterium* spp., *Lb. sakei, Leuconostoc* spp., *St. xylosus*	Pini *et al.* (2020)
	Piacentino	*Lb. acidophilus, Lb. antri, Lb. brevis, Lb. coryneformis, Lb. curvatus, Lb. frumenti, Lb. graminis, Lb. helveticus, Lb. oris, Lb. panis, Lb. paracasei, Lb. paralimentarius, Lb. plantarum, Lb. reuteri, Lb. sakei, Lb. vaginalis, Lb. versmoldensis, Lb zeae, St. arlettae, St. auricularis, St. carnosus, St. caseolyticus, St. cohnii, St. epidermidis, St. equorum, St. gallinarum, St. kloosii, St. saprophyticus, St. sciuri, St. succinus, St. xylosus*	Di Cagno *et al.* (2008), Polka *et al.* (2015)

		Microorganisms	References
	Salami (Torino)	*C. divergens, Lb. curvatus, Lb. sakei, Ln. carnosum, Ln. mesenteroides, St. cohnii, St. equorum, St. saprophyticus, St. succinus, St. xylosus*	Greppi *et al.* (2015)
	Salami (Friuli Venezia Giulia region)	*E. pseudoavium, Lc. garvieae, Lc. lactis, Lb. brevis, Lb. casei, Lb. curvatus, Lb. paraplantarum, Lb. plantarum, Lb. sakei, Ln. carnosum, Ln. citreum, Ln. mesenteroides, W. hellenica, W. paramesenteroides, St. cohnii, St. epidermidis, St. intermedius, St. saprophyticus, St. simulans, St. warneri, St. xylosus*	Comi *et al.* (2005), Urso *et al.* (2006), Iacumin *et al.* (2006)
	Salame Milano	*Lb. plantarum, Lb. sakei, St. sciuri, St. xylosus*	Rebecchi *et al.* (1998)
	Salame Mantovano	*Ko. salsicia, Ko. varians, Lb. curvatus, Lb. fermentum, Lb. paracasei, Lb. plantarum, Lb. sakei, Lb. salivarius, St. epidermidis, St. equorum, St. pasteuri, St. saprophyticus, St. simulans, St. xylosus*	Pisacane *et al.* (2015)
	Soppressata molisana	*Lb. brevis, Lb. corynyformis, Lb. curvatus, Lb. graminis, Lb. paracasei, Lb. paralimentarius, Lb. plantarum, Lb. viridescens, M. kristinae, M. roseus, M. varians, St. equorum, St. kloosii, St. simulans, St. xylosus*	Coppola *et al.* (1997, 1998)
	Salami (Basilicata)	*Lb. curvatus, Lb. plantarum, Ln. carnosum, Ln. gelidum, Ln. pseudomesenteroides, St. epidermidis, St. equorum, St. intermedius, St. lentus, St. pasteuri, St. pulvereri, St. saprophyticus, St. succinus, St. vitulus, St. warneri*	Parente *et al.* (2001), Blaiotta *et al.* (2004), Bonomo *et al.* (2008, 2009)
	Varzi	*Lb. curvatus, Lb. sakei, St. xylosus*	Di Cagno *et al.* (2008)
Portugal	Alheira	*E. gallinarum, Lb. curvatus, Lb. brevis, Lb. sakei, Ln. lactis, Ln. mesenteroides, Pd. acidilactici, W. paramesenteroides*	Albano *et al.* (2008)
Spain	Androlla	*Ko. kristinae, Ko. varians, Lb. alimentarius, Lb. curvatus, Lb. plantarum, Lb. sakei, M. luteus, M. lylae, St. capitis, St. epidermidis, St. equorum, St. saprophyticus, St. xylosus*	Garcia Fontan *et al.* (2007a)
	Botillo	*Lb. alimentarius, Lb. curvatus, Lb. farciminis, Lb. plantarum, Ln. mesenteroides, Micrococcus spp., St. capitis, St. cohnii, St. epidermidis, St. lentus, St. sciuri, St. xylosus*	Garcia Fontan *et al.* (2007b)

(Contd.)

Table 2: *(Comtd.)*

Origin	Product	Bacteria Species	References
	Chorizo	*Lb. brevis, Lb. curvatus, Lb. sakei, Lc. lactis, Ln. mesenteroides, St. aureus, St. carnosus, St. epidermidis, St. hominis, St. intermedius, St. saprophyticus, St. warneri, St. xylosus, Pd. acidilactici, Pd. pentosaceus*	Benito *et al.* (2007), Aymerich *et al.* (2006), Martin *et al.* (2006), Garcia-Varona *et al.* (2000)
	Asturian chorizo	*Lb. curvatus, Lb. futsai, Lb. graminis, Lb. plantarum, Lb. sakei*	Prado *et al.* (2019)
	Galician chorizo	*Lb. sakei, St. equorum*	Fonseca *et al.* (2013)
	Fuet	*Lb. curvatus, Lb. sakei, Ln. mesenteroides, St. epidermidis, St. carnosus, St. warneri, St. xylosus*	Aymerich *et al.* (2006), Martin *et al.* (2006)
	Salchichon	*Ko. varians, Lb. brevis, Lb. curvatus, Lb. plantarum, Lb. sakei, Lc. lactis, Ln. mesenteroides, Pd. acidilactici, Pd. pentosaceus, St. epidermidis, St. warneri, St. xylosus*	Benito *et al.* (2007), Aymerich *et al.* (2006), Martin *et al.* (2006)

C.: Carnobacterium; E.: Enterococcus; Ko.: Kocuria; Lb.: Lactobacillus; Lc.: Lactococcus; Ln.: Leuconostoc; M.: Micrococcus; Pd.: Pediococcus; St.: Staphylococcus; V.: Vagococcus; W.: Weissella by whole genome sequencing (Chaillou *et al.*, 2005). More accurately, adaptation in meat microenvironment is indicated by the ability to efficiently utilise substrates that are abundant in the specific substrate, namely nucleosides. The nucleosides that *Lb. sakei* is capable of utilising but *Lb. curvatus* and *Lb. plantarum* are not, and therefore are not considered as adapted to the specific niche, are arginine and N-acetyl-D-neuraminic acid (Hebert *et al.*, 2012; Siezen *et al.*, 2012).

Regarding staphylococci, *Staphylococcus xylosus* and *St. carnosus* are the most frequently isolated species (Paramithiotis *et al.*, 2010b; Vignolo *et al.*, 2010). Rosenstein *et al.* (2009) attributed the persistence of *St. carnosus* to the ability to catabolise carbon sources abundantly present in meat and to grow at low-water activity and temperature. All these metabolic features have been revealed through genomic analysis of *St. carnosus* strain TM300. As far as yeasts were concerned, the most frequently isolated species is *Debaryomyces hansenii*.

2.1.3 Sourdough Fermentation

The microecosystem of a wide variety of spontaneously fermented sourdoughs has been thoroughly assessed. In the majority of the cases, these sourdoughs were made with wheat and rye flours. However, several other cereals, pseudocereals and legumes, including amaranth, barley, bean, buckwheat, chickpea, einkorn, oat, lentil, millet, quinoa, rice and sorghum have been exploited for this purpose (Sterr *et al.*, 2009; Vogelmann *et al.*, 2009; Weckx *et al.*, 2010; Flander *et al.*, 2011; Moroni *et al.*, 2012; Rieder *et al.*, 2012; Mariotti *et al.*, 2014; Rizzello *et al.*, 2014; Farahmand *et al.*, 2015; Ogunsakin *et al.*, 2015; Kuligowski *et al.*, 2016; Slukova *et al.*, 2016; Korcari *et al.*, 2019; Cakir *et al.*, 2020).

The sourdough microecosystem is dominated by LAB; yeasts are very often present at lower populations, most of the times approximately at 1-2 log CFU/g less. The lactic acid microbiota that drives sourdough fermentation depends upon the preparation procedure. Solid sourdoughs prepared at ambient temperatures through frequent back-slopping (termed type I sourdoughs) are usually dominated by *Lb. sanfransiscensis* with *Lb. plantarum*, *Lb. brevis* and *Lb. alimentarius* being frequently isolated. On the other hand, liquid sourdoughs prepared at elevated temperatures (> 30°C) (termed type II sourdoughs) are usually dominated by *Lb. pontis* with *Lb. fermentum*, *Lb. reuteri*, *Lb. frumenti* and *Lb. amylovorus* being frequently part of the microecosystem (Van Kerrebroeck *et al.*, 2017; Gaenzle and Zheng, 2019). The yeast species most frequently isolated in both cases are *Saccharomyces cerevisiae* and *Kazachstania humilis*. In Table 3, the composition of the bacterial microbiota of sourdoughs from around the world is exhibited.

Dominance of *Lb. sanfranciscensis* has been attributed to the ability for rapid growth within the sourdough microecosystem, despite the relatively small genome size (Vogel *et al.*, 2011; Gaenzle *et al.*, 1998). Similarly, *Lb. pontis* growth seems to be favoured by elevated incubation temperature and is more acid resistant and thus dominates Type II sourdough fermentations. From a technological perspective, LAB hold a pivotal role in sourdough breadmaking. Apart from acidification, production of homo- and hetero-exopolysaccharides with positive effect on the rheological properties, texture, volume and shelf-life (Korakli *et al.*, 2001; Tieking *et al.*, 2003;

Table 3: Lactic Acid Bacteria and Yeasts Involved in Spontaneous Sourdough Fermentation

Origin	Microbial Species	References
Wheat sourdoughs		
Albania	*Lb. plantarum, Lc. lactis, Ln. citreum, Ln. mesenteroides, Pd. pentosaceus*	Nionelli *et al.* (2014)
Belgium	*Lb. acidifarinae, Lb. brevis, Lb. buchneri, Lb. crustorum, Lb. hammesii, Lb. helveticus, Lb. nantensis, Lb. parabuchneri, Lb. paracasei, Lb. paralimentarius, Lb. plantarum, Lb. pontis, Lb. rossiae, Lb. sakei, Lb. sanfranciscensis, Lb. spicheri, Ln. mesenteroides, Pd. pentosaceus, W. cibaria, W. confusa. K. barnetti, K. unispora, S. cerevisiae, T. delbrueckii, Wi. anomalus*	Scheirlinck *et al.* (2007a, b, 2008), Vrancken *et al.* (2010)
Brazil	*E. durans, E. faecalis, E. faecium, E. gilvus, E. hirae, Lb. brevis, Lb. farciminis, Lc. lactis, Ln. citreum*	Menezes *et al.* (2020)
China	*E. durans, E. faecium, Lb. brevis, Lb. casei, Lb. crustorum, Lb. curvatus, Lb. farciminis, Lb. fermentum, Lb. guizhouensis, Lb. helveticus, Lb. mindensis, Lb. paralimentarius, Lb. paraplantarum, Lb. plantarum, Lb. pontis, Lb. rossiae, Lb. sanfranciscensis, Lb. zeae, Lc. lactis, Ln. citreum, Ln. mesenteroides, Pd. pentosaceus, W. cibaria, W. confusa. Ca. humilis, Ca. parapsilosis, K. exigua, Mz. guilliermondii, P. kudriavzevii, S. cerevisiae, T. delbrueckii, Wi. anomalus*	Zhang *et al.* (2011), Fu *et al.* (2020), Xing *et al.* (2020)
France	*E. hirae, Lb. acidophilus, Lb. brevis, Lb. casei, Lb. curvatus, Lb. delbrueckii, Lb. diolivorans, Lb. farraginis, Lb. frumenti, Lb. hammesii, Lb. hilgardii, Lb. kimchii, Lb. koreensis, Lb. nantensis, Lb. panis, Lb. paracasei, Lb. paralimentarius, Lb. paraplantarum, Lb. pentosus, Lb. plantarum, Lb. pontis, Lb. sakei, Lb. sanfranciscensis, Lb. spicheri, Lb. xiangfangensis, Lc. lactis, Ln. citreum, Ln. mesenteroides, Pd. pentosaceus, W. cibaria, W. confusa. Ca. carpophila, Ca. humilis, H. pseudoburtonii, K. bulderi, K. exigua, K. servazzii, K. unispora, R. mucilaginosa, S. cerevisiae, T. delbrueckii*	Infantes and Tourneur, 1991, Valcheva *et al.* (2005), Ferchichi *et al.* (2008), Robert *et al.* (2009), Lhomme *et al.* (2015a, b, 2016)
Germany	*Lb. brevis, Lb. buchneri, Lb. casei, Lb. delbrueckii, Lb. fermentum, Lb. plantarum*	Spicher (1959)
Greece	*E. faecium, Lb. brevis, Lb. paralimentarius, Lb. plantarum, Lb. sanfranciscensis, Lb. zymae, Pd. pentosaceus, W. cibaria, D. hansenii, P. membranifaciens, S. cerevisiae, T. delbrueckii, Y. lipolytica*	Paramithiotis *et al.* (2000, 2010c), De Vuyst *et al.* (2002), Vancanneyt *et al.* (2005)

Italy	E. faecalis, Lb. acidophilus, Lb. alimentarius, Lb. belbrueckii, Lb. brevis, Lb. casei, Lb. cellobiosus, Lb. curvatus, Lb. farciminis, Lb. fermentum, Lb. fructivorans, Lb. gallinarum, Lb. graminis, Lb. helveticus, Lb. paracasei, Lb. paralimentarius, Lb. paraplantarum, Lb. pentosus, Lb. plantarum, Lb. rhamnosus, Lb. rossiae, Lb. rossii, Lb. sakei, Lb. salivarius, Lb. sanfranciscensis, Lc. lactis, Ln. citreum, Ln. durionis, Ln. fructosus, Ln. mesenteroides, Ln. pseudomesenteroides, Pd. pentosaceus, Pd. argentinicus, Pd. inopinatus, Pd. parvulus, W. cibaria, W. confusa, W. paramesenteroides, Ca. glabrata, Ca. humilis, Ca. milleri, Ca. stellata, K. exigua, Me. pulcherrima, P. kudriavzevii, P. membranifaciens, S. cerevisiae, S. pastorianus, T. delbrueckii, Wi. anomalus	Galli et al. (1988), Gobbetti et al. (1994), Ottogalli et al. (1996), Corsetti et al. (2001, 2005), Gullo et al. (2003), Succi et al. (2003), Foschino et al. (2004), Vernocchi et al. (2004a, b), Randazzo et al. (2005), Garofalo et al. (2008), Zotta et al. (2008), Iacumin et al. (2009), Osimani et al. (2009), Minervini et al. (2012, 2015), Ventimiglia et al. (2015)
Poland	Enterococcus sp., Lb. plantarum, Lb. brevis, Pd. pentosaceus, Lactobacillus spp.	Boreczek et al. (2020)
Spain	Lb. brevis, Lb. plantarum, S. cerevisiae	Barber et al. (1983)
Rye sourdoughs		
Belgium	Lb. brevis, Lb. hammesii, Lb. nantensis, Lb. paralimentarius, Lb. plantarum, S. cerevisiae, Wi. anomalus	Scheirlinck et al. (2007b, 2008), Vrancken et al. (2010)
Bulgaria	Lb. kimchii, Lb. paralimentarius, Lb. sanfranciscensis, Lb. spicheri	Ganchev et al. (2014)
Denmark	Lb. amylovorus, Lb. panis, Lb. reuteri, S. cerevisiae	Rosenquist and Hansen (2000)
Finland	Lb. acidophilus, Lb. casei, Lb. plantarum, Ca. humilis, Ca. stellata, K. exigua, K. unispora, S. cerevisiae, Wi. anomalus	Salovaara and Katunpaa (1984), Mantynen et al. (1999)
Germany	Lb. amylovorus, Lb. frumenti, Lb. pontis, Lb. reuteri	Mueller et al. (2001)
Poland	Enterococcus sp., Lb. brevis, Lb. plantarum, Lactobacillus spp., Pd. pentosaceus	Boreczek et al. (2020)

Lb.: *Lactobacillus*; *Lc.*: *Lactococcus*; *Ln.*: *Leuconostoc*; *Pd.*: *Pediococcus*; *W.*: *Weissella*; *Ca.*: *Candida*; *D.*: *Debaryomyces*; *H.*: *Hyphopichia*; *K.*: *Kazachstania*; *Me.*: *Metschnikowia*; *Mz.*: *Meyerozyma*; *P.*: *Pichia*; *R.*: *Rhodotorula*; *S.*: *Saccharomyces*; *T.*: *Torulaspora*; *Wi.*: *Wickerhamomyces*; *Y.*: *Yarrowia*

Di Cagno *et al.*, 2006; Kaditzky *et al.*, 2008), enrichment of flavour and aroma through their metabolic activities (Vogel *et al.*, 2011), degradation of phytic acid with concomitant increase of mineral bioavailability (De Angelis *et al.*, 2003), gluten degradation, especially of toxic gliadin epitopes (Di Cagno *et al.*, 2004) as well as production of metabolites with antimold activity (Magnusson *et al.*, 2003; Dal Bello *et al.*, 2007; Gerez *et al.*, 2009; Ndagano *et al.*, 2011; Ryan *et al.*, 2011; Mu *et al.*, 2012; Black *et al.*, 2013; Demirbas *et al.*, 2017' Russo *et al.*, 2017) are among the most important technological properties of sourdough LAB.

The role of sourdough yeasts, on the other hand, was thought to be restricted to CO_2 production and leavening. However, this is currently revisited and yeast's role is enriched with functions that affect organoleptic qualities, nutritional value as well as the shelf-life through the production of antimold compounds (Heitmann *et al.*, 2017; Axel *et al.*, 2017).

2.1.4 Dairy Fermentation

Milk produced by healthy animals is considered sterile. However, it can be easily contaminated by microorganisms residing in the rearing environment, namely soil, bedding, feed, faeces, etc. Further contamination may take place through the equipment and utensils used for milking and milk storage. Milking under proper hygienic conditions will result in the enumeration of a few Gram-positive cocci and bacilli as well as pseudomonads and yeasts in milk. Occurrence of these microorganisms at elevated populations or presence of pathogens, such as salmonelae, *Listeria monocytogenes*, clostridia, etc. may indicate either a diseased animal or improper rearing conditions.

Lactic acid fermentation is employed for the production of cheese as well as various types of fermented milks. Both constitute an important part of the human diet since ancient times; nowadays a wide range of products exist, most of which are characteristic of certain geographical regions. Regarding cheesemaking, this variability is attributed to the type of milk used as well as the coagulation, acidification and ripening conditions. The microorganisms that are used as starter cultures are mesophilic and thermophilic lactic acid bacteria. The former belongs mainly to the *Lactococcus* and *Leuconostoc* genera and are applied at fermentation temperatures of 20-40°C. Thermophilic starters, namely *Streptococcus salivarius* subsp. *thermophilus*, *Lb. delbrueckii* subsp. *bulgaricus*, *Lb. delbrueckii* subsp. *lactis*, etc. are used when fermentation takes place at higher temperatures (40-50°C). Acidification and texture development are the primary functions of these microorganisms. Flavour development may also be assigned to microbiota added as secondary or adjunct microbiota. The latter may consist of LAB, yeasts or moulds.

Several LAB belonging to the genera *Lactobacillus*, *Lactococcus*, *Leuconostoc*, *Pediococcus, Streptococcus* and *Bifidobacterium*, yeasts of the genera *Candida, Klyuveromyces, Saccharomyces* and *Torulaspora* and moulds have been reported to contribute to milk fermentation. Generally, products prepared with thermophilic starters (e.g. yoghurt) are characteristic of southeast Europe and South and Central Asia, while products prepared with mesophilic starters (e.g. sour milk) are characteristic of northeast Europe (Oberman and Libudzisz, 1998). Marshall

(1984) proposed a classification scheme, according to the prevailing microbiota, and classified traditional fermented milks into four categories: (i) products, in which mesophilic lactococci and leuconostocs dominate; these products have characteristic consistency and flavour typical of central and eastern Europe as well as Scandinavia; (ii) products, in which *Lactobacillus* spp. dominate; these are typical of eastern Europe; (iii) products, in which thermophilic bacteria dominate; these are typical of central and eastern Europe with yoghurt being the most known example; and (iv) products, in which LAB and yeasts dominate; such products are widely produced in Asia.

In Table 4, recent studies on the characterisation of the spontaneous microbiota of cheese and fermented milk are shown. In most of the cases, dominance of the aforementioned species is reported. In addition, dominance is indicated already from the beginning of the fermentation. However, there are cases in which species succession during fermentation is reported. More accurately, production of *lait caille* was characterised by dominance of *Ln. mesenteroides* during the first few hours of fermentation with *Pd. pentosaceus* and *W. paramesenteroides*, forming a secondary microbiota, while *Lc. lactis* and *Enterococcus* spp. dominated after seven hours of fermentation. Similarly, *Candida parapsilopsis* prevailed over the yeast microbiota during the first 18 hours of fermentation, followed by *S. cerevisiae* that dominated until the end of fermentation (Bayili *et al.*, 2019).

2.2 Acetic Acid Fermentation

Acetic acid fermentation is driven by acetic acid bacteria, despite the fact that acetic acid may be produced during fermentation by other microorganisms as well, such as heterofermentative LAB and yeasts. Acetic acid bacteria consist of as many as 19 genera; the ones associated with food fermentation belong to the genera *Acetobacter*, *Gluconobacter*, *Gluconoacetobacter* and *Komagataeibacter*. Species belonging to these genera are constantly isolated during spontaneous vinegar, kombucha, kefir, cocoa and lambic beer fermentation (Table 5).

Acetic acid is produced through oxidative fermentation during which dehydrogenases of the respiratory chain catalyse the oxidation of organic molecules, such as ethanol, carbohydrates and sugar alcohols to a wide variety of products including organic acids, aldehydes and ketones. Acetic acid bacteria are characterised by their strict aerobic metabolism. In addition, *Gluconobacter oxydans* seem to lack two important components of the respiratory chain, namely the genes encoding for cytochrome c oxidase and the proton-translocating NADH: ubiquinone oxidoreductase, compromising thus the proton translocation leading to poor growth capacity (Prust *et al.*, 2005). However, this is not evident for *Acetobacter pasteurianus* (Illeghems *et al.*, 2013). As far as their acetic acid production capacity was concerned, *Komagataeibacter* spp. seems to be more tolerant than *Acetobacter* spp. Indeed, *Ko. europaeus* and *Ko. oboediens* have been isolated from vinegar fermentation with acidity up to 20 per cent while *A. aceti* and *A. pasteurianus* up to 10 per cent (Andres-Barrao and Barja, 2017).

A wide variety of raw materials are currently used for vinegar production. Fruits are most commonly used in Europe, while cereals in Asia (Lynch *et al.*, 2019). Despite

Table 4: Bacteria and Yeasts Involved in Representative Spontaneous Dairy Product Fermentation

Product	Microbial Species	References
Cheeses		
Herve (Belgium)	*Arthrobacter* spp., *Brevibacterium* spp., *Co. casei, Lc. lactis* subsp. *cremoris, Microbacterium* spp., *Staphylococcus* spp.,	Delcenserie *et al.* (2014)
Poro (Mexico)	*Acinetobacter* spp., *Bacillus* spp., *Chryseobacterium* spp., *Enterococcus* spp., *Lb. delbrueckii, Lactococcus* spp., *Sediminibacter* spp., *Staphylococcus* spp., *Str. thermophilus*	Aldrete-Tapia *et al.* (2014)
Istrian (Croatia)	*Lc. lactis* subsp. *lactis, Enterococcus* spp.	Fuka *et al.* (2010)
Kazak (China)	*Lc. garvieae, Lc. lactis, Lb. casei, Lb. helveticus, Lb. paracasei, Lb. plantarum, Lb. rhamnosus, Ln. lactis, Str. thermophilus, W. confusa*	Li J. *et al.* (2020)
Feta (Greece)	*Ab. johnsonii, Ab. junii, E. canis, E. faecalis, Lb. brevis, Lb. mali, Lb. plantarum, Lc. garvieae, Lc. lactis, Ln. citreum, Pseudomonas* spp., *Str. macedonicus, Str. uberis*	Bozoudi *et al.* (2016)
Bryndza (Slovakia)	*Lc. garvieae, Lc. lactis* subsp. *cremoris, Lc. lactis* subsp. *lactis, Str. parauberis*	Pangallo *et al.* (2014)
Fermented milks		
Gioddu (Italy)	*Lb. delbrueckii, Lb. kefiri, Str. thermophilus, G. candidum, G. geotrichum, Kl. marxianus*	Maoloni *et al.* (2020)
Pendidaam (Cameroon)	*E. faecalis, E. faecium, Lb. casei, Lb. delbrueckii, Lb. fermentum, Lb. helveticus, Lb. plantarum, Ln. mesenteroides, Ln. paramesenteroides, Str. thermophilus,*	Mbawala *et al.* (2013)
Nunu (Ghana)	*E. faecium, E. italicus, Lb. fermentum, Lb. helveticus, Lb. plantarum, Lactococcus* spp., *Ln. mesenteroides, W. confusa, Ca. parapsilosis, Ca. rugosa, Ca. tropicalis, G. geotrichum, P. kudriavzevii, S. cerevisiae*	Akabanda *et al.* (2013)
Madila (Botswana)	*Lb. acidophilus, Lb. brevis, Lb. delbrueckii, Lb. fermentum, Lb. plantarum, Lc. lactis*	Ohenhen *et al.* (2013)
Fene (Mali)	*Enterococcus* spp., *Lb. fermentum, Lb. plantarum, Lc. lactis* subsp. *lactis, Pd. pentosaceus, W. confusa*	Wullschleger *et al.* (2013)
Fermented cow milk (China)	*E. durans, E. faecalis, E. faecium, E. italicus, Lb. brevis, Lb. coryniformis* subsp. *torquens, Lb. delbrueckii* subsp. *bulgaricus, Lb. fermentum, Lb. graminis, Lb. helveticus, Lb. kefiranofaciens* subsp. *kefiranofaciens, Lb. kefiri, Lb. paracasei, Lb. plantarum, Lb. sakei, Lc. garvieae, Lc. lactis* subsp. *cremoris, Lc. lactis* subsp. *lactis, Ln. citreum, Ln. lactis, Ln. mesenteroides, Ln. pseudomesenteroides, Pd. pentosaceus, Str. thermophilus, W. cibaria, W. confusa*	Mo *et al.* (2019)

Fermented camel milk (China)	E. faecium, Lb. helveticus, Lb. kefranofaciens subsp., kefranofaciens, Lb. kefiri, Lb. plantarum, Lc. lactis subsp. lactis, Ln. lactis, Ln. mesenteroides, Str. thermophilus,	Mo et al. (2019)
Fermented horse milk (China)	Lb. delbrueckii subsp. bulgaricus, Lb. fermentum, Lb. helveticus, Lb. kefranofaciens subsp. kefranofaciens, Lb. paracasei, Ln. lactis, Ln. mesenteroides, Str. thermophilus	Mo et al. (2019)
Dadih (West Sumatra)	Bifidobacterium spp., Klebsiella spp., Lactococcus spp., Leuconostoc spp., Streptococcus spp.	Venema and Surono (2018)
Matsoni (Georgia)	Enterococcus spp., Lb. delbrueckii subsp. bulgaricus, Lb. delbrueckii subsp. lactis, Lb. rhamnosus, Lb. vaginalis, Str. thermophilus	Kakabadge et al. (2019)
Kefir (Belgium)	A. orientalis, A. lovaniensis, Glu. frateurii, Lb. kefiri, Lb. kefranofaciens, Lc. lactis ssp. cremoris, Ln. mesenteroides, Kl. marxianus, Ka. kefir, Naumovozyma sp.	Korsak et al. (2015)
Kefir (Italy)	Ac. fabarum, Ac. lovaniensis, Ac. orientalis, Bacillus sp., Enterococcus sp., Lb. kefranofaciens, Lc. lactis, Str. thermophilus, D. anomala	Garofalo et al. (2015)
Kefir (Tibet)	Lb. casei, Lb. kefranofaciens, Lb. kefiri, Lb. helveticus, Lb. paracasei, Lc. lactis, Ln. mesenteroides, Str. thermophilus, K. marxianus, Ka. exigua, K. unispora, S. cerevisiae	Zhou et al. (2009), Gao and Zhang, 2018
Kefir (Russia)	Lb. casei, Lb. paracasei, Lb. kefiri, Lb. kefranofaciens subsp. kefirgranum, Lc. lactis ssp. cremoris/ lactis, Ln. pseudomesenteroides, K. unispora, S. cerevisiae	Kotova et al. (2016)
Lait caillé (Burkina Faso)	Lc. lactis, Ln. mesenteroides, Pd. pentosaceus, W. paramesenteroides, Enterococcus spp., Ca. parapsilosis, S. cerevisiae	Bayili et al. (2019)
Fermented Milks from Messinese Goat Breed	E. durans, E. faecalis, E. faecium, E. hirae, E. lactis, Lc. lactis, Ln. lactis	Palmeri et al. (2019)

A.: Acetobacter; Ab.: Acinetobacter; Co.: Corynebacterium; E.: Enterococcus; Glu.: Gluconobacter; Lb.: Lactobacillus; Lc.: Lactococcus; Ln.: Leuconostoc; Pd.: Pediococcus; Str.: Streptococcus; W.: Weissella; Ca.: Candida; D.: Dekkera; G.: Galactomyces; K.: Kazachastania; Kl.: Kluyveromyces; P.: Pichia; S.: Saccharomyces

Table 5: Acetic Acid Bacteria Involved in Spontaneous Fermentations

Product	Acetic Acid Bacteria Species	References
Balsamic vinegar	*A. aceti, A. malorum, A. pasteurianus, Ga. europaeus, Ga. hansenii, Ga. xylinus*	Gullo and Giudici (2008)
Shanxi aged vinegar	*A. indonesiensis, A. malorum, A. orientalis, A. pasteurianus, A. senegalensis, Glu. oxydans*	Wu *et al.* (2012)
Diospyros kaki vinegar	*A. malorun, A. pasteurianus, A. syzygii, Ga. europaeus, Ga. intermedius, Ga. saccharivorans*	Hidalgo *et al.* (2012)
Blueberry vinegar	*A. pasteurianus*	Hidalgo *et al.* (2013)
Lambic beer	*A. fabarum, A. lambici, A. orientalis, Glu. cerevisiae, Glu. cerinus*	Spitaels *et al.* (2015)
Milk kefir	*A. syzygii, Glu. japonicus*	Miguel *et al.* (2010)
Sugary kefir	*A. fabarum, A. orientalis, A. lovaniensis, Glu. liquefaciens*	Magalhaes *et al.* (2010), Miguel *et al.* (2011), Gulitz *et al.* (2011), Laureys and De Vuyst (2014)
Cocoa beans	*A. cerevisiae, A. cibinongensis, A. fabarum, A. ghanaensis, A. lovaniensis, A. malorum/cerevisiae, A. malorum/indonesiensis, A. orientalis, A. pasteurianus, A. peroxydans, A. pomorum, A. senegalensis, A. syzygii, Glu. europaeus, Ga. entanii, Ga. persimonis, Ga. saccharivorans*	Camu *et al.* (2007), Garcia-Armisen *et al.* (2010), Lefeber *et al.* (2011) Meersman *et al.* (2013), Papalexandratou *et al.* (2011)
Haipao	*A. aceti, A. pasteurianus, A. xylium, Glu. oxydans*	Liu *et al.* (1996)
Kombucha	*Acetobacter* spp., *Gluconobacter* spp.	Marsh *et al.* (2014)

A.: Acetobacter; Glu.: Gluconobacter; Ga.: Gluconoacetobacter

that, vinegar production is performed in two steps, namely alcoholic fermentation driven in most of the cases by *Sa. cerevisiae* and acetic acid fermentation driven in most of the cases by a consortium of acetic acid bacteria. A saccharification step may also be included, especially in cereal vinegars. The relatively small amount of studies on acetic acid bacteria species dynamics during fermentation, does not allow generalisations regarding their relative fitness. However, it seems that cocoa fermentation is dominated by *A. pasteurianus* and Kombucha fermentation by *Ko. xylinus* (formerly *A. xylinum* and *Ga. xylinus*). Species succesion has also been reported. More accurately, Hidalgo *et al.* (2012) reported that during acetic acid fermentation of *Diospyros kaki*, *A. malorum* dominated the initial and mid-stages and was replaced by *Ga. saccharivorans* in the latter stages.

2.3 Propionic Acid Fermentation

Propionic acid fermentation is a fermentation in which the main end product is propionic acid. Several microbial species can produce propionic acid through fermentation pathways (acrylate, succinate and 1,2-propanediol pathways), amino acid catabolic pathways and biosynthetic pathways (Gonzalez-Garcia *et al.*, 2017). Regarding microorganisms associated with food fermentations, this is carried out by *Propionibacterium freudenreichii* and *Pro. acidipropionici* through the Wood-Werkman cycle and the decarboxylation of succinate (Fig. 2). Both species are used as starters for ripening hard cheeses and contribute to their characteristic flavour and texture, although they are not considered to be adapted to milk as growth substrate. Indeed, Pivateau *et al.* (2000) reported that both species were not able to grow in milk and whey when the inoculum was less than 6 log CFU/mL. This was explained by the inability to utilise milk proteins due to the greater peptidase than protease activity (Gagnaire *et al.*, 1999) and to the inability, at least of some strains, to utilise lactose (Pivateau, 1999). Thus, during Emmental production, presence of LAB is necessary in order to provide lactate and peptides, and thus allow propionic acid bacteria development (Cousin *et al.*, 2012). Research is currently focused on revealing the probiotic potential of these species. This has been attributed to their ability to adapt in various niches, which was confirmed by whole genome sequencing (Falentin *et al.*, 2010; Parizzi *et al.*, 2012) as well as their capacity to produce biofunctional metabolites. Indeed, modulation of intestinal microbiota, immunomodulation as well as anticancer activities have been attributed to a series of metabolites, including short chain fatty acids, 1,4-dihydroxy-2-naphthoic acid, 2-amino-3-carboxy-1,4-naphthoquinone, conjugated fatty acids and surface proteins (Altieri, 2016; Rabah *et al.*, 2017).

3. Alkaline Fermentation

Alkaline fermentation, although of great importance for local populations, mostly of certain parts of Asia and Africa, is not as studied as acid or alcoholic ones. Alkaline fermentation is the microbial-driven bio-transformation, during which the pH value of the substrate increases to values above 7. This increase is assigned to the ammonia released during enzymatic protein degradation. The latter occurs due to extracellular

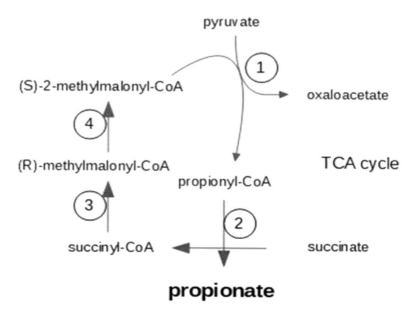

Fig. 2: Wood-Werkman cycle for the production of propionate. The reactions are catalysed by the enzymes: 1: methylmalonyl-CoA carboxyltransferase; 2: propionyl-CoA:succinyl-CoA transferase; 3: methylmalonyl-CoA mutase; and 4: methylmalonyl-CoA epimerase

proteolytic enzymes produced by the dominating microbiota, which consists mostly of *Bacillus* spp. In Table 6, the dominating microbiota of a variety of alkaline-fermented products is exhibited. In most of the cases, dominance of *B. pumilus* sensu lato, *B. cereus* sensu lato and *B. subtilis* is reported. In general, the production procedure involves 1-2 boiling steps at time/temperature that depends upon the raw material used. The fermentation step usually takes place after draining and cooling of the raw materials in ambient conditions; therefore time depends upon temperature. The most studied product and the one that starter cultures are commercially available is the Japanese Natto and consists of *B. subtilis*. It is believed that starter cultures will be soon available for the Korean cheonggukjang as well (Kimura and Yokoyama, 2019).

The raw materials that are mostly used as substrate for alkaline fermentation are protein-rich seeds and grains of wild and domesticated plants and trees (Parkouda *et al.*, 2009). The valorisation of *Parkia biglobosa* (Jacq.) Benth, *Adansonia digitata* L., *Pentaclethra macrophylla* Benth, *Prosopis africana*, *Glycine max* (L.) Merr, *Vigna subterranea*, *Hibiscus sabdariffa*, *Gossypium hirsutum* L. and *Colocynthis citrullus* seeds through alkaline fermentation has been reported (Ogbadu and Okagbue, 1988; Sanni and Ogbonna, 1991; Omafuvbe *et al.*, 1999, 2000; Sarkar *et al.*, 2002; Parkouda *et al.*, 2008, 2010; Ahaotu *et al.*, 2013; Chettri and Tamang, 2015; Adedeji *et al.*, 2017). However, alkaline fermentation of other plant material may not be excluded. Indeed, Mbozo *et al.* (2017) studied the microecosystem of Ntoba Mbodi, a product based on the alkaline fermentation of cassaca leaves.

Table 6: Bacteria Involved in Alkaline Fermentations

Product	Basic Ingredient	Bacteria Species	References
Ntoba Mbodi	Cassava leaves	*B. altitudinis, B. amyloliquefaciens, B. cereus sensu lato, B. licheniformis, B. megaterium, B. pumilus sensu lato, B. safensis, B. siamensis, B. subtilis, L. louembei*	Mbozo *et al.* (2017)
Maari[a]	Baobab seed (*Adansonia digitata* L.)		Parkouda *et al.* (2010)
	Mansila	*Ae. viridians, B. circulans, B. licheniformis, B. pumilus, B. subtilis, Corynebacterium spp., E. avium, E. faecium, Gc. sulfidifaciens, L. sphaericus, Lf. shinshuensis, Ma. caseolyticus*	
	Toulfe	*Ae. viridians, B. licheniformis, B. subtilis, Corynebacterium spp., E. avium, E. faecium, L. fusiformis, L. sphaericus, Ma. caseolyticus, St. hominis*	
	Gorgadji	*Ae. viridians, B. cereus, B. circulans, B. coagulans, B. endophyticus, B. licheniformis, B. megaterium, B. pumilus, B. subtilis, Br. borstelensis, E. avium, E. faecium, Ku. gibsonii, P. polymyxa, Pd. acidilactici, Pr. mirabilis, St. hominis, St. sciuri*	
	Ouagadougou	*Ab. calcoaceticus, B. cereus, B. circulans, B. licheniformis, B. megaterium, B. pumilus, B. subtilis, Corynebacterium spp., E. casseliflavus, E. faecium, L. sphaericus, St. hominis, St. gallinarum, St. sciuri*	
Ugba	Oil beans (*Pentaclethra macrophylla* Benth)	*B. cereus sensu lato, B. clausii, B. licheniformis, B. safensis, B. subtilis, L. xylanilyticus*	Ahaotu *et al.* (2013)
Iru	Locust bean seeds (*Parkia biglobosa*)	*B. anthracis, Eb. clocae, Pr. mirabilis, St. sciuri*	Adedeji *et al.* (2017)

Table 6: (*Contd.*)

Product	Basic Ingredient	Bacteria Species	References
Ogiri	Melon seeds (*Colocynthis citrullus*)	*Al. faecalis, B. licheniformis, P. dendritiformis, Pr. mirabilis*	Adedeji *et al.* (2017)
Kinema	Soybeans (*Glycine max*)	*B. cereus, B. circulans, B. licheniformis, B. sphaericus, B. subtilis, B. thuringiensis*	Sarkar *et al.* (2002)
Soumbala	Locust bean seeds (*Parkia biglobosa*)	*B. badius, B. cereus, B.firmus, B. licheniformis, B. megaterium, B. mycoides, B. sphaericus, B. subtilis, B. thuringiensis, Br. laterosporus, Pb. alvei, Pb. larvae*	Sarkar *et al.* (2002)
Kpaye	*Prosopis africana* seeds	*B. licheniformis, B. pumilus, B. subtilis*	Omafuvbe *et al.* (1999)
Soy-daddawa	Soybeans (*Glycine max*)	*B. subtilis*	Omafuvbe *et al.* (2000)
Afitin	African locust bean (*Parkia biglobosa*)	*Bacillus* spp., *Staphylococcus* spp.	Azokpota *et al.* (2006)
Sonru	African locust bean (*Parkia biglobosa*)	*Bacillus* spp., *Staphylococcus* spp.	Azokpota *et al.* (2006)
Tungrymbai	Soybeans (*Glycine max*)	*B. licheniformis, B. pumilus, B. subtilis*	Chettri and Tamang (2015)
Bekang	Soybeans (*Glycine max*)	*B. brevis, B. coagulans, B. circulans, B. licheniformis, B. pumilus, B. sphaericus, B. subtilis, L.fusiformis*	Chettri and Tamang (2015)

[a] The microbiota of the final product from four different production sites of Burkina Faso indicated was reported
Ab.: Acinetobacter; Ae.: Aerococcus; Al.: Alcaligenes; B.: Bacillus; Br.: Brevibacillus; E.: Enterococcus; Eb.: Enterobacter; Gc.: Globicatella; Ku.: Kurthia; L.: Lysinibacillus; Lf.: Leifsonia; Ma.: Macrococcus; Pb.: Paenibacillus; Pd. Pediococcus; Pr.: Proteus; St.: Staphylococcus

Since soybean is most often the raw material used, the fate of nutrients during alkaline fermentation has been studied to some extent. Increase of free fatty acids, particularly linoleic acid and free amino acids, particularly hydrophobic, apolar and acidic ones has been reported as a result of extensive lipolysis and proteolysis, respectively. Regarding vitamins and polyphenols, their content is the outcome of a balance between the production capacity of the *Bacillus* strains and the depleting action of the soaking and cooking step as well as the presence of other microbiota. However, increase of thiamine, riboflavin, niacin, total phenolic compounds, anthocyanins, daidzein and genistein during fermentation of cooked soybeans, has been reported (Shrestha *et al.*, 2010).

4. Alcoholic Fermentation

An extended variety of fruits, vegetables and grains has been used as substrates for alcoholic fermentation. Liberation of fermentable carbohydrates is the first necessary step which may involve crushing, in the case of fruits, or more elaborate procedures which, in the case of grains, may include procedures, such as germination, precooking and saccharification through mold fermentation. Ethanol is produced through the decarboxylation of pyruvate produced by the glycolytic pathway to acetaldehyde, which is subsequently reduced to ethanol. The reactions are catalysed by puryvate decarboxylase and alcohol dehydrogenase, respectively. Alcoholic fermentation is, in the majority of cases, carried out by strains belonging to the *Saccharomyces* genus. This dominance has been principally attributed to the ability to tolerate elevated ethanol concentration in the fermenting medium (Stanley *et al.*, 2010). During alcoholic fermentation, species succession is evident, at least in the majority of cases; non-*Saccharomyces* species prevail during the initial stages and *Saccharomyces* spp. during the last. Indeed, Bougreu *et al.* (2019) reported that *Aureobasidium pullulans* and *Hanseniaspora* spp. dominated the yeast microecosystem during the first days of Tempranillo must fermentation while *S. cerevisiae* and *Lachancea thermotolerans* completed the fermentation. Li *et al.* (2020) reported dominance of *Mrakia* spp. during the first days of raspberry must fermentation being gradually replaced by *Saccharomyces* spp. Satora and Tyszynski (2005) reported that during the first days of plum must fermentation, the yeast microecosystem was dominated by *Aureobasidium* spp. and *Klockera apiculata*, which were replaced by *S. cerevisiae*. This succession at species level is accompanied by a respective at strain level. Indeed, Pateraki *et al.* (2014), Sabate *et al.* (1998); Egli *et al.* (1998) reported this during grape must fermentation. In Table 7, the yeast species associated with alcoholic fermentation of a variety of substrates is exhibited. In most of the cases, dominance of *Saccharomyces* spp. is reported. However, in some cases, this was not achieved, mainly due to the rather restriced amount of ethanol produced.

Yeasts may also drive spontaneous vegetable fermentation and play an important technological role during sausage fermentation. The former was reported by Paramithiotis *et al.* (2016) during spontaneous fermentation of globe artichoke immature inflorescense. In this case, dominance of *Hanseniaspora* spp. throughout fermentation was suggested by both culture-dependent and -independent approaches, most probably due to the preparation procedure that resulted in a rather low initial

Table 7: Yeasts Involved in Representative Spontaneous Alcoholic Fermentations

Substrate	Yeast Species	References
Tempranillo Grapes	*S. cerevisiae, La. thermotolerans, Hanseniaspora* spp.	Bougreau *et al.* (2019)
Grapes of Priorat region	*Au. pullulans, Ca. intermedia, Ca. zemplinina, Ha. guilliermondii, Ha. uvarum, Ha. valbyensis, I. terricola, La. thermotolerans, S. cerevisiae, Sm. ludwigii, Z. fermentati*	Padila *et al.* (2016)
Plums	*Aureobasidium* sp., *Kc. apiculata, C. pulcherrima, S. cerevisiae*	Satora and Tuszyński (2005)
Rapsberries	*Mrakia* spp., *Saccharomyces* spp., *Guehomyces* spp., *Cladosporium* spp., *Cystofilobasidium* spp.	Li H. *et al.* (2020)
Apples	*H. osmophila, H. uvarum, H. valbyensis, Me. pulcherrima, P. guilliermondii, S. bayanus, S. cerevisiae*	Valles *et al.* (2007)
Oenocarpus bacaba	*P. caribbica, P. guilliermondii*	Puerari *et al.* (2015)
Pineapple	*Ha. uvarum, P. guilliermondii*	Chanprasartsuk *et al.* (2010)
Campomanesia pubescens	*Ca. quercitrusa, I. terricola, S. cerevisiae*	Duarte *et al.* (2009)
Palm	*Ca. fermentati, Ca. parapsilopsis, Ha. uvarum, P. fermentans, S. cerevisiae, Sm. ludwigii, Z. bailii*	Stringini *et al.* (2009)
Papaya	*S. bayanus, S. cerevisiae, S. italicus, S. pastorianus, S. uvarum, Sc. pombe, Zygosaccharomyces* sp.	Maragatham and Panneerselvam (2011)
Ziziphus mauritiana	*Au. pullulans, Ca. glabrata, Ca. parapsilosis, Ca. pyralidae, Cr. flavus, Cr. magnus, Ha. opuntiae, I. orientalis, P. ciferrii, P. fabianii, R. mucilaginosa, S. cerevisiae, Sp. fibuligera, Za. hellenicus*	Nyanga *et al.* (2007)
Rice	*Ha. guillermondii, Ha. opuntiae, Ha. uvarum, Ko. ohmeri, Mz. guilliermondii, P. kluyveri, P. kudriavzevii, R. mucilaginosa, R. slooffiae, S. cerevisiae, Wi. anomalus*	Grijalva-Vallejos *et al.* (2020)

Oat	*Candida* spp., *Cryptococcus* spp. *Ha. guillermondii, Ha. opuntiae, Ha. uvarum, P. kluyveri, R. mucilaginosa*	Grijalva-Vallejos et al. (2020)
Seven corn mix	*Candida* spp., *S. cerevisiae, Y. lipolytica*	Grijalva-Vallejos et al. (2020)
Sorghum	*Ca. tropicalis, Cr. albidus, D. hansenii, I. orientalis, Kc. apiculata, Kl. africanus, Kl. marxianus, P.anomala, S. cerevisiae, S. pastorianus, Sc. pombe, T. delbrueckii*	Lyumugabe et al. (2010), Kayode et al. (2007), Maoura et al. (2005), Sefa-Dedeh et al. (1999), Van der Aa Kuhle et al. (2001)
Honey by-products	*La. fermentati, P. anomala, P. kudriavzevii, S. cerevisiae, Wi. anomalus, Z. bailii, Z. rouxii*	Gaglio et al. (2017)
Agave	*Kl. marxianus, P. kudriavzevii, S. cerevisiae*	Aldrete-Tapia et al. (2020)

Au.: Aureobasidium; Ca.: Candida; Cr.: Cryptococcus; D.: Debaryomyces; Ha.: Hanseniaspora; I.: Issatchenkia; Kc.: Kloeckera; Kl.: Kluyveromyces; La.: Lachancea; Me.: Metschnikowia; Mz.: Meyerozyma; P.: Pichia; R.: Rhodotorula; S.: Saccharomyces; Sc.: Schizosaccharomyces; Sm.: Saccharomycodes; Sp.: Saccharomycopsis; Wi.: Wickeramomyces; Y.: Yarrowia; Z.: Zygosaccharomyces; Za.: Zygoascus pH value. During sausage fermentation, yeasts may grow on the surface and thus contribute to colour development and protection against rancidity through O_2 depletion. At the same time, through their enzymatic activities, they may contribute to the development of characteristic organoleptic properties of the product. However, their use should be optimised as they may inhibit growth of *Staphylococcus* spp. and generate off-flavours. The yeast species most frequently isolated from spontaneous sausage fermentation belong to the genera *Candida, Debaryomyces, Hansenula, Pichia, Rhodotorula* and *Torulopsis* (Selgas and Garcia, 2007).

5. Mold Fermentation

The principal role of moulds is degradation of macromolecules through their pectinolytic, amylolytic, proteolytic and lipolytic enzymes. As a result, softening of the texture, developing of the characteristic taste and aroma as well as improving of digestibility and nutritional value occur. A wide range of fermented foods require the participation of moulds. In Table 8, the fermented products in which moulds possess an important role are exhibited. When cereals are the basic raw materials, the main objective of mold fermentation is saccharification and may be performed by a variety of species of the genera *Rhizopus, Aspergillus, Mucor*, etc. Indeed, this is evident for the production of rice, sugarcane and finger millet alcoholic beverages, e.g. *Aspergillus oryzae* is critical in the first stage of sake production, i.e. *Koji* production (Akaike *et al.*, 2020). Tempe is a product from Indonesia that is made principally from precooked soybeans that are glued together by the mycelium of *Rhizopus* spp. During mold growth, nearly 25 per cent of the initial protein gets degraded, enriching the product with peptides and amino acids. In addition, more than 30 per cent of the glycerides is hydrolysed, increasing the free fatty acid content, while arabinogalactan and pectin fractions are solubilized. In addition, stachyose, raffinose, trypsin inhibitor and phytic acid levels decrease, partially due to pretreatment of the beans (Nout and Kiers, 2005).

In the case of meat products and cheese, in which mold contribution is important, this takes place during ripening. In the case of cheese, mold ripening contributes to the characteristic appearance and texture as well as the typical aroma and taste. Regarding meat products, apart from the development of the characteristic taste and aroma, through proteolytic and lipolytic enzymes, surface mold growth reduces O_2 levels, resulting in colour improvement and prevention of lipid oxidation and thus rancidity (Spotti *et al.*, 2008).

6. Conclusion

The production procedure and the dominating microbiota of a wide diversity of fermented products throughout the world have been described. Currently, research is focused on the technological and functional properties of the dominating as well as the secondary microbiota that compose the micro-community of these products. Technological and conceptual advancements have enabled accurate assessment of the microbial dynamics during fermentation at species and in many cases at strain level. In addition, our understanding of the genetic basis that accounts for the persistence

Table 8: Molds Involved in Spontaneous Fermentations

Product	Major Ingredient	Mold Species	References
Sufu	Soybean	*Actinomucor* spp., *Mucor* spp., *Rhizopus* spp.	Han *et al.* (2001)
Tempeh	Soybean	*Mucor* spp. *N. intermedia*, *N. sitophila*, *Rh. microsporus*, *Rh. oligosporus*, *Rh. oryzae*	Nout and Kiers (2005)
Rice wine	Rice	*Am. rouxii*, *Rh. oligosporus*, *Rh. oryzae*	Dung *et al.* (2007)
Kodo ko jaanr	Finger millet	*Rhizopus* spp.	Thapa and Tamang (2006)
Bhaati jaanr	Rice	*Rhizopus* spp.	Tamang and Thapa (2006)
Tapuy	Rice	*Mc. cirnelloides*, *Mc. grisecyanus*, *Rh. cohnii*	Kozaki and Uchimura (1990)
Tape	Rice	*Rhizopus* spp.	Suprianto *et al.* (1989)
Miso	Soybean, wheat, barley, rice	*Aspergillus* sp.	Asahara *et al.* (2006)
Shochu	Sweet potato	*As. luchuensis*	Setoguchi *et al.* (2019)
Sake	Rice	*As. oryzae*	Murakami (1971)
Dairy products (Camembert, Brine, Roquefort, Gorgonzola)	Milk	*Gt. candidum*, *Pe. camemberti*, *Pe. roqueforti*	Marcellino and Benson (2013)
Fermented sausages	Meat	*Penicillium* spp., *Aspergillus* spp., *Eurotium* spp.	Spotti *et al.* (2008)

Am.: Amylomyces; As.: Aspergillus; Gt.: Geotrichum; Mc.: Mucor; N.: Neurospora; Pe.: Penicillium; Rh.: Rhizopus of some species in specific environments and the capacity of others to prevail over a wide range of conditions inflicted by several biotic and abiotic stimuli, has been improved. Furthermore, the production of a wide range of functional microbial metabolites and the assessment of conditions under which this is facilitated, allow us to design products with enhanced nutritional value and functional potential. As more such data are generated over time, our understanding of microbial physiology will constantly improve, enabling us to effectively utilise microorganisms not only for food production, but also for additional biotechnological interventions.

Abbreviations

CFU – Colony-forming units
LAB – Lactic acid bacteria
NAD – Nicotinamide adenine dinucleotide
NADH – Nicotinamide adenine dinucleotide (reduced)
RH – Relative humidity

References

Adedeji, B.S., Ezeokoli, O.T., Ezekiel, C.N., Obadina, A.O., Somorin, Y.M., Sulyok M., Adeleke, R.A., Warth, B., Nwangburuka, C.C., Omemu, A.M., Oyewole, O.B. and Krska, R. (2017). Bacterial species and mycotoxin contamination associated with locust bean, melon and their fermented products in south-western Nigeria. *International Journal of Food Microbiology*, 258: 73-80.

Ahaotu, I., Anyogu, A., Njoku, O.H., Odu, N.N., Sutherland, J.P. and Ouoba, L.I.I. (2013). Molecular identification and safety of *Bacillus* species involved in the fermentation of African oil beans (*Pentaclethra macrophylla* Benth) for production of Ugba. *International Journal of Food Microbiology*, 162: 95-104.

Akabanda, F., Owusu-Kwarteng, J., Tano-Debrah, K., Glover, R.L.K., Nielsen, D.S. and Jespersen, L. (2013). Taxonomic and molecular characterisation of lactic acid bacteria and yeasts in nunu, a Ghanaian fermented milk product. *Food Microbiology*, 34: 277-283.

Akaike, M., Miyagawa, H., Kimura, Y., Terasaki, M., Kusaba, Y., Kitagaki, H. and Nishida, H. (2020). Chemical and bacterial components in Sake and Sake production process. *Current Microbiology*, 77: 632-637.

Akanni, G.B., De Kock, H.L., Naudé, Y. and Buys, E.M. (2018). Volatile compounds produced by *Bacillus* species alkaline fermentation of bambara groundnut (*Vigna subterranean* (L.) Verdc) into a dawadawa-type African food condiment using headspace solid-phase microextraction and GC × GC-TOFMS. *International Journal of Food Properties*, 21: 930-942.

Albano, H., Henriques, I., Correia, A., Hogg, T. and Teixeira, P. (2008). Characterisation of microbial population of 'Alheira' (a traditional Portuguese fermented sausage) by PCR-DGGE and traditional cultural microbiological methods. *Journal of Applied Microbiology*, 105: 2187-2194.

Aldrete-Tapia J.A., Escalante-Minakata, P., Martínez-Peniche, R.A., Tamplin, M.L. and Hernández-Iturriaga, M. (2020). Yeast and bacterial diversity, dynamics and fermentative kinetics during small-scale tequila spontaneous fermentation. *Food Microbiology*, 86: 103339.

Aldrete-Tapia, A., Escobar-Ramirez, M.C., Tamplin, M.L. and Hernandez-Iturriaga, M. (2014). High-throughput sequencing of microbial communities in Poro cheese, an artisanal Mexican cheese. *Food Microbiology*, 44: 136-141.

Altieri, C. (2016). Dairy propionibacteria as probiotics: Recent evidences. *World Journal of Microbiology and Biotechnology*, 32: 172.

Ammor, S., Rachman, C., Chaillou, S., Prévost, H., Dousset, X., Zagorec, M., Dufour, E. and Chevallier, I. (2005). Phenotypic and genotypic identification of lactic acid bacteria isolated from a small-scale facility producing traditional dry sausages. *Food Microbiology*, 22: 373-382.

Andres-Barrao, C. and Barja, F. (2017). Acetic acid bacteria strategies contributing to acetic acid resistance during oxidative fermentation. pp. 92-119. *In*: I.Y. Sengun (Ed.). *Acetic Acid Bacteria: Fundamentals and Food Applications*. CRC Press, Boca Raton, FL, USA.

Aquilanti, L., Santarelli, S., Silvestri, G., Osimani, A., Petruzzelli, A. and Clementi, F. (2007). The microbial ecology of a typical Italian salami during its natural fermentation. *International Journal of Food Microbiology*, 120: 136-145.

Arici, M. and Coskun F. (2001). Hardaliye: Fermented grape juice as a traditional Turkish beverage. *Food Microbiology*, 18: 417-421.

Asahara, N., Zhang, X.B. and Ohta, Y. (2006). Antimutagenicity and mutagen-binding activation of mutagenic pyrolyzates by microorganisms isolated from Japanese *miso*. *Journal of Science of Food Agriculture*, 58: 395-401.

Axel, C., Zannini, E. and Arendt, E.K. (2017). Mold spoilage of bread and its biopreservation: A review of current strategies for bread shelf life extension. *Critical Reviews of Food Science and Nutrition*, 57: 3528-3542.

Aymerich, T., Martin, B., Garriga, M., Vidal-Carou, M.C., Bover-Cid, S. and Hugas, M. (2006). Safety properties and molecular strain typing of lactic acid bacteria from slightly fermented sausages. *Journal of Applied Microbiology*, 100: 40-49.

Azokpota, P., Hounhouigan, D.J. and Nago, M.C. (2006). Microbiological and chemical changes during the fermentation of African locust bean (*Parkia biglobosa*) to produce afitin, iru and sonru, three traditional condiments produced in Benin. *International Journal of Food Microbiology*, 107: 304-309.

Barber, S., Baguena, R., Martinez-Anaya, M.A. and Torner, M.J. (1983). Microflora de la masa madre panaria. I. Identificacion y propiedades funcionales de microorganismos de masas madre industriales, elaboradas con harina de trigo. *Revista de agroquímica y tecnología de alimentos*, 23: 552-562.

Barrangou, R., Yoon, S.S., Breidt Jr., F., Fleming, H.P. and Klaenhammer, T.R. (2002). Identification and characterisation of *Leuconostoc fallax* strains isolated from an industrial sauerkraut fermentation. *Applied and Environmental Microbiology*, 68: 2877-2884.

Baruzzi, F., Matarante, A., Caputoa, L. and Morea, M. (2006). Molecular and physiological characterisation of natural microbial communities isolated from a traditional Southern Italian processed sausage. *Meat Science*, 72: 261-269.

Bayili, G.R., Johansen, P., Nielsen, D.S., Sawadogo-Lingani, H., Ouedraogo, G.A., Diawara, B. and Jespersen, L. (2019). Identification of the predominant microbiota during production of *lait caillé*, a spontaneously fermented milk product made in Burkina Faso. *World Journal of Microbiology and Biotechnology*, 35: 100.

Beganovic, J., Kos, B., Lebos Pavunc, A., Uroic, K., Jokic, M. and Suskovi, J. (2014). Traditionally produced sauerkraut as source of autochthonous functional starter cultures. *Microbiological Research*, 169: 623-632.

Benítez-Cabello, A., Bautista-Gallego, J., Garrido-Fernández, A., Rantsiou, K., Cocolin, L., Jiménez-Díaz, R. and Arroyo-López, F.N. (2016). RT-PCR–DGGE analysis to elucidate the dominant bacterial species of industrial Spanish-style green table olive fermentations. *Frontiers in Microbiology*, 7: 1291.

Benito, M.J., Martin, A., Aranda, E., Perez-Nevado, F., Ruiz-Moyano, S. and Cordoba, M.G. (2007). Characterisation and selection of autochthonous lactic acid bacteria isolated from traditional Iberian dry-fermented Salchichon and Chorizo sausages. *Journal of Food Science*, 72: M193-M201.

Black, B.A., Zannini, E., Curtis, J.M. and Gaenzle, M.G. (2013). Antifungal hydroxy fatty acids produced during sourdough fermentation: Microbial and enzymatic pathways, and antifungal activity in bread. *Applied and Environmental Microbiology*, 79: 1866-1873.

Blaiotta, G., Pennacchia, C., Villani, F., Ricciardi, A., Tofalo, R. and Parente, E. (2004). Diversity and dynamics of communities of coagulase–negative staphylococci in traditional fermented sausages. *Journal of Applied Microbiology*, 97: 271-284.

Bonomo, M.G., Ricciardi, A., Zotta, T., Parente, E. and Salzano, G. (2008). Molecular and technological characterisation of lactic acid bacteria from traditional fermented sausages of Basilicata region (southern Italy). *Meat Sci.*, 80: 1238-1248.

Bonomo, M.G., Ricciardi, A., Zotta, T., Sico, M.A. and Salzano, G. (2009). Technological and safety characterisation of coagulase-negative staphylococci from traditionally fermented sausages of Basilicata region (southern Italy). *Meat Science*, 83: 15-23.

Boreczek, J., Litwinek, D., Żylińska-Urban, J., Izak, D., Buksa, K., Gawor, J., Gromadka, R., Bardowski, J.K. and Kowalczyk, M. (2020). Bacterial community dynamics in spontaneous sourdoughs made from wheat, spelt, and rye wholemeal flour. *Microbiology Open*, 9: e1009.

Bougreau, M., Ascencio, K., Bugarel, M., Nightingale, K. and Loneragan, G. (2019). Yeast species isolated from Texas high plains vineyards and dynamics during spontaneous fermentations of Tempranillo grapes. *PLoS ONE*, 14(5): e0216246.

Bozoudi, D., Torriani, S., Zdragas, A. and Litopoulou-Tzanetaki, E. (2016). Assessment of microbial diversity of the dominant microbiota in fresh and mature PDO Feta cheese made at three mountainous areas of Greece. *LWT-Food Science and Technology*, 72: 525-533.

Cakır, E., Arici, M., Durak, M.Z. and Karasu, S. (2020). The molecular and technological characterisation of lactic acid bacteria in einkorn sourdough: Effect on bread quality. *Food Measure*, https://doi.org/10.1007/s11694-020-00412-5

Camu, N., De Winter, T., Verbrugghe, K., Cleenwerck, I., Vandamme, P., Takrama, J.S., Vancanneyt, M. and De Vuyst, L. (2007). Dynamics and biodiversity of populations of lactic acid bacteria and acetic acid bacteria involved in spontaneous heap fermentation of cocoa beans in Ghana. *Applied and Environmental Microbiology*, 73: 1809-1824.

Chaillou, S., Champomier-Vergès, M.C., Cornet, M., Crutz Le Coq, A.M., Dudez, A.M., Martin, V., Beaufils, S., Darbon-Rongère, E., Bossy, R., Loux, V. and Zagorec, M. (2005). Complete genome sequence of the meat-borne lactic acid bacterium *Lactobacillus sakei* 23K. *Nature Biotechnology*, 23: 1527-1533.

Chang, H.C. (2018). Healthy and safe Korean traditional fermented foods: Kimchi and chongkukjang. *Journal of Ethnic Foods*, 5: 161-166.

Chang, H.W., Kim, K.H., Nam, Y.D., Roh, S.W., Kim, M.S., Jeon, C.O., Oh, H.M. and Bae, J.W. (2008). Analysis of yeast and archaeal population dynamics in kimchi using denaturing gradient gel electrophoresis. *International Journal of Food Microbiology*, 126: 159-166.

Chanprasartsuk, O.O., Prakitchaiwattana, C., Sanguandeekul, R. and Fleet, G.H. (2010). Autochthonous yeasts associated with mature pineapple fruits, freshly crushed juice and their ferments and the chemical changes during natural fermentation. *Bioresource Technology*, 101: 7500-7509.

Chao, S.H., Wu, R.J., Watanabe, K. and Tsai, Y.C. (2009). Diversity of lactic acid bacteria in suantsai and fu-tsai, traditional fermented mustard products of Taiwan. *International Journal of Food Microbiology*, 135: 203-210.

Chen, Y.S., Yanagida, F. and Hsu, J.S. (2006). Isolation and characterisation of lactic acid bacteria from suan-tsai (fermented mustard), a traditional fermented food in Taiwan. *Journal of Applied Microbiology*, 101: 125-130.

Chettri, R. and Tamang, J.P. (2015). *Bacillus* species isolated from tungrymbai and bekang, naturally fermented soybean foods of India. *International Journal of Food Microbiology*, 197: 72- 76.

Cho, J., Lee, D., Yang, C., Jeon, J., Kim, J. and Han, H. (2006). Microbial population dynamics of kimchi, a fermented cabbage product. *FEMS Microbiology Letters*, 257: 262-267.

Comi, G., Urso, R., Iacumin, L., Rantsiou, K., Cattaneo, P., Cantoni, C. and Cocolin, L. (2005). Characterisation of naturally fermented sausages produced in the north-east of Italy. *Meat Science*, 69: 381-392.

Coppola, R., Giagnacovo, B., Iorizzo, M. and Grazia, L. (1998). Characterisation of Lactobacilli involved in the ripening of soppressata molisana, a typical southern Italy fermented sausage. *Food Microbiology*, 15: 47-353.

Coppola, R., Iorizzo, M., Saotta, R., Sorrentino, E. and Grazia, L. (1997). Characterisation of micrococci and staphylococci isolated from soppressata molisana, a southern Italy fermented sausage. *Food Microbiology*, 14: 47-53.

Corsetti, A., Settanni, L., van Sinderen, D., Felis, G.E., Dellaglio, F. and Gobbetti M. (2005). *Lactobacillus rossii* sp. nov., isolated from wheat sourdough. *International Journal of Systematics and Evolutionary Microbiology*, 55: 35-40.

Corsetti, A., Lavermicocca, P., Morea, M., Baruzzi, F., Tosti, N. and Gobbetti, M. (2001). Phenotypic and molecular identification and clustering of lactic acid bacteria and yeasts from wheat (species *Triticum durum* and *Triticum aestivum*) sourdoughs of southern Italy. *International Journal of Food Microbiology*, 64: 95-104.

Coton, E., Desmonts, M.H., Leroy, S., Coton, M., Jamet, E., Christieans, S., Donnio, P.Y., Lebert, I. and Talon R. (2010). Biodiversity of coagulase-negative staphylococci in French cheeses, dry fermented sausages, processing environments and clinical samples. *International Journal of Food Microbiology*, 137: 221-229.

Cousin, F.J., Louesdon S., Maillard, M.B., Parayre, S., Falentin, H., Deutsch, S.M., Boudry, G. and Jan, G. (2012). The first dairy product exclusively fermented by *Propionibacterium freudenreichii*: A new vector to study probiotic potentialities *in vivo*. *Food Microbiology*, 32: 135-146.

Daeschel, M.A., Andersson, R.E. and Fleming, H.P. (1987). Microbial ecology of fermenting plant materials. *FEMS Microbiology Reviews*, 46: 357-367.

Dal Bello, F., Clarke, C.I., Ryan, L.A.M., Ulmer, H., Schober, T.J., Strom, K., Sjogren, J., van Sinderen, D., Schnurer, J. and Arendt, E.K. (2007). Improvement of the quality and shelf-life of wheat bread by fermentation with the antifungal strain *Lactobacillus plantarum* FST 1.7. *Journal of Cereal Science*, 45: 309-318.

De Angelis, M., Gallo, G., Corbo, M.R., McSweeney, P.L., Faccia, M., Giovine, M. and Gobbetti, M. (2003). Phytase activity in sourdough lactic acid bacteria: Purification and characterisation of a phytase from *Lactobacillus sanfranciscensis* CB1. *International Journal of Food Microbiology*, 87: 259-270.

De Vuyst, L., Schrijvers, V., Paramithiotis, S., Hoste, B., Vancanneyt, M., Swings, J., Kalantzopoulos, G., Tsakalidou, E. and Messens, W. (2002). The biodiversity of lactic acid bacteria in Greek traditional wheat sourdough is reflected in both composition and metabolite formation. *Applied and Environmental Microbiology*, 68: 6059-6069.

Delcenserie, V., Taminiau, B., Delhalle, L., Nezer, C., Doyen, P., Crevecoeur, S., Roussey, D., Korsak, N. and Daube G. (2014). Microbiota characterisation of a Belgian protected designation of origin cheese, Herve cheese, using metagenomic analysis. *Journal of Dairy Science*, 9: 6046-6056.

Demirbas, F., Ispirli, H., Kurnaz, A.A., Yilmaz, M.T. and Dertli, E. (2017). Antimicrobial and functional properties of lactic acid bacteria isolated from sourdoughs. *LWT–Food Science and Technology*, 79: 361-366.

Di Cagno, R., Chaves Lopez, C., Tofalo, R., Gallo, G., De Angelis, M., Paparella, A., Hammes, W. and Gobbetti, M. (2008). Comparison of the compositional, microbiological, biochemical and volatile profile characteristics of three Italian PDO fermented sausages. *Meat Science*, 79: 224-235.

Di Cagno, R., De Angelis, M., Limitone, A., Minervini, F., Carnevali, A., Corsetti, P., Gaenzle, M., Ciati, R. and Gobbetti, M. (2006). Glucan and fructan production by sourdough *Weissella cibaria* and *Lactobacillus plantarum*. *Journal of Agriculture and Food Chemstry*, 54: 9873-9881.

Di Cagno, R., De Angelis, M., Auricchio, S., Greco, L., Clarke, C., De Vincenzi, M., Giovannini, C., D'Archivio, M., Landolfo, F., Parrilli, G., Minervini, F., Arendt, E. and Gobbetti, M. (2004). Sourdough bread made from wheat and nontoxic flours and started with selected Lactobacilli is tolerated in celiac sprue patients. *Applied and Environmental Microbiology*, 70: 1088-1096.

Drosinos, E.H., Mataragas, M., Xiraphi, N., Moschonas, G., Gaitis, F. and Metaxopoulos, J. (2005). Characterisation of the microbial flora from a traditional Greek fermented sausage. *Meat Science*, 69: 307-317.

Drosinos, E.H., Paramithiotis, S., Kolovos, G., Tsikouras, I. and Metaxopoulos, I. (2007). Phenotypic and technological diversity of lactic acid bacteria and staphylococci isolated from traditionally fermented sausages in southern Greece. *Food Microbiology*, 24: 260-270.

Duarte, W.F., Dias, D.R., de Melo Pereira, G.V., Gervásio, I.M. and Schwan, R.F. (2009). Indigenous and inoculated yeast fermentation of gabiroba (*Campomanesia pubescens*) pulp for fruit wine production. *Journal of Industrial Microbiology and Biotechnology*, 36: 557-569.

Dung, N.T.P., Rombouts, F.M. and Nout, M.J.R. (2007). Characteristics of some traditional Vietnamese starch-based rice wine starters (men), *LWT–Food Science and Technology*, 40: 130-135.

Egli, C.M., Edinger, W.D., Mitrakul, C.M. and Henick-Kling, T. (1998). Dynamics of indigenous and inoculated yeast populations and their effects on the sensory character of Riesling and Chardonnay wines. *Journal of Applied Microbiology*, 85: 779-789.

Falentin, H., Deutsch, S.M., Jan, G., Loux, V., Thierry, A., Parayre, S., Maillard, M.B., Dherbécourt, J., Cousin, F.J., Jardin, J., Siguier, P., Couloux, A., Barbe, V., Vacherie, B., Wincker, P., Gibrat, J.F., Gaillardin, C. and Lortal, S. (2010). The complete genome of *Propionibacterium freudenreichii* CIRM-BIA1T, a hardy actinobacterium with food and probiotic applications. *PLoS ONE*, 5: e11748.

Farahmand, E., Razavi, S.H., Yarmand, M.S. and Morovatpour, M. (2015). Development of Iranian rice-bran sourdough breads: Physicochemical, microbiological and sensorial characterisation during the storage period. *Quality Assurance and Safety of Crops and Foods*, 7: 295-303.

Federici, S., Ciarrocchi, F., Campana, R., Ciandrini, E., Blasi, G. and Baffone, W. (2014). Identification and functional traits of lactic acid bacteria isolated from Ciauscolo salami produced in central Italy. *Meat Science*, 98: 575-584.

Ferchichi, M., Valcheva, R., Oheix, N., Kabadjova, P., Prevost, H., Onno, B. and Dousset, X. (2008). Rapid investigation of French sourdough microbiota by restriction fragment length polymorphism of the 16S-23S rRNA gene intergenic spacer region. *World Journal of Microbiology and Biotechnology*, 24: 2425-2434.

Flander, L., Suortti, T., Katina, K. and Poutanen, K. (2011). Effects of wheat sourdough process on the quality of mixed oat-wheat bread. *LWT–Food Science and Technology*, 44: 656-664.

Fleming, H.P., Kyung, K.H. and Breidt, F. (1995). Vegetable fermentations. pp. 629-662. *In*: G. Reed and T.W. Nagodawithana (Eds.). Biotechnology (second completely revised edition), vol. 9, *Enzymes, Biomass, Food and Feed*, VCH, Weinheim, Germany.

Fonseca, S., Cachaldora, A., Gómez, M., Franco, I. and Carballo, J. (2013). Monitoring the bacterial population dynamics during the ripening of Galician chorizo, a traditional dry fermented Spanish sausage. *Food Microbiology*, 33: 77-84.

Fontana, C., Cocconcelli, P.S. and Vignolo, G. (2005). Monitoring the bacterial population dynamics during fermentation of artisanal Argentinean sausages. *International Journal of Food Microbiology*, 103: 131-142.

Foschino, R., Gallina, S., Andrighetto, C., Rossetti, L. and Galli, A. (2004). Comparison of cultural methods for the identification and molecular investigation of yeasts from sourdoughs for Italian sweet baked products. *FEMS Yeast Research*, 4: 609-618.

Fu, W., Rao, H., Tian, Y. and Xue W. (2020). Bacterial composition in sourdoughs from different regions in China and the microbial potential to reduce wheat allergens. *LWT–Food Science and Technology*, 117: 108669.

Fuka, M.M., Engel, M., Skelin, A., Redzepovic, S. and Schloter M. (2010). Bacterial communities associated with the production of artisanal Istrian cheese. *International Journal of Food Microbiology*, 142: 19-24.

Gaglio, R., Alfonzo, A., Francesca, N., Corona, O., Di Gerlando, R., Columba, P. and Moschetti G. (2017). Production of the Sicilian distillate 'Spiritu re fascitrari' from honey byproducts: An interesting source of yeast diversity. *International Journal of Food Microbiology*, 261: 62-72.

Gagnaire, V., Molle, D., Sorhaug, T. and Leonil, J. (1999). Peptidases of dairy propionic acid bacteria. *Lait*, 79: 43-57.

Galli, A., Franzetti, L. and Fortina, M.G. (1988). Isolation and identification of sourdough microflora. *Microbiologie Aliments Nutrition*, 6: 345-351.

Ganchev, I., Koleva, Z., Kizheva, Y., Moncheva, P. and Hristova, P. (2014). Lactic acid bacteria from spontaneously fermented rye sourdough. *Bulgarian Journal of Agricultural Science*, 20: 69-73.

Gaenzle, M.G. and Zheng, J. (2019). Lifestyles of sourdough Lactobacilli – Do they matter for microbial ecology and bread quality? *International Journal of Food Microbiology*, 302: 15-23.

Gaenzle, M.G., Ehmann, M. and Hammes, W.P. (1998). Modelling of growth of *Lactobacillus sanfranciscensis* and *Candida milleri* in response to process parameters of the sourdough fermentation. *Applied and Environmental Microbiology*, 64: 2616-2623.

Gao, W. and Zhang, L. (2018). Genotypic diversity of bacteria and yeasts isolated from Tibetan kefir. *International Journal of Food Microbiology*, 53: 1535-1540.

Garcia Fontan, M.C., Lorenzo, J.M., Parada, A., Franco, I. and Carballo, J. (2007a). Microbiological characteristics of 'androlla', a Spanish traditional pork sausage. *Food Microbiology*, 24: 52-58.

Garcia Fontan, M.C., Lorenzo, J.M., Martinez, S., Franco, I. and Carballo, J. (2007b). Microbiological characteristics of Botillo, a Spanish traditional pork sausage. *LWT–Food Science and Technology*, 40: 610-622.

Garcia-Armisen, T., Papalexandratou, Z., Hendryckx, H., Camu, N., Vrancken, G., De Vuyst, L. and Cornelis, P. (2010). Diversity of the total bacterial community associated with Ghanaian and Brazilian cocoa bean fermentation samples as revealed by a 16S rRNA gene clone library. *Applied Microbiology Biotechnology*, 87: 2281-2292.

Garcia-Varona, M., Santos, E.M., Jaime, I. and Rovira, J. (2000). Characterisation of Micrococcaceae isolated from different varieties of chorizo. *International Journal of Food Microbiology*, 54: 189-195.

Gardini, F., Suzzi, G., Lombardi, A., Galgano, F., Crudele, M.A., Andrighetto, C., Schirone, M. and Tofalo, R.A. (2001). Survey of yeasts in traditional sausages of southern Italy. *FEMS Yeast Research*, 1: 161-167.

Garofalo, C., Osimani, A., Milanovic, V., Aquilanti, L., De Filippis, F., Stellato, G., Di Mauro, S., Turchetti, B., Buzzini, P., Ercolini, D. and Clementi, F. (2015). Bacteria and yeast microbiota in milk kefir grains from different Italian regions. *Food Microbiology*, 49: 123-133.

Garofalo, C., Silvestri, G., Aquilanti, L. and Clementi, F. (2008). PCR-DGGE analysis of lactic acid bacteria and yeast dynamics during the production processes of three varieties of Panettone. *Journal of Applied Microbiology*, 105: 243-254.

Gashe, B.A. (1987). Kocho fermentation. *Journal of Applied Bacteriology*, 62: 473-477.

Gerez, C.L., Torino, M.I., Rollan, G. and Font de Valdez, G. (2009). Prevention of bread mold spoilage by using lactic acid bacteria with antifungal properties. *Food Control*, 20: 144-148.

Gill, C.O. (2005). Sources of bacterial contamination at slaughtering plants. pp. 231-243. *In*: J.N. Sofos (Ed.). *Improving the Safety of Fresh Meat*. CRC/Woodhead Publishing Limited, Cambridge, UK.

Gobbetti, M., Corsetti, A., Rossi, J., La Rosa, F. and De Vincenzi, S. (1994). Identification and clustering of lactic acid bacteria and yeasts from wheat sourdoughs of central Italy. *International Journal of Food Science*, 6: 85-94.

Gonzalez-Garcia, R.A., McCubbin, T., Navone, L., Stowers, C., Nielsen, L.K. and Marcellin, E. (2017). Microbial propionic acid production. *Fermentation*, 3: 21.

Goswami, G., Bora, S.S., Parveen, A., Boro, R.C. and Barooah, M. (2017). Identification and functional properties of dominant lactic acid bacteria isolated from Kahudi, a traditional rapeseed fermented food product of Assam, India. *Journal of Ethnic Foods*, 4: 187-197.

Greco, M., Mazzette, R., De Santis, E.P.L., Corona, A. and Cosseddu, A.M. (2005). Evolution and identification of lactic acid bacteria isolated during the ripening of Sardinian sausages. *Meat Science*, 69: 733-739.

Greppi, A., Ferrocino, I., La Storia, A., Rantsiou, K., Ercolini, D. and Cocolin, L. (2015). Monitoring of the microbiota of fermented sausages by culture independent rRNA-based approaches. *International Journal of Food Microbiology*, 212: 67-75.

Grijalva-Vallejos, N., Aranda, A. and Matallana, E. (2020). Evaluation of yeasts from Ecuadorian chicha by their performance as starters for alcoholic fermentations in the food industry. *International Journal of Food Microbiology*, 317: 108462.

Gulitz, A., Stadie, J., Wenning, M., Ehrmann, M.A. and Vogel, R.F. (2011). The microbial diversity of water kefir. *International Journal of Food Microbiology*, 151: 284-288.

Gullo, M. and Giudici, P. (2008). Acetic acid bacteria in traditional balsamic vinegar: Phenotypic traits relevant for starter cultures selection. *International Journal of Food Microbiology*, 125: 46-53.

Gullo, M., Romano, A.D., Pulvirenti, A. and Giudici, P. (2003). *Candida humilis* – Dominant species in sourdoughs for the production of durum wheat bran flour bread. *International Journal of Food Microbiology*, 80: 55-59.

Han, B.Z., Rombouts, F.M. and Nout M.J.R. (2001). A Chinese fermented soybean food. *International Journal of Food Microbiology*, 65: 1-10.

Harris, L.J. (1998). The microbiology of vegetable fermentations. pp. 45-72. *In*: B.J.B. Wood (Ed.). *Microbiology of Fermented Foods*. Blackie Academic and Professional, London, UK.

Hebert, E.M., Saavedra, L., Taranto, M.P., Mozzi, F., Magni, C., Nader, M.E., de Valdez, G.F., Sesma, F., Vignolo, G. and Raya, R.R. (2012). Genome sequence of the bacteriocin producing *Lactobacillus curvatus* strain CRL705. *Journal of Bacteriology*, 194: 538-539.

Heitmann, M., Zannini, E. and Arendt, E. (2017). Impact of *Saccharomyces cerevisiae* metabolites produced during fermentation on bread quality parameters: A review. *Critical Reviews of Food Science and Nutrition*, 58: 1152-1164.

Hidalgo, C., Mateo, E., Mas, A. and Torija, M.J. (2012). Identification of yeast and acetic acid bacteria isolated from the fermentation and acetification of persimmon (*Diospyros kaki*), *Food Microbiology*, 30: 98-104.

Hidalgo, C., Garcıa, D., Romero, J., Mas, A., Torija, M.J. and Mateo, E. (2013). *Acetobacter* strains isolated during the acetification of blueberry (*Vaccinium corymbosum* L.) wine. *Letters in Applied Microbiology*, 57: 227-232.

Hurtado, A., Reguant, C., Bordons, A. and Rozès, N. (2012). Lactic acid bacteria from fermented table olives. *Food Microbiology*, 31: 1-8.

Iacumin, L., Cecchini, F., Manzano, M., Osualdini, M., Boscolo, D., Orlic, S. and Comi, G. (2009). Description of the microflora of sourdoughs by culture-dependent and culture independent methods. *Food Microbiology*, 26: 128-135.

Iacumin, L., Comi, G., Cantoni, C. and Cocolin, L. (2006). Ecology and dynamics of coagulase-negative cocci isolated from naturally fermented Italian sausages. *Systemic and Applied Microbiology*, 29: 480-486.

Illeghems, K., De Vuyst, L. and Weckx, S. (2013). Complete genome sequence and comparative analysis of *Acetobacter pasteurianus* 386B, a strain well-adapted to the cocoa bean fermentation ecosystem. *BMC Genomics*, 14: 526.

Infantes, M. and Tourneur, C. (1991). Etude de la flore lactique de levains naturels de panification provenant de differentes regions francaises. *Sciences des Aliments*, 11: 527-545.

Jeong, S.H., Jung, J.Y., Lee, S.H., Jin, H.M. and Jeon, C.O. (2013a). Microbial succession and metabolite changes during fermentation of dongchimi, traditional Korean watery kimchi. *International Journal of Food Microbiology*, 164: 46-53.

Jeong, S.H., Lee, S.H., Jung, J.Y., Choi, E.J. and Jeon, C.O. (2013b). Microbial succession and metabolite changes during long-term storage of kimchi. *Journal of Food Science*, 78: M763-M769.

Jung, J.Y., Lee, S.H., Jin, H.M., Hahn, Y., Madsen, E.L. and Jeon, C.O. (2013). Metatranscriptomic analysis of lactic acid bacterial gene expression during kimchi fermentation. *International Journal of Food Microbiology*, 163: 171-179.

Kaditzky, S., Seitter, M., Hertel, C. and Vogel, R.F. (2008). Performance of *Lactobacillus sanfranciscensis* TMW 1.392 and its levansucrase deletion mutant in wheat dough and comparison of their impact on bread quality. *European Food Research and Technology*, 227: 433-442.

Kakabadze, E., Zago, M., Rossetti, L., Bonvini, B., Tidona, F., Carminati, D., Chanishvili, N. and Giraffa, G. (2019). Characterisation of lactic acid bacteria isolated from the Georgian, yoghurt-like matsoni. *International Journal of Dairy Technology*, 72: 373-380.

Kayanush, J.A. and Olson, D.W. (2017). A 100-year review: Yogurt and other cultured dairy products. *Journal of Dairy Science*, 100: 9987-10013.

Kayodé, A.P.P., Hounhouigana, J.D., Nout, M.J.R. and Niehof, A. (2007). Household production of sorghum beer in Benin: Technological and socio-economic aspects. *International Journal of Consumer Studies*, 31: 258-264.

Kimura, K. and Yokoyama. S. (2019). Trends in the application of *Bacillus* in fermented foods. *Current Opinion in Biotechnology*, 56: 36-42.

Korakli, M., Rossmann, A., Gaenzle, M.G. and Vogel, R.F. (2001). Sucrose metabolism and exopolysaccharide production in wheat and rye sourdoughs by *Lactobacillus sanfranciscensis*. *Journal of Agriculture and Food Chemistry*, 49: 5194-5200.

Korcari, D., Ricci, G., Quattrini, M. and Fortina, M.G. (2019). Microbial consortia involved in fermented spelt sourdoughs: Dynamics and characterisation of yeasts and lactic acid bacteria. *Letters in Applied Microbiology*, 70: 48-54.

Korsak, N., Taminiau, B., Leclercq, M., Nezer, C., Crevecoeur, S., Ferauche, C., Detry, E., Delcenserie, V. and Daube, G. (2015). Short communication: Evaluation of the microbiota of kefir samples using metagenetic analysis targeting the 16S and 26S ribosomal DNA fragments. *Journal of Dairy Science*, 98: 3684-3689.

Kotova, I.B., Cherdyntseva, T.A. and Netrusov, A.I. (2016). Russian kefir grains microbial composition and its changes during production process. *Advances in Experimental Medicine and Biology*, 932: 93-121.

Kozaki, M. and Uchimura, T. (1990). Micro-organisms in Chinese starter 'bubod' and rice wine 'tapuy' in the Philippines. *Journal of the Brewing Society of Japan*, 85: 818-824.

Kuligowski, M., Nowak, J. and Jasinska-Kuligowska, I. (2016). Fermentation process and bioavailability of phytochemicals from sourdough bread. pp. 68-78. *In*: C.M. Rosell, J. Bajerska and A.F. El Sheikha (Eds.). *Bread and its Fortification*. CRC Press, Boca Raton,

Laureys, D. and De Vuyst, L. (2014). Microbial species diversity, community dynamics, and metabolite kinetics of water kefir fermentation. *Applied and Environmental Microbiology*, 80: 2564-2572.

Lee, J.S., Heo, G.Y., Lee, J.W., Oh, Y.J., Park, J.A., Park, Y.H., Pyun, Y.R. and Ahn, J.S. (2005). Analysis of *kimchi* microflora using denaturing gradient gel electrophoresis, *International Journal of Food Microbiology*, 102: 143-150.

Lefeber, T., Gobert, W., Vrancken, G., Camu, N. and De Vuyst, L. (2011). Dynamics and species diversity of communities of lactic acid bacteria and acetic acid bacteria during spontaneous cocoa bean fermentation in vessels. *Food Microbiology*, 28: 457-464.

Lhomme, E., Lattanzi, A., Dousset, X., Minervini, F., De Angelis, M., Lacaze, G., Onno, B. and Gobbetti, M. (2015a). Lactic acid bacterium and yeast microbiotas of sixteen French traditional sourdoughs. *International Journal of Food Microbiology*, 215: 161-170.

Lhomme, E., Urien, C., Legrand, J., Dousset, X., Onno, B. and Sicard, D. (2016). Sourdough microbial community dynamics: An analysis during French organic bread-making processes. *Food Microbiology*, 53: 41-50.

Lhomme, E., Orain, S., Courcoux, P., Onno, B. and Dousset, X. (2015b). The predominance of *Lactobacillus sanfranciscensis* in French organic sourdoughs and its impact on related bread characteristics. *International Journal of Food Microbiology*, 213: 40-48.

Li, H., Jiang, D., Liu, W., Yang, Y., Zhang, Y., Jin, C. and Sun, S. (2020). Comparison of fermentation behaviors and properties of raspberry wines by spontaneous and controlled alcoholic fermentations. *Food Research International*, 128: 108801.

Li, J., Huang, Q., Zheng, X., Ge, Z., Lin, K., Zhang, D., Chen, Y., Wang, B. and Shi, X. (2020). Investigation of the lactic acid bacteria in Kazak cheese and their contributions to cheese fermentation. *Frontiers in Microbiology*, 11: 228.

Liu, C.H., Hsu, W.H., Lee, F.L. and Liao, C.C. (1996). The isolation and identification of microbes from a fermented tea beverage, Haipao, and their interactions during Haipao fermentation. *Food Microbiology*, 13: 407-415.

Lizaso, G., Chasco, M. and Beriain, J. (1999). Microbiological and biochemical changes during ripening of salchichon, a Spanish dry cured sausage. *Food Microbiology*, 16: 219-228.

Lynch, K.M., Zannini, E., Wilkinson, S., Daenen, L. and Arendt, E.K. (2019). Physiology of acetic acid bacteria and their role in vinegar and fermented beverages. *Comprehensive Reviews of Food Science and Food Safety*, 18: 587-625.

Lyumugabe, L., Kamaliza, G., Bajyana, E. and Thonart, Ph. (2010). Microbiological and physico-chemical characteristics of Rwandese traditional beer 'Ikigage'. *African Journal of Biotechnology*, 9: 4241-4246.

Magalhaes, K.T., Pereira, G.V.D., Dias, D.R. and Schwan, R.F. (2010). Microbial communities and chemical changes during fermentation of sugary Brazilian kefir. *World Journal of Microbiology and Biotechnology*, 26: 1241-1250.

Magnusson, J., Strom, K., Roos, S., Sjogren, J. and Schnurer, J. (2003). Broad and complex antifungal activity among environmental isolates of lactic acid bacteria. *FEMS Microbiology Letters*, 219: 129-135.

Mantynen, V.H., Korhola, M., Gudmundsson, H., Turakainen, H., Alfredsson, G.A., Salovaara, H. and Lindstrom, K. (1999). A polyphasic study on the taxonomic position of industrial sour dough yeasts. *Systemic and Applied Microbiology*, 22: 87-96.

Maoloni, A., Blaiotta, G., Ferrocino, I., Mangia, N.P., Osimani, A., Milanović, V., Cardinali, F., Cesaro, C., Garofalo, C., Clementi, F., Pasquini, M., Trombetta, M.F., Cocolin, L. and Aquilanti, L. (2020). Microbiological characterisation of Gioddu, an Italian fermented milk. *International Journal of Food Microbiology*, 323: 108610.

Maoura, N., Mbaiguinam, M., Nguyen, H.V., Gaillardin, C. and Pourquie, J. (2005). Identification and typing of the yeast strains isolated from bili bili, a traditional sorghum beer of Chad. *African Journal of Biotechnology*, 4: 646-656.

Maragatham, C. and Panneerselvam, A. (2011). Isolation, identification and characterisation of wine-yeast from rotten papaya fruits for wine production. *Advances in Applied Science Research*, 2: 93-98.

Marcellino, N. and Benson, D.R. (2013). The good, the bad, and the ugly: Tales of mold-ripened cheese. *Microbiology Spectrum*, 1: CM-0005-2012.

Mariotti, M., Garofalo, C., Aquilanti, L., Osimani, A., Fongaro, L., Tavoletti, S., Hager, A.S. and Clementi, F. (2014). Barley flour exploitation in sourdough bread-making: A technological, nutritional and sensory evaluation. *LWT–Food Science and Technology*, 59: 973-980.

Marsh, A.J., O'Sullivan, O., Hill, C., Ross, R.P. and Cotter, P.D. (2014). Sequence-based analysis of the bacterial and fungal compositions of multiple kombucha (tea fungus) samples. *Food Microbiology*, 38: 171-178.

Marshall, V.M. (1984). Flavour development in fermented milks. pp. 153-186. *In*: F.L. Davies and B.A. Law (Eds.). *Advances in the Microbiology and Biochemistry of Cheese and Fermented Milk*. Elsevier Applied Science, London, UK.

Martin, B., Garriga, M., Hugas, M., Bover-Cid, S., Veciana-Nogués, M.T. and Aymerich, T. (2006). Molecular, technological and safety characterisation of Gram-positive catalase-positive cocci from slightly fermented sausages. *International Journal of Food Microbiology*, 107: 148-158.

Mauriello, G., Casaburi, A., Blaiotta, G. and Villani, F. (2004). Isolation and technological properties of coagulase negative staphylococci from fermented sausages of southern Italy. *Meat Science*, 67: 149-158.

Mbawala, A., Mahbou, P.Y., Mouafo, H.T. and Tatsadjieu, L.N. (2013). Antibacterial activity of some lactic acid bacteria isolated from a local fermented milk product (pendidam) in ngaoundere, Cameroon. *Journal of Animal and Plant Sciences*, 23: 157-166.

Mbozo, A.B.V., Kobawila, S.C., Anyogu, A., Awamaria, B., Louembe, D., Sutherland, J.P. and Ouoba, L.I.I. (2017). Investigation of the diversity and safety of the predominant *Bacillus pumilus* sensu lato and other *Bacillus* species involved in the alkaline fermentation of cassava leaves for the production of Ntoba Mbodi. *Food Control*, 82: 154-162.

Meersman, E., Steensels, J., Mathawan, M., Wittocx, P.J., Saels, V., Struyf, N., Bernaert, H., Vrancken, G. and Verstrepen, K.J. (2013). Detailed analysis of the microbial population in Malaysian spontaneous cocoa pulp fermentations reveals a core and variable microbiota. *PLoS ONE*, 8: e81559.

Menezes, L.A.A., Savo Sardaro, M.L., Duarte, R.T.D., Mazzon, R.R., Neviani, E., Gatti, M. and De Dea Lindner, J. (2020). Sourdough bacterial dynamics revealed by metagenomic analysis in Brazil. *Food Microbiology*, 85: 103302.

Metaxopoulos, J., Samelis, J. and Papadelli, M. (2001). Technological and microbiological evaluation of traditional processes as modified for the industrial manufacturing of dry fermented sausage in Greece. *International Journal of Food Science*, 13: 3-18.

Miguel, M.G.D.P., Cardoso, P.G., Magalhaes, K.T. and Schwan, R.F. (2011). Profile of microbial communities present in Tibico (sugary kefir) grains from different Brazilian states. *World Journal of Microbiology and Biotechnology*, 27: 1875-1884.

Miguel, M.G.D.P., Cardoso, P.G., Lago, L.D. and Schwan, R.F. (2010). Diversity of bacteria present in milk kefir grains using culture-dependent and culture-independent methods. *Food Research International*, 43: 1523-1528.

Minervini, F., Lattanzi, A., De Angelis, M., Celano, G. and Gobbetti, M. (2015). House microbiotas as sources of lactic acid bacteria and yeasts in traditional Italian sourdoughs. *Food Microbiology*, 52: 66-76.

Minervini, F., Di Cagno, R., Lattanzi, A., De Angelis, M., Antonielli, L., Cardinali, G., Cappelle, S. and Gobbetti, M. (2012). Lactic acid bacterium and yeast microbiotas of 19 sourdoughs used for traditional/typical Italian breads: Interactions between ingredients and microbial species diversity. *Applied and Environmental Microbiology*, 78: 1251-1264.

Mo, L., Jin, H., Pan, L., Hou, Q., Li, C., Darima, I., Zhang, H. and Yu, J. (2019). Biodiversity of lactic acid bacteria isolated from fermented milk products in Xinjiang, China. *Food Biotechnology*, 33: 174-192.

Moroni, A.V., Zannini, E., Sensidoni, G. and Arendt, E.K. (2012). Exploitation of buckwheat sourdough for the production of wheat bread. *European Food Research and Technology*, 235: 659-668.

Mu, W., Yu, S., Zhu, L., Jiang, B. and Zhang, T. (2012). Production of 3-phenyllactic acid and 4-hydroxyphenyllactic acid by *Pediococcus acidilactici* DSM 20284 fermentation. *European Food Research and Technology*, 235: 581-585.

Mueller, M.R.A., Wolfrum, G., Stolz, P., Ehrmann, M.A. and Vogel, R.F. (2001). Monitoring the growth of *Lactobacillus* species during rye flour fermentation. *Food Microbiology*, 18: 217-227.

Murakami, H. (1971). Classification of the koji mould. *The Journal of General and Applied Microbiology*, 17: 281-309.

Ndagano, D., Lamoureux, T., Dortu, C., Vandermoten, S. and Thonart, P. (2011). Antifungal activity of 2 lactic acid bacteria of the *Weissella* genus isolated from food. *Journal of Food Science*, 76: M305-M311.

Nguyen, D.T.L., Van Hoorde, K., Cnockaert, M., De Brandt, E., Aerts, M., Thanh, L. and Vandamme, P. (2013). A description of the lactic acid bacteria microbiota associated with the production of traditional fermented vegetables in Vietnam. *International Journal of Food Microbiology*, 163: 19-27.

Nionelli, L., Curri, N., Curiel, J.A., Di Cagno, R., Pontonio, E., Cavoski, I., Gobbetti, M. and Rizzello, C.G. (2014). Exploitation of Albanian wheat cultivars: Characterisation of the flours and lactic acid bacteria microbiota, and selection of starters for sourdough fermentation. *Food Microbiology*, 44: 96-107.

Nout, M.J.R. and Kiers, J.L. (2005). Tempe fermentation, innovation and functionality: Update into the 3rd millennium. *Journal of Applied Microbiology*, 98: 789-805.

Nyanga, L.K., Nout, M.J.R., Gadaga, T.H., Theelen, B., Boekhout, T. and Zwietering, M.H. (2007). Yeasts and lactic acid bacteria microbiota from masau (*Ziziphus mauritiana*) fruits and their fermented fruit pulp in Zimbabwe. *International Journal of Food Microbiology*, 120: 159-166.

Nychas, G.J.E., Skandamis, P.N., Tassou, C.C. and Koutsoumanis, K.P. (2008). Meat spoilage during distribution. *Meat Science*, 78: 77-89.

Oberman, H. and Libudzisz, Z. (1998). Fermented milks. pp. 308-350. *In*: B.J.B. Woods (Ed.). *Microbiology of Fermented Foods*. Blackie Academic/Professional, London, UK.

Ogbadu, L.J. and Okagbue, R.N. (1988). Fermentation of African locust bean *Parkia biglobosa*, seeds: Involvement of different species of *Bacillus*. *Food Microbiology*, 5: 195-199.

Ogunsakin, O.A., Banwo, K., Ogunremi, O.R. and Sanni, A.I. (2015). Microbiological and physicochemical properties of sourdough bread from sorghum flour. *International Food Research Journal*, 22: 2610-2618.

Ohenhen, R.E., Imarenezor, E.P.K. and Kihuha, A.N. (2013). Microbiome of madila – A southern African fermented milk product. *International Journal of Basic and Applied Science*, 2: 170-175.

Omafuvbe, B.O., Shonukan, O.O. and Abiose, S.H. (2000). Microbiological and biochemical changes in the traditional fermentation of soybean for 'soy-daddawa' – Nigerian food condiment. *Food Microbiology*, 17: 469-474.

Omafuvbe, B.O., Abiose S.H. and Adaraloye (1999). The production of 'Kpaye' – A fermented condiment from *Prosopis africana* (O.O. Guill and Perr) Taub. Seeds. *International Journal of Food Microbiology*, 51: 183-186.

Osimani, A., Zannini, E., Aquilanti, L., Mannazzu, I., Comitini, F. and Clementi, F. (2009). Lactic acid bacteria and yeasts from wheat sourdoughs of the Marche region. *International Journal of Food Science*, 21: 269-286.

Ottogalli, G., Galli, A. and Foschino, R. (1996). Italian bakery products obtained with sour dough: Characterisation of the typical microflora. *Advances in Food Science*, 18: 131-144.

Padilla, B., Garcia- Fernández, D., González, B., Izidoro, I., Esteve-Zarzoso, B., Beltran, G. and Mas, A. (2016). Yeast biodiversity from DOQ priorat uninoculated fermentations. *Frontiers in Microbiology*, 7: 930.

Palmeri, M., Mancuso, I., Barbaccia, P., Cirlincione, F. and Scatassa, M. (2019). Dominant lactic acid bacteria in naturally fermented milks from Messinese Goat's breed. *Journal of Food Quality and Hazards Control*, 6: 66-72.

Pangallo, D., Šaková, N., Koreňová, J., Puškárová, A., Kraková, L., Valík, L. and Kuchta, T. (2014). Microbial diversity and dynamics during the production of May bryndza cheese. *International Journal of Food Microbiology*, 170: 38-43.

Papalexandratou, Z., Falony, G., Romanens, E., Jimenez, J.C., Amores, F., Daniel, H.M. and De Vuyst, L. (2011). Species diversity, community dynamics, and metabolite kinetics of the microbiota associated with traditional Ecuadorian spontaneous cocoa bean fermentations. *Applied and Environmental Microbiology*, 77: 7698-7714.

Papamanoli, E., Tzanetakis, N., Litopoulou-Tzanetaki, E. and Kotzekidou, P. (2003). Characterisation of lactic acid bacteria isolated from a Greek dry-fermented sausage in respect of their technological and probiotic properties. *Meat Science*, 65: 859-867.

Papamanoli, E., Kotzekidou, P., Tzanetakis, N. and Litopoulou-Tzanetaki, E. (2002). Characterisation of micrococcaceae isolated from dry fermented sausage. *Food Microbiology*, 19: 441-449.

Paramithiotis, S., Doulgeraki, A.I., Vrelli, A., Nychas, G.J.E. and Drosinos, E.H. (2016). Evolution of the microbial community during traditional fermentation of globe artichoke immature inflorescence. *International Journal of Clinical and Medical Microbiology*, 1: 117.

Paramithiotis, S., Kouretas, K. and Drosinos, E.H. (2014a). Effect of ripening stage on the development of the microbial community during spontaneous fermentation of green tomatoes. *Journal of Science of Food and Agriculture*, 94: 1600-1606.

Paramithiotis, S., Doulgeraki, A.I., Karahasani, A. and Drosinos, E.H. (2014b). Microbial population dynamics during spontaneous fermentation of *Asparagus officinalis* L. young sprouts. *European Food Research and Technology*, 239: 297-304.

Paramithiotis, S., Hondrodimou, O.L. and Drosinos, E.H. (2010a). Development of the microbial community during spontaneous cauliflower fermentation. *Food Research International*, 43: 1098-1103.

Paramithiotis, S., Drosinos, E.H., Sofos, J. and Nychas, G.J.E. (2010b). Fermentation: Microbiology and biochemistry. pp. 185-198. *In*: F. Toldra (Ed.). *Handbook of Meat Processing*. Springer, New York, USA.

Paramithiotis, S., Tsiasiotou, S. and Drosinos, E.H. (2010c). Comparative study of spontaneously fermented sourdoughs originating from two regions of Greece: Peloponnesus and Thessaly. *European Food Research and Technology*, 231: 883-890.

Paramithiotis, S., Gioulatos, S., Tsakalidou, E. and Kalantzopoulos, G. (2006). Interactions between *Saccharomyces cerevisiae* and lactic acid bacteria in sourdough. *Process Biochemistry*, 41: 2429-2433.

Paramithiotis, S., Kagkli, D.M., Blana, V.A., Nychas, G.J.E. and Drosinos, E.H. (2008). Identification and characterisation of *Enterococcus* spp. in Greek spontaneous sausage fermentation. *Journal of Food Protection*, 71: 1244-1247.

Paramithiotis, S., Mueller, M.R.A., Ehrmann, M.A., Tsakalidou, E., Seiler, H., Vogel, R.F. and Kalantzopoulos, G. (2000). Polyphasic identification of wild yeast strains isolated from Greek sourdoughs. *Systemic and Applied Microbiology*, 23: 156-164.

Pardali, E., Paramithiotis, S., Papadelli, M., Mataragas, M. and Drosinos, E.H. (2017). Lactic acid bacteria population dynamics during spontaneous fermentation of radish (*Raphanus sativus* L.) roots in brine. *World Journal of Microbiology and Biotechnology*, 33: 110.

Parente, E., Grieco, S. and Crudele, M.A. (2001). Phenotypic diversity of lactic acid bacteria isolated from fermented sausages produced in Basilicata (southern Italy). *Journal of Applied Microbiology*, 90: 943-952.

Parizzi, L.P., Grassi, M.C.B., Llerena, L.A., Carazzolle, M.F., Queiroz, V.L., Lunardi, I., Zeidler, A.F., Teixeira, P.J., Mieczkowski, P., Rincones, J. and Pereira, G.A. (2012). The genome sequence of *Propionibacterium acidipropionici* provides insights into its biotechnological and industrial potential. *BMC Genomics*, 13: 562.

Parkouda, C., Diawara, B. and Ouoba, L.I.I. (2008). Technology and physico-chemical characteristics of Bikalga, alkaline fermented seeds of *Hibiscus sabdariffa*. *African Journal of Biotechnology*, 7: 916-922.

Parkouda, C., Nielsen, D.S., Azokpota, P., Ouoba, L.I.I., Amoa-Awua, W.K., Thorsen, L., Hounhouigan, J.D., Jensen, J.S., Tano-Debrah, K., Diawara, B. and Jakobsen, M. (2009). The microbiology of alkaline-fermentation of indigenous seeds used as food condiments in Africa and Asia. *Critical Reviews in Microbiology*, 35: 139-156.

Parkouda, C., Thorsen, L., Compaoré, C.S., Nielsen, D.S., Tano-Debrah, K., Jensen, J.S., Diawara, B. and Jakobsen, M. (2010). Microorganisms associated with Maari, a Baobab seed fermented product. *International Journal of Food Microbiology*, 142: 292-301.

Pateraki, C., Paramithiotis, S., Doulgeraki, A.I., Kallithraka, S., Kotseridis, G. and Drosinos, E.H. (2014). Effect of sulfur dioxide addition in wild yeast population dynamics and polyphenolic composition during spontaneous red wine fermentation from *Vitis vinifera* cultivar Agiorgitiko. *European Food Research and Technology*, 239: 1067-1075.

Pérez-Díaz, M., Hayes, J., Medina, E., Anekella, K., Daughtry, K., Dieck, S., Levi, M., Price, R., Butz, N., Lu, Z. and Azcarate-Peril, M.A. (2017). Reassessment of the succession of lactic acid bacteria in commercial cucumber fermentations and physiological and genomic features associated with their dominance. *Food Microbiology*, 63: 217-227.

Pini, F., Aquilani, C., Giovannetti, L., Viti, C. and Pugliese, C. (2020). Characterisation of the microbial community composition in Italian Cinta Senese sausages dry-fermented with natural extracts as alternatives to sodium nitrite. *Food Microbiology*, 89: 103417.

Pisacane, V., Callegari, M.L., Puglisi, E., Dallolio, G. and Rebecchi, A. (2015). Microbial analyses of traditional Italian salami reveal microorganisms transfer from the natural casing to the meat matrix. *International Journal of Food Microbiology*, 207: 57-65.

Piveteau, P. (1999). Metabolism of lactate and sugars by dairy propionibacteria: A review. *Lait*, 79: 23-41.

Piveteau, P., Condon, S. and Cogan, T.M. (2000). Inability of dairy propionibacteria to grow in milk from low inocula. *Journal of Dairy Research*, 67: 65-71.

Polka, J., Rebecchi, A., Pisacane, V. and Morelli, L. (2015). Bacterial diversity in typical Italian salami at different ripening stages as revealed by high-throughput sequencing of 16s rRNA amplicons. *Food Microbiology*, 46: 342-356.

Prado, N., Sampayo, M., González, P., Lombó, F. and Díaz, J. (2019). Physicochemical, sensory and microbiological characterisation of Asturian Chorizo, a traditional fermented sausage manufactured in northern Spain. *Meat Science*, 156: 118-124.

Prust, C., Hoffmeister, M., Liesegang, H., Wiezer, A., Fricke, W. F., Ehrenreich, A., Gottschalk, G. and Deppenmeier, U. (2005). Complete genome sequence of the acetic acid bacterium *Gluconobacter oxydans*. *Nature Biotechnology*, 23: 195-200.

Puerari, C., Magalhães-Guedes, K.T. and Schwan, R.F. (2015). Bacaba beverage produced by Umutina Brazilian Amerindians: Microbiological and chemical characterisation. *Brazilian Journal of Microbiology*, 46: 1207-1216.

Rabah, H., Carmo, F.L.R. and Jan, G. (2017). Dairy propionibacteria: Versatile probiotics. *Microorganisms*, 5: 24.

Randazzo, C.L., Heilig, H., Restuccia, C., Giudici, P. and Caggia, C. (2005). Bacterial population in traditional sourdough evaluated by molecular methods. *Journal of Applied Microbiology*, 99: 251-258.

Rantsiou, K., Drosinos, E.H., Gialitaki, M., Metaxopoulos, I., Comi, G. and Cocolin, L. (2006). Use of molecular tools to characterise *Lactobacillus* spp. isolated from Greek traditional fermented sausages. *International Journal of Food Microbiology*, 112: 215-222.

Rantsiou, K., Drosinos, E.H., Gialitaki, M., Urso, R., Krommer, J., Gasparik-Reichardt, J., Toth, S., Metaxopoulos, I., Comi, G. and Cocolin, L. (2005). Molecular characterisation of *Lactobacillus* species isolated from naturally fermented sausages produced in Greece, Hungary and Italy. *Food Microbiology*, 22: 19-28.

Rebecchi, A., Crivori, S., Sarra, P.G. and Cocconcelli, P.S. (1998). Physiological and molecular techniques for the study of bacterial community development in sausage fermentation. *Journal Applied Microbiology*, 84: 1043-1049.

Rieder, A., Holtekjolen, A.K., Sahlstrom, S. and Moldestad, A. (2012). Effect of barley and oat flour types and sourdoughs on dough rheology and bread quality of composite wheat bread. *Journal of Cereal Science*, 55: 44-52.

Rizzello, C.G., Calasso, M., Campanella, D., De Angelis, M. and Gobbetti, M. (2014). Use of sourdough fermentation and mixture of wheat, chickpea, lentil and bean flours for enhancing the nutritional, texture and sensory characteristics of white bread. *International Journal of Food Microbiology*, 180: 78-87.

Robert, H., Gabriel, V. and Fontagnı-Faucher, C. (2009). Biodiversity of lactic acid bacteria in French wheat sourdough as determined by molecular characterisation using species specific PCR. *International Journal of Food Microbiology*, 135: 53-59.

Rosenquist, H. and Hansen, A. (2000). The microbial stability of two bakery sourdoughs made from conventionally and organically grown rye. *Food Microbiology*, 17: 241-250.

Rosenstein, R., Nerz, C., Biswas, L., Resch, A., Raddaz, G., Schuster, S.C. and Gotz, F. (2009). Genome analysis of the meat starter culture bacterium *Staphylococcus carnosus* TM300. *Applied and Environmental Microbiology*, 75: 811-822.

Rossello, C., Barbas, J.I., Berna, A. and Lopez, N. (1995). Microbial and chemical changes in 'Sobrasada' during ripening. *Meat Science*, 40: 379-385.

Russo, P., Fares, C., Longo, A., Spano, G. and Capozzi, V. (2017). *Lactobacillus plantarum* with broad antifungal activity as a protective starter culture for bread production. *Foods*, 6: E110.

Ryan, L.A., Zannini, E., Dal Bello, F., Pawlowska, A., Koehler, P. and Arendt, E.K. (2011). *Lactobacillus amylovorus* DSM 19280 as a novel food-grade antifungal agent for bakery products. *International Journal of Food Microbiology*, 29: 276-283.

Sabate, J., Cano, J., Querol, A. and Guillamon, J.M. (1998). Diversity of *Saccharomyces cerevisiae* strains in wine fermentations: Analysis for two consecutive years. *Letters in Applied Microbiology*, 26: 452-455.

Sagdic, O., Ozturk, I., Yapar, N. and Yetim, H. (2014). Diversity and probiotic potentials of lactic acid bacteria isolated from gilaburu, a traditional Turkish fermented European cranberry bush (*Viburnum opulus* L.) fruit drink. *Food Research International*, 64: 537-545.

Salovaara, H. and Katunpaa, H. (1984). An approach to the classification of Lactobacilli isolated from Finnish sour rye dough ferments. *Acta Scientiarum Polonorum Technologia Alimentaria*, 10: 231-239.

Samelis, J., Maurogenakis, F. and Metaxopoulos, J. (1994). Characterisation of lactic acid bacteria isolated from naturally fermented Greek dry salami. *International Journal of Food Microbiology*, 23: 179-196.

Samelis, J., Metaxopoulos, J., Vlassi, M. and Pappa, A. (1998). Stability and safety of traditional Greek salami – A microbiological ecology study. *International Journal of Food Microbiology*, 44: 69-82.

Sanni, A.I. and Ogbonna, D.N. (1991). The production of owoh – A Nigerian fermented seasoning agent from cotton seed (*Gossypium hirsutum* L.). *Food Microbiology*, 8: 223-229.

Sarkar, P.K., Hasenack, B. and Nout, M.J.R. (2002). Diversity and functionality of Bacillus and related genera isolated from spontaneously fermented soybeans (Indian Kinema) and locust beans (African Soumbala). *International Journal of Food Microbiology*, 77: 175-186.

Satora, P. and Tuszyński, T. (2005). Biodiversity of yeasts during plum Wegierka Zwykla spontaneous fermentation. *Food Technology and Biotechnology*, 43: 277-282.

Scheirlinck, I., Van der Meulen, R., Van Schoor, A., Huys, G., Vandamme, P., De Vuyst, L. and Vancanneyt, M. (2007a). *Lactobacillus crustorum* sp. nov., isolated from two traditional Belgian wheat sourdoughs. *International Journal of Systemic and Evolutionary Microbiology*, 57: 1461-1467.

Scheirlinck, I., Van der Meulen, R., Van Schoor, A., Vancanneyt, M., De Vuyst, L., Vandamme, P. and Huys, G. (2007b). Influence of geographical origin and flour type on diversity of lactic acid bacteria in traditional Belgian sourdoughs. *Applied and Environmental Microbiology*, 73: 6262-6269.

Scheirlinck, I., Van der Meulen, R., Van Schoor, A., Vancanneyt, M., De Vuyst, L., Vandamme, P. and Huys, G. (2008). Taxonomic structure and stability of the bacterial community in Belgian sourdough ecosystems as assessed by culture and population fingerprinting. *Applied and Environmental Microbiology*, 74: 2414-2423.

Sefa-Dedeh, S., Sanni, A.I., Tetteh, G. and Sakyi-Dawson, E. (1999). Yeasts in the traditional brewing of pito in Ghana. *World Journal of Microbiology and Biotechnology*, 15: 593-597.

Selgas, M.D. and Garcia, M.L. (2007). Starter cultures: Yeasts. pp. 159-170. *In*: F. Toldra (Ed.). *Handbook of Fermented Meat and Poultry*. Blackwell Publishing Professional. Iowa, USA.

Setoguchi, S., Mizutani, O., Yamada, O., Futagami, T., Iwai, K., Takase, Y. and Tamaki, H. (2019). Effect of *pepa* deletion and overexpression in *Aspergillus luchuensis* on sweet potato shochu brewing. *Journal of Bioscience and Bioengineering*, 128: 456-462.

Shrestha, A.K., Dahal, N.R. and Ndungutse, V. (2010). *Bacillus* fermentation of soybean: A review. *Journal of Food Science and Technology, Nepal*, 6: 1-9.

Siezen, R.J., Francke, C., Renckens, B., Boekhorst, J., Wels, M., Kleerebezem, M. and van Hijum, S.A. (2012). Complete resequencing and reannotation of the *Lactobacillus plantarum* WCFS1 genome. *Journal of Bacteriology*, 194: 195-196.

Silvestri, G., Santarelli, S., Aquilanti, L., Beccaceci, A., Osimani, A., Tonucci, F. and Clementi, F. (2007). Investigation of the microbial ecology of Ciauscolo, a traditional Italian salami, by culture-dependent techniques and PCR-DGGE. *Meat Science*, 77: 413-423.

Slukova, M., Hinkova, A., Henke, S., Smrz, F., Lukacikova, M., Pour, V. and Bubnik, Z. (2016). Cheese whey treated by membrane separation as a valuable ingredient for barley sourdough preparation. *Journal of Food Engineering*, 172: 38-47.

Spicher, G. (1959). Die Mikroflora des Sauerteiges. I. Mitteilung: Untersuchungen uber die Art der in Sauerteigen anzutreffenden stabchenformigen Milchsaurebakterien (Genus *Lactobacillus* Beijerinck). *Zeitblatt fur Bakteriologie II Abt*, 113: 80-106.

Spitaels, F., Wieme, A.D., Janssens, M., Aerts, M., Van Landschoot, A., De Vuyst, L. and Vandamme, P. (2015). The microbial diversity of an industrially produced lambic beer shares members of a traditionally produced one and reveals a core microbiota for lambic beer fermentation. *Food Microbiology*, 49: 23-32.

Spotti, E., Berni, E. and Cacchioli, C. (2008). Characteristics and applications of molds. pp. 181-198. *In*: F. Toldra (Ed.). *Meat Biotechnology*. Springer, New York, USA.

Stamer, J.R., Stoyla, B.O. and Dunckel, B.A. (1971). Growth rates and fermentation patterns of lactic acid bacteria associated with the sauerkraut fermentation. *Journal of Milk Food Technology*, 34: 521-525.

Stanley, D., Bandara, A., Fraser, S., Chambers, P.J. and Stanley, G.A. (2010). The ethanol stress response and ethanol tolerance of *Saccharomyces cerevisiae*. *Journal of Applied Microbiology*, 109: 13-24.

Sterr, Y., Weiss, A. and Schmidt, H. (2009). Evaluation of lactic acid bacteria for sourdough fermentation of amaranth. *International Journal of Food Microbiology*, 136: 75-82.

Stringini, M., Comitini, F., Taccari, M. and Ciani, M. (2009). Yeast diversity during tapping and fermentation of palm wine from Cameroon. *Food Microbiology*, 26: 415-420.

Succi, M., Reale, A., Andrighetto, C., Lombardi, A., Sorrentino, E. and Coppola, R. (2003). Presence of yeasts in southern Italian sourdoughs from *Triticum aestivum* flour. *FEMS Microbiology Letters*, 225: 143-148.

Suprianto, Ohba, R., Koga, T. and Ueda, S. (1989). Liquefaction of glutinous rice and aroma formation in tapé preparation by ragi. *Journal of Fermentation and Bioengineering*, 64: 249-252.

Tamang, J.P. and Thapa, S. (2006). Fermentation dynamics during production of bhaati jaanr, a traditional fermented rice beverage of the eastern Himalayas. *Food Biotechnology*, 20: 251-261.

Tamang, J.P., Tamang, B., Schillinger, U., Franz, C.M.A.P., Gores, M. and Holzapfel, W.H. (2005). Identification of predominant lactic acid bacteria isolated from traditionally fermented vegetable products of the Eastern Himalayas. *International Journal of Food Microbiology*, 105: 347-356.

Tanguler, H. and Erten, H. (2012). Chemical and microbiological characteristics of shalgam (şalgam): A traditional Turkish lactic acid fermented beverage. *Journal of Food Quality*, 35: 298-306.

Thapa, S. and Tamang, J.P. (2006). Microbiological and physico-chemical changes during fermentation of kodo ko jaanr, a traditional alcoholic beverage of the Darjeeling hills and Sikkim. *Indian Journal of Microbiology*, 46: 333-341.

Tieking, M., Korakli, M., Ehrmann, M.A., Gaenzle, M.G. and Vogel, R.F. (2003). *In situ* production of exopolysaccharides during sourdough fermentation by cereal and intestinal isolates of lactic acid bacteria. *Applied and Environmental Microbiology*, 69: 945-952.

Urso, R., Comi, G. and Cocolin, L. (2006). Ecology of lactic acid bacteria in Italian fermented sausages: Isolation, identification and molecular characterisation. *Systemics and Applied Microbiology*, 29: 671-680.

Valcheva, R., Korakli, M., Onno, B., Prevost, H., Ivanova, I., Ehrmann, M.A., Dousset, X., Gaenzle, M.G. and Vogel, R.F. (2005). *Lactobacillus hammesii* sp. nov., isolated from French sourdough. *International Journal of Systemics and Evolutionary Microbiology*, 55: 763-767.

Valles, B.S., Bedrinana, R.P., Tascon, N.F., Simon, A.Q. and Madrera, R.R. (2007). Yeast species associated with the spontaneous fermentation of cider. *Food Microbiology*, 24: 25-31.

Van der Aa Kühle, A., Jespersen, L., Glover, R.L., Diawara, B. and Jakobsen, M. (2001). Identification and characterisation of *Saccharomyces cerevisiae* strains isolated from West African sorghum beer. *Yeast*, 18: 1069-1079.

Van Kerrebroeck, S., Maes, D. and De Vuyst, L. (2017). Sourdoughs as a function of their species diversity and process conditions, a meta-analysis. *Trends in Food Science and Technology*, 68: 152-159.

Vancanneyt, M., Neysens, P., De Wachter, M., Engelbeen, K., Snauwaert, C., Cleenwerck, I., Van der Meulen, R., Hoste, B., Tsakalidou, E., De Vuyst, L. and Swings, J. (2005). *Lactobacillus acidifarinae* sp. nov. and *Lactobacillus zymae* sp. nov., from wheat sourdoughs. *International Journal of Systemic and Evolutionary Microbiology*, 55: 615-620.

Venema, K. and Surono, I.S. (2018). Microbiota composition of dadih – A traditional fermented buffalo milk of west Sumatra. *Letters in Applied Microbiology*, 68: 234-240.

Ventimiglia, G., Alfonzo, A., Galluzzo, P., Corona, O., Francesca, N., Caracappa, S., Moschetti, G. and Settanni, L. (2015). Codominance of *Lactobacillus plantarum* and obligate heterofermentative lactic acid bacteria during sourdough fermentation. *Food Microbiology*, 51: 57-68.

Vernocchi, P., Valmorri, S., Dalai, I., Torriani, S., Gianotti, A., Suzzi, G., Guerzoni, M.E., Mastrocola, D. and Gardini, F. (2004a). Characterisation of the yeast population involved in the production of a typical Italian bread. *Journal of Food Science*, 69: 182-186.

Vernocchi, P., Valmorri, S., Gatto, V., Torriani, S., Gianotti, A., Suzzi, G., Guerzoni, M.E. and Gardini, F. (2004b). A survey on yeast microbiota associated with an Italian traditional sweet leavened baked good fermentation. *Food Research International*, 37: 469-476.

Vignolo, G., Fontana, C. and Fadda, S. (2010). Fermentation: Microbiology and biochemistry. pp. 379-398. *In*: F. Toldra (Ed.). *Handbook of Meat Processing*. Springer, New York, USA.

Vogel, R.F., Pavlovic, M., Ehrmann, M.A., Wiezer, A., Liesegang, H., Offschanka, S., Voget, S., Angelov, A., Böcker, G. and Liebl, W. (2011). Genomic analysis reveals *Lactobacillus sanfranciscensis* as stable element in traditional sourdoughs. *Microbial Cell Factories*, 10: S6.

Vogelmann, S.A., Seitter, M., Singer, U., Brandt, M.J. and Hertel, C. (2009). Adaptability of lactic acid bacteria and yeasts to sourdoughs prepared from cereals, pseudocereals and cassava and use of competitive strains as starters. *International Journal of Food Microbiology*, 130: 205-212.

von Holy, A., Holzapfel, W.H. and Dykes, G.A. (1992). Bacterial populations associated with Vienna sausage packaging. *Food Microbiology*, 9: 45-53.

Vrancken, G., De Vuyst, L., van der Meulen, R., Huys, G., Vandamme, P. and Daniel, H.M. (2010). Yeast species composition differs between artisan bakery and spontaneous laboratory sourdoughs. *FEMS Yeast Research*, 10: 471-481.

Weckx, S., Van der Meulen, R., Maes, D., Scheirlinck, I., Huys, G., Vandamme, P. and De Vuyst, L. (2010). Lactic acid bacteria community dynamics and metabolite production of rye sourdough fermentations share characteristics of wheat and spelt sourdough fermentations. *Food Microbiology*, 27: 1000-1008.

Wouters, D., Bernaert, N., Conjaerts, W., Van Droogenbroeck, B., De Loose, M. and De Vuyst, L. (2013a). Species diversity, community dynamics and metabolite kinetics of spontaneous leek fermentations. *Food Microbiology*, 33: 185-196.

Wouters, D., Grosu-Tudor, S., Zamfir, M. and De Vuyst, L. (2013b). Bacterial community dynamics, lactic acid bacteria species diversity and metabolite kinetics of traditional Romanian vegetable fermentations. *Journal of Science of Food and Agriculture*, 93: 749-760.

Wu, J.J., Ma, Y.K., Zhang, F.F. and Chen, F.S. (2012). Biodiversity of yeasts, lactic acid bacteria and acetic acid bacteria in the fermentation of 'Shanxi aged vinegar', a traditional Chinese vinegar. *Food Microbiology*, 30: 289-297.

Wullschleger, S., Lacroix, C., Bonfoh, B., Sissoko-Thiam, A., Hugenschmidt, S., Romanens, E., Baumgartner, S., Traore, I., Yaffee, M., Jans, C. and Meile, L. (2013). Analysis of lactic acid bacteria communities and their seasonal variations in a spontaneously fermented dairy product (Malian fene) by applying a cultivation/genotype based binary model. *International Dairy Journal*, 29: 28-35.

Xing, X., Ma, J., Fu, Z., Zhao, Y., Ai, Z. and Suo, B. (2020). Diversity of bacterial communities in traditional sourdough derived from three terrain conditions (mountain, plain and basin) in Henan province, China. *Food Research International*, 133: 109139.

Yeun, H., Yang, H.S., Chang, H.C. and Kim, H.Y. (2013). Comparison of bacterial community changes in fermenting kimchi at two different temperatures using a denaturing gradient gel electrophoresis analysis. *Journal Microbiology and Biotechnology*, 23: 76-84.

Zhang, J., Liu, W., Sun, Z., Bao, Q., Wang, F., Ju, Y., Chen, W. and Zhang, H. (2011). Diversity of lactic acid bacteria and yeasts in traditional sourdoughs collected from western region in Inner Mongolia of China. *Food Control*, 22: 767-774.

Zhou, J., Liu, X., Jiang, H. and Dong, M. (2009). Analysis of the microflora in Tibetan kefir grains using denaturing gradient gel electrophoresis. *Food Microbiology*, 26: 770-775.

Zotta, T., Piraino, P., Parente, E., Salzano, G. and Ricciardi, A. (2008). Characterisation of lactic acid bacteria isolated from sourdoughs for Cornetto, a traditional bread produced in Basilicata (southern Italy). *World Journal of Microbiology and Biotechnology*, 24: 1785-1795.

Bioactive Compounds in Fermented Foods

Swati Sharma[1], Srichandan Padhi[1], Megha Kumari[1], Amit Kumar Rai[1] and Dinabandhu Sahoo[1,2]*

[1] Institute of Bioresources and Sustainable Development, Regional Centre, Tadong - 737102, Sikkim, India
[2] Department of Botany, University of Delhi, Delhi - 110007, India

1. Introduction

Fermentation has been in practice since long, not just in food preservation, but also in improving the texture, appearance, nutrition value and masking of the taste (Sourabh et al., 2015; Rai et al., 2017a). Fermented foods are produced traditionally since many generations by adopting unique methods of production in different countries across the globe (Rai and Kumar, 2015; Sanjukta et al., 2017). The traditional knowledge associated with food fermentation is passed from one generation to another without much knowledge about the specific role of individual microorganisms involved in the production of bioactive compounds (Sourabh et al., 2015). During fermentation there are possibilities of production of bioactive compounds or degradation of anti-nutritional factors due to microorganisms associated with specific food fermentation (Alzate et al., 2008; Rai and Appaiah, 2014; Bunnoy et al., 2015; Hati et al., 2020). Due to the increasing interest of consumers and research in recent past on the role of individual microorganisms in production of bioactive compounds in some of the fermented foods has resulted in development of controlled processes for food production with specific health benefits (Rai et al., 2017a; Salmeron, 2017).

The bioactive compounds in different fermented foods are either produced by the starter culture (bacteria, yeast and filamentous fungi) or on biotransformation of food components by enzymes produced by the starter culture (Cui et al., 2020; Rai et al., 2019; Sanjukta and Rai, 2016). The enzymes involved are produced by microorganisms associated with food fermentation and include enzymes such as proteases, amylases, β-glucosidases, xylanases, lipases, cellulases, glucoamylases and β-glucanases (Hur et al., 2014; Rai et al., 2017b). These enzymes act on specific substrates (proteins, carbohydrates, lipids and polyphenols), resulting in production

---

*Corresponding author: dbsahoo@hotmail.com

of transformed metabolites in the final product (Cho *et al.*, 2011; Hur *et al.*, 2014; Hati *et al.*, 2019). Substrate-derived bioactive compounds will depend on the origin of the raw material used for fermentation and their biochemical composition (Sourabh *et al.*, 2015; Barla *et al.*, 2016).

The source of raw material used for fermented food production can be either plant based or animal based. Plant-based fermented food products are mainly rich in carotenoids, polyphenols including isoflavones, flavanols and phenolics acids (Cho *et al.*, 2011; Hur *et al.*, 2014). Animal-based fermented food products are rich in bioactive compounds, such as peptides, enzymes, lipids and microbial metabolites produced by starter culture (Lee *et al.*, 2003; Barla *et al.*, 2016; Majumdar *et al.*, 2016). Bioactive compounds produced by the starter culture are reported in both plant- and animal-based products, including vitamins, bacteriocins, γ-aminobutyric acid (GABA), exopolysaccharides and cell-wall constituents, including β-glucan (Barla *et al.*, 2016; Cui *et al.*, 2020). Bioactive compounds produced during food fermentation are responsible for several health benefits, including antimicrobial, antioxidant, antiaging, anticancer, antidiabetic, antihypertensive and immunomodulatory properties (Lee *et al.*, 2003; Rai and Jeyaram, 2015; Barla *et al.*, 2016; Sanjukta and Rai, 2016; Rai *et al.*, 2019). Some of the health benefits exhibited by bioactive compounds in fermented food products are proven by animal studies as well as clinical trials (Pouliot-Mathieu *et al.*, 2013; Nilsen *et al.*, 2014; Rai *et al.*, 2017a; Santiago-López *et al.*, 2018). The different types of bioactive compounds produced during fermentation and their health benefits are presented in Fig. 1. This chapter gives an overview of the different types of bioactive compounds produced during food fermentation.

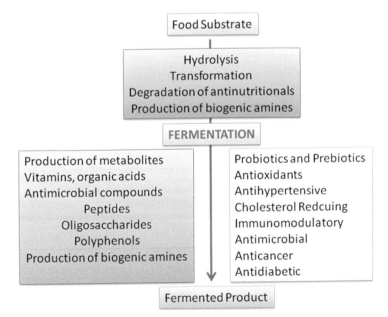

Fig. 1: Production of bioactive compounds during food fermentation and their potential health benefits

2. Bioactive Metabolites Produced by Starter Culture

Different types of bioactive compounds are produced by starter cultures themselves and their presence in the fermented product results in specific health benefits. These bioactive compounds include (i) exopolysaccharides, (ii) γ-aminobutyric acid, (iii) vitamins, (iii) biopigments and lipids, (iv) bacteriocins and (v) selenium. In this section, the different types of bioactive compounds produced by the starter cultures are discussed.

2.1   Exopolysaccharides

Exopolysaccharides (EPS) are high-molecular-weight polymers that are composed of sugar residues, present on many bacterial surfaces and secreted into the surrounding environment. EPS production is an important characteristic of many starter cultures involved in the production of traditional and fermented foods (Parente *et al.*, 2017). Further, these polymeric compounds also contribute to improving the texture, flavour and rheology of fermented food products (Nampoothiri *et al.*, 2017, Zhou *et al.*, 2019). Besides, EPS have also been reported to exhibit various health benefits such as prebiotic, antidiabetic, antioxidant, cholesterol-lowering, immunomodulatory, and anticancer properties (Nampoothiri *et al.*, 2017; Lynch *et al.*, 2018; Perez-Ramos *et al.*, 2018). β-glucans are common examples of EPS produced in fermented foods and have been associated mainly with hypocholesterolemic activity (Rai *et al.*, 2019). Other examples include dextran, levan and pullulan which are also considered as prebiotic as they are selectively produced by gut microorganisms having a positive impact on host health (Patel and Prajapati, 2013; Perez-Ramos *et al.*, 2018).

EPS-producing strains have been isolated from a wide range of fermented products including dairy products, fermented vegetables, bakery and cereal-based beverages and meat products (Behera *et al.*, 2018). LABs have particularly gained attention for the production of EPS. Among them genera *Lactobacillus*, *Leuconostoc*, *Pediococcus*, *Lactococcus* and *Streptococcus* have been extensively investigated as in many instances they enhance the biofunctionality (Patel and Prajapati, 2013). For example, cheese produced by using *Lactobacillus. plantarum* JLK0142 as an adjunct culture resulted in a product having significantly higher antioxidant properties and the ability to inhibit key enzymes related to diabetes and hypertension (Wang *et al.*, 2019). A few types of yeast, such as *Saccharomyces cerevisiae* and *Kluyveromyces marxianus* are also reported to produce functional EPS, including β-glucan and α-D mannan (Rai *et al.*, 2019). Furthermore, co-cultures of LAB and yeasts have also been found successful in improving the quality as well as quantity of EPS production. Higher EPS production was observed when co-culturing *S. cerevisiae* and *Lactobacillus rhamnosus* in comparison to monoculture, suggesting a positive impact of yeast and LAB interaction on biosynthesis of EPS (Bertsch *et al.*, 2019).

2.2   γ-aminobutyric Acid (GABA)

Gamma aminobutyric acid (GABA), a naturally occurring amino acid, functions as a neurotransmitter in the brain of widely divergent animal species including humans (Wu and Shah, 2017). GABA inhibits or blocks certain brain signals and reduces activity in the nervous system. In addition, GABA has several other beneficial roles, such as antihypertensive, antidiabetic and immunomodulatory (Aoki *et al.*, 2003; Cui *et al.*, 2020; Zhang *et al.*, 2020). In recent years, microbial fermentation has been widely recognised for GABA production. Many bacteria associated with food fermentation, particularly LAB produces GABA as a defence mechanism to maintain viability under acidic conditions. Functional fermented dairy products including cheese, soymilk and *kimchi* have been manufactured using GABA-producing LAB (Dhakal *et al.*, 2012). Glutamic acid decarboxylases produced by LAB are the key biocatalyst responsible for production of GABA (Kumar *et al.*, 2019; Cui *et al.*, 2020; Cui *et al.*, 2020). Some of the studies have shown the production of GABA using *Lactobacillus brevis, Lb. paracasei, Lactococcus lactis, Lb.buchneri* and *Weissella hellenica* isolated from different fermented food products (Barla *et al.*, 2016; Cui *et al.*, 2020).

Production of GABA varies in different species of same genus as well as different strains of same species isolated from fermented foods (Cui *et al.*, 2020). Dairy products, such as yoghurt, cheese and fermented milk contain GABA due to a high concentration of glutamic acid in cow's milk (Santos-Espinosa *et al.*, 2018). GABA production was reported in milk fermented by using specific two *L. lactis* strains isolated from Mexican cheese product (Santos-Espinosa *et al.*, 2020). Production of soya yoghurt using *Lb. brevis* OPY-1 resulted in enhanced levels of GABA (Park *et al.*, 2007). Fermentation of black raspberry using *Lb. brevis* was carried out to obtain fermented juice enriched with high GABA concentration (Kim *et al.*, 2009). Similarly, fermented fish sauce was prepared from Giant Masu salmon and enrichment of GABA was observed when *Lactobacillus plantarum* strain N10 was added before fermentation (Taoka *et al.*, 2019).

Efficiency of GABA production depends on the microbial strain used for fermentation. *Lactobacillus futsaii* CS3 isolated from Thai fermented shrimp (*Kung-Som*) showed higher efficiency of bioconversion of monosodium glutamate to GABA with yield more than 99 per cent conversion rate within three days (Sanchart *et al.*, 2017). Mulberry beverage enriched with GABA was produced using co-culturing specific combination of LAB and yeast strains (Zhang *et al.*, 2020). Feeding GABA-enriched *tempeh* to spontaneously hypertensive rats resulted in lowering of blood pressure in comparison to control group (Aoki *et al.*, 2003). GABA-enriched milk consumption has also shown to reduce blood pressure in mild hypertensive subjects (Inoue *et al.*, 2003). Application of a specific combination of lactic acid bacteria and yeast can be applied for production of fermented product, enriched with GABA.

2.3 Biopigments and Lipids

Colour is an important characteristic of food as that creates impressions on the taste, flavour and its safety as well. The colours of the food are due to the natural pigments or colourants added to them while processing. Pigments are also considered as important constituents in fermented food as these are associated with the improvement of its functional properties (Bunnoy *et al.*, 2015; Majumdar *et al.*, 2016). Carotenoids are the common pigments encountered in various fermented food products which could either be from plant-based substrate or microbial metabolites produced during fermentation (Cerrillo *et al.*, 2014; Giuffrida *et al.*, 2016). Apart from adding colour, these pigments are also responsible for the enhancement of several health benefits including antioxidant, antidiabetic and anti-inflammatory activities, prevention of cancer and atherosclerosis (Giuffrida *et al.*, 2016; Hernández-Almanza *et al.*, 2017). Cheese ripening with *Arthrobacter arilaitensis* resulted in identification of carotenoids, including 9-Z-decaprenoxanthin, 15-Z-decaprenoxanthin, all-E-sarcinaxanthin, E-decaprenoxanthin, decaprenoxanthin monoglucoside, decaprenoxanthin diglucoside, sarcinaxanthin monoglucoside pentaacetate and decaprenoxanthin-C16:0 (decaprenoxanthin-palmitate) (Giuffrida *et al.*, 2016). In a study, total carotenoid content increased significantly during orange juice fermentation, which might be due to the enzymatic action resulting in higher carotenoids extractability from the food matrix (Cerrillo *et al.*, 2014). Carotenoid-enriched yoghurt was produced by adding fresh biomass of *Spirulina* in milk and fermented by using *Lb. acidophilus* (Patel *et al.*, 2019). Besides the carotenoids, some flavonoids and polyketides have been identified as colourants in fermented foods and reported to enhance the food functionality. *Monascus* species, especially *M. pilosus, M. purpureus, M. ruber* and *M. froridanus* associated with fermented food products, produce pigments, which are responsible for the food colour, increasing health benefits as well as function as preservatives (Hsu and Pan, 2012). These include yellow pigments monascin and ankaflavin, a lemon-yellow citrinin and a red pigment monacolin K, which are responsible for a wide range of health benefits, such as antioxidant, anticancer, antihypertensive, antidiabetic, anti-inflammatory and cholesterol-reducing properties (Shi and Pan, 2011; Hsu and Pan, 2012; Bunnoy *et al.*, 2015).

Apart from pigments, beneficial lipids, including polyunsaturated fatty acids (PUFA) and conjugated linoleic acid (CLA), contribute to functional properties of fermented food products (Gursoy *et al.*, 2012; Majumdar *et al.*, 2016; Ilyasoglu and Yilmaz, 2018; Chourasia *et al.*, 2020). *Shidals,* a salt-free fermented fish product of northeast India was found to be rich in PUFA, including eicosapentaenoic, docosahexaenoic and linolenic acid (Majumdar *et al.*, 2016). PUFA are responsible for several health benefits, including lipid-lowering effect, reducing risk of cardiovascular diseases and neuropsychiatric disorders (Rai *et al.*, 2013). PUFA, consisting of linoleic and linolenic acids, was observed to increase in yoghurt by adding walnut slurry (10-50 per cent) and skimmed milk (Ilyasoglu and Yilmaz, 2018). Increased contents of *cis*-9, *trans*-11-conjugated linoleic acid (CLA) and PUFA were witnessed in milk fermented by using a potential probiotic strain *L. lactis* MRS47, isolated from *kefir* grains (Vieira *et al.*, 2017). A proper combination of CLA-producing bacteria and yeast can result in cheese product having enhanced CLA content (Gursoy *et al.*, 2012). CLA has been associated with several health benefits including antioxidative, antidiabetic, anti-obesity, antiadipogenic, antihypertensive, anticarcinogenic, and anti-inflammatory properties (Florence *et al.*, 2009; Bassaganya-Riera *et al.*, 2012; Koba and Yanagita, 2014; Murru *et al.*, 2018).

2.4   Vitamins

Vitamins are essential nutrients that are not synthesised in the human body but are needed for several biochemical reactions in the body (Walther and Schmid, 2016). Microorganisms associated with food fermentation can be one of the potential sources of vitamins, such as cyanocobalamin (B_{12}), riboflavin (B_2), pyridoxine (B_6), niacin (B_3), folates (B_{11}) and vitamin K (Patel *et al.*, 2013; Gu and Li, 2016; Walther and Schmid, 2016). Several LAB associated with fermented foods have been reported to produce vitamins and include *Lactobacillus acidophilus, L. lactis, Lactobacillus reuteri, Lb. bulgaricus, Lb. plantarum, Streptococcus thermophilus* and *Bifidobacterium longum* (LeBlanc *et al.*, 2011, 2015; Walther and Schmid; 2016). LAB-fermented food products, such as yoghurt, different varieties of cheese, *kefir*, curd, *tempeh*, fermented skimmed milk, yoghurt, bamboo shoot and sourdough have been reported to have a higher content of vitamins (Patel *et al.*, 2013; Bhushan *et al.*, 2016; Thakur *et al.*, 2016).

Specific LAB has been applied for enhanced production of a particular type of vitamin in fermented food product (Bhushan *et al.*, 2016; Thakur *et al.*, 2016). Strains, such as *Lactobacillus fermentum* MTCC 8711, *Lb. lactis* NZ900, *Lb. plantarum* CRL725 and BBC32B have been established for the production of riboflavin in dairy products and fermented breads (Russo *et al.*, 2014; Thakur *et al.*, 2016). Similarly, *Lb. plantarum* BCF20 has been applied for production of fermented soymilk with fortified vitamin B_{12} (Bhushan *et al.*, 2016). Several species of yeast have also been reported for the production of vitamins (Moslehi-Jenabian *et al.*, 2010). High vitamin K and B_9 contents were observed in several fermented foods produced by using yeasts, *Kluyveromyces marxianus, Saccharomyces bayanus, S. cerevisiae* and *Metschnikowia lochheadii* (Hjortmo *et al.*, 2008; Moslehi-Jenabian *et al.*, 2010). Application of specific strain having the ability for production of a particular vitamin can be applied for production of fermented foods, enriched with the desired vitamin.

2.5 Bacteriocins

Bacteriocins are generally defined as peptides or proteins synthesised ribosomally by bacteria that kill or inhibit other related or unrelated microorganisms (Vuyst and Leroy, 2007). Most bacteriocins belong to LAB, a group present naturally in foods and bear Qualified Presumption of Safety (QPS) and Generally Recognised as Safe (GRAS) status. LAB isolated from traditional fermented foods including fermented fish product, fermented milk product, fermented vegetables, fermented cereal products have been reported for production of bacteriocins active against spoilage causing food-borne pathogens *Listeria monocytogenes, Escherichia coli* and *Staphylococcus aureus* (Castilho *et al.*, 2009; Vuyst and Leroy, 2007). *Bifidobacterium* spp., *Leuconostoc* spp., *Lactococcus* spp., *Enterococcus* spp., *Lactobacillus* spp., and *Pediococcus* spp. (Martinez *et al.*, 2009; Rai *et al.*, 2009; Diep *et al.*, 2009) are some of the bacteriocin producers. The bacteriocins produced by wide range of LAB include bifidin, acidophilin or acidocin, pediocin, nisin, bulgarin, plantaricin, etc. (Deegan *et al.*, 2006; Abo-Amer, 2011; Cheikhyoussef *et al.*, 2010; Dussault *et al.*, 2016; Devi and Halami, 2014; Simova *et al.*, 2008).

LAB have been used for production of bacteriocins for food application as preservative effect (Alvarez-Sieiro, 2016). Bacteriocins enhance the shelf-life of food products as they are active in broad pH range and can resist high temperature, increasing its scope of application (Ahmad *et al.*, 2017). Due to their proteinaceous nature, they are easily digestible proteases and have no negative impact on gut microbiota (Egan *et al.*, 2016). Apart from LAB, *Bacillus* spp., isolated from traditional fermented food products, have also been reported for the production of antimicrobial compounds (Zheng and Slavik, 1999). *Bacillus* sp. LM 7 isolated from *chungkookjang*-fermented soybean product in Korea, have been found to produce antimicrobial liopeptides having activity against pathogenic bacterial and fungal strains (Lee *et al.*, 2016). Many *Bacillus* bacteriocins isolated from different sources have been shown to have antibacterial effect with a broad spectrum of pathogenic bacteria, particularly at alkaline pH (Abriouel *et al.*, 2011). However, detailed studies on exploring novel strains of *Bacillus* spp. from fermented food products are needed for application in food preservation.

2.6 Selenium

Selenium is an important micronutrient, which helps in improving the immune system and protecting against oxidative stress and cancer (Kieliszek and Blazejak, 2016; Constantinescu-Aruxandei *et al.*, 2018; Rai *et al.*, 2019). Selenium has been transformed by many lactic acid bacteria from its inorganic form to organic form (Pophaly *et al.*, 2014). *Lb. acidophilus* CRL636 and *Lb. reuteri* CRL1101 reportedly have the ability to transform selenium into selenomethionine and selenocysteine (Pescuma *et al.*, 2017). *Lb. brevis* strain CGMCC6683 was used to prepare yoghurt which was found to be rich in selenium (Deng *et al.*, 2015). In recent years, selenium-enriched probiotic LAB exhibited several health benefits, such as antioxidant, antimicrobial, anticancer, antimutagenic and anti-inflammatory properties (Pophaly *et al.*, 2014; Yang *et al.*, 2018).

Yeast-based fermented products, such as alcoholic beverages and bread are shown to be the source of selenium as they have the property to accumulate selenium in organic form (Perez-Corona *et al.* 2011; Rai *et al.*, 2019). These products are also helpful in reducing cancer progression and improving immunity and antioxidative properties in several animal models (Abedi *et al.*, 2018; Perez-Corona *et al.*, 2011; Rai *et al.*, 2019). Yeasts, such as *S. cerevisiae* and *Saccharomyces bayanus*, have been involved in the biotransformation of selenium in white wine without affecting alcoholic fermentation (Lyons *et al.*, 2007). Selection and application of selected yeast for food fermentation can result in products rich in selenium.

3. Substrate-derived Bioactive Compounds

Substrate-derived bioactive compounds depend on the biochemical composition of the food components and specificity of the enzymes produced by the starter cultures. These bioactive compounds are present in unfermented foods in bound/inactive form, which, on enzymatic reaction, are released in their bioactive form. The substrate-derived bioactive compounds include bioactive peptides, oligosaccharides, polyphenols, phytoestrogens, isothiocyanates and polyols. In this section we will discuss the bioactive compounds produced on biotransformation of food components.

3.1 Polyphenols

Polyphenols have been remaining a subject of interest and gained research attention due to their important health benefits, and are characterised by at least one aromatic ring (C_6) with one hydroxyl group. Based on the number of phenol rings and the structural elements binding these rings, polyphenols may be classified into different groups including flavonoids, phenolic acids, polyphenolic amides and other polyphenols. Being the largest, flavonoids make up 60 per cent of all polyphenols and are further classified into flavonols, flavones, isoflavones, flavanones, anthocyanidins and flavanols (Springob and Kutchan, 2009). Around one-third of the total polyphenols account for phenolic acids and notable examples include stilbenes and lignans, and are mostly found in fruits, vegetables, whole grains and seeds. Polyphenols are abundantly produced in plants as secondary metabolites to the stressed conditions and several of such compounds derived from the crude extracts of fruits, vegetables, cereals, legumes are efficient bioactive agents (Gharras, 2009). Moreover, the foods rich in polyphenols are promoted as functional foods because of their potential impact on human health and long-term consumption of which may have preventive effects against cardiovascular diseases, obesity and neurodegenerative disorders (Xiao and Hogger, 2015). Production of bioactive compounds in fermented foods is presented in Table 1.

Fermentation of plant-based foods results in breakdown of cell-wall constituents and releases phenolic compounds (Lee *et al.*, 2008; Hur *et al.*, 2014). Polyphenols are present in the raw material in bound form and are released in free form during the fermentation process, which improves its bioavailability in the intestine and contributes to beneficial health impacts (Sourabh *et al.*, 2015; Sanjukta *et al.*, 2017). For example, some isoflavones in its aglycone forms like genistein, daidzein, and glycitein, identified in fermented soy products, were initially present in its bound forms (glycone) genistin, daidzin and glycitin respectively in the unfermented soy (Cho *et al.*, 2011). Similarly, polyphenol enhancement of 144 per cent was observed in soybean fermented by using *Bacillus subtilis* (Moktan *et al.*, 2008). Similarly, antioxidant activity due to polyphenols increased by 5.2 -7.4 folds in *Monascus-*fermented soybean (Pyo and Lee, 2007).

The families of phenolic compounds that are considered as phytoestrogens and present in plant-based fermented foods are Isoflavones, stilbenes and lignans (Cornwell *et al.*, 2004). Isoflavones are of important bioactive compounds present in fermented soybean products (Hati *et al.*, 2020; Sanjukta *et al.*, 2017). The aglycone forms of isoflavones are more bioavailable and known for several health benefits, such as antioxidant, anticancer and protection against cardiovascular diseases (Sanjukta *et al.*, 2017; Wong *et al.*; 2008). Apart from isoflavones, fermented soybean products are also rich in flavanols and phenolic acids, which also increase during fermentation (Cho *et al.*, 2011; Sanjukta *et al.*, 2015).

Polyphenols conversion into free form has also been reported in fermented beverages due to enzymes produced by starter cultures (Rai *et al.*, 2014, 2019). Garcinia wine fermented with yeast species (*S. cerevisiae* and *Hansiniaspora* sp.) has demonstrated enhancement of free polyphenols and antioxidant properties (Rai *et al.*, 2010, 2014). Antioxidant properties due to polyphenols in a fermented beverage depend on the concentration of must used for fermentation and enzyme-producing ability of the yeast strain (Rai *et al.*, 2014). Caffeic acid bioconversion to dihydrocaffeic acid occurs during fermentation of cherry juice by *Lb. plantarum*

Table 1: Health Benefits of Polyphenols in Fermented Foods and Beverages

Fermented Foods	Starter Used	Bioactivity	Bioactive Compounds	References
Fermented black soybean	*Bacillus subtilis*	Antioxidative	Total phenolic, flavonoids and vitamin K2 increased and aglycone contents	Juan and Chou (2010)
Soybean *kinema*	*Bacillus subtilis*	Antioxidative	Polyphenols	Moktan *et al.* (2008)
Korean soybean paste (*doenjang*)	*Bacillus subtilis* CS90	Antioxidative	Daidzein (aglycone type) and acetyl Daidzin (glycoside type)	Cho *et al.* (2011)
Fermented soymilk	Lactic acid bacteria and *Saccharomyces boulardii*	Antioxidative	daidzein + genistein	Rekha and Vijayalakshmi (2008)
Bread wheat doughs	*Lactobacillus plantarum*	Antioxidative	Phenolic acid and carotenoid	Antognoni *et al.* (2018)
Boza (fermented cereals)	*Saccharomyces uvarum* *Saccharomyces cerevisiae*	Antioxidative	Protocatechuic acid Ferulic acid	Blandino *et al.* (2003), Dogan and Ozpinar (2017)

along with increase in antioxidant properties (Shiferaw and Augustin, 2019). Increase in antioxidant activity and changes in the phenolic profile were also observed in fermented apple juice produced by using *Lb. plantarum* (Li *et al.*, 2019).

3.2 Bioactive Peptides

Food-derived bioactive peptides are chains of 2-20 amino acids and capable of preventing or ameliorating many disease conditions (Rai and Jeyaram, 2015). Besides their biological activity, they impart nutritional as well as several other functional properties and their efficiency certainly depends on their sequence length and structural characteristics (Rai *et al.*, 2017a; Sarmadi and Ismail, 2010; Zhou *et al.*, 2016). The peptide having biological activity was first reported by Marcuse in 1960. Since then, the peptides from several protein ingredients have been extensively studied and considered as potential and feasible dietary elements for human health promotion. Protein-rich fermented foods are one of the potential sources of bioactive peptides, which are formed on hydrolysis of parent food proteins (Sanjukta and Rai, 2016; Rai and Jeyaram; 2015). A schematic representation on production of bioactive peptides is presented in Fig. 2.

Research in the past two decades has resulted in discovery of several bioactive peptides from protein-rich foods, such as fermented milk, fermented soybean, fermented meat and fish products (Lee *et al.*, 2003; Rai and Jeyaram, 2015; Sanjukta and Rai, 2016). Depending on the parent protein and specificity of proteases, different types of bioactive peptides are produced, which differ in their length, amino acid composition and their sequence (Sanjukta and Rai, 2016; Rai *et al.*, 2017a). Among the different biological activities exhibited by these peptides properties like

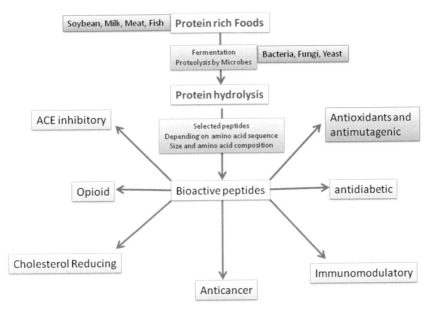

Fig. 2: Production of bioactive peptides in fermented foods and their potential health benefits

antioxidant, antihypertensive, antidiabetic, anticancer, immunomodulatory are most frequent (Sanjukta and Rai, 2016; Ayyash *et al.*, 2018). Production of bioactive peptides and their physical characteristics is also dependent on the starter culture used. Several studies have shown that different strains of same species may result in production of different types of bioactive peptides with specific health benefits (Quiros *et al.*, 2007; Rodriguez-Figueroa *et al.*, 2010; Chen *et al.*, 2014; Sanjukta *et al.*, 2015).

Most of the bioactive peptides have been reported in fermented milk and cheese products, which are produced using (i) lactic acid bacteria and (ii) a combination of LAB and yeast (Rai *et al.*, 2017a, 2019). Bioactive peptides in fermented milk and cheese products are produced during fermentation, storage and ripening process (in case of cheese) (Rai *et al.*, 2016, 2017). Milk products have been reported as a source of antioxidant and antihypertensive peptides (Rai *et al.*, 2017). Antihypertensive peptides in milk products range from 2-12 amino acids having the ability to inhibit angiotensin-converting enzymes (ACE) (Chen *et al.*, 2014; Rodriguez-Figueroa *et al.*, 2010; Rai *et al.*, 2017a). Soybean products fermented by using *Bacillus* spp., LAB and fungi have also been reported as a source of bioactive peptides with antioxidant, immunomodulatory, antidiabetic, anticancer, antimicrobial and ACE-inhibitory properties (Sanjukta and Rai, 2016; Rai *et al.*, 2017a). The fermented soybean products reported to have bioactive peptides are *natto, cheonggukjang, doenjang,* soy yoghurt, *tempeh, douchi, sufu* and *meju* (Cho *et al.* 2003; Donkor *et al.*, 2005; Ibe *et al.*, 2009; Sanjukta and Rai, 2016). Bioactive peptides have also been reported in fermented fish and meat products with antioxidant and ACE inhibitory properties (Je *et al.*, 2005; Ohata *et al.*, 2016; Choksawangkarn *et al.*, 2018). Peptide fraction from anchovy sauce has been reported to induce apoptosis, which is considered as one of the key players for cancer chemoprevention (Lee *et al.*, 2003). The details of different types of bioactive peptides produced in fermented foods using different starter culture is presented in Table 2.

Antioxidant property of the peptides depends on the composition and sequence of the amino acids (Zhang *et al.*, 2020; Rai *et al.*, 2016). Peptides having amino acids, such as cysteine, histidine, tryptophan, methionine, tyrosine and phenyl alanine have been reported for high antioxidant activity (Zhang *et al.*, 2020). Peptides have also been reported to inhibit key enzymes responsible for stimulating oxidative stress, such as myeloperoxidases (Zhang *et al.*, 2020). Production of bioactive peptides was also observed in fermented cereal products (Coda *et al.*, 2012; Rizzello *et al.*, 2008). Peptides having ACE-inhibitory properties were produced during sourdough fermentation using selected LAB (Rizzello *et al.*, 2008). Fermented milk, including cheese and soybean products, has been reported to possess antidiabetic properties (Tamam *et al.*, 2019; Sohri *et al.*, 2020). Water-soluble extract of LAB-fermented sourdough using various cereal flours resulted in production of antioxidative peptides (Coda *et al.*, 2012). Despite the above beneficial roles, stability of bioactive peptides during prolonged fermentation, storage, ripening and gastrointestinal digestion is still a challenge and needs to be addressed. There are possibilities of production of novel bioactive peptides on degradation during these processes, which can add significant value to the product (Rai *et al.*, 2017a, b).

Table 2: Bioactive Peptides in Fermented Milk and Their Potential Health Benefits

Product	Bioactivity	Starter	Bioactive Peptides	References
Sour milk (Calpis)	Antihypertensive	*Lactobacillus helveticus, Saccharomyces cerevisiae*	VPP, IPP	Nakamura *et al.* (1995)
Fermented soybean	ACE inhibitory	*Pediococcus pentosaceus* SDL 1409	EDEVSFSP, SRPFNL, RSPFNL, ENPFNL	Daliri *et al.* (2018)
Fermented milk	Antihypertensive	*Lb. helveticus* CM4	VPP, IPP	Wakai *et al.* (2012)
Fermented camel milk	ACE inhibitory	*Lb. rhamnosus* MTCC 5945	AHTPDB	Solanki and Hati (2018)
Honey based fermented milk	ACE inhibitory	*Lb. helveticus* MTCC 5463	VPP, IPP	Hati *et al.* (2016)
Fermented milk	ACE inhibitory	*Lb. casei* IMAU10408, *Lb. casei* IMAU 20411	VPP, IPP	Li *et al.* (2017)
Fermented goat milk	ACE inhibitory	*Lb. casei* NK9, *Lb. fermentum*	AFPEHK	Parmar *et al.* (2018)
Fermented fish (*Pekasam*)	Antioxidant	*Lb. plantarum*	AIPPHPYP, IAGVFLITDPK	Najafian and Banji (2018)
Fermented yoghurt	Antidiabetic	*Lactococcus lactis*	GHWYYRCW	D'Souza *et al.* (2012)
Fermented milk	Antioxidant, ACE-inhibitory	Commercial	GPVRGPFPII	Hernandez-Ledesma *et al.* (2005)
Fermented milk	Antihypertensive peptides	*Enterococcus faecalis* CECT 5727	LHLPLP LVYPFPGPIPNSLPQNIPP	Quiros *et al.* (2007)

3.3   Oligosaccharides

Oligosaccharides are low-molecular-weight molecules comprising 2-10 monosaccharides linked by glycosidic bonds, intermediate in nature between simple sugars and polysaccharides (Weijers *et al.*, 2008; Patel and Goyal, 2011). The anomeric C atoms (C1 or C2) of the monosaccharide units in the non-digestible oligosaccharides have configuration that makes their bonds non-susceptible to the hydrolysis of human digestive enzymes (Roberfroid and Slavin, 2000). Fermented foods prepared by using specific carbohydrate active enzymes can result in production of oligosaccharides with functional properties (Basinskiene *et al.*, 2016; Zhou *et al.*, 2019). Soybean dregs, a traditional fermented food product of China produced by using *Neurospora crassa* showed enhancement in oligosaccharide production, which can improve the probiotic potential (Zhou *et al.*, 2019).

A traditional non-alcoholic beverage fermented by using *Lactobacillus sakei* and addition of xylanolytic enzyme resulted in a product with a higher yield of oligosaccharides from 300-1100 mg/L (Basinskiene *et al.*, 2016). Fermented milk produced by using *Leuconostoc* spp. was found to be rich in oligosaccharides, specially panose (Seo *et al.*, 2007). The oligosaccharides which have been reported as prebiotic agents are xylooligosaccharides (XOS), mannan oligosaccharides (MOS), isomaltooligosaccharides (IMO), pectic-oligosaccharides, fructooligosaccharides (FOS), arabinoxylanoligosaccharides and galactooligosaccharides (Qiang *et al.*, 2009; Rai *et al.*, 2019). Cashew apple juice fermented with *Lb. plantarum* and *Lb. acidophilus* resulted in a product having higher content of FOS and raffinose-family oligosaccharides (Kaprasob *et al.*, 2018). The product fermented with *Lb. acidophilus* resulted in high FOS of nystose, 1-kestose, and 1F-β-fructofuranosyl nystose whereas *Lb. plantarum* fermentation resulted in high raffinose and stachyose.

Oligosaccharides have also been incorporated as food additives in some of the fermented foods and beverages to improve the prebiotic potential of the product. Juice blend of orange juice and malt extract was fermented by using *Weissella cibaria* for the production of IMO (Rolim *et al.*, 2018). Prebiotics oligosaccharides are known for several health benefits including risk reduction of diabetes and colon cancer, cholesterol lowering, treatment of obesity and enhancement of immunity (Qiang *et al.*, 2009; Patel and Goyal, 2011). Fermented foods with enriched functional oligosaccharides need to be explored further for development of functional fermented food products.

4.   Conclusion

In recent years, fermented foods have been reported as a source of bioactive compounds with a wide range of health benefits. Some of the fermented foods enriched with specific bioactive compounds have been produced at industrial level and commercialised. Due to increasing awareness of health benefits of fermented foods, novel bioactive compounds and their production have been investigated in different food products. However, microbial fermentation of unexplored fermented foods using specific strains can result in production of novel bioactive compounds with specific health benefits. Application of a foodomic approach on health benefits of bioactive compounds during food fermentation could give more scientific knowledge in functional food development with select health benefits.

Acknowledgement

The authors would like to thank the Institute of Bioresources and Sustainable Development and Department of Biotechnology, Government of India for the financial support.

Abbreviations

ACE – Angiotensin Converting Enzyme
CLA – Conjugated Linoleic Acid
EPS – Exopolysaccharides
FOS – Fructooligosaccharides
GABA – γ-aminobutyric Acid
GRAS – Generally Recognised as Safe
MOS – Mannan Oligosaccharides
IMO – Isomaltooligosaccharides
LAB – Lactic Acid Bacteria
PUFA – Polyunsaturated Fatty Acids
QPS – Qualified Presumption of Safety
XOS – Xylooligosaccharides

References

Abedi, J., Saatloo, M.V., Nejati, V., Hobbenaghi, R., Tukmechi, A., Nami, Y. and Khosroushahi, A.Y. (2018). Selenium-enriched *Saccharomyces cerevisiae* reduces the progression of colorectal cancer. *Biological Trace Elemental Research*, 185: 424-432.

Abo-Amer, A.E. (2011). Optimisation of bacteriocin production by *Lactobacillus acidophilus* AA11, a strain isolated from Egyptian cheese. *Annals of Microbiology*, 61: 445-452.

Abriouel, H., Franz, C.M., Ben Omar, N. and Gálvez, A. (2011). Diversity and applications of *Bacillus* bacteriocins. *FEMS Microbiology Review*, 35: 201-232.

Ahmad, V., Khan, M.S., Jamal, Q.M.S., Alzohairy, M.A., Al Karaawi, M.A. and Siddiqui, M.U. (2017). Antimicrobial potential of bacteriocins: In therapy, agriculture and food preservation. *International Journal of Antimicrobial Agents*, 49: 1-11.

Alvarez-Sieiro, P., Montalbán-López, M., Mu, D. and Kuipers, O.P. (2016). Bacteriocins of lactic acid bacteria: Extending the family. *Applied Microbiology Biotechnology*, 100: 2939-2951.

Alzate, A., Fernández-Fernández, A., Pérez-Conde, M.C., Gutiérrez, A.M. and Cámara, C. (2008). Comparison of biotransformation of inorganic selenium by *Lactobacillus* and *Saccharomyces* in lactic fermentation process of yoghurt and *kefir*. *Journal of Agricultural and Food Chemistry*, 56: 8728-8736.

Antognoni, F., Mandrioli, R., Potente, G., Saa, D.L.T. and Gianotti, A. (2019). Changes in carotenoids, phenolic acids and antioxidant capacity in bread wheat dough fermented with different lactic acid bacteria strains. *Food Chemistry*, 292: 211-216.

Aoki, H., Furuya, Y., Endo, Y. and Fujimoto, K. (2003). Effect of gamma-aminobutyric acid-enriched tempeh-like fermented soybean (GABA-Tempeh) on the blood pressure of spontaneously hypertensive rats. *Bioscience Biotechnology and Biochemistry*, 67: 1806-1808.

Ayyash, M., Al-Nuaimi, A.K., Al-Mahadin, S. and Liu, S.Q. (2018). *In vitro* investigation of anticancer and ace-inhibiting activity, α-amylase and α-glucosidase inhibition, and antioxidant activity of camel milk fermented with camel milk probiotic: A comparative study with fermented bovine milk. *Food Chemistry*, 239: 588-597.

Barla, F., Koyanagi, T., Tokuda, N., Matsui, H., Katayama, T., Kumagai, H., Michihata, T., Sasaki, T., Tsuji, A. and Enmoto, T. (2016). The γ-aminobutyric acid-producing ability under low pH conditions of lactic acid bacteria isolated from traditional fermented foods of Ishikawa Prefecture, Japan, with a strong ability to produce ACE-inhibitory peptides. *Biotechnology Reports*, 10: 105-110.

Basinskiene, L., Juodeikiene, G., Vidmantiene, D., Tenkanen, M., Makaravicius, T. and Bartkiene, E. (2016). Non-alcoholic beverages from fermented cereals with increased oligosaccharide content. *Food Technology and Biotechnology*, 54: 36-44.

Bassaganya-Riera, J., Viladomiu, M., Pedragosa, M., De Simone, C., Carbo, A., Shaykhutdinov, R. and Hontecillas, R. (2012). Probiotic bacteria produce conjugated linoleic acid locally in the gut that targets macrophage PPAR γ to suppress colitis. *PLoS ONE*, 7: e31238.

Behera, S.S., Ray, R.C. and Zdolec, N. (2018). *Lactobacillus plantarum* with functional properties: An approach to increase safety and shelf-life of fermented foods. *BioMed. Reserch International*, 9361614.

Bertsch, A., Roy, D. and LaPointe, G. (2019). Enhanced exopolysaccharide production by *Lactobacillus rhamnosus* in co-culture with *Saccharomyces cerevisiae*. *Applied Sciences*, 9: 4026.

Bhushan, B., Tomar, S.K. and Chauhan, A. (2017). Techno-functional differentiation of two vitamin B_{12} producing *Lactobacillus plantarum* strains: An elucidation for diverse future use. *Applied Microbiology and Biotechnology*, 101: 697-709.

Blandino, A., Al-Aseeri, M.E., Pandiella, S.S., Cantero, D. and Webb. C. (2003). Cereal-based fermented foods and beverages. *Food Research International*, 36: 527-543.

Bunnoy, A., Saenphet, K., Lumyong, S., Saenphet, S. and Chomdej, S. (2015). *Monascus purpureus*-fermented Thai glutinous rice reduces blood and hepatic cholesterol and hepatic steatosis concentrations in diet-induced hypercholesterolemic rats. *BMC Complementary and Alternative Medicine*, 15: 88.

Castilho, N.P.A., Colombo, M., de Oliveira, L.L., Todorov, S.D. and Nero, L.A. (2019). *Lactobacillus curvatus* UFV-NPAC1 and other lactic acid bacteria isolated from Calabresa, a fermented meat product, present high bacteriocinogenic activity against *Listeria monocytogenes*. *BMC Animal Health Sciences*, 19: 63.

Cerrillo, I., Escudero-López, B., Hornero-Méndez, D., Martín, F. and Fernández-Pachón, M.S. (2014). Effect of alcoholic fermentation on the carotenoid composition and provitamin A content of orange juice. *Journal of Agricultural and Food Chemistry*, 62: 842-849.

Cheikhyoussef, A., Cheikhyoussef, N., Chen, H., Zhao, J., Tang, J., Zhang, H. and Chen, W. (2010). Bifidin I – A new bacteriocin produced by *Bifidobacterium infantis* BCRC 14602: Purification and partial amino acid sequence. *Food Control*, 21: 746-753.

Chen, Y., Liu, W., Xue, J., Yang, J., Chen, X., Shao, Y., Kwok, L., Bilige, M., Mang, L. and Zhang, H. (2014). Angiotensin-converting enzyme inhibitory activity of *Lactobacillus helveticus* strains from traditional fermented dairy foods and antihypertensive effect of fermented milk of strain H9. *Journal of Dairy Science*, 10: 2014-7962.

Cho, K.M., Hong, S.Y., Math, R.K., Lee, J.H., Kambiranda, D.M., Kim, J.M., Islam, S.M.A., Yun, M.G., Cho, J.J. and Lim, W.J. (2009). Biotransformation of phenolics (isoflavones, flavanols and phenolic acids) during the fermentation of cheonggukjang by *Bacillus pumilus* HY1. *Food Chemistry*, 114: 413-419.

Cho, K.M., Lee, J.H., Yun, H.D., Ahn, B.Y., Kim, H. and Seo, W.T. (2011). Changes of phytochemical constituents (isoflavones, flavanols, and phenolic acids) during cheonggukjang soybeans fermentation using potential probiotics *Bacillus subtilis* CS90. *Journal of Food Composition and Analysis*, 24: 402-410.

Choksawangkarn, W., Phiphattananukoon, S., Jaresitthikunchai, J. and Roytrakul, S. (2018). Antioxidative peptides from fish sauce by-product: Isolation and characterization. *Agriculture and Natural Resources*, 52: 460-466.

Chourasia, R., Phukon L.C., Abedin, M., Sahoo, D., Singh, S.P. and Rai, A.K. (2020). Biotechnological approaches for the production of designer cheese with improved functionality. *Comprehensive Reviews in Food Science and Food Safety*. doi: 10.1111/1541-4337.12680

Coda, R., Rizzello, C.G., Pinto, D. and Gobbetti, M. (2012). Selected lactic acid bacteria synthesise antioxidant peptides during sourdough fermentation of cereal flours. *Applied and Environmental Microbiology*, 78: 1087-1096.

Constantinescu-AruxandeiCornwell, T., Cohick, W. and Raskin, I. (2004). Dietary phytoestrogens and health. *Phytochemistry*, 65: 995-1016.

Cui, Y., Miao, K., Niyaphorn, S. and Qu, X. (2020). Production of gamma-aminobutyric acid from lactic acid bacteria: A systematic review. *International Journal of Molecular Science*, 21: 995.

Daliri, E.B.M., Lee, B.H., Park, M.H., Kim, J.H. and Oh, D.H. (2018). Novel angiotensin I-converting enzyme inhibitory peptides from soybean protein isolates fermented by Pediococcus pentosaceus SDL1409. *LWT Food Science and Technology*, 93: 88-93.

De Vuyst, L. and Leroy, F. (2007). Bacteriocins from lactic acid bacteria: Production, purification, and food applications. *Journal of Molecular Microbiology and Biotechnology*, 13: 194-199.

Deegan, L.H., Cotter, P.D., Colin, H. and Ross, P. (2006). Bacteriocins: Biological tools for bio-preservation and shelf-life extension. *International Dairy Journal*, 16: 1058-1071.

Devi, S.M. and Halami, P.M. (2014). Detection and characterisation of pediocin PA-1/AcH like bacteriocin-producing lactic acid bacteria. *Current Microbiology*, 63: 181-185.

Dhakal, R., Bajpai, V.K. and Baek, K.H. (2012). Production of GABA by microorganisms: A review. *Brazilian Journal of Microbiology*, 43: 1230-1241.

Diep, D.B., Straume, D., Kjos, M. and Torres, C. (2009). An overview of the mosaic bacteriocin pln loci from *Lactobacillus plantarum*. *Peptides*, 30: 1562-1574.

Donkor, O.N., Henriksson, A., Vasiljevic, T. and Shah, N.P. (2005). Probiotic strains as starter cultures improve angiotensin-converting enzyme inhibitory activity in soy yogurt. *Journal of Food Science*, 70: 375-381.

D'Souza, R., Pandeya, D.R., Rahman, M., Lee, H.S., Jung, J.K. and Hong, S.T. (2012). Genetic engineering of *Lactococcus lactis* to produce an amylase inhibitor for development of an anti-diabetes biodrug. *The New Microbiologica*, 35: 35-42.

Dussault, D., Vu, K.D. and Lacroix, M. (2016). Enhancement of nisin production by *Lactococcus lactis* sub sp. *Lactis*, *Probiotics* and *Antimicrobial Proteins*, 8: 170-175.

Egan, K., Field, D., Rea, M.C., Ross, R.P., Hill, C. and Cotter, P.D. (2016). Bacteriocins: Novel solutions to age old spore-related problems. *Frontiers in Microbiology*, 7: 461.

Florence, A.C.R., Da Silva, R.C., Do Espírito Santo, A.P., Gioielli, L.A., Tamime, A.Y. and De Oliveira, M.N. (2009). Increased CLA content in organic milk fermented by bifidobacteria or yoghurt cultures. *Dairy Science and Technology*, 89: 541-553.

Gharras, H.E. (2009). Polyphenols: Food sources, properties and applications – A review. *International Journal of Food Science and Technology*, 44: 2512-2518.

Giuffrida, D., Sutthiwong, N., Dugo, P., Donato, P., Cacciola, F., Girard-Valenciennes, E. and Dufossé, L. (2016). Characterisation of the C50 carotenoids produced by strains of the cheese-ripening bacterium *Arthrobacter arilaitensis*. *International Dairy Journal*, 55: 10-16.

Gu, Q. and Li, P. (2016). Biosynthesis of vitamins by probiotic bacteria. pp. 135-148. *In*: Rao, V., Rao, L.G. (Eds.), Probiotics and Prebiotics in Human Nutrition and Health. *Tech Open*, London.

Gursoy, O., Seckin, A.K., Kinik, O. and Karaman, A.D. (2012). The effect of using different probiotic cultures on conjugated linoleic acid (CLA) concentration and fatty acid composition of white pickle cheese. *International Journal of Food Sciences and Nutrition*, 63: 610-615.

Hati, S., Ningtyas, D.W., Khanuja, J.K. and Prakash, S. (2020). β-glucosidase from almonds and yoghurt cultures in the biotransformation of isoflavones in soy milk. *Food Bioscience*, 34: 100542.

Hati, S., Patel, M., Mishra, B.K. and Das, S. (2019). Short-chain fatty acid and vitamin production potentials of *Lactobacillus* isolated from fermented foods of Khasi tribes, Meghalaya, India. *Annals of Microbiology*, 69: 1191-1199.

Hernández-Almanza, A., Muñiz-Márquez, D.B., de la Rosa, O., Navarro, V., Martínez-Medina, G., Rodríguez-Herrera, R. and Aguilar, C.N. (2017). Microbial production of bioactive pigments, oligosaccharides, and peptides. pp. 95-134. *In*: A. Grumezescu and A.M. Holban (Eds.). *Food Biosynthesis*. Elsevier.

Hernandez-Ledesma, B., Miralles, B., Amigo, L., Ramos, M. and Recio, I. (2005). Identification of antioxidant and ACE-inhibitory peptides in fermented milk. *Journal of Science of Food and Agriculture*, 85: 1041-1048.

Hjortmo, S., Patring, J., Jastrebova, J. and Andlid, T. (2008). Biofortification of folates in white wheat bread by selection of yeast strain and process. *International Journal of Food Microbiology*, 127: 32-36.

Hsu, W.H. and Pan, T.M. (2012). *Monascus purpureus*-fermented products and oral cancer: A review. *Applied Microbiology and Biotechnology*, 93: 1831-1842.

Hur, S.J., Lee, S.Y., Kim, Y.C., Choi, I. and Kim, G.B. (2014). Effect of fermentation on the antioxidant activity in plant-based foods. *Food Chemistry*, 160: 346-356.

Ibe, S., Yoshida, K., Kumada, K., Tsurushiin, S., Furusho, T. and Otobe, K. (2009). Antihypertensive effects of natto, a traditional Japanese fermented food, in spontaneously hypertensive rats. *Food Science and Technology Research*, 15: 199-202.

Ilyasoglu, H. and Yilmaz, F. (2018). Characterisation of yoghurt enriched with polyunsaturated fatty acids by using walnut slurry. *International Journal of Dairy Technology*, 72: 110-119.

Inoue, K., Shirai, T., Ochiai, H., Kasao, M., Hayakawa, K., Kimura, M. and Sansawa, H. (2003). Blood-pressure-lowering effect of a novel fermented milk containing gamma-aminobutyric acid (GABA) in mild hypertensives. *European Journal of Clinical Nutrition*, 57: 490-495.

Je, J.Y., Park, P.J., Byun, H.G., Jung, W.K. and Kim, S.K. (2005). Angiotensin I converting enzyme (ACE) inhibitory peptide derived from the sauce of fermented blue mussel, Mytilus edulis. *Bioresource Technology*, 96: 1624-1629.

Juan, M.Y. and Chou, C.C. (2010). Enhancement of antioxidant activity, total phenolic and flavonoid content of black soybeans by solid state fermentation with *Bacillus subtilis* BCRC 14715. *Food Microbiology*, 27: 586-591.

Kaprasob, R., Kerdchoechuen, O., Laohakunjit, N. and Somboonpanyakul, P.B. (2018). B vitamins and prebiotic fructooligosaccharides of cashew apple fermented with probiotic strains *Lactobacillus* spp., *Leuconostoc mesenteroides* and *Bifidobacterium longum*. *Process Biochemistry*, 70: 9-19.

Kim, J.Y., Lee, M.Y., Ji, G.E., Lee, Y.S. and Hwang, K.T. (2009) Production of gamma-aminobutyric acid in black raspberry juice during fermentation by *Lactobacillus brevis* GABA100. *International Journal of Food Microbiology*, 130: 12-16.

Koba, K. and Yanagita, T. (2014). Health benefits of conjugated linoleic acid (CLA). *Obesity Research and Clinical Practice*, 8: 525-532.

Kumar, J., Sharma, N., Kaushal, G., Samurailatpam, S., Sahoo, D., Rai, A.K. and Singh, S.P. (2019). Metagenomic Insights into the taxonomic and functional features of Kinema, a traditional fermented soybean product of Sikkim Himalaya. *Frontiers in Microbiology*, 10: 1744.

LeBlanc, J.G., Laiño, J.E., del Valle, M.J., Giori, F. and deTaranto, M.P. (2015). B-group vitamins production by probiotic lactic acid bacteria. pp. 279-296. *In*: Fernanda Mozzi, Raúl R. Raya and Graciela M. Vignolo (Eds.). *Biotechnology of Lactic Acid Bacteria: Novel Applications*, 2nd ed. John Wiley & Sons, Ltd.

LeBlanc, J.G., Laino, J.E., delValle, M.J., Vannini, V., van, S.D., Taranto, M.P., de Valdez, G.F., deGiori, G.S. and Sesma, F. (2011). B-group vitamin production by lactic acid bacteria – Current knowledge and potential applications. *Journal of Applied Microbiology*, 111: 1297-1309.

Lee, Y.G., Kim, J.Y., Lee, K.W., Kim, K.H. and Lee, H.J. (2003). Peptides from anchovy sauce in a human lymphoma cell (U937) through the increase of Caspase-3 and -8 activities. *Annals of the New York Academy of Sciences*, 1010: 399-404.

Lee, I. H., Hung, Y.-H. and Chou, C.C. (2008). Solid-state fermentation with fungi to enhance the antioxidative activity, total phenolic and anthocyanin contents of black bean. *International Journal of Food Microbiology*, 121: 150-156.

Li, Z., Teng, J., Lyu, Y., Hu, X., Zhao, Y. and Wang, M. (2019). Enhanced antioxidant activity for apple juice fermented with *Lactobacillus plantarum* ATCC14917. *Molecules*, 24: 51.

Lynch, K.M., Zannini, E., Coffey, A. and Arendt, E.K. (2018). Lactic acid bacteria exopolysaccharides in foods and beverages: Isolation, properties, characterisation and health benefits. *Annual Reviews of Food Science and Technology*, 9: 155-176.

Lyons, M.P., Papazyan, T.T. and Surai, P.F. (2007). Selenium in food chain and animal nutrition: Lessons from nature. *Asian-Australasian Journal of Animal Sciences*, 20: 1135-1155.

Majumdar, R.K., Roy, D., Bejjanki, S. and Bhaskar, N. (2016). Chemical and microbial properties of shidal, a traditional fermented fish of northeast India. *Journal of Food Science and Technology*, 53: 401-410.

Martinez, F.A.C., Balciunas, E.M., Converti, A., Cotter, P.D. and de Souza Oliveira, R.P. (2013). Bacteriocin production by *Bifidobacterium* spp.: A review. *Biotechnology Advances*, 31: 482-488.

Moktan, B., Saha, J. and Sarkar, P.K. (2008). Antioxidant activities of soybean as affected by *Bacillus*-fermentation to kinema. *Food Research International*, 41: 586-593.

Moslehi-Jenabian, S., Lindegaard, L. and Jespersen, L. (2010). Beneficial effects of probiotic and food borne yeasts on human health. *Nutrients*, 2: 449-473.

Murru, E., Carta, G., Cordeddu, L., Melis, M.P., Desogus, E., Ansar, H., Chilliard, Y., Ferlay, A., Stanton, C., Coakley, M., Ross, P.R., Piredda, G., Addis, M., Mele, M.C., Cannelli, G., Banni, S. and Manca, C. (2018). Dietary conjugated linoleic acid-enriched cheeses influence the levels of circulating n-3 highly unsaturated fatty acids in humans. *International Journal of Molecular Sciences*, 19: 1730.

Najafian, L. and Babji, A.S. (2018). Purification and identification of antioxidant peptides from fermented fish sauce (Budu). *Journal of Aquatic Food Product Technology*, 8850: 1-12.

Nakamura, Y., Yamamoto, N., Sakai, K., Okubo, A., Yamazaki, S. and Takano, T. (1995).

Purification and characterization of angiotensin I converting enzyme inhibitors from sour milk. *Journal of Dairy Science*, 78: 777-783.

Nampoothiri, K.M., Beena, D.J., Vasanthakumari, D.S. and Ismail, B. (2017). Health benefits of exopolysaccharides in fermented foods. pp. 49-62. *In*: Juana Frías, Cristina Martínez-Villaluenga and Elena Peña (Eds.). *Fermented Foods in Health and Disease Prevention*. Academic Press, Oxford.

Nilsen, R., Pripp, A.H., Høstmark, A.T., Haug, A. and Skeie, S. (2014). Short communication: Is consumption of a cheese rich in angiotensin converting enzyme-inhibiting peptides, such as the Norwegian cheese Gamalost, associated with reduced blood pressure. *Journal of Dairy Science*, 97: 2662-2668.

Ohata, M., Uchida, S., Zhou, L. and Arihara, K. (2016). Antioxidant activity of fermented meat sauce and isolation of an associated antioxidant peptide. *Food Chemistry*, 194: 1034-1039.

Parente, E., Cogan, T.M. and Powell, I.B. (2017). Starter cultures: General aspects. pp. 201-226. *In*: P.L.H. McSweeney, P.F. Fox., P.D. Cotter., D.W. Everett (Eds.). *Cheese*. Elsevier.

Park, K.B. and Oh S.H. (2007). Production of yogurt with enhanced levels of gamma aminobutyric acid and valuable nutrients using lactic acid bacteria and germinated soybean extract. *Bioresource Technology*, 98: 1675-1679.

Patel, A. and Prajapati, J.B. (2013). Food and health applications of exopolysaccharides produced by lactic acid bacteria. *Advances in Dairy Research*, 1: 2.

Patel, A., Shah, N. and Prajapati. J. (2013). Biosynthesis of vitamins and enzymes in fermented foods by lactic acid bacteria and related genera: A promising approach. *Croatian Journal of Food Science and Technology*, 5: 85-91.

Patel, P., Jethani, H., Radha, C., Vijayendra, S.V.N., Mudlier, S.N., Sarada, R. and Chauhan, V.S. (2019). Development of a carotenoid enriched probiotic yogurt from fresh biomass of Spirulina and its characterisation. *Journal of Food Science and Technology*, 56: 3721-3731.

Patel, S. and Goyal, A. (2011). Functional oligosaccharides: Production, properties and applications. *World Journal of Microbiology and Biotechnology*, 27: 1119-1128.

Parmar, H., Hati, S. and Sakure, A. (2018). *In vitro* and in silico analysis of novel ACE-inhibitory bioactive peptides derived from fermented goat milk. *International Journal of Peptide Research and Therapeutics*, 24: 441-453.

Perez-Corona, M.T., Sánchez-Martínez, M., Valderrama, M.J., Rodríguez, M.E., Cámara, C. and Madrid, Y. (2011). Selenium biotransformation by *Saccharomyces cerevisiae* and *Saccharomyces bayanus* during white wine manufacture: Laboratory-scale experiments. *Food Chemistry*, 124: 1050-1055.

Perez-Ramos, A., Mohedano, M.L., Pardo, M.A. and Lopez, P. (2018). Beta-glucan-producing *Pediococcus parvulus* 2.6: Test of probiotic and immunomodulatory properties in zebrafish models. *Frontiers in Microbiology*, 9: 1684.

Pescuma, M., Gomez-Gomez, B., Perez-Corona, T., Font, G., Madrid, Y. and Mozzi, F. (2017). Food prospects of selenium-enriched *Lactobacillus acidophilus* CRL 636 and *Lactobacillus reuteri* CRL 1101. *Journal of Functional Foods*, 35: 466-473.

Pophaly, S.D., Poonam, Singh, P., Kumar, H., Tomar, S.K. and Singh, R. (2014). Selenium enrichment of lactic acid bacteria and bifidobacteria: A functional food perspective. *Trends in Food Science and Technology*, 39: 135-145.

Pouliot-Mathieu, K., Gardner-Fortier, C., Lemieux, S., St-Gelais, D., Champagne, C.P. and Vuillemard, J.C. (2013). Effect of cheese containing gamma-aminobutyric acid-producing lactic acid bacteria on blood pressure in men. *Pharma Nutrition*, 1: 141-148.

Pyo, Y. H. and Lee, T. C. (2007). The potential antioxidant capacity and angiotensin I converting enzyme inhibitory activity of Monascus-fermented soybean extracts: Evaluation of

Monascus-fermented soybean extracts as multifunctional food additives. *Journal of Food Science*, 72: 218-223.

Qiang, X., YongLie, C. and QianBing, W. (2009). Health benefit application of functional oligosaccharides. *Carbohydrate Polymer*, 77: 435-441.

Quiros, A., Ramos, M., Muguerza, B., Delgado, M., Miguel, M., Alexaindre, A. and Recio, I. (2007). Identification of novel antihypertensive peptides in milk fermented with *Enterococcus faecalis*. *International Dairy Journal*, 17: 33-41.

Rai, A.K., Pandey, A. and Sahoo, D. (2019). Biotechnological potential of yeasts in functional food industry. *Trends in Food Science and Technology*, 83: 129-137.

Rai, A.K., Sanjukta, S. and Jeyaram, K. (2017a). Production of Angiotensin I converting enzyme inhibitory (ACE-I) peptides during milk fermentation and its role in treatment of hypertension. *Critical Reviews in Food Science and Nutrition*, 57: 2789-2800.

Rai, A.K., Sanjukta, S., Chourasia, R., Bhat, I., Bhardwaj, P.K. and Sahoo, D. (2017b). Production of soybean bioactive hydrolysate using protease, amylase and β-glucosidase from novel *Bacillus* spp. strains isolated from *kinema*. *Bioresource Technology*, 235: 358-365.

Rai, A.K., Kumari, R., Sanjukta, S. and Sahoo, D. (2016). Production of bioactive protein hydrolysate using the yeasts isolated from soft *chhurpi*. *Bioresource Technology*, 219: 239-245.

Rai, A.K. and Jeyaram, K. (2015). Health benefits of functional proteins in fermented foods. pp. 455-476. *In*: J.P. Tamang (Ed.). *Health Benefits of Fermented Foods and Beverages*. CRC Press, Taylor and Francis Group of USA.

Rai, A.K. and Kumar, R. (2015). Potential of microbial bio-resources of Sikkim Himalayan region. *ENVIS Himalayan Bulletin*, 23: 99-105.

Rai, A.K. and Appaiah, A.K. (2014). Application of native yeast from Garcinia (*Garcinia xanthochumus*) for the preparation of fermented beverage: Changes in biochemical and antioxidant properties. *Food Bioscience*, 5: 101-107.

Rai, A.K., Bhaskar, N. and Baskaran, V. (2013). Bioefficacy of EPA-DHA from lipids recovered from fish processing wastes through biotechnological approaches. *Food Chemistry*, 136: 80-86.

Rai, A.K., Prakash, M. and Appaiah, A.K. (2010). Production of *Garcinia* wine: Changes in biochemical parameters, organic acids and free sugars during fermentation of *Garcinia* must. *International Journal of Food Science and Technology*, 45: 1330-1336.

Rizzello, C.G., Cassone, A., Di Cagno, R. and Gobbetti, M. (2008). Synthesis of angiotensin I-converting enzyme (ACE)-inhibitory peptides and γ-aminobutyric acid (GABA) during sourdough fermentation by selected lactic acid bacteria. *Journal of Agricultural and Food Chemistry*, 16: 6936-6943.

Roberfroid, M.B. and Slavin, J. (2000). Nondigestible oligosaccharides. *Critical Review in Food Science and Nutrition*, 40: 461-480.

Rodriguez-Figueroa Rolim, P., Hu, Y. and Geanzle, M. (2018). Sensory analysis of juice blend containing isomaltooligosaccharides produced by fermentation with Weissella cibaria. *Food Research International*, 124: 86-92.

Russo, P., Capozzi, V., Arena, M.P., Spadaccino, G., Teresa, D.M., Lopez, P., Fiocco, D. and Spano, G. (2014). Riboflavin-overproducing strains of *Lactobacillus fermentum* for riboflavin-enriched bread. *Applied Microbiology and Biotechnology*, 98: 3691-3700.

Salmerón, I. (2017). Fermented cereal beverages: From probiotic, prebiotic and synbiotic towards nanoscience designed healthy drinks. *Letters in Applied Microbiology*, 65: 114-124.

Sanchart, C., Rattanaporn, O., Haltrich, D., Phukpattaranont, P. and Maneerat, S. (2017). *Lactobacillus futsaii* CS3, a New GABA-producing strain isolated from Thai fermented shrimp (*Kung-Som*). *Indian Journal of Microbiology*, 57: 211-217.

Sanjukta, S., Rai, A.K. and Sahoo, D. (2017). Bioactive metabolites in fermented soybean products and their potential health benefits. *In*: Ramesh C. Ray and Didier Montet (Eds.). *Fermented Foods: Part II: Technological Interventions*, CRC Press, Florida.

Sanjukta, S. and Rai, A.K. (2016). Production of bioactive peptides during soybean fermentation and their potential health benefits. *Trends in Food Science and Technology*, 50: 1-10.

Sanjukta, S., Rai, A.K., Ali, M.M., Jeyaram, K. and Talukdar, N.C. (2015). Enhancement of antioxidant properties of two soybean varieties of Sikkim Himalayan region by proteolytic *Bacillus subtilis* fermentation. *Journal of Functional Foods*, 14: 650-658.

Santiago-López, L., Aguilar-Toalá, J.E., Hernández-Mendoza, A., Vallejo-Cordoba, B., Liceaga, A.M. and González-Córdova, A.F. (2018). Invited review: Bioactive compounds produced during cheese ripening and health effects associated with aged cheese consumption. *Journal of Dairy Science*, 101: 3742-3757.

Santos-Espinosa, A., Beltrán-Barrientos, L.M., Reyes-Díaz, R., Mazorra, M.A.M., Hernandez-Mendoza, A., Gonalez-Aguilar, G.A., Sayago-Ayerdi, S.G., Vallejo-Cordoba, B. and Gonzalez-Cordova, A.F. (2020). Gamma-aminobutyric acid (GABA) production in milk fermented by specific wild lactic acid bacteria strains isolated from artisanal Mexican cheeses. *Annual Microbiology*, 70: 12.

Sarmadi, B. and Ismail, A. (2010). Antioxidative peptides from food proteins: A review. *Peptides*, 31: 1949-1956.

Seo, M.D., Kim, S.Y., Eom, H.J. and Han, N.S. (2007). Synbiotic synthesis of oligosaccharides. *Journal of Microbiology Biotechnology*, 17: 1758-1764.

Shi, Y.C. and Pan, T.M. (2011). Beneficial effects of *Monascus purpureus* NTU 568-fermented products: A review. *Applied Microbiology and Biotechnology*, 90: 1207-1217.

Shiferaw, Terefe, N. and Augustin, M.A. (2019). Fermentation for tailoring the technological and health related functionality of food products. *Critical Reviews in Food Science and Nutrition*, 60: 2887-2913.

Simova, E.D., Beshkova, D.M., Angelov, M.P. and Dimitrov, Zh.P. (2008). Bacteriocin production by strain *Lactobacillus delbrueckii* ssp. *bulgaricus* BB18 during continuous pre-fermentation of yogurt starter culture and subsequent batch coagulation of milk. *Journal of Industrial Microbiology and Biotechnology*, 35: 559-567.

Sohri, A.B. (2020). Proteolytic activity, antioxidant, and α-Amylase inhibitory activity of yogurt enriched with coriander and cumin seeds. *LWT–Food Science and Technology*, 133: 109912.

Solanki, D. and Hati, S. (2018). Considering the potential of *Lactobacillus rhamnosus* for producing Angiotensin I-Converting Enzyme (ACE) inhibitory peptides in fermented camel milk (Indian breed). *Food Science*, 23: 16-22.

Sourabh, A., Rai, A.K., Chauhan, A., Jeyaram, K., Taweechotipatr, M., Panesar, P.S., Sharma, R., Panesar, R., Kanwar, S.S., Walia, S., Tanasupawat, S., Sood, S., Joshi, V.K., Bali, V., Chauhan, V. and Kumar, V. (2015). Health-related issues and indigenous fermented products. pp. 303-343. *In*: V.K. Joshi (Eds.). *Indigenous Fermented Foods of South Asia*. Taylor and Francis Group of USA.

Springob, K. and Kutchan, T.M. (2009). Introduction to the different classes of natural products. pp. 3-50. *In*: A. Osbourn, E. and V. Lanzotti (Eds.). *Plant-derived Natural Products, Synthesis, Function, and Application*. Springer Science + Business Media.

Tamam, B., Syah, D., Suhartono, M.T., Kusuma, W.A., Tachibana, S. and Lioe, H.N. (2019). Proteomic study of bioactive peptides from *tempe*. *Journal of Bioscience and Bioengineering*, 128: 241-248.

Taoka, Y., Nakamura, M., Nagai, S., Nagasaka, N., Tanaka, R. and Uchida, K. (2019). Production of anserine-rich fish sauce from giant masu salmon. *Oncorhynchus Masou*

masou and γ-aminobutyric acid (GABA)-enrichment by *Lactobacillus plantarum* strain N10. *Fermentation*, 5: 45.

Thakur, K., Tomar, S.K. and De, S. (2016). Lactic acid bacteria as a cell factory for riboflavin production. *Microbiology Biotechnology*, 9: 441-451.

Vieira, C.P., Cabral, C.C., da Costa Lima, B.R., Paschoalin, V.M.F., Leandro, K.C. and Conte-Junior, C.A. (2017). *Lactococcus lactis* ssp. *cremoris* MRS47, a potential probiotic strain isolated from *kefir* grains, increases cis-9, trans-11-CLA and PUFA contents in fermented milk. *Journal of Functional Foods*, 31: 172-178.

Wakai, T., Yamaguchi, N., Hatanaka, M., Nakamura, Y. and Yamamoto, N. (2012). Repressive processing of antihypertensive peptides, Val-Pro-Pro and Ile-Pro-Pro, in *Lactobacillus helveticus* fermented milk by added peptides. *Journal of Bioscience and Bioengineering*, 114: 133-137.

Walther, B. and Schmid, A. (2016). Effect of fermentation on vitamin content in food. *In*: J. Frías, C. Martínez-Villaluenga and E. Penas (Eds.). *Fermented Foods in Health and Disease Prevention*. Oxford, UK: Elsevier Science.

Wang, J., Wu, T., Fang, X. and Yang, Z. (2019). Manufacture of low-fat Cheddar cheese by exopolysaccharide-producing *Lactobacillus plantarum* JLK0142 and its functional properties. *Journal of Dairy Science*, 102: 3825-3838.

Weijers, C.A., Franssen, M.C. and Visser, G.M. (2008). Glycosyltransferase catalyzed synthesis of bioactive oligosaccharides. *Biotechnology Advances*, 26: 436-456.

Wong, M.C., Emery, P.W., Preedy, V.R and Wiseman, H. (2008). Health benefits of isoflavones in functional foods? Proteomic and metabonomic advances. *Inflammopharmacology*, 16: 235-239.

Wu, Q. and Shah, N.P. (2016). High γ-aminobutyric acid production from lactic acid bacteria: Emphasis on *Lactobacillus brevis* as a functional dairy starter. *Critical Reviews of Food, Science and Nutrition*, 17: 3661-3672.

Xiao, J.B. and Hogger, P. (2015). Dietary polyphenols and type 2 diabetes: Current insights and future perspectives. *Current Medical Chemistry*, 22: 23-38.

Yang, J., Wang, J., Yang, K., Liu, M.Y., Qi, Y., Zhang, T., Fan, M. and Wei, X. (2018). Antibacterial activity of selenium-enriched lactic acid bacteria against common food-borne pathogens *in vitro*. *Journal of Dairy Science*, 101: 1930-1942.

Zhang, Y., He, S., Bonneil, E. and Simpson, B.K. (2020). Generation of antioxidative peptides from Atlantic sea cucumber using alcalase versus trypsin: *In vitro* activity, *de novo* sequencing, and in silico docking for *in vivo* function prediction. *Food Chemistry*, 306: 125581.

Zheng, G. and Slavik, M. (1999). Isolation, partial purification and characterisation of a bacteriocin produced by a newly isolated Bacillus subtilis strain. *Letters in Applied Microbiology*, 28: 363-367.

Zhou, R., Ren, Z., Ye, J., Fan, Y., Liu, X., Yang, J., Deng, Z.Y. and Li, J. (2019). Fermented soybean dregs by *Neurospora crassa*: A traditional prebiotic food. *Applied Biochemistry and Biotechnology*, 189: 608-625.

Zhou, Y., Zheng, J., Li, Y., Xu, D.P., Li, S., Chen, Y.M. and Li, H.B. (2016). Natural polyphenols for prevention and treatment of cancer. *Nutrients*, 8: 515.

Part II

Health Promoting Components in Fermented Foods

Polyphenols in Fermented Foods and Their Potential Health Benefits

E. Sener-Aslay[1a] and Z. Tacer-Caba[1]*

[1] Department of Gastronomy and Culinary Arts, Bahcesehir University, Ihlamur Yildiz Caddesi, No:8 Gayrettepe, 34353 Besiktas, Istanbul, Turkey

[a] Current address: Department of Gastronomy and Culinary Arts, Izmir University of Economics, Fevzi Çakmak, Sakarya Cd. No:156, 35330 Balçova/İzmir, Turkey

1. Introduction

Several attempts have been made through ages to make stable and safe food products from food materials. Fermentation process is defined as utilisation of growth and/or activity of microorganisms to transform simple raw products into numerous types of value-added products (Farnworth, 2003). Fermentation has been the subject of various cultures for thousands of years to create desired, acceptable and stable food products from perishable food sources. Moreover, fermentation plays a critical role in the local culinary traditions, as they evoked through the ages and modified technically as products of cultural practices to create the unique flavours, aromas and textures (Shiferaw et al., 2019; Xiang et al., 2019). Fermentation results in many biochemical alterations that result in a changed ratio of nutritive and anti-nutritive constituents. Therefore, although geographical location is the main factor to direct the differences in the type of fermented foods developed, salt and other condiments used with the substrate perishable food commodity affects the inherent bacteria through fermentation. This alteration consequently affects the properties of products, such as digestibility, taste, texture and bioactivity (Chouhan et al., 2019; Min et al., 2019).

Fermentation is classified into three main groups, of which two are traditional applications: (1) spontaneous fermentation, (2) continuous usage of food-adapted microorganisms, back-slopping and the last group is the modern approach for the fermented food industry, and (3) usage of starter cultures, controlled fermentation (Hutkins, 2008). This invention and wide use of pure starter cultures contributed to the construction of fermented food industry and enabled the production of safe, high-yield, high-scale and versatile fermented products around the world (Mapelli-Brahm et al., 2020; Szutowska, 2020). Classification among industrial fermentation applications

*Corresponding author: zeynep.tacercaba@sad.bau.edu.tr

is made depending on the type of microorganisms: (1) alcoholic fermentation driven by yeast (most commonly *Brettanomyces, Candida, Cryptococcus, Debaryomyces, Dekkera, Galactomyces, Geotrichum, Hansenula, Hanseniaspora, Hyphopichia, Issatchenkia, Kazachstania, Kluyveromyces, Metschnikowia, Pichia, Rhodotorula, Rhodosporidium, Saccharomyces, Saccharomycodes, Saccharomycopsis, Schizosaccharomyces, Sporobolomyces, Torulaspora, Torulopsis, Trichosporon, Yarrowia,* and *Zygosaccharomyces*) and (2) lactic fermentation driven by lactic acid bacteria (most commonly *Alkalibacterium, Carnobacterium, Enterococcus, Lactobacillus, Lactococcus, Leuconostoc, Oenococcus, Pediococcus, Streptococcus, Tetragenococcus, Vagococcus,* and *Weissella*) (Tamang *et al.*, 2016). Moreover, other types of fermentation, such as propionic, malolactic, butyric, etc. also share the similar basic mechanism of fermentation of being 'an anaerobic catabolism of organic compounds whose yield is lower than respiratory chain in term of energy production' that produces a variety of products (Septembre-Malaterre *et al.*, 2018).

2. Polyphenols

Polyphenols, with more than 8,000 species, are among the most significant bioactive compounds to be included in the secondary metabolites of plants. They have diverse chemical structures and exert health benefits, particularly antioxidant activity. Antioxidant activity may be defined as the total capacity of antioxidants for eliminating free radicals in the cell and in food. Their ability of scavenging free radicals via their hydroxyl substituents and their H-donating ability, therefore, enable antioxidants to protect DNA or cell walls from oxidative stress (Bakir *et al.*, 2016, Khosravi and Razavi, 2020). This ability is of vital significance as these reactive oxygen species, produced during aerobic cell respiration, are widely proposed to play a key role in many degenerative diseases, such as cancer, diabetes, autoimmune conditions, various respiratory diseases, eye diseases, schizophrenia, osteoporosis and aging (Balasundram *et al.*, 2006). The antioxidants also act as metal chelators or hydrogen donors to radicals (Hur *et al.*, 2014). Moreover, this pivotal function keeps the body's oxidative stress level below the critical point (Georgetti *et al.*, 2009).

Polyphenols range from simple phenolic molecules with one aromatic ring to larger and complex polymerised molecules. Polyphenols are grouped into five major families including phenolic acids, flavonoids, lignans, stilbenes and curcuminoids according to their structure. Phenolic compounds have two subgroups of hydroxybenzoic acids (such as gallic, p-hydroxybenzoic, protocatechuic, syringic, and vanillic acids) and hydroxycinnamic acids (caffeic, ferulic, p-coumaric, and sinapic acids). Flavonoids are larger structures with two aromatic rings (A and B) joined by a three-carbon bridge (C6–C3–C6 structure) (Adebo and Medina-Meza, 2020). Different eight subclasses, including, flavones, flavonols, flavanones, isoflavones, flavan-3-ols, proanthocyanidins, anthocyanins and chalcones/dihydrochalcones, belong to flavonoids. Among them, anthocyanins are the naturally occurring pigments in plants and are widely distributed in Nature in their glycosylated form. Sugars are attached to the 3-hydroxyl position of the anthocyanidin (sometimes to the 5 or 7 position of flavynium ion) (McGhie and Walton, 2007). The antioxidant activity of phenolic acids depends on the number and orientation of the hydroxyl groups

(Rice-Evans *et al.*, 1996). In this regard, flavonoids are the most effective radical scavengers because of their chemical structure (Hur *et al.*, 2014; Lee *et al.*, 2004). The main families of polyphenols with compound structures are presented in Fig. 1.

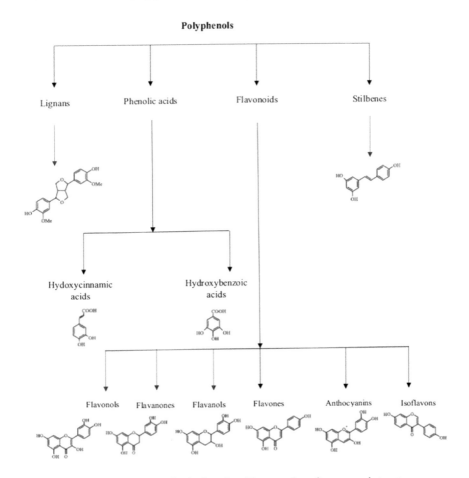

Fig. 1: Main families of polyphenols with examples of compound structures (adapted from Gómez-Caravaca *et al.*, 2014)

Foods, particularly fruits and vegetables, are rich in polyphenols. Therefore, they are widely ingested as a part of daily diet. However, due to the arrangement of the gastrointestinal tract and its structures, their bioavailability for the body may be affected by the food matrix. The gastrointestinal system with different environments in the stomach, small intestine and colon in gastrointestinal tract has different conditions to affect the polyphenols. Moreover, since many of the phenolics are found as covalently bound to sugars, proteins or cell wall components, absorption of these moieties is related with the characteristics and also the liberation of these molecules (Acosta-Estrada *et al.*, 2014). Variations in phenolics is, therefore, mostly related to the nature and number of sugar moiety attached to the phenolic molecule (McGhie and Walton, 2007; Tacer-Caba, 2015). For absorption through the gastrointestinal tract, this bound structure is required to be liberated (Du and Myracle, 2018). This phenomenon might be explained with the concept of bioavailability, since it is basically defined as the extent of absorption and transportation of nutrients to body tissues and affected by factors like the biological characteristics of food, components and genotypes (Oksuz *et al.*, 2019).

3. Food Fermentation and Bioavailability of Polyphenols

Recent studies related to phenolics have shown that food processing methods might depend on the increased bioavailability of polyphenols. In this respect, food processes, such as 'fermentation, malting, thermoplastic extrusion or enzymatic, alkaline and acid hydrolysis occasionally assisted with microwave or ultrasound' are suggested as methods to increase the bioavailability by releasing the polyphenols (Du and Myracle, 2018). During fermentation, complex substrates are decomposed and/ or biotransformed into smaller compounds by the aid of microbial enzymes, such as 'glycosidases, esterases, phenolic acid decarboxylases, phenolic acid reductase and tannase' (Rodríguez *et al.*, 2009; Xiang *et al.*, 2019). In this regard, hydrolysis of the β-glycosidic bonds in the conjugated phenolic compounds structured as one or more sugar residues linked to the hydroxyl groups has been reported as the main mechanism to increase the amount of free polyphenols and it is mainly attributed to the activity of the enzymes (Georgetti *et al.*, 2009; Hur *et al.*, 2014). Mainly glucosidases are significant enzymes in this hydrolysation activity that may have metal ion chelation activity (Hur *et al.*, 2014). Microbial enzymes such as glycoside hydrolase, cellulose or xylan-degrading ones and esterases also have a significant role in acting on complex polysaccharides during fermentation, resulting in change in properties and quantity of certain bioactive compounds (Chouhan *et al.*, 2019; Shiferaw *et al.*, 2019). Moreover, the activity of microorganisms breaking down the complex phenolics leading to the liberation of phenolic metabolites has been proposed as another approach for increasing the bioavailability of phenolics (Du and Myracle, 2018).

These enzymes release the bioactive esterified and insoluble bound nutrients by softening of the kernel structure and breakdown of cell walls of cereals (Hur *et al.*, 2014). Moreover, the production of protease, α-amylase and some other enzymes can be influenced by fermentation

In a previous study on rice vinegar, microbial enzymes, such as glucosidase, amylase, cellulase, chitinase, inulinase, phytase, xylanase, tannase, esterase, invertase or lipase were attributed not only to glucoside hydrolysis but also to plant cell-wall matrix disintegration. This cell wall disintegration has been causally associated with the increase in the extraction of flavonoids (Hur *et al.*, 2014). In contrast, it has been detected that the total amount of some high molecular weight phenolic compounds, such as oleuropein in olives decreased during fermentation, while there was an increase in the simpler phenolic acids (Othman *et al.*, 2009). This dramatic change has been attributed to hydrolysis (acidic and enzymatic) or the lactic acid bacterial activity to depolymerise high molecular weight phenolic compounds, such as oleuropein. Fermentation has also been shown to facilitate the increase of some distinct phenolics, such as hydroxytyrosol and caffeic acid, with high antioxidant activity (Hur *et al.*, 2014; Othman *et al.*, 2009). Therefore, further studies focusing on the relation between fermentation and different food products are required.

4. Fermented Dairy Products

Since many generations, nearly every culture around the world has developed fermented milk of some kind. Most people are familiar with the words *dahi*, buttermilk, yoghurt, *leben* and acidophilus milk but those who first created such foods had not actually realised that they were fermenting with bacteria. All around the world, there are traces of this early fermentation, like buttermilk and *dahi* from India, *kımız* and *keffir* from Central Asia, *laban zeer* from Egypt, *taetta* from Scandinavia and many more (Hui *et al.*, 2004). Although there are many different varieties of fermented milk products, these products can be divided into two categories, according to the microorganisms that are responsible for fermentation. Firstly, there is a large group of lactic fermented products, which is the result of lactic acid bacteria, breaking up protein and carbohydrate molecules and releasing lactic acid in the process. The fermentation is led by different types of lactic acid bacteria and produces thermophilic, probiotic and mesophilic products. The second group of fermented milk products also rely on lactic acid bacteria with the addition of yeasts. Yeast produces alcohol in the milk and along with lactic acid bacteria, results in slightly alcoholic fermented milk products, such as koumiss and *kefir* (Mayo *et al.*, 2010). Polyphenol content in fermented milk products is mainly due to the extracts or polyphenol-rich additives added to milk prior to fermentation (Du and Myracle, 2018). The knowledge of milk fermentation led to one of the most important food products: cheese and products derived from cheese. Introducing fermented milk products with phenolic compounds can be done in different ways which are summarised in Table 1.

Table 1: Approaches to Increase in Total Phenols and/or Antioxidant Activity in Fermented Dairy Products

Combination of dairy products with natural antioxidant sources (generally plant extracts) (Wang *et al.*, 2020).

Use of milk from different animals such as goat, sheep, buffalo, camel, etc. with added phenolic/bioactive profiles in fermented food production added before fermentation (Chávez-Servín *et al.*, 2018).

Utilisation of non-conventional starter base with better phenolic/bioactive profile (Caric *et al.*, 2019).

4.1 Yoghurt

4.1.1 *Product Properties*

Yoghurt is defined as a semi-solid dairy product made by adding specific bacteria strains (*Streptococcus salivarius* subsp. *thermophilus* and *Lactobacillus delbrueckii* subsp. *Bulgaricus* often cocultured with other lactic acid bacteria for taste or health effects) into milk, fermenting under controlled temperatures (42-43°C). Produced lactic acid after bacterial ingestion of natural milk sugars increases acidity, causing milk proteins to coagulate into a solid mass (curd) (Yildiz, 2016). Yoghurt has many different types around the world – set yoghurt, stirred yoghurt, fruit yoghurt, etc. being the most famous ones. *Dadiah, labneh, tarator, Rahmjoghurt, jameed,* etc. are among the other variants of yoghurt.

4.1.2 Polyphenols and Potential Health Benefits

Yoghurt is a significant dairy product to fulfil many current nutritional requirements. One distinguishable feature of yoghurt is its being a useful matrix for adding high polyphenol from plants sources to enhance the polyphenols. In dairy products, the effect of fermentation is also related with the proteolytic activity of microbial strains, especially yoghurt, in which lactic acid bacteria enhance the release of bonded peptides via fermentation. Previous research revealed that different fragments of β-casein in addition to a few N-terminal fragments of κ-casein exhibit antioxidant properties (Sah *et al.*, 2014). Addition of green tea rich in catechins and phenolic acids has been suggested to improve functional properties of fermented dairy beverages (Caric *et al.*, 2019). Addition of green tea has resulted in more than 30 times enhancement of antioxidant properties (Caric *et al.*, 2019). In another study, yoghurt was fortified with polyphenol extracts from strawberry press residues followed by fermentation using probiotic lactic acid bacteria and this resulted in enhancement of antioxidant activity (Ivanov and Dimitrova, 2019). Ingredients of milk, such as fat, vitamins A and D, polyunsaturated fatty acids, riboflavin, retinol, tocopherol, ascorbic acid, lactoferrin, and antioxidant enzymes also contribute to antioxidative capacity (Caric *et al.*, 2019).

4.2   Cheese

4.2.1 Product Properties

The most common belief about the origins of cheese is that in Central Asia, pastoral nomads of the day used animal stomach bags as containers, and since the inner sides of animal's stomach contain a curdling enzyme, rennin, it is possible that sour milk turned into cheese under the heat of the sun. This advancement is believed to have occurred around 7,000 BC, probably spread via nomadic routes through the Fertile Crescent (between the Euphrates and Tigris rivers in today's Iraq) and Babylonian, Indian and Egyptian records prove that there were milk, butter and cheese products all along this region. It is believed that cheese-making practices widely spread from the Mediterranean to the rest of the world. The most common cultures that are used for development of flavour and texture during ripening of cheese are summarised as *Brevibacterium linens, Propionibacterium freudenreichii, Debaryomyces hansenii, Geotrichum candidum, Penicillium camemberti,* and *Penicillium roqueforti* (Tamang *et al.*, 2016) in cheese-making practices using different milks, formulas and techniques. Despite this, only little progress was made technologically until the 19th century, and the knowledge of cheese making was only limited to old traditions.

In 1878, Sir Joseph Lister, managed to isolate milk bacterium known as *Lactococcus lactis* and thus selected cultures for different kinds of fermented milk products. And later on in 1890, a Danish scientist, Wilhelm Storch used the world's first selected culture for butter making (Hui *et al.*, 2004).

4.2.2   Polyphenols and Potential Health Benefits

Cheese is a nutritionally rich product possessing healthy nutrients, such as proteins, lipids, vitamins and minerals. Moreover, beneficial effects of cheese on calcium for osteoporosis and dental caries, in addition to its inherent conjugated linoleic acid with anticarcinogenic and antiatherogenic effects and peptides with various biological activities have been reported (Xiang *et al.*, 2019). Supplementation of cheese with high phenolic sources is the general approach applied in order to increase their polyphenolic content and antioxidant capacity. Different cheese types have been produced with the addition of organic Bordeaux grape (Pasini Deolindo *et al.*, 2019), mint leaf extract (Fezea *et al.*, 2017) and many other fruits, vegetables, extracts or herbs.

4.3   Kefir

4.3.1   Product Properties

Kefir is an ancient fermented milk product, composed of up to 30 microorganisms including lactic acid bacteria, yeast and occasionally acetic acid bacteria. *Kefir's* functional and biological properties, such as antibacterial, anticarcinogenic and anti-inflammatory effects are well known. *Kefir* can also protect the polyphenols from degradation in the small intestine and diverse microorganisms found in *kefir* may metabolise phenolic compounds and increase bioactivity (Du and Myracle, 2018).

4.3.2   Polyphenols and Potential Health Benefits

A study on *kefir*, using phenolic-rich sources for fermentation, was the common approach. In a recent study, the juice of aronia (*Aronia melanocarpa*), a berry rich in proanthocyanins and anthocyanins as phenolic compounds, has been used to ferment cow's milk to make *aronia kefir* (Du and Myracle, 2018). According to the findings of the study, the antioxidant capacity and bioavailable phenolics were higher after digestion of *aronia kefir*. Although the total amount of anthocyanin has decreased, new metabolites were detected after fermentation, probably being brokendown anthocyanins with higher antioxidant activity. The mechanism of fermentation action on anthocyanins has been related with the acidic pH of *kefir* which is a significant factor to protect anthocyanins from degradation (Du and Myracle, 2018). In literature, many other studies also support the positive effect of fermentation on increased antioxidant activity (Aiello *et al.*, 2020; Setiyoningruma and Priadi, 2019; Du and Myracle, 2018).

5.   Fermented Meat Products and Polyphenols

5.1   Product Properties

Fermented meat products are unique and often represented as an element of culinary heritage and identity. Since ancient times, meat fermentation is used to preserve meat and it is actually related to many changes summarised as acidification (carbohydrate catabolism), solubilisation and gelation of myofibrilla and sarcoplasmic proteins, degradation of proteins and lipids, reduction of nitrate into nitrite, formation of nitroso-myoglobin and dehydration in muscles in muscle-based-fermented products (Ojha *et al.*, 2015).

There are mainly two groups of fermented meat products: (1) those made from whole meat pieces or slices, such as dried meat and jerky; and (2) those made by chopping or comminuting the meat, usually called sausages (Adams, 2010). Fermented sausages and *salami, sucuk,* jerky, *nham* and *nemchua*, etc. are among the most famous fermented meat products from different parts of the world, having variable properties of size, shape, texture, appearance and flavour. The main microbial groups involved in meat fermentation are indigenous lactic acid bacteria, followed by coagulase-negative staphylococci, micrococci and *Enterobacteriaceae*, and depending on the product, some species of yeasts and molds, which may play a role in meat ripening (Tamang *et al.*, 2016). Modern use of industrial starter cultures provides additional functionalities, technological, nutritional or health advantages in terms of the final product composition whereas traditional artisanal fermentation techniques give more desired and usual organoleptic quality with limited safety profile and shelf-life.

5.2   Polyphenols and Potential Health Benefits

Intrinsic and extrinsic factors, such as enzymatic activity, pH, and temperature, compositions of the protein and lipid fractions significantly affect oxidative stability of meat products. Polyphenols in meat products have been reported as antioxidants through scavenging reactive species, decreasing oxidative degradation of proteins, inhibiting lipoxygenase activity and functioning as reducing agents for metmyoglobin to preserve colour (Papuc *et al.*, 2017). Therefore, extracts rich in polyphenols (mostly from spices, fruit and vegetable sources) function in a similar way to commonly used synthetic antioxidants in meat products, such as butylhydroxyanisole (BHA), butylhydroxytoluene (BHT), tert-butylhydroquinone (TBHQ), and propyl gallate (PG) (Cunha *et al.*, 2018). Use of plant sources, such as radish powder with chitosan (Ozaki *et al.*, 2020) and celery powder and tocopherol extract (Magrinya *et al.*, 2009) in fermented cooked sausages was studied to test their potential as inhibitors of lipid oxidation instead of nitrites and gave promising results to produce nitrite-free products. Trials on *Sucuk* (traditional fermented Turkish sausage) have also been conducted with green tea and shrub extract and according to the findings, the used extracts prevented lipid oxidation better than synthetic antioxidants (Bozkurt, 2006). Therefore, polyphenol-rich extracts have the potential as significant antioxidants against the lipid and protein oxidation in fermented meat products.

6.   Fermented Grain, Soy and Legume Products

Cereals have been the staple food for ages by being either consumed as individual foods or incorporated as ingredients into other food products. Fermented cereals contain numerous products, including mainly bread, beverages, biscuits and breakfast cereals, in addition to fermented beverages with many distinct properties. Some of the common ones are bread, fermented oat, *tempe, koji, akamu*, sourdough bread, porridge, etc. Sourdough fermentation decreases the glycemic response of baked goods, improves the properties against oxidative stress, bioavailability fibre complex and phytochemicals, and may increase the uptake of minerals (Gobbetti *et al.*, 2020). They are also good alternatives for people with specific dietary requirements, such as vegans, vegetarians or individuals with lactose intolerance (Xiang *et al.*, 2019).

The plant cell wall has ether or glycosidic linkages that increase its mechanical strength. Since most of the polyphenols are found esterified within mainly the cell-wall components in the bound form, they need additional processes to be released. In general, fermentation enables this breakdown of ester bonds between cell walls, dietary fibre-phenolic compounds and polysaccharides (Vitaglione *et al.*, 2008). Although fermentation is attributed to cereal products rather than grains, it has also been utilised as a method to improve the functional properties in some grains. For example, the bioavailability of free phenolics was improved in barley samples fermented with three probiotic microorganisms (*Lactobacillus johnsonii* LA1, *Lb. reuteri* SD2112, and *Lb. acidophilus* LA-5) (Hole *et al.*, 2012). Moreover, many other grains like rice have many polyphenols identified in light brown pericarp, such as derivatives of hydroxybenzoic acids and hydroxycinnamic acids, such as ferulic acid, ρ-coumaric acid, sinapic acid, protocatechuic acid, chlorogenic acid, hydroxybenzoic acid, and flavonoid compounds and also anthocyanins in some types. During fermentation, microorganisms produce enzymes, such as α-amylase, β-glucosidase and xylanase which break down the linkages between bound polyphenols and cell-wall components, thereby increasing the extractable polyphenols (Khosravi and Razavi, 2020;Wang *et al.*, 2014). Different mechanisms related to increase of total phenols and/or antioxidant activity in fermented grain soy and legume products are summarised in Table 2.

Table 2: Approaches to Increase in Total Phenols and/or Antioxidant Activity in Fermented Grain Soy and Legume Products (Xiang et al., 2019)

Bioconversion of the conjugated forms of phenolic compounds into their free forms by action of biocatalyst produced by starter culture.

Structural breakdown of the cereal cell walls potentially liberating and/or induce the synthesis of various bioactive compounds.

Changes in the content and the type of free polyphenols due to transformation due to specific enzymes.

6.1 Bread and Other Bakery Products

Breadmaking is an ancient skill practiced in almost every civilisation in many forms. First ever breads probably were unleavened, cake-like breads made with germinated, roasted and crushed grains, mostly from wheat, barley and spelt. Around 4000 BC, people started to process the grain raw, which enables proteins and carbohydrates to stay intact and help in leavening the bread. This grounded grain was mixed with water and salt, left in breweries and benefited from the natural occurring yeast (Sonnenfeld *et al.*, 1999). From the accidental discovery of bread until the commercialisation of baker's yeast (*Saccharomyces cerevisiae*) in the 19th century, people mostly used wild yeasts and lactic acid bacteria as leavening agents (Corsetti and Settanni, 2007).

Most cereals, like wheat, rye, barley and maize are fermented by natural fermentation process or by adding commercial baker's yeast into the batter for dough breads/loaves. Bread formulae generally contain a fair percentage of wheat flour. Wheat, being one of the ancient crops used in bread making, has the specific elastic protein of gluten which is responsible for expansion of the dough and most of the time for bread's texture. The production specifications are mostly associated with the formation of gluten, which requires both the hydration of proteins in the flour and the application of energy through the process of kneading (Cauvain, 2003). Modern bread-making practices do not differ from the ancient techniques. In many cultures, breadmaking is a combination of similar basic processes, although resulting in a wide range of breads. Mixing flour (not only from wheat, but also from other grains, such as rye, barley, maize, etc.), water, yeast, salt and proofing this mixture for yeast to produce enough carbon dioxide gas, expanding the dough and baking it were the basics of breadmaking. It was actually a piece of dough from the previous baking that was left aside until the next baking, when it was mixed with flour, salt and water to make the bread dough (Hui *et al.*, 2004). Sourdough is characterised by a complex microbial ecosystem, mainly represented by lactic acid bacteria and yeasts, whose fermentation confers to the resulting bread its characteristic features, such as palatability and high sensory quality (Corsetti and Settanni, 2007).

The most important factors that influence sourdough fermentation include the microbial starter, fermentation conditions (time and temperature; number of back sloppings) and flour composition (especially protein, carbohydrate and ash contents) (Abedfar and Sadeghi, 2019). Yeasts and some species of lactic acid bacteria (*Enterococcus, Lactococcus, Lactobacillus, Leuconostoc, Pediococcus, Streptococcus*, and *Weissella*) are widely used in sourdough fermentation. Besides, sourdough bread also enables significant increase in bread shelf-life, improvement of nutritional value and organoleptic properties (Plessas *et al.*, 2011). Among the mentioned parameters, fermentation time as well as flour ash and sugar contents have the largest effects on sourdough acidification. Sourdough fermentation is one of the oldest food biotechnologies, which has been studied and recently rediscovered for its effect on the sensory, structural, nutritional and shelf-life properties of leavened baked goods. Acidification, proteolysis and activation of a number of enzymes as well as the synthesis of microbial metabolites cause several changes during sourdough fermentation, which affect the dough and baked good matrix and influence the nutritional/functional quality.

6.1.1 Polyphenols and Potential Health Benefits

A recent study has aimed to reveal the different effects of fermentation among bread and sourdough bread. They revealed that activity of LAB with yeasts phenolic acid esterase's during fermentation increased the content of free phenolic acids and the amount of free phenolic acids in bread samples correlated with the antioxidant activity measured (Skrajda-Brdak *et al.*, 2019). Overall they also mentioned that the indicated final content of phenolic compounds in bread samples was the result of rather simultaneous activity of their release, decomposition by native flour and/ or enzymes in the microflora (to cause cell-wall degradation), together with thermal decomposition and repeated binding by surrounding chemicals (Skrajda-Brdak *et al.*, 2019). Therefore, the matrix effect in the bread significantly affects the phenomenon between fermentation and polyphenols.

6.2   Fermented Soybean and Polyphenols

6.2.1   Product Properties

Fermented soy products have high amounts of amino acids and fatty acids to give them their unique flavour, high nutritional value and good texture (Steinkraus *et al.*, 2004). Fermented soy products are categorised into two main groups – soybean foods fermented by *Bacillus* spp. (mostly *B. subtilis*), and soybean foods fermented by filamentous molds (mostly *Aspergillus, Mucor, Rhizopus*) (Tamang *et al.*, 2016; Sanjukta *et al.*, 2017). Fermented tofu, among the most famous traditional soy foods, is produced in many different countries with different names: China and Taiwan (*sufu* and *furu*), Malaysia (*tau ju*), Philippines (*tafuri*), Thailand (*tau-fu yee*) and Japan (*tofuyo*). Other fermented soy products, such as *meju*, soybean sauce, soybean paste, *miso, tempeh, natto* and *Cheonggukjang* are also commonly consumed (Lee *et al.*, 2018). In a recent review, fermented soybean products are generally considered as being a part of local identities as endogenous microorganisms are very commonly and traditionally used during their fermentation processes (Licandro *et al.*, 2020).

6.2.2   Polyphenols and Potential Health Benefits

Soybean and soy products have many phenolic compounds having antioxidant activity. A significant class of flavonoids, isoflavone, has phenolic hydroxyl radicals with potent antioxidant properties. Isoflavones have anticancerogenic properties, particularly limiting the onset of breast carcinogenesis. Genistein, daidzein and glycitein are the three main isoflavones which contribute the total phenolic content in soybean and soy products and isoflavonoids are the carbohydrate-conjugated compounds of isoflavones (Lee *et al.*, 2018; Septembre-Malaterre *et al.*, 2018). *Doenjang* is a protein-rich traditional fermented soybean product in Korea. Traditionally, *Bacillus* species are used as starter culture for *Doenjang* and then secondary microflora of *Aspergillus* species, such as *A. oryzae* and *A. niger* are used for fermentation process (Shukla *et al.*, 2019).

Isoflavone aglycones are defined as estrogen-like compounds. Previous studies revealed that isoflavone aglycones, being similar to many other polyphenolic compounds, have higher bioavailability than their conjugated, highly-polar glycosidic forms (Chen *et al.*, 2018; Georgetti *et al.*, 2009; Lin *et al.*, 2006). Fermentation basically affects the soybean products in a similar way as other plants, by breaking down phenolic glucosides (Georgetti *et al.*, 2009). It is an effective method in the transformation mechanism of glycosylated isoflavones into their aglycones (Hur *et al.*, 2014). Specifically for isoflavone glucosides, released lipophilic compounds by fermentation have been related with the increased antioxidant activity (Chouhan *et al.*, 2019). The catalytic action of β-glucosidase during fermentation may be either by *Bacillus* or *Aspergillus* species. However, studies revealed that although all produce β-glucosidases, different filamentous fungi (*Aspergillus niger*, *Aspergillus niveus* and *Aspergillus awamori*) in defatted soy flour samples caused differences in total antioxidant activities (Georgetti *et al.*, 2009). According to the findings of the study, *A. niveus* exhibited the highest enzymatic activity and specificity to genistein to increase the genistein content in fermented samples (Georgetti *et al.*, 2009).

Freeze-thawing has been proposed as an additional process to aid fermentation of tofu with *B. subtilis.* As freeze-thawing denatured the proteins and created sponge-like pores, microbial strain and enzymes better penetrate the tofu to decrease isoflavone glycosides (daidzin and genistin) while gradually increasing isoflavone aglycones (daidzein and genistein) with increased number of hydroxyl groups having higher antioxidant activity (Lee *et al.*, 2018). Recent studies revealed further effects of polyphenols on soybean products. In a recent study, extracts of lotus, ginkgo and garlic added later during the fermentation of *Doenjang* made this fermented product safer by reducing *B. cereus* counts, biological amines, aflatoxins and other food-borne pathogens in addition to increasing its antioxidant activity (Shukla *et al.*, 2019). Another enzymatic activity attributed to some of the lactic acid bacteria, such as *Lb. plantarum, Lb. paraplantarum* and *Lb. pentosus,* is the tannase activity. These species hydrolyse ester bounds of tannic acid, thus releasing glucose and gallic acid (Rodríguez *et al.*, 2009). Fermentation of soybean flour by *Lb. plantarum* CECT 748 led to the β-glucosidase catalysed conversion of most of the isoflavone glucosides and the malonyl and the acetyl glucosides to aglycones, which are absorbed faster and in a larger amount through the gut wall (María Landete *et al.*, 2015; Shiferaw Terefe and Augustin, 2019).

6.3   Legumes

Legumes are from the Fabaceae family and are a group of plants that yield around one to 12 seeds or grains which are protected inside a pod. According to FAO, common beans (*Phaseolus vulgaris*), broad beans (*Vicia faba*), dry peas (*Pisum sativum*), chickpeas (*Cicer arietinum*), cowpeas (*Vigna sinensis*), lentils (*Lens culinaris*) and *mung* bean (*Vigna radiata*) are considered as legumes having different sizes, forms and colours, and they are used as significant sources of nutrition for humans, animal feed and the production of plant-based oils (FAO, 1994). As being rich nutritional sources with relatively low-cost, legumes have been significant parts of the diets around the world for ages. Most of the legumes have a high protein and dietary fibre contents in addition to containing different types of phenolic acids, anthocyanins, proanthocyanidins and flavonols (Kan *et al.*, 2018). In contrast, legumes also contain antinutritional factors of protease inhibitors, α-amylase inhibitors, phytic acid, lectins and saponins that bind to proteins, carbohydrates, vitamins, minerals, etc. and decrease their bioavailability. Different processes, such as cooking, autoclaving, soaking, enzyme processing, fermentation and germination aim to minimise the effect of these antinutrients in legumes (Moreno-Valdespino *et al.*, 2019).

6.3.1 Polyphenols and Potential Health Benefits

Different legumes have a variety of phenolic acids, anthocyanins, proanthocyanidins and flavonols which are distributed generally between the cotyledon and seed coats (Oomah *et al.*, 2011), Fermentation of legumes is a natural and traditional process, since indigenous bacteria (mostly the lactic acid bacteria*)* are responsible for legume fermentation. The fermentation of legumes has many benefits, which primarily concern enhancing the digestibility of proteins and related nutritional values and the biological quality of fibres and phenolic compounds. Therefore, legume fermentation can be used to increase the content of bioactive phenolic compounds in legumes and thereby increase the antioxidant activity (Gobbetti *et al.*, 2020; Limón *et al.*, 2015). In general, strains of lactic acid bacteria and mainly *Lactobacillus genera*, emphasise the production of bioactive compounds, providing health benefits beyond basic nutrition (Rocchetti *et al.*, 2019). Moreover, bioconversion of the conjugated forms of phenolic compounds into their free forms during fermentation is another significant mechanism for legumes (Balasundram *et al.*, 2006; Hur *et al.*, 2014; Torino *et al.*, 2013). Polyphenols content in soybean fermented products have shown to increase during fermentation due to the enzymes produced by the starter culture (Rai *et al.*, 2017; Sanjukta *et al.*, 2015).

Another recent study on legume phenolics also revealed that fermentations made qualitative and quantitative changes in phenolic compounds (Limón *et al.*, 2015). Apart from microbiological enzymatic activities to increase free phenolic contents and form hydroxycinnamic acids, such as p-coumaric and ferulic acids similar to previous studies (Dueñas *et al.*, 2007), they detected significant increase in the amount of p-hydroxybenzoic acids (84 per cent). According to their findings, this increase seemed to be due to the hydroxybenzoic acid synthesis from bean hemicelluloses by *Bacillus subtilis* (Limón *et al.*, 2015; Reddy and Krishnan, 2013). They also reported decreases in (+)-catechin (catechin monomer) while more polymeric procyanidins (although not detected individually with analytical methods) are formed during fermentation (Limón *et al.*, 2015). A significant change in the phenolic profile of *mung* bean flour was also observed during fermentation with substantial increase in the concentration of apigenin (40,5,7-trihydroxyflavone) derivatives, having anti-inflammatory, antioxidant and antimicrobial properties (María Landete *et al.*, 2015; Shiferaw *et al.*, 2019). In general, the unique structure of legumes in terms of microbiological content, fermentation cause significant changes in legumes to increase the polyphenols and thus need further investigations.

7. Fermented Fruits and Vegetables

High content of carbohydrates, vitamins, minerals and dietary fibres, in addition to their phenolic contents, make fruits and vegetables very good matrices for fermentation (Septembre-Malaterre *et al.*, 2018; Szutowska, 2020). In a very recent review, positive health effects, such as probiotic delivery, antidiabetic, anticancer and anti-inflammatory properties, in addition to their contribution to decrease obesity, have been related with fermented juices of beetroots, carrots, pomegranates or noni fruits (Szutowska, 2020).

Since the group of fermented fruits and vegetables include numerous different products, the summarised approaches are given in Table 3.

Table 3: Approaches to Increase in Total Phenols and/or Antioxidant Activity in Plants

---

Various microbial ligninolytic and carbohydrate-metabolizing enzymes in the hydrolysis of glucosides leading to increase free phenolics (Georgetti *et al.*, 2009).

Enzymatical hydrolysis of glucosides and breakdown of plant cell walls or starch (Đorđević *et al.*, 2010).

Conversion of phenolic acids by lactic acid bacteria enzyme activity from a complex to a simpler form.

7.1 Pickles, Kimchi and Other Products

7.1.1 Product Properties

It is a traditional and very common approach to ferment perishable and seasonal vegetables into edible fermented products. Traditionally, pickling is as an ancient craft to make desirable changes in flavour, texture and colour in fermented pickles (Behera *et al.*, 2020). *Kimchi* and *sauerkraut* from cabbage, in addition to radish, cucumbers, carrots, cauliflowers, tomatoes, olives, green peas and peppers are fermented at home or produced industrially and generally called as pickles (Lu *et al.*, 2003; Septembre-Malaterre *et al.*, 2018). Species of *Lactobacillus* and *Pediococcus*, followed by *Leuconostoc, Weissella, Tetragenococcus*, and *Lactococcus* are among the most commonly used bacteria during the fermentation of vegetables (Tamang *et al.*, 2016).

Sauerkraut, a popular fermented vegetable is made via lactic acid fermentation of salted white cabbage (*Brassicaoleracea* var. capitata). Korean *kimchi* products include ordinary (without added water) and also many other types that are produced by LAB fermentation of baechu cabbage (diced Chinese cabbage) or its different types along with other vegetables, such as radish (some common ones as *tongbaechu, yeolmoo, kakdugi,* dongchimi, *nabakkimchi* etc.). Turkish *tursu* is defined as a combination made of cabbage, cucumber, carrot, beet, pepper, turnip, eggplant, and beans, most commonly fermented by *Lb. plantarum, Lb. brevis, Leuconostoc mesenteroides* and *Pediococcus pentosaceus* (Behera *et al.*, 2020). Fermented cucumbers are produced via fermentation in brine (usually 5-8 per cent NaCl) in open-faced tanks to convert saccharides into acids and other products. Fermented vegetables have significant amounts of LAB (10^7–10^9 CFU/g), dietary fibres, vitamins and minerals (Fleming *et al.*, 1978; Lu *et al.*, 2003; Xiang *et al.*, 2019).

7.1.2 Polyphenols and Potential Health Benefits

The high amount of phenolics in fruits and vegetables has presented them as specific substrates and fermentation processed have been reported to help in effective preservation and enhancement of their antioxidant capacity (Sayın and Alkan, 2015). During fermentation of fruits and vegetables, microorganisms produce glycosidases, esterases, phenolic acid decarboxylases, phenolic acid reductase and tannase that enables polyphenolic compounds into different degradation products with higher antioxidant capacity and bioavailability (Rodríguez *et al.*, 2009; Shiferaw Terefe and Augustin, 2019). Previously fermented vegetables have been reported to include flavonoids, phenols and sterols as antioxidants (Behera *et al.*, 2020). More specifically, *kimchi* products have been reported to have probiotic, antioxidant, antiaging, anti-inflammatory, antibacterial, antiobesity and anticancer effects (Xiang *et al.*, 2019). Hydrolysis of phenolic compounds into different phenolics is also observed among different fermented products from vegetables. For example, hydrolysis of chlorogenic acid into caffeic acid having better absorption in the stomach and the small intestine compared to chlorogenic acid, was observed during fermentation of broccoli puree by *Lb. reuteri* FUA3168 (Filannino *et al.*, 2015; Shiferaw Terefe and Augustin, 2019). Moreover, *Lb plantarum* fermentation has been proposed to mediate production of phenyllactic acid from phenylalanine, a phenolic acid phytochemical with antimicrobial (mostly antifungal) activity in Chinese pickles (Li *et al.*, 2015).

7.2 Vinegar

7.2.1 Product Properties

Vinegar is another fruit product with distinct properties. Its production might be summarized as 'aerobic conversion of sugar or ethanol containing substrates and hydrolysed starchy materials to acetic acid' (Solieri and Giudici, 2009). Although it might be produced from grain sources, as well, fruits are significant sources for fruit vinegar. The history of vinegar goes back to 2000 BC and fruit vinegars are made from grapes, apples, tomatoes, persimmons and pineapple by liquid-state fermentation (Duan *et al.*, 2019). Fruit vinegars are among the most commonly used acidic condiments globally. Other uses of vinegars include preservatives and healthy drinks. However, it is not considered as food in some sources as it does not have any great nutritional value (Solieri and Giudici, 2009) because in common industrial fruit vinegar-making practice, wine is produced through fermentation by yeasts and then transformed into acetic acid by acetic acid bacteria (Bakir *et al.*, 2016).

7.2.2 Polyphenols and Potential Health Benefits

Vinegars made from fruit sources have different polyphenols. Apple, cherry, grape, oak, chestnut or strawberry are among the most commonly used fruit sources in making fruit vinegars. In addition to the raw materials, also the processing method and nature of strains used for fermentation are significant parameters for sensory properties and bioactives in vinegars. Different vinegar products may contain different amounts of flavanols (e.g. catechin), hydroxybenzoic acids, hydroxycinnamic acids and tartaric esters of hydroxycinnamic acids (Xiang *et al.*, 2019).

7.3 Fruit Juices and Other Products

7.3.1 Product Properties

Fruit juices are popular among all people because of their desirable properties of freshness ,and sweetness mainly. In recent years, increasing consumer interest in functional uses, fermented juices are proposed as alternatives to traditional fermented dairy-based products with functional properties (Panda *et al.*, 2017). Other fermented beverages, like non-dairy *kefir* and *kombucha* (fermented brewed tea) are examples of famous products comprising lactic acid bacteria and yeasts (Chakravorty *et al.*, 2016).

7.3.2 Polyphenols and Potential Health Benefits

Among different types of foods, fruits and vegetables constitute the group with the highest amount of phenolics. Therefore, they are also very significant sources of antioxidants (Rodríguez *et al.*, 2009). The vitamin C in fruits also depicts significant antioxidant activity in addition to its role in the normal functioning of cultured endothelial cells, the increasing endothelial nitric oxide synthase activity and in the total antioxidant potential of fruits from phenolic compounds (Michels and Frei, 2013). Although, the general acidic structure of fermented fruit juices has been indicated as an additional advantage on the stability of polyphenols (Panda *et al.*, 2017), significant decreases in total phenolic compounds have also been observed in some fermented fruit juices, such as sweet lemon juices (Hashemi *et al.*, 2017) in previous studies. This phenomenon has been related to the metabolism of phenolic compounds starter of *Lb. plantarum*.

Bioconversion of phenolic compounds was also common during fermentation of fruits, in a similar manner to fermented vegetables. Fermentation of cherry juice by different strains of *Lb. plantarum* resulted in the reduction of caffeic acid into dihydrocaffeic acid, with a higher antioxidant capacity compared to its precursor (Shiferaw *et al.*, 2019). Similarly, after fermentation with *Lb. plantarum* LS5, an increased amount of antioxidant activity was measured in sweet lemon juice (Hashemi *et al.*, 2017). Moreover, fermentation also affected the aromatic profile of elderberry juice when *Lb. plantarum*, *Lb. rhamnosus* and *Lb. casei* strains were used in lactic acid fermentation (Ricci *et al.*, 2018). Garcinia beverage fermented with selected yeast species increased polyphenol content and its antioxidant activity (Rai *et al.*, 2010, 2014). Fermentation of fruits with specific yeast strains may improve the functional properties of final product due to increase in polyphenols.

8. Conclusion

Polyphenols in fermented food products have gained attention in recent years due to increasing awareness about health foods. Polyphenols are either fortified in fermented food products or their free forms are enhanced by the action of biocatalysts produced by the microorganisms involved in fermented food production. Food fermentation with specific enzyme-producing bacteria has the potential to add value to fermented products by enhancement of polyphenols and their bioconversion to highly active form. The type of polyphenols will depend on the food substrate or the extract added to the fermented food product. The health benefits associated with fermented foods, rich in polyphenols and their transformed products, need to be validated by animal studies and clinical trials. Further studies on effect of gut microbes on transformation of specific polyphenols can lead to production of novel metabolites with specific health benefits.

Abbreviations

BC – Before Christ
BHA – Butylhydroxyanisole
BHT – Butylhydroxytoluene
CFU – Colony forming units
LAB – Lactic acid bacteria
PG – Propyl gallate
TBHQ – Tert-butylhydroquinone

References

Abedfar, A. and Sadeghi, A. (2019). Response surface methodology for investigating the effects of sourdough fermentation conditions on Iranian cup bread properties. *Heliyon*, 5: e02608.

Acosta-Estrada, B.A., Gutiérrez-Uribe, J.A. and Serna-Saldívar, S.O. (2014). Bound phenolics in foods: A review. *Food Chemistry*, 152: 46-55.

Adebo, O.A. and Medina-Meza, I.G. (2020). Impact of fermentation on the phenolic compounds and antioxidant activity of whole cereal grains: A mini review. *Molecules*, 25: 927.

Aiello, F., Restuccia, D., Spizzirri, U.G., Carullo, G., Leporini, M. and Loizzo, M.R. (2020). Improving *kefir* bioactive properties by functional enrichment with plant and agro-food waste extracts. *Fermentation*, 6: 83.

Bakir, S., Toydemir, G., Boyacioglu, D., Beekwilder, J. and Capanoglu, E. (2016). Fruit antioxidants during vinegar processing: Changes in content and *in vitro* bio-accessibility. *International Journal of Molecular Sciences*, 17: 1658.

Balasundram, N., Sundram, K. and Samman, S. (2006). Phenolic compounds in plants and agri-industrial by-products: Antioxidant activity, occurrence and potential uses. *Food Chemistry*, 99: 191-203.

Behera, S.S., El Sheikha, A.F., Hammami, R. and Kumar, A. (2020). Traditionally fermented pickles: How the microbial diversity associated with their nutritional and health benefits? *Journal of Functional Foods*, 70: 103971.

Bozkurt, H. (2006). Utilisation of natural antioxidants: Green tea extract and Thymbra spicata oil in Turkish dry-fermented sausage. *Meat Science*, 73: 442-450.

Caric, M., Milanovic, S. and Ilicic, M. (2019). Novel trends in fermented dairy technology, *Zbornik Matice Srpske Za Prirodne Nauke Matica Srpska. Journal for Natural Sciences*, 136: 9-21.

Cauvain, S.P. (2003). *Bread making: Improving Quality*. CRC Press, Boca Raton, USA.

Chakravorty, S., Bhattacharya, S., Chatzinotas, A., Chakraborty, W., Bhattacharya, D. and Gachhui, R. (2016). Kombucha tea fermentation: Microbial and biochemical dynamics. *International Journal of Food Microbiology*, 220: 63-72.

Chávez-Servín, J.L., Andrade-Montemayor, H.M., Vázquez, C.V., Barreyro, A.A., García-Gasca, T., Martínez, R.A.F., Olvera Ramirez, A.M. and de la Torra-Carbot, K. (2018). Effects of feeding system, heat treatment and season on phenolic compounds and antioxidant capacity in goat milk, whey and cheese. *Small Ruminant Research*, 160: 54-58.

Chen, W., Yang, Y., Guo, G., Chen, C. and Huang, Y. (2018). β-Glucosidase in a *Yarrowia lipolytica* transformant to degrade soybean isoflavones. *Catalysts*, 8: 1-10.

Chouhan, S., Sharma, K. and Guleria, S. (2019). Augmenting bioactivity of plant-based foods using fermentation. pp. 165-184. *In*: Saran, S., Babu, V., Chaubey, A. (Eds.). *High Value Fermentation Products*. Scrivener Publishing LLC, Beverly.

Corsetti, A. and Settanni, L. (2007). Lactobacilli in sourdough fermentation. *Food Research International*, 40: 539-558.

Cunha, L.C.M., Monteiro, M.L.G., Lorenzo, J.M., Munekata, P.E.S., Muchenje, V., de Carvalho, F.A.L. and Conte-Junior, C.A. (2018). Natural antioxidants in processing and storage stability of sheep and goat meat products. *Food Research International*, 111: 379-390.

Đorđević, T.M., Šiler-Marinković, S.S. and Dimitrijević-Branković, S.I. (2010). Effect of fermentation on antioxidant properties of some cereals and pseudo cereals. *Food Chemistry*, 119: 957-963.

Du, X. and Myracle, A.D. (2018). Fermentation alters the bioaccessible phenolic compounds and increases the alpha-glucosidase inhibitory effects of aronia juice in a dairy matrix following: *In vitro* digestion. *Food and Function*, 9: 2998-3007.

Duan, W., Xia, T., Zhang, B., Li, S., Zhang, C., Zhao, C., Song, J. and Wang, M. (2019). Changes of physicochemical, bioactive compounds and antioxidant capacity during the brewing process of Zhenjiang aromatic vinegar. *Molecules*, 24: 3935.

Dueñas, M., Hernández, T. and Estrella, I. (2007). Changes in the content of bioactive polyphenolic compounds of lentils by the action of exogenous enzymes: Effect on their antioxidant activity. *Food Chemistry*, 101: 90-97.

FAO (1994). *Definition and Classification of Commodities*. http://www.fao.org/es/faodef/fdef04e.htm

Farnworth, E. (2003). *Handbook of Fermented Functional Food*. CRC Press, Boca Raton, USA.

Fezea, A.F., Al-Zobaidy, H.N. and Al-Quraishi, M.F. (2017). Total phenolic content, microbial content and sensory attributes evaluation of white soft cheese incorporated with mint (*Mentha spicata*) leaf extract. *IOSR Journal of Agriculture and Veterinary Science*, 10: 36-40.

Filannino, P., Bai, Y., Di Cagno, R., Gobbetti, M. and Gänzle, M.G. (2015). Metabolism of phenolic compounds by *Lactobacillus* spp. during fermentation of cherry juice and broccoli puree. *Food Microbiology*, 46: 272-279.

Fleming, H.P., Thompson, R.L., Bell, T.A. and Hontz, L.H. (1978). Controlled fermentation of sliced cucumbers. *Journal of Food Science*, 43: 888-891.

Georgetti, S.R., Vicentini, F.T.M. de C., Yokoyama, C.Y., Borin, M. de F., Spadaro, A.C.C. and Fonseca, M.J.V. (2009). Enhanced *in vitro* and *in vivo* antioxidant activity and mobilization of free phenolic compounds of soybean flour fermented with different β-glucosidase-producing fungi. *Journal of Applied Microbiology*, 106: 459-466.

Gobbetti, M., De Angelis, M., Di Cagno, R., Polo, A. and Rizzello, C.G. (2020). The sourdough fermentation is the powerful process to exploit the potential of legumes, pseudo-cereals and milling by-products in baking industry. *Critical Reviews in Food Science and Nutrition*, 60(13): 2158-2173.

Hashemi, S.M.B., Khaneghah, A.M., Barba, F.J., Nemati, Z., Shokofti, S.S. and Alizadeh, F. (2017). Fermented sweet lemon juice (*Citrus limetta*) using *Lactobacillus plantarum* LS5: Chemical composition, antioxidant and antibacterial activities. *Journal of Functional Foods*, 38: 409-414.

Hole, A.S., Rud, I., Grimmer, S., Sigl, S., Narvhus, J. and Sahlstrøm, S. (2012). Improved bioavailability of dietary phenolic acids in whole grain barley and oat groat following fermentation with probiotic *Lactobacillus acidophilus, Lactobacillus johnsonii*, and *Lactobacillus reuteri*. *Journal of Agricultural and Food Chemistry*, 60: 6369-6375.

Hui, Y.H., Hansen, A.S., Stanfield, P.S. and Told, F. (2004). *Handbook of Food and Beverage Fermentation Technology*. CRC Press, Boca Raton, USA.

Hur, S.J., Lee, S.Y., Kim, Y.-C., Choi, I. and Kim, G.B. (2014). Effect of fermentation on the antioxidant activity in plant-based foods. *Food Chemistry*, 160: 346-356.

Hutkins, R.W. (2008). *Microbiology and Technology of Fermented Foods* (vol.. 22). John Wiley and Sons.

Ivanov, G.Y. and Dimitrova, M.R. (2019). Functional yoghurt fortified with phenolic compounds extracted from strawberry press residues and fermented with probiotic lactic acid bacteria. *Pakistan Journal of Nutrition*, 18: 530-537.

Kan, L., Nie, S., Hu, J., Wang, S., Bai, Z., Wang, J., Zhou, Y., Jiang, J., Zeng, Q. and Song, K. (2018). Comparative study on the chemical composition, anthocyanins, tocopherols and carotenoids of selected legumes. *Food Chemistry*, 260: 317-326.

Khosravi, A. and Razavi, S.H. (2020). The role of bioconversion processes to enhance bioaccessibility of polyphenols in rice. *Food Bioscience*, 35: 100605.

Lee, J., Koo, N. and Min, D.B. (2004). Reactive oxygen species, aging, and antioxidative nutraceuticals. *Comprehensive Reviews in Food Science and Technology*, 3: 21-33.

Lee, M., Kim, J. and Lee, S. (2018). Effects of fermentation on SDS-PAGE patterns, total peptide, isoflavone effects of fermentation on SDS-PAGE patterns, e contents and antioxidant activity of freeze-thawed tofu fermented with *Bacillus subtilis*. *Food Chemistry*, 249: 60-65.

Li, X., Ning, Y., Liu, D., Yan, A., Wang, Z., Wang, S., Miao, M., Zhu, H. and Jia, Y. (2015). Metabolic mechanism of phenyllactic acid naturally occurring in Chinese pickles. *Food Chemistry*, 186: 265-270.

Licandro, H., Ho, P.H., Nguyen, T.K.C., Petchkongkaew, A., Nguyen, H. Van, Chu-Ky, S., Nguyen, T.V.A., Lorn, D. and Waché, Y. (2020). How fermentation by lactic acid bacteria can address safety issues in legumes food products. *Food Control*, 110: 106957.

Limón, R.I., Peñas, E., Torino, M.I., Martínez-Villaluenga, C., Dueñas, M. and Frias, J. (2015). Fermentation enhances the content of bioactive compounds in kidney bean extracts. *Food Chemistry*, 172: 343-352.

Lin, C., Lin, Y. and Wei, C. (2006). Enhanced antioxidative activity of soybean koji prepared with various filamentous fungi. *Food Microbiology*, 23: 628-633.

Lu, Z., Breidt, F., Plengvidhya, V. and Fleming, H.P. (2003). Bacteriophage ecology in commercial sauerkraut fermentations. *Applied and Environmental Microbiology*, 69: 3192-3202.

Magrinya, N., Bou, R., Tres, A., Rius, N., Codony, R. and Guardiola, F. (2009). Effect of tocopherol extract, Staphylococcus carnosus culture, and celery concentrate addition on quality parameters of organic and conventional dry-cured sausages. *Journal of Agricultural and Food Chemistry*, 57: 8963-8972.

Mapelli-Brahm, P., Barba, F.J., Remize, F., Garcia, C., Fessard, A., Mousavi Khaneghah, A., Sant'Ana, A.S., Lorenzo, J.M., Montesano, D. and Meléndez-Martínez, A.J. (2020). The impact of fermentation processes on the production, retention and bioavailability of carotenoids: An overview. *Trends in Food Science and Technology*, 99: 389-401.

María Landete, J., Hernández, T., Robredo, S., Duenas, M., de las Rivas, B., Estrella, I. and Munoz, R. (2015). Effect of soaking and fermentation on content of phenolic compounds of soybean (*Glycine max* cv. *Merit*) and mung beans (*Vigna radiata* [L] Wilczek). *International Journal of Food Sciences and Nutrition*, 66: 203-209.

Mayo, B., Aleksandrzak-Piekarczyk, T., Fernández, M., Kowalczyk, M., Álvarez-Martín, P. and Bardowski, J. (2010). Updates in the metabolism of lactic acid bacteria. pp. 3-33. *In*: *Biotechnology of Lactic Acid Bacteria*. Blackwell Publishing.

McGhie, T.K. and Walton, M.C. (2007). The bioavailability and absorption of anthocyanins: Towards a better understanding. *Molecular Nutrition and Food Research*, 51: 702-713.

Michels, A.J. and Frei, B. (2013). Myths, artifacts, and fatal flaws: Identifying limitations and opportunities in vitamin C research. *Nutrients*, 5: 5161-5192.

Min, M., Bunt, C.R., Mason, S.L. and Hussain, M.A. (2019). Non-dairy probiotic food products: An emerging group of functional foods. *Critical Reviews in Food Science and Nutrition*, 59: 2626-2641.

Moreno-Valdespino, C.A., Luna-Vital, D., Camacho-Ruiz, R.M. and Mojica, L. (2019). Bioactive proteins and phytochemicals from legumes: Mechanisms of action preventing obesity and type-2 diabetes. *Food Research International*, 130: 108905.

Ojha, Kumarı S., Kerry, J.P., Duffy, G., Beresford, T. and Tiwari, B.K. (2015). Technological advances for enhancing quality and safety of fermented meat products. *Trends in Food Science and Technology*, 44: 105-116.

Oksuz, T., Tacer-Caba, Z., Nilufer-Erdil, D. and Boyacioglu, D. (2019). Changes in bioavailability of sour cherry (*Prunus cerasus* L.) phenolics and anthocyanins when consumed with dairy food matrices. *Journal of Food Science and Technology*, 56: 4177-4188.

Oomah, B.D., Caspar, F., Malcolmson, L.J. and Bellido, A. (2011). Phenolics and antioxidant activity of lentil and pea hulls. *Food Research International*, 44: 436-441.

Othman, N. Ben, Roblain, D., Chammen, N., Thonart, P. and Hamdi, M. (2009). Antioxidant phenolic compounds loss during the fermentation of Chétoui olives. *Food Chemistry*, 116: 662-669.

Ozaki, M.M., Munekata, P.E.S., de Souza Lopes, A., da Silva do Nascimento, M., Pateiro, M., Lorenzo, J.M. and Pollonio, M.A.R. (2020). Using chitosan and radish powder to improve stability of fermented cooked sausages. *Meat Science*, 167: 108165.

Panda, S.K., Behera, S.K., Qaku, X.W., Sekar, S., Ndinteh, D.T., Nanjundaswamy, H.M., Ray, R.C. and Kayitesi, E. (2017). Quality enhancement of prickly pears (*Opuntia* sp.) juice through probiotic fermentation using *Lactobacillus fermentum* – ATCC 9338, *LWT – Food Science and Technology*, 75: 453-459.

Papuc, C., Goran, G.V., Predescu, C.N., Nicorescu, V. and Stefan, G. (2017). Plant Polyphenols as antioxidant and antibacterial agents for shelf-life extension of meat and meat products: Classification, structures, sources, and action mechanisms. *Comprehensive Reviews of Food Science and Food Safety*, 16: 1243-1268.

Pasini Deolindo, C.T., Monteiro, P.I., Santos, J.S., Cruz, A.G., Cristina da Silva, M. and Granato, D. (2019). Phenolic-rich Petit Suisse cheese manufactured with organic Bordeaux grape juice, skin, and seed extract: Technological, sensory, and functional properties. *LWT–Food Science and Technology*, 115: 108493.

Plessas, S., Alexopoulos, A., Mantzourani, I., Koutinas, A., Voidarou, C., Stavropoulou, E. and Bezirtzoglou, E. (2011). Application of novel starter cultures for sourdough bread production. *Anaerobe*, 17(6): 486-489.

Rai, A.K., Pandey, A. and Sahoo, D. (2019). Biotechnological potential of yeasts in functional food industry. *Trends in Food Science and Technology*, 83: 129-137.

Rai, A.K., Sanjukta, S., Chourasia, R., Bhat, I., Bhardwaj, P.K. and Sahoo, D. (2017). Production of soybean bioactive hydrolysate using protease, amylase and β-glucosidase from novel *Bacillus* spp. strains isolated from *kinema. Bioresource Technology*, 235: 358-365.

Rai, A.K. and Appaiah, A.K. 2014. Application of native yeast from Garcinia (*Garcinia xanthochumus*) for the preparation of fermented beverage: Changes in biochemical and antioxidant properties. *Food Bioscience*, 5: 101-107.

Rai, A.K., Prakash, M. and Appaiah, A.K. (2010). Production of Garcinia wine: Changes in biochemical parameters, organic acids and free sugars during fermentation of Garcinia must. *International Journal of Food Science and Technology*, 45: 1330-1336.

Reddy, S.S. and Krishnan, C. (2013). Characterisation of enzyme released antioxidant phenolic acids and xylooligosaccharides from different Graminaceae or Poaceae members. *Food Biotechnology*, 27: 357-370.

Ricci, A., Cirlini, M., Levante, A., Dall'Asta, C., Galaverna, G. and Lazzi, C. (2018). Volatile profile of elderberry juice: Effect of lactic acid fermentation using *Lb. plantarum, Lb.. rhamnosus* and *Lb. casei* strains. *Food Research International*, 105: 412-422.

Rice-Evans, C.A., Miller, N.J. and Paganga, G. (1996). Structure-antioxidant activity relationships of flavonoids and phenolic acids. *Free Radical Biology and Medicine*, 20: 933-956.

Rocchetti, G., Lucini, L., Rodriguez, J.M.L., Barba, F.J. and Giuberti, G. (2019). Gluten-free flours from cereals, pseudocereals and legumes: Phenolic fingerprints and *in vitro* antioxidant properties. *Food Chemistry*, 271: 157-164.

Rodríguez, H., Curiel, J.A., Landete, J.M., de las Rivas, B., de Felipe, F.L., Gómez-Cordovés, C., Mancheño, J. M. and Muñoz, R. (2009). Food phenolics and lactic acid bacteria. *International Journal of Food Microbiology*, 132: 79-90.

Sah, B.N.P., Vasiljevic, T., McKechnie, S. and Donkor, O.N. (2014). Effect of probiotics on antioxidant and antimutagenic activities of crude peptide extract from yoghurt. *Food Chemistry*, 156: 264-270.

Sanjukta, S., Rai, A.K., Ali, M.M., Jeyaram, K. and Talukdar, N.C. (2015). Enhancement of antioxidant properties of two soybean varieties of Sikkim Himalayan region by proteolytic *Bacillus subtilis* fermentation. *Journal of Functional Foods*, 14: 650-658.

Sayın, F.K. and Alkan, S.B. (2015). The effect of pickling on total phenolic contents and antioxidant activity of 10 vegetables. *Food and Health*, 1: 135-141.

Setiyoningruma, F. and Priadi, G.A.F. (2019). Supplementation of ginger and cinnamon extract into goat milk *kefir*. *AIP Conference Proceedings*, 2175: 020069.

Septembre-Malaterre, A., Remize, F. and Poucheret, P. (2018). Fruits and vegetables, as a source of nutritional compounds and phytochemicals: Changes in bioactive compounds during lactic fermentation. *Food Research International*, 104: 86-99.

Shiferaw, Terefe, N. and Augustin, M.A. (2019). Fermentation for tailoring the technological and health related functionality of food products. *Critical Reviews in Food Science and Nutrition*. https://doi.org/10.1080/10408398.2019.1666250

Shukla, S., Suk, J., Bajpai, V.K., Khan, I., Suk, Y., Han, Y. and Kim, M. (2019). Toxicological evaluation of lotus, ginkgo, and garlic tailored fermented Korean soybean paste (*Doenjang*) for biogenic amines, aflatoxins and microbial hazards. *Food and Chemical Toxicology*, 133: 110729.

Skrajda-Brdak, M., Konopka, I., Tańska, M. and Czaplicki, S. (2019). Changes in the content of free phenolic acids and antioxidative capacity of wholemeal bread in relation to cereal species and fermentation type. *European Food Research and Technology*, 245: 2247-2256.

Solieri, L. and Giudici, P. (2009). Vinegars of the world. pp. 1-16. *In*: Solieri, L. and Giudici, P. (Eds.). *Vinegars of the World*, Springer, New York.

Sonnenfeld, A., Flandrin, J.-L. and Montanari, M. (1999). *Food: A Culinary History*. Columbia University Press.

Steinkraus, K.H., Hui, Y.H., Meunier-Goddik, L., Hansen, A.S., Josephsen, J., Nip, W.K., Stanfeld, P.S. and Toldra, F. (2004). Origin and history of food fermentations. *Food Science and Technology*, 1-8. New York, Marcel Dekker.

Szutowska, J. (2020). Functional properties of lactic acid bacteria in fermented fruit and vegetable juices: A systematic literature review. *European Food Research and Technology*, 246: 357-372.

Tacer-Caba, Z. (2015). Functional Properties and Quality Parameters of Grape Extract Powder Substituted Bread and Extruded Products. PhD. Thesis. İstanbul Technical University, Institute of Science and Technology.

Tamang, J.P., Watanabe, K. and Holzapfel, W.H. (2016). Review: Diversity of microorganisms in global fermented foods and beverages. *Frontiers in Microbiology*, 7: 377.

Torino, M.I., Limón, R.I., Martínez-Villaluenga, C., Mäkinen, S., Pihlanto, A., Vidal-Valverde, C. and Frias, J. (2013). Antioxidant and antihypertensive properties of liquid and solid state fermented lentils. *Food Chemistry*, 136: 1030-1037.

Vitaglione, P., Napolitano, A. and Fogliano, V. (2008). Cereal dietary fibre: A natural functional ingredient to deliver phenolic compounds into the gut. *Trends in Food Science and Technology*, 19: 451-463.

Wang, T., He, F. and Chen, G. (2014). Improving bioaccessibility and bioavailability of phenolic compounds in cereal grains through processing technologies: A concise review. *Journal of Functional Foods*, 7: 101-111.

Wang, X., Kristo, E. and LaPointe, G. (2020). Adding apple pomace as a functional ingredient in stirred-type yoghurt and yoghurt drinks. *Food Hydrocolloids*, 100: 105453.

Xiang, H., Sun-Waterhouse, D., Waterhouse, G.I.N., Cui, C. and Ruan, Z. (2019). Fermentation-enabled wellness foods: A fresh perspective. *Food Science and Human Wellness*, 8: 203-243.

Yildiz, F. (2016). *Development and Manufacture of Yoghurt and Other Functional Dairy Products*. CRC Press, Boca Raton.

Bioactive Peptides in Fermented Food Products: Production and Functionality

Sanjukta Samurailatpam[1], Reena Kumari[1], Dinabandhu Sahoo[1,2] and Amit Kumar Rai[1]*

[1] Institute of Bioresources and Sustainable Development, Regional Centre, Tadong, Sikkim, India
[2] Department of Botany, University of Delhi, Delhi 110007, India

1. Introduction

Fermentation is one of the ancient techniques of food preservation, which results in improvement of aroma and texture to the fermented form (Ray and Joshi, 2014; Sourabh *et al.*, 2015). Fermentation improves the digestion and bioavailability of nutritional constituents (protein, carbohydrates, etc.) in food and enhances biosynthesis of vitamins along with degradation of antinutritional factors in relatively cost-effective and low-energy preservation mode (Liu *et al.*, 2011; Sanjukta *et al.*, 2017). Apart from conventional usage, fermented food products can also be consumed for various health benefits (Fitzgerald and Murray, 2006; LeBlanc *et al.*, 2004; Nakamura *et al.*, 2013; Rai and Jeyaram, 2015). Fermentation criteria of different foods differ and so do their applications, though some of the basic criteria are duration of fermentation period, temperature, selection of starter culture and moisture content in raw material (Sourabh *et al.*, 2015). The enzymes produced by the starter culture act on the substrate during food fermentation, resulting in different types of metabolites (Chancharoonpong *et al.*, 2012; Rai *et al.*, 2019).

Protein-rich food fermentation results in breakdown of protein in the raw material into smaller peptides and free amino acids (Sanjukta and Rai, 2016). Proteins are macromolecules that help in nourishing the growth and development of the human body and which are composed of one or more chains of amino acids. Transformation of proteins into small peptides in fermented foods with improved functionality occurs by (i) proteolytic enzymes produced by starter culture and (ii) gastrointestinal enzymes on consumption. After ingestion, proteins are hydrolysed

---

*Corresponding author: amitraikvs@gmail.com

by the gastrointestinal enzymes, resulting in several small peptides that may or may not be in their functional forms (Rai *et al.*, 2017a). Microbial hydrolysis that occurs during fermentation is the most efficient and cost-effective way of obtaining functionally active compounds (Hernandez *et al.*, 2004; Sanjukta *et al.*, 2017). In this process, complex food components are broken down into their simpler forms by microbial enzymatic activities and further biotransformation by microbes and digestive enzymes in the gastrointestinal tract (Rai and Jeyaram, 2017).

Food-derived bioactive peptides are chains of amino acids formed on hydrolysis of food protein and exhibit health benefits, depending on amino acid sequence and composition (Sanjukta and Rai, 2016). Peptide with therapeutic activities binds to specific receptors in the body to perform specific function on the target cells. The starter culture plays an important role in rendering the fate of the proteins into different functional peptides in the fermented products via hydrolysis with acidic, neutral and alkaline proteases (Chancharoonpong *et al.*, 2012). The change in starter culture at strain level leads to production of altered or totally new structures and composition of the bioactive peptide(s) that may exhibit different function(s) (Gobbetti *et al.*, 2000; Rai *et al.*, 2017b). Bioactive peptides in fermented foods are known to exhibit many health-beneficial properties, such as angiotensin converting enzyme (ACE) inhibitory (Gobbetti *et al.*, 2000; Okamoto *et al.*, 1995), antioxidant (Gupta *et al.*, 2010; Sanjukta *et al.*, 2015), antidiabetic (Islam *et al.*, 2009, Rajasekaran *et al.*, 2009), immunomodulatory (Ichimura *et al.*, 2003; LeBlanc *et al.*, 2002), anticancer (Cao *et al.*, 2009) and antimicrobial properties (Benkerroum *et al.*, 2004; Lee *et al.*, 2016). The chapter presents details of bioactive peptides formed in fermented foods and their potential health benefits.

2. Biochemical Changes on Proteins during Fermentation

The properties of a fermented food product depend on the composition of bioactive components formed during/after the process of fermentation. The microbes associated during fermentation are found to possess different enzymes-producing ability, such as amylase, glucosidase, lipase, protease, invertase, pectinase, cellulase, etc. which hydrolyse the macromolecules to simple, active and easily absorbable forms (Chancharoonpong *et al.*, 2012; Fitzgerald and Murray, 2006; Rai *et al.*, 2017). Protein-rich fermented foods can be acidic or alkaline, based on the metabolites produced during fermentation. Foods fermented with lactic acid bacteria (LAB) are generally acidic due to the production of organic acids. Fermentation by *Bacillus* species are generally neutral or alkaline due to the biogenic amines produced during the process (Chancharoonpong *et al.*, 2012). Food fermentation can also be categorised as solid state fermentation (i.e. soybean-fermented products, including *natto*, *kinema*, etc.) and submerged or liquid state fermentation (i.e., curd, yoghurt, etc.) (Rai *et al.*, 2017a). Functionality of the final product also depends on the type of fermentation and starter culture (Torino *et al.*, 2013). Lentil fermentation was carried out bt using *Lactobacillus plantarum* (submerged fermentation) and *Bacillus subtilis* (solid state fermentation). The result of this study showed that biofunctionality of

Lb. plantarum (liquid state fermentation) exhibited good Angiotensin converting enzyme (ACE) inhibitory and antioxidant activities (Torino *et al.*, 2013).

Production of bioactive peptides during food fermentation depends on the specificity of the proteases and composition of food proteins (Rai and Jeyaram, 2015). The variation in the types of peptides and free amino acid composition in final fermented product depends on the starter culture at species and strain level (Rai and Jeyaram, 2015; Sanjukta *et al.*, 2015). The composition and sequence of amino acids in the peptide defines the functional properties of the product (Sanjukta and Rai, 2016). Proteolysis of proteins to smaller peptides is carried out by the microbes involved in the fermentation process as starter, which is mainly due to the protease (neutral, acid and alkaline) produced by specific microbes. Apart from functional properties, presence of specific amino acids also affects the final taste of the product (Kabelova *et al.*, 2008). A study on the free amino acids in Czech beer revealed that among all amino acids detected, only nine were more concentrated as compared to the amino acids in other beer brands (Kabelova *et al.*, 2008). These amino acids contribute to the bitter taste in beer which includes isoleucine, leucine, lysine, phenylalanine and histidine found in high concentrations followed by bitter sweet amino acids valine and proline; salty umami amino acids glutamic acid and aspartic acid (Kabelova *et al.*, 2008). Apart from microbial species involved during fermentation, the degree of proteolysis and production of bioactive peptides greatly depends on physical parameters of fermentation, such as temperature, pH, moisture content, salt content, duration, etc. (Chancharoonpong *et al.*, 2012; Marshall and Tamime, 1997). As for example, fermented anchovy sauce was found to possess ACE-I peptides, whose activity increased with the high salt concentration and extension in time of fermentation (Kim *et al.*, 2016). Therefore, several conditions need to be optimised to obtain product with desired bioactivity.

3. Protein-rich Foods as Source of Bioactive Peptides

Protein-rich fermented food products are a rich source of bioactive peptides, which are produced on hydrolysis of specific food proteins. The production of these peptides mainly depends on the protein composition of the food material. The major proteins in most popular protein-rich foods are discussed below.

3.1 Fermented Milk Products

Fermented milk products, including cheese, are a good source of bioactive peptides, which are produced during fermentation and ripening process (Rai *et al.*, 2017a). During milk fermentation, milk proteins α_{s1}, α_{s2}, β, κ-casein, which are considered as the precursor of bioactive peptides along with whey proteins (α-lactalbumin, β-lactoglobulin) (Rai *et al.*, 2017a). These proteins are hydrolysed into several small amino acid chains of which a few of them exhibit health-beneficial properties. In milk products, LAB are the most prominent microorganisms used as starter culture, including *Lactococcus lactis*, *Lactobacillus delbrueckii*, *Leuconostoc* spp, *Lactobacillus helveticus* and *Streptococcus thermophilus* (Settanni and Moschetti,

2010). Many health-beneficial peptides are formed during fermentation which are given in Table 1. In case of cheese ripening, non-starter microbes (LAB and yeast) are involved, which further contribute to proteolysis of milk proteins (Settanni and Moschetti, 2010). Apart from microbial enzymes milk enzymes plasmin and acid protease are also reported to act on β- and α_{S1}-casein and release small peptides (Bastian and Brown, 1996). Cathepsin D and cell-envelope proteinases (CEP) secreted by thermophilic Lactobacilli species hydrolyse Rs1-casein and α-casein in Emmental cheese (Pihlanto *et al.*, 1994). These enzymes exhibit potential participation in proteolysis of milk protein with specific cleavage site at Phe24-Val25 and Phe32-Gly33 on Rs1-casein and at Phe52-Ala53 and Pro81-Val82 on α-casein. Such cleavages are found to be associated with the formation of peptides, such as α-CN(1-6) and α-CN(7- 28). Apart from composition of different types of milk protein, the type of enzymes also affects the production of specific bioactive peptides.

3.2   Fermented Legumes

Unlike milk, legumes are fermented with bacteria (*Bacillus* sp. and lactic acid bacteria), filamentous fungi (*Aspergillus* sp., *Mucor* sp. and *Rhizopus* sp.), and in some cases fermented using mix cultures. Among legumes, soybean is considered to be the richest source of protein (35 – 40 per cent D/W). In fermented soybean products peptides are also released either by the hydrolysis of soybean proteins (Glycinin and β-conglycinin) or produced by the microbes associated during fermentation. It is reported that glycinin and β-conglycinin are the major soy proteins that act as precursors for most of the bioactive peptides isolated from fermented soybean (Sanjukta and Rai, 2016; Yang *et al.*, 2000). There are several peptides in soybean reported for a wide range of health benefits (Gonzalez *et al.*, 2004; Sanjukta and Rai, 2016). Lunasin (43 amino acid peptide) found in different soybean varieties exhibits antioxidant, anticancer and anti-inflammatory properties (Gonzalez *et al.*, 2004; Hernandez *et al.*, 2009; Lule *et al.*, 2015). In fermented soybean products, application of different startera has resulted in products with differences in free amino acids composition and bioactive properties (Ibe *et al.*, 2006; Sanjukta and Rai, 2016; Sanjukta *et al.*, 2015). It is reported that fermentation of soybean using the proteolytic strains of *Bacillus subtilis* resulted in increase of free amino acid content of soybean up to 10-20 folds (Sanjukta *et al.*, 2015). Further, screening of protease, amylase and β-glucosidase producing *Bacillus* species from *kinema* produced in different seasons resulted in selection of strains with better fermentative property and proteolytic property for the production of bioactive peptides (Rai *et al.*, 2017b). Similarly, proteolytic strains of filamentous fungi *Aspergillus oryzae* S. were used for koji fermentation, which produced both protease and amylase during 48 hours of fermentation (Chancharoonpong *et al.*, 2012).

In Chinese soybean cheese like products (*sufu, tofu, pehtze*), both salted and non-salted products exhibited different patterns of increase in free amino acids (FAA). In China, *pehtze* is an *Actinomucor elegans* fermented *tofu* product. FAA content in *pehtze* was observed to be 1.3 to 15.6 mg/g dry matter, while salted *pehtze* exhibited an increased to 11.9 mg/g dry matter (Han *et al.*, 2004). It is also known from this

Table 1: Fermented Foods, Bioactive Peptides and Their Specific Health Benefits

Fermented Food	Peptide	Microbes Involved	Property	References
Calpis sour milk	VPP IPP	*Lactobacillus helveticus* and *Saccharomyces cerevisiae* *Lb. helveticus* and *S. cerevisiae*	Antihypertensive	Nakamura *et al.* (1995) Yamamoto *et al.* (1994)
Fermented milk	VPP IPP	*Lb. helveticus* LBK-16H	Antihypertensive	Seppo *et al.* (2002), Seppo *et al.* (2003)
Fermented milk	LHLPLP and LVYPFPGPIPNSLPQNIPP SKVYPFPGPI	*Enterococcus faecalis* CECT 5727	ACE inhibitory	Quiros *et al.* (2007)
Fermented milk	LHLPLP, LVYPFPGPIPNSLPQ-NIPP, VLGPVRGPFP, and VRGPFPIIV	*E. faecalis*	Antihypertensive	Muguerza *et al.* (2006)
Fermented milk	LPYPY	*Lb. delbrueckii* QS306	ACE inhibitory, Antihypertensive	Wu *et al.* (2019)
Dahi	SLVTP	*Lb. delbrueckii* ssp. *bulgaricus*	ACE inhibitory	Ashar and Chand, 2004
Yoghurt	VPP, IPP, SLVTP, PPGPI, SLV, TP	*Lb. delbrueckii* ssp. *bulgaricus* Lb1466, *Streptococcus thermophilus* St1342, *Lb. acidophilus* L10, *L. casei* L26 and *Bifidobacterium lactis* B94	ACE inhibitory	Donkar *et al.* (2007)

(Contd.)

Table 1: (*Contd.*)

Fermented Food	Peptide	Microbes Involved	Property	References
Gouda cheese	VPP IPP	*Lb. helveticus*	ACE inhibitory	Butikofer *et al.* (2007), Meyer *et al.* (2009)
Manchego Cheese	KKYNVPQL, VRYL, VRGPFP	*Lactobacillus, Lactococcus* and *Leuconostoc*	ACE inhibitory	Saito *et al.* (2000), Gomez-Ruiz *et al.* (2002)
Mozzarella, Italico, Crescenza, and Gorgonzola	LVYPFPGPINSLPQ	*Lactobacillus, Lactococcus, Streptococcus*	ACE inhibitory and antimicrobial	Smacchi and Gobbetti (1998)
Festivo cheese	αs1-cn f(1–9), f(1–7), f(1–6)	*Lactobacillus acidophilus* and Bifidobacteria	ACE inhibitory	Ryhanen *et al.* (2001)
Emmental cheese	αs1- and β-Casein fragments	*Endothia parasitica*	Immunomodulatory and antimicrobial	Gagnaire *et al.* (2001)
Fermented meat sauce	GTP	*Aspergillus*	Antioxidant	Ohata *et al.* (2016)
Fermented soybean	AT, AT, GT, ST, GT, VP, AP, AI and VG.	*Aspergillus sojae*	ACE inhibitory	Nakahara *et al.* (2010)
Doenjang	AP	*Bacillus subtilis* CSY191	Anticancer Antihypertension	Lee *at el.* (2012) Kim *et al.* (1999)
Fermented *Acetes chinensis* sauce	AP, GTG, ST	*Lb. fermentum* SM 605	ACE inhibitory	Wang *et al.* (2008)

Product	Peptide sequences	Microorganism/Fermentation	Functionality	Reference
Fermented milk	ALG, GPF, TPT, YPS, PEP, VSL, LKD, KAL, PEP	Lb. fermentum M2	Antidiabetic	Kinariwala et al. (2019)
Fermented casein	DELQDKIHPF	Lb. fermentum M2	Antihypertensive	Fan et al. (2018)
Fermented bovine skim milk	FSDIPNPIGSENSEKTTMPLW	Lb. helveticus CICC 6024 Lactococcus lactis SL6	Antioxidant	Kim et al. (2017)
Fermented goat milk	DERF, REF, FFD, FPE FFL, FLV, ILA, LTL, QRQ, QQR	Lb. casei KR732325 Lb. fermentum TDS030603	ACE inhibitory	Parmar et al. (2017)
Fermented milk	VPP, IPP	Lb. casei IMAU10408, Lb. casei IMAU20411	ACE inhibitory	Li et al. (2017)
Fermented cow milk (thayir/ dahi)	VPP, IPP	Lb. helveticus KII13, Lb. helveticus KHI1,	Antimicrobial, Anticancer, Cholesterol lowering	Damodaran et al. (2016)
Fermented fish (Fermented blue mussel, Mytilus edulis)	EVMAGNLYPG	Natural fermentation (25% NaCl w/w)	ACE inhibitory	Je et al. (2005)
Fermented fish budu (Ilisha melastoma)	LDDPVFIH VAAGRTDAGVH	Natural fermentaion	Antioxidant	Najafian and Babji (2019)
Fermented fish (pekasam)	AIPPHPYP IAEVFLITDPK	Lb. plantarum IFRPD P15	Antioxidant	Najafian and Babji (2018)
Fermented fish (bekasam)	KGENYNTGVTPNLRPKA AEVVAFLNKEAIEAIADT MKK	Lb. acidophilus	HGM CoA reductase inhibitor	Rinto et al. (2017)
Fermented anchovy sauce	PK, GCK, NPH, DGGP	Natural fermentation (20-25% salt concentration)	ACE inhibitory	Kim et al. (2016)

(Contd.)

Table 1: (*Contd.*)

Fermented Food	Peptide	Microbes Involved	Property	References
Fermented soybean (*tempe*)	TY, EF, PS, SV, AE, SI, EP, HV, VH, PF, RN, NR, HF AV, GL, GF, PL, AF, DM, PAP, DY, IAK, ALEP, VIKP, RIY	*Rhizopus* spp.	Antioxidant, CaMPDE inhibitory, DPP IV Inhibitory, ACE inhibitory	Tamam *et al.* (2019)
Fermented soybean	EDEVSFSP, SRPFNL, RSPFNL, ENPFNL	*Pediococcus pentosaceus* SDL 1409	ACE inhibitory	Daliri *et al.* (2018)
Fermented soybean (*natto*)	KL, KI	*Aspergillus oryzae*	DPP IV Inhibitory	Sato *et al.* (2018)
Fermented soybean (*natto*)	SMA, TPHVAGAAAL, ILSKHPTWTN, AQVRDRLEST, ATYLGNSFYY	*B. subtilis*	ACE inhibitory	Kitagawa *et al.* (2017)

study that during ripening up to 80 days, the FAA content in red *sufu* was highest in 8 per cent salt concentration rather than 11 or 14 per cent (Han *et al.*, 2004). Proteolysis during *tempeh* production has also resulted in production of wide ranges of bioactive peptides, such as antioxidant, antimicrobial, anticancer, antihypertensive, antithrombotic and hypocholesterolaemic effect (Gibbs *et al.*, 2004; Moreno *et al.*, 2002; Nout and Kiers, 2005). Soybean fermentation using combination of bacteria and filamentous fungi has also resulted in production of bioactive peptides. Two-step fermentation using *Rhizopus oligosporus* and *B. subtilis* is reported to increase the protein hydrolysis with higher content of aspartic acid and glutamic acid after fermentation. Such fermentation is accompanied with many small peptides of molecular weight <20 kD without affecting the essential amino acid profile (Weng and Chen, 2011). Bioactive peptides produced during legume fermentation possess several health-beneficial properties which will be discussed later.

3.3   Fermented Fish and Meat Products

Traditionally fermented fish and meat products are popularly consumed in many Asian countries (Jemil *et al.*, 2016; Vastag *et al.*, 2010). Fermentation and ripening of meat and sausage bring several changes to the end product by changing the amino acid composition, particularly free essential amino acids, such as threonine, valine, leucine and lysine (Beriain *et al.*, 2000; Campbell-Platt and Cook 1995). Meat, poultry and fish are a great source of histidine-containing dipeptides, carnosine (β-alanyl-L-histidine) and anserine (N-β-alanyl-1-methyl-L-histidine) (Nagasawa *et al.*, 2001; Sarmadi and Ismail, 2010; Young *et al.*, 2013). Unlike other fermented products, meat and fish products are fermented in different ways, like dry curing, ageing, salting, etc. Fermentation of meat is mostly done with the muscle proteins with a very few reports on direct fermentation of seafood and sausagea using specific starter (Jemil *et al.*, 2016; Vastag *et al.*, 2010). A report on casein-derived peptides from dry fermented sausages added with casein is found in which the dry fermented sausages were inoculated with a starter culture C-P-77S bactoferm containing *Lactobacillus pentosus* and *Staphylococcus carnosus* (Mora *et al.*, 2015). Peptidomic study showed that intense hydrolysis took place due to the action of endopeptidases and exopeptidases, which are found to be from both endogenous muscle origin and LAB added as the starter.

In their study, Fernendez *et al.* (2016) also studied the importance of microbial proteases from proteolytic starters *Pediococcus acidilactici* MS200 and *Staphylococcus vitulus* RS34 along with the protease EPg222 in dry fermentation of sausage 'salchichon' for obtaining higher nitrogenous compounds that exhibit ACE inhibitory and antioxidant activities. In comparison to fermented milk and legume products, fermented meat and fish products are least explored for bioactive peptide production.

4.   Health Benefits of Bioactive Peptides in Fermented Foods

Bioactive peptides in various fermented foods exhibit different health-beneficial properties. Some of the health-beneficial properties reported worldwide by the bioactive peptides in fermented food products are antihypertensive peptides that inhibit or lower hypertension, antioxidative peptides that scavenge or quench the free radicals and prevent several age-related problems, antidiabetic peptides and amino acids that control Type II diabetes, anticancer peptides that prevents cancerous cells and mutations, antimicrobial peptides that fight against toxin-producing microbes and immunomodulatory peptides that boost up the immune system. The biological activities of these peptides are discussed in this section.

4.1   Antihypertensive Peptides

Hypertension is a condition when blood pressure is higher than normal rate (120/80 mmHg), which is an alarming factor that leads to coronary heart disease, stroke and paralysis. Hypertension occurs due to poor diet condition, unhealthy lifestyle and lack of cardiovascular activities that gradually facilitates the conversion of Angiotensin I to Angiotensin II, which is a potent vasoconstrictor (Skeggs *et al.*, 1956). Excessive production of vasoconstrictor leads to thickening of the inner wall of the blood vessels and hinders blood circulation that results in cardiac arrest. Many fermented food products are known to exhibit ACE inhibitory properties, which are in high demand as functional foods and nutraceuticals. It is known from studies that peptides with high ACE inhibitory activity generally have higher quantity of hydrophobic and positively charged amino acids in C-terminal, alkaline or aromatic amino acids in N-terminal (Li and Yu 2015; Rai *et al.*, 2017a). Fermented milk products are found to be a potent source of ACE inhibitory peptides, which is also observed in spontaneously hypertensive rats (SHR) and clinical trials (Rai *et al.*, 2017a).

The ACE inhibitory activity in fermented milk is mainly due to the microbial proteolytic activity along with the factors involved during the fermentation process, like temperature, duration, pH and the probiotic used (Donkor *et al.*, 2007a,b; Nielsen *et al.*, 2009). In their study, Nielsen *et al.* (2009) reported that milk fermented by using proteolytic the strain of *Lactobacillus helveticus* exhibited highest ACE inhibitory activity under specific pH of 4.6 to 4.3. It is also found that cold storage of the fermented milk significantly increased the ACE inhibitory activity. In their study, Donkor *et al.* (2007) discussed the proteolytic activities of the bacterial strains with intracellular and extracellular specific peptidases, like X-prolyl-dipeptidyl aminopeptidase that specifically cleaves the proline-containing sequences that are considered for liberation of various peptides and improving cell growth. Probiotic yoghurts prepared with *Lactobacillus acidophilus* L10, *Lactobacillus casei* L26 and *Bifidobacterium lactis* B94 exhibited good ACE inhibitory activity as compared to control yoghurt (Donkor *et al.*, 2007a,b). Highest ACE inhibitory activity (IC_{50} of 103.30–27.79 μg mL^{-1}) was observed during first and third week of storage, which was correlated with the liberation of peptide through casein degradation. ACE inhibiting peptides, VPP and IPP were found with other casein originated small peptides (Donkor *et al.*, 2007a,b).

Fermented milk products (curd, cheese, sour milks etc.) are sources of ACE inhibitory peptides that inhibit ACE in the rennin angiotensin system (RAS) and are resistant to gastrointestinal digestion (De-Leo *et al.*, 2009; Korhonen and Pihlanto,

2006; Vermeirssen *et al.*, 2004). In a comparative study of bovine and camel milk fermented by *Lactobacillus rhamnosus* PTCC 1637, camel milk exhibited higher ACE inhibitory activity which was assumed to be due to the higher proline content in the primary structure of camel-milk caseins (Moslehishad *et al.*, 2013). Purified ACE inhibitory peptide was found in a Japanese soft drink called Calpis, made from skim milk fermented by *Lb. helveticus* and *Saccharomyces cerevisiae* (Nakamura *et al.*, 1995). A study on peptides derived from α_{s1}- and β-caseins by hydrolysis of the *Lb. helveticus* CP790 proteinase was done to check the inhibitory activities of ACE and check the activity in SHR by oral administration of the dosage of 15 mg/kg of body weight. Result of this study highlighted the potentiality of *Lb. helveticus* CP790 in milk fermentation thereby attributing 0.3 per cent peptides of which ACE inhibitory peptide was found (15 mg of peptide/kg) with its activity in SHR rats with 5 ml/kg of body weight (Yamamoto *et al.*, 1994a). Good ACE inhibitory activity was observed in peptide fractions from bovine yoghurt with starter culture containing probiotic organisms, including *Lb. acidophilus* L10 and *Bifidobacterium longum* Bl 536 with IC_{50} values of 0.196 and 0.151 mg/mL, respectively (Donkor *et al.*, 2007a,b). ACE inhibitory peptides are found in milk fermented with *Lactobacillus delbrueckii* ssp. *bulgaricus* with the sequence Ser-Lys-Val-Tyr-Pro, Phe-Pro-Gly-Pro-Ile along with Ser-Lys-Val, Tyr-Pro, which were derived from casein when fermented with *Streptococcus salivarius* ssp. *thermophilus* and *Lactococcus lactis* biovar *diacetylactis*. Further study on these peptides revealed that both the peptides were manifestly stable to digestive enzymes at a wide range of pH and also sustained during storage at 5 and 10°C for four days.

Mung bean milk fermentation with *Lb. plantarum* B1-6 supplemented with sucrose exhibited hydrolysis percentages between 49-64 per cent. Several larger peptides were found to be disappeared and many smaller hydrophilic peptides were observed with reverse phase high-performance liquid chromatography (RP-HPLC). Presence of such small peptides was correlated with the significantly high ACE inhibitory activity of 67.5 per cent at the end of fermentation (Wu *et al.*, 2015). Navy bean milk (NBM) fermentation with LAB strains, *Lactobacillus bulgaricus, Lb. helveticus* MB2-1, *Lb. plantarum* B1-6, and *Lb. plantarum* 70810 were studied for the production of ACE-I peptides (Rui *et al.*, 2015). RP-HPLC resulted in many small peptides which showed high ACE inhibitory activity compared to the unfermented NBM with higher activities observed at 2, 3, and 5h when fermented with *Lb. plantarum* 70810 (IC_{50} - 109±5.1 μg protein/ml), *Lb. plantarum* B1-6 (IC_{50} - 108±1.1 μg protein/ml) and *Lb. bulgaricus* (IC_{50} - 101±2.2 μg protein/ml) respectively.

ACE inhibitory activity in SHR was observed due to bioactive peptides in *doenjang* by a dipeptide Ala-Pro (Kim *et al.*, 1999) and in *miso* by tripeptides Val-Pro-Pro and Ile-Pro-Pro (Inoue *et al.*, 2009). Traditional fermented shrimp pastes, *Kapi Ta Dam* and *Kapi Ta Deang*, are also found to exhibit high ACE inhibitory activity (Kleekayai *et al.*, 2015). Two dipeptides, Ser-Val and Ile-Phe, were identified to possess ACE-I activity with IC_{50} values of 60.68 ± 1.06 and 70.03 ± 1.45 μM, respectively. ACE-I peptide with IC_{50} value of 19.34 lg/ml was purified from fermented blue mussel (*Mytilus edulis*) sauce purchased from a local shellfish market using Sephadex G-75 gel chromatography, SP-Sephadex C-25 ion exchange chromatography and RP-HPLC using C_{18} column (Je *et al.*, 2005). Further, 10 amino acid residues of N-terminal sequence were found to be EVMAGNLYPG and possess antihypertensive effect on oral administration of SHR. Positive effect of the purified peptide was observed, thereby lowering the blood pressure significantly and reduced hypertension.

ACE inhibitory activity was observed in fermented oyster sauce, which on purification and fractionation gave the purified inhibitor an IC_{50} value of 0.0874 mg/ml that exhibited competitive inhibition against ACE. Further evaluation of the purified peptide for antihypertensive effect was performed through oral administration of SHRs and found significant decrease in blood pressure (Je *et al.*, 2005a). Fermented sauce of *Acetes chinensis*, a shrimp species found in China, using *Lb. fermentum* SM 605 exhibits good ACE inhibitory activity with minimum IC_{50} value of 3.37 ± 0.04 mg/mL (Wang *et al.*, 2008). From this study three novel ACE inhibitory peptides were purified with amino acid sequences, Asp-Pro (IC_{50} values 2.15 ± 0.02 µM), Gly-Thr-Gly (IC_{50} values 5.54 ± 0.09 µM), and Ser-Thr (IC_{50} values 4.03 ± 0.10 µM). Fermented anchovy sauce fermented with solar sea-salt was found to possess ACE inhibitory peptides. On purification and characterisation, four short ACE inhibitory peptides, Pro-Lys (IC_{50} value-164 µM), Gly-Cys-Lys (IC_{50} value 178 µM), Asn- His-Pro (IC_{50} value-1172 µM) and Asp-Gly-Gly-Pro (IC_{50} value 4092 µM), were identified and analysed for the activity (Kim *et al.*, 2016).

In their study, Vastag *et al.* (2010) studied ACE inhibitory activities in fermented Petrovac sausage (*Petrovska Kolbasa*) from Backi Petrovac (province Vojvodina, Serbia) in which the protein extracts exhibited *in vitro* ACE inhibitory activity that increased with the ripening process. The protein extracts of 90 days of fermented sausage exhibited good ACE inhibitory activity of 73.74 ± 3.29 per cent as compared to 27.11 ± 2.16 per cent (day 0). Buckwheat sprouts fermented by *L. plantarum* KT, termed as 'neo-fermented buckwheat sprouts' (neo-FBS) are found to be a potent blood-pressure-lowering effect than traditionally prepared products by decreasing both systolic and diastolic blood pressure in SHRs at a dose of 0.01 mg/kg (Nakamura *et al.*, 2013). It was noticed that when the amount of FBS increased to 10 mg/kg, there was considerable decrease in ACE activity in lung, thoracic aorta, heart, kidney, and liver of SHRs. These effects on lowering blood pressure and ACE inhibitory effect in the fermented sprouts is due to the bioactive peptides and γ-aminobutyric acid (GABA) released during lactic acid fermentation (Nakamura *et al.*, 2013).

4.2   Antioxidant Peptides

Antioxidants are compounds that either scavenge or quench the free radicals released by our body during biochemical processes performed for normal body functioning. Fermented foods possess many small peptides that possess the antioxidant property and help in scavenging free radicals and delaying the ageing process (Sanjukta *et al.*, 2017; Sanjukta and Rai, 2016). Antioxidant peptides are found in many fermented foods, such as fermented milk products, fermented soybean and fermented sea foods (Rai and Jeyaram, 2015). Antioxidant activity in commercially fermented milk from Europe was evaluated by using ABTS radical-scavenging capacity in which many small peptides were recovered in fractions obtained by RP-HPLC and were sequenced using RP-HPLC–tandem mass spectrometry (Hernandez *et al.*, 2005). *Lb. plantarum* 55-fermented milk was found to exhibit high antioxidant activity of 282.8-362.3 μmol of trolox equivalent when determined by ABTS and ORAC methods. It is known from this study that both crude extracts and peptide fractions of sizes <3 kDa and 3-10 kDa of the fermented milk possess the capability to neutralise Reactive Oxygen Species (ROS) (Aguilar-Toalá *et al.*, 2017).

Crude peptides from yoghurt fermented with proteolytic strains of *Lactobacillus* were separated by high-speed centrifugation and examined for antioxidant activity. In this study, it was observed that many small peptides were liberated that exhibited high radical scavenging activity of IC_{50} value 1.51 and 1.63 mg/ml for 1,1-diphenyl-2-picrylhydrazyl (DPPH) and ABTS radicals, respectively (Sah *et al.*, 2014). Fresh yoghurt purchased from local supermarket of Denmark were studied for antioxidant activity in which it was found that peptides and free amino acids released during fermentation result in antioxidant activity (Farvin *et al.*, 2010). With ultra filtration using different membrane sizes, the antioxidant peptide was fractionated and identified with few N-terminal fragments of αs_1-, αs_2-, and κ-casein with one proline residue (Farvin *et al.*, 2010).

Fermentation of marine blue mussel (*M. edulis*) released many bioactive peptides in the fractions obtained during purification with ion exchange, gel filtration and high performance liquid chromatographic techniques exhibiting radical scavenging activity. Out of many fractions, significant radical scavenging activity was observed in the hepta-peptide sequence, HFGBPFH with MW of 962 kDa. This antioxidant sequence scavenged free radicals with remarkable IC_{50} values and exhibited strong lipid peroxidation inhibition, higher than the standard α-tocopherol. (Rajapakse *et al.*, 2005). Fish *miso*, a fermented fish paste prepared by mixing horse mackerel meat and traditional Japanese *koji* fermented with *Aspergillus oryzae*, exhibited remarkable antioxidant activity. During the early stages of fermentation, hydrolysis of the protein was observed by SDS-PAGE profiling and many low-molecular-weight peptides were also observed which were found responsible for radical scavenging and reducing power potential. These peptides were further partially purified and characterised and based on molecular weight study, it was found that low-molecular-weight peptides (<500 Da) were responsible for the antioxidant activity (Giri *et al.*, 2012).

Antioxidant activities were also observed in water-soluble fraction of fermented Loma fish, *Pekasam* and on further analysis of the extract, many novel peptides were found (Najafian and Babji, 2018). The antioxidant activities of the crude extract and further purification of the extract was done through chromatography techniques. The IC_{50} values of the strongest radical scavenging activity were observed in purified fraction FIII3 with the value of 0.636 mg/ml that generated two novel peptides, Ala-Ile-Pro-Pro-His-Pro-Tyr-Pro and Ile-Ala-Glu-Val-Phe-Leu-Ile-Tre-Asp-Pro-Lys. It is assumed from this study that the presence of hydrophobic and basic amino acids, Ile, Ala, Pro and Lys in the peptide (IAEVFLITDPK) resulted in high antioxidant activity as compared to the other peptide.

Traditional fermented shrimp pastes, *Kapi Ta Dam* and *Kapi Ta Deang*, were studied for antioxidant activity and identified a peptide, Trp-Pro, with high 2,2'-azino-bis (3-ethylbenzthiazoline-6-sulphonic acid) radical scavenging activity (EC50

17.52 ± 0.46 µM) (Kleekayai *et al.*, 2015). Another traditional fermented shrimp and krill product of Thai, namely *Jaloo, Koong-Som* and *Kapi*, were studied for the antioxidant activity, which was resistant to heat treatment (100°C) (Faithong *et al.*, 2010). Antioxidant activity was also observed in peptides isolated from fermented *Ganoderma lucidum* obtained from Guoyao Co. (Wuhan, China). The antioxidant activity of *G. lucidum* peptide was also observed in the rat liver tissue homogenates and mitochondrial membrane peroxidation systems (Sun *et al.*, 2004). Antioxidant activity was observed in water-soluble (pH 7.4) protein extracts of fermented Petrovac sausage (*P. Kolbasa*) from Backi Petrovac (province Vojvodina, Serbia) in which the degree of hydrolysis is found to be 17.71 ± 0.76 per cent after 90 days as compared to 6.07 ± 0.84 per cent on 0 day. DPPH radical scavenging activity and reducing activity of the 90-day extract showed scavenging activity of 27.61 ± 0.73 per cent and 0.493 ± 0.016 per cent, respectively as compared to activities of 0 days extract of 27.61 ± 0.73 per cent and 0.493 ± 0.016 per cent respectively. Fermented meat sauce (FMS) exhibited radical scavenging and antioxidant activities against OH-radicals (61.2 per cent) and DPPH (0.55 per cent µmol Trolox eq/ml) (Ohata *et al.*, 2016). The activity was found to be due to proteolysis of substrate by the protease of *Aspergillus*. Further, on LC-MS analysis, a tripeptide Gln-Tyr-Pro was detected that scavenged >90 per cent of OH-radical.

4.3   Anticancer Peptides

Cancer is one of the most common diseases in world and invention of drugs against it is a great challenge to science. However, many fermented foods are found to possess anticancer property, which can be applied as a preventive measure. Anticancer property was observed in milk fermented by *Lb. helveticus* R389 or *Lb. helveticus* L89 (LeBlanc *et al.*, 2005). Fermented milks were fed to mice for two or seven consecutive days and given subcutaneous injection of tumor cells in the mammary gland after the feeding period was over. After four days of injection, the mice were given fermented milk on a cyclical basis after which it was observed that there was delay and termination in tumor development and maintenance of the balance between immune and endocrine systems, which is significant in oestrogen-dependent tumour and induced cellular apoptosis. Crude extract of milk fermented with *Lb. plantarum* exhibited high antimutagenic activity (Aguilar-Toala *et al.*, 2017). Anticancer activity is found in many fermented soybean products, like *natto*, *cheonggukjang*, *doenjang* and many soy sauces. *Bacillus natto* TK-1 fermented *natto* is reported to exhibit antitumour activity in which the cytotoxic potential was tested with purified lipopeptide by treating on K-562 and BEL-7402 cells by using different concentrations for 48 hours (Cao *et al.*, 2009). *Doenjang* fermented with *B. subtilis* CSY191 is found to produce a surfactin-like compound with anticancerous property, which can survive on highly acidic condition of pH 3 for three hours on gastrointestinal digestion (Lee *et al.*, 2012).

On purification, three purified isoforms were isolated with identical amino acid sequence GLLVDLL that inhibit the growth of human breast cancer cells (MCF-7) when delivered in a dose-dependent manner (IC$_{50}$- 10 mg/ml at 24 h). *Cheonggukjang* fermented with CSY191 showed rise in surfactin contents from 0.3 to 48.2 mg/kg in

48 hours fermentation with the rise in anticancer activity from 2.6 to 5.1 folds (Lee *et al.*, 2012). Anticancer activity was observed in Lunasin, an anticancer peptide found in fermented soy products, such as *tempeh, tofu,* soymilk and soy infant formula (Cavazos *et al.*, 2012; Hernandez *et al.*, 2009). Lunasin peptide showed prevention in loci formation in mouse fibroblast NIH3T3 cells induced by oncogene (EIA) (Lam, 2003), which also reduced ~70 per cent skin cancer in sensitivity to carcinogenesis mouse skin tumour model induced by chemical carcinogens DMBA (Galvez *et al.*, 2001). Peptide fraction from anchovy sauce is found to possess anticancer property on being studied by using human lymphoma cell line U937 (Lee *et al.*, 2003). Apoptotic DNA fragmentation and increased in Caspase 3 and 8 were observed, with the positive outcome of apoptosis in cancer cells and hence resulting in a promising anticancer peptide fraction. Protein-rich fermented products need to be explored for peptides with anticancer properties, which can be applied as nutraceuticals.

4.4 Antidiabetic Peptides

Diabetes (Type II) is most common affecting 90 per cent people with diabetic problems (Bhat *et al.*, 2008). Several fermented foods possess antidiabetic properties and many of them are studied by using animal models. Aqueous extracts of *Monascus perpureus* MTCC1090 fermented Indian variety rice IR-532-E-576 (10 days) exhibited antidiabetic property (Rajasekaran *et al.*, 2009). Rat fed with fermented rice significantly reduced glucose level, as also total cholesterol and triglycerides (Rajasekaran *et al.*, 2009). In a comparative study of fermented (using *Mardi Rhizopus* sp. strain 5351) and non-fermented *mung* bean extracts for antidiabetic activity, it was observed that fermented *mung* bean extracts significantly reduced blood sugar levels of hyperglycemic mice (Yeap *et al.*, 2012). Water extract of *touchi*, soybean fermentation using *koji* fermented with *Aspergillus* sp. showed good inhibitory activity against diabetic mellitus (Fujita and Yamagami, 2001). Examination of the long-term effects of inhibitory activity was done by feeding genetically-modified eight-week old male diabetic model KKA^y mice for 60 days (Fujita and Yamagai, 2001). Water extract of *meju*, fermented soybean, is found to possess antidiabetic property. It approximately has 70 per cent of the rosiglitazone activity and smaller peptides in *meju* increase cell proliferation in Min6 cells, thereby exhibiting insulinotropic activity and have greater impact on PPAR-γ activation (Kwon *et al.*, 2011). Peptides in fermented food products still need to be explored for antidiabetic properties.

4.5 Antimicrobial Peptides

Antimicrobial peptides in fermented foods play a vital role in prevention of growth of undesirable microorganisms in food products (Sanjukta and Rai, 2016). Antimicrobial properties are due to peptides produced by microorganisms involved in the process of fermentation as well as food-derived peptides formed on hydrolysis of food proteins (Rai and Jeyaram, 2015). Moderate antimicrobial activity against many pathogenic microorganisms, such as *Escherichia coli, Listeria innocua, Salmonella typhimurium, Staphylococcus aureus* and *Salmonella choleraesuis* was observed in crude extract of *Lb. plantarum* 55 fermented milk (Aguilar-Toala *et al.*,

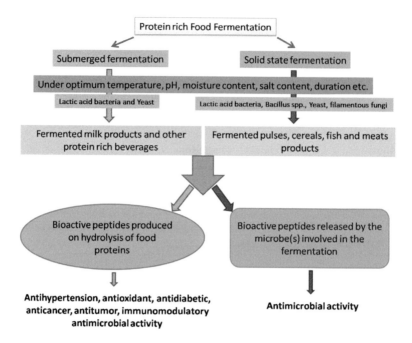

Fig. 1: Types of fermentation and production of bioactive peptides

2017). Antimicrobial activity was observed in protein-rich crude extract of milk fermented by using *Lb. plantarum* (Aguilar-Toala *et al.*, 2017; Atanasova *et al.*, 2014; Peres *et al.*, 2014). Heat-stable bacteriocins were characterised from LAB isolated from Malaysian mold-fermented soybean product, *tempeh*. The bacteriocins were produced by *E. faecium* LMG 19827 as well as *E. faecium* LMG 19828 and inhibited *Listeria monocytogenes* (Moreno *et al.*, 2002). It was found that bacteriocins were produced at late exponential/early stationary growth with molecular mass 3.4 kDa for B1 bacteriocin, and 5.8 kDa for B2 bacteriocins.

Bacillus mojavensis A21, a proteolytic bacterium-fermented zebra blenny (*Salaria basilisca*) exhibits antibacterial activity against both Gram negative and Gram positive pathogens, like *E. coli*, *Enterobacter* sp., *Pseudomonas aeruginosa*, *Klebsiella pneumoniae*, *Salmonella enterica*, *S. typhimurium*, *Bacillus cereus*, *Micrococcus luteus* and *S. aureus* (Jemil *et al.*, 2016). Further on purification, two different fractions, A (GLPPYPYAG, ETPGGTPLAPEPD, AGCAGVGGAG) and B (NVLSGGTTMYPGIAD; NRRIQLVEEELDRAQER), were obtained with high proline, glycine and arginine residues. Presence of such amino acids is correlated with prominent antimicrobial activity of hydrolysate fractions (Jemil *et al.*, 2016). Bacteriocin were also found to be produced in LAB-fermented cucumber, including pediocin, plantaricin A, mesentericin, enterocin A and nisin at different time intervals when fermented at 35°C in 4 per cent saline solution (Singh and Ramesh, 2008). There are many studies on bacteriocins-producing LAB, yeast and filamentous fungi. Future studies are needed to explore food-derived antimicrobial peptides produced during fermentation.

4.6   Immunomodulatory Peptides

Immunomodulatory peptides are reported to be present in many protein-rich fermented foods. These peptides can enhance the immune cell functions, like antibody synthesis, natural killer (NK) cell activity, lymphocyte proliferation, cytokine regulation (Korhonen and Pihlanto, 2003). Immunomodulatory activity was observed in casein peptides that stimulate the proliferation of human lymphocytes and the phagocytic activities of macrophages (Migliore-Samour *et al.*, 1988). Immunomodulatory activity of milk fermented with *Lb. helveticus* R389 was observed in BALB/c mice by analysing the effects on IL-6 production by small intestine epithelial cells (Vinderola *et al.*, 2007). Peptidic fraction of milk fermented by *Lb. helveticus* was studied for its immunomodulatory and anti-infectious effects (IgA immune response) in the mucosa-associated lymphoid tissue through administration of the fraction in *E. coli* O157:H7 infected BALB/c murine model (LeBlanc *et al.*, 2004). An increase in the number of IgA-secreting B lymphocytes in the intestinal lamina propria was observed along with an enhancement in total secretory and systemic IgA response. There was a visible difference in cytokine profiling where stimulation of a Th2 (T helper) response was observed in fraction-fed mice contrary to the pro-inflammatory Th1 response in infected control mice, which indicates that the peptides released by LAB have immunomodulatory effects. Peptidic compounds released by *Lb. helveticus*-fermented milk exhibited good proteolytic activity and three fractions were studied after size-exclusion HPLC elution profiles for its immunomodulatory property (LeBlanc *et al.*, 2002). Double dose of fraction II showed highest number of IgA-secreting cells and antitumor activity, thereby modulating the cellular immune response. Fermented food products can be a good source of immunomodulatory peptides.

5. Conclusion

Fermented foods are consumed in different part of the world and these include fermented milk, soybean and pulses, meat and fish products as a source of bioactive peptides. These fermented foods are found to possess bioactive peptides that have enormous health-beneficial properties, such as ACE inhibitory, antioxidant, antitumor, antidiabetic and antimicrobial which can be used as a source of functional foods and nutraceuticals. Some of the peptides produced during fermentation are also studied to retain their active (functional) form even after gastrointestinal digestion. Thus many researchers around the globe have reported fermented food products as a healthy substitute for prevention and treatment of many degenerative diseases. Future research on effect of novel proteolytic microbial strains on fermented products can be explored, which can lead to production of novel bioactive peptides with improved functional properties.

Abbreviations

ABTS – 2, 2'-azino-bis (3-ethylbenzothiazoline-6-sulfonic acid)
ACE – Angiotensin converting enzyme

ACE-I – Angiotensin converting enzyme inhibitory
CEP – Cell-envelope proteinases
DPPH – 1, 1-diphenyl-2-picrylhydrazyl
FAA – Free amino acids
FMS – Fermented meat sauce
GABA – γ-aminobutyric acid
HPLC – High Performance Liquid Chromatography
LAB – Lactic Acid Bacteria
NBM – Navy bean milk
ROS – Reactive oxygen species
RP-HPLC – Reverse phase high-performance liquid chromatography

References

Aguilar-Toala, J.E., Santiago-Lopez, L., Peres, C.M., Peres, C., Garcia, H.S., Vallejo-Cordoba, B., Gonzalez-Cordova, A.F. and Hernandez-Mendoza, A. (2017). Assessment of multifunctional activity of bioactive peptides derived from fermented milk by specific *Lactobacillus plantarum* strains. *Journal of Dairy Science*, 100: 65-75.

Atanasova, J., Moncheva, P. and Ivanova, I. (2014). Proteolytic and antimicrobial activity of lactic acid bacteria grown in goat milk. *Biotechnology and Biotechnological Equipment*, 28(6): 1073-1078.

Ashar, M.N. and Chand, R. (2004). Antihypertensive peptides purified from milks fermented with *Lactobacillus delbrueckii* spp. *Bulgaricus, Milchwissenschaft*, 59(1): 14-17.

Bastian, E.D. and Brown, R.J (1996). Plasmin in milk and dairy products: An update. *International Dairy Journal*, 6(5): 435-457.

Benkerroum, N., Mekkaoui, M., Bennani, N. and Hidane, K. (2004). Antimicrobial activity of camel's milk against pathogenic strains of *Escherichia coli* and *Listeria monocytogenes*. *International Dairy Technology*, 57(1): 39-43.

Beriain, M.J., Lizaso, G. and Chasco, J. (2000). Free amino acids and proteolysis involved in 'salchichon' processing. *Food Control*, 11: 41-47.

Bhat, M., Zinjarde, S.S., Bhargava, S.Y., Kumar, A.R. and Joshi, B.N. (2008). Antidiabetic Indian plants: A good source of potent amylase inhibitors. *Evidence-based Complementary and Alternative Medicine*, 1: 1-6.

Butikofer, U., Meyer, J., Sieber, R. and Wechsler, D. (2007). Quantification of the angiotensin converting enzyme-inhibiting tripeptides Val Pro-Pro and Ile-Pro-Pro in hard, semi-hard and soft cheeses. *International Dairy Journal*, 17(8): 968-975.

Campbell-Platt, G. and Cook, P.E. (1995). *Fermented Meats*. Blackie Academic and Professional, London.

Cao, X.H., Liao, Z.Y., Wang, C.L., Yang, W.Y. and Lu, M.F. (2009). Evaluation of a lipopeptide biosurfactant from *Bacillus natto* TK-1 as a potential source of anti-adhesive, antimicrobial and antitumor activities. *Brazilian Journal of Microbiology*, 40: 373-379.

Chancharoonpong, C., Hsieh, P.C. and Sheu, S.C. (2012). Production of enzyme and growth of *Aspergillus oryzae* S. on soybean Koji. *International Journal of Bioscience, Biochemistry and Bioinformatics*, 2(4): 228-231.

Cavazos, A., Morales, E., Dia, V.P. and Gonzalez de Mejia, E. (2012). Analysis of lunasin in commercial and pilot plant produced soybean products and an improved method of lunasin purification. *Journal of Food Science*, 77(5): 539-545.

Daliri, E.M.V., Lee, B.H., Kim, J.K. and Oh, D.H. (2018). Novel angiotensin I-converting enzyme inhibitory peptides from soybean protein isolates fermented by *Pediococcus pentosaceus* SDL1409. *LWT–Food Science and Technology*, 93: 88-93.

Damodaran, K., Palaniyandi, S.A., Yang, S.H. and Suh, J.W. (2016). Functional probiotic characterisation and *in vivo* cholesterol-lowering activity of *Lactobacillus helveticus* isolated from fermented cow milk. *Journal of Microbial Biotechnology*, 26: 1675-1686.

De-Leo, F., Panarese, S., Gallerani, R. and Ceci, L.R. (2009). Angiotensin-converting enzyme (ACE) inhibitory peptides: Production and implementation of functional food. *Current Pharmaceutical Design*, 15(31): 3622-3643.

Donkor, O.N. (2007). Influence of probiotic organisms on release of bioactive compounds in yoghurt and soy yoghurt. Ph.D. thesis. Victoria University, Melbourne, Australia.

Donkor, O.N., Henriksson, A., Singh, T.K., Vasiljevic, T. and Shah, N.P. (2007a). ACE-inhibitory activity of probiotic yoghurt. *International Dairy Journal*, 17(11): 1321-1331.

Donkor, O.N., Henriksson, A., Vasiljevic, T. and Shah, N.P. (2007b). Proteolytic activity of dairy lactic acid bacteria and probiotics as determinant of growth and in vitro angiotensin-converting enzyme inhibitory activity in fermented milk. *Le Lait*, 87(1): 21-38.

Faithong, N., Benjakul, S., Phatcharat, S. and Binsan, W. (2010). Chemical composition and antioxidative activity of Thai traditional fermented shrimp and krill products. *Food Chemistry*, 119: 133-140.

Fan, M., Guo, T., Li, W., Chen, J., Li, F., Wang, C., Shi, Y., Xi, D., Li, A. and Zhang, S. (2019). Isolation and identification of novel casein-derived bioactive peptides and potential functions in fermented casein with *Lactobacillus helveticus*. *Food Science and Human Wellness*, 8(2): 156-176.

Farvin, K.H.S., Baron, C.P., Nielsen, N.S., Otte, J. and Jacobse, C. (2010). Antioxidant activity of yoghurt peptides: Part 2 – Characterisation of peptide fractions. *Food Chemistry*, 123: 1090-1097.

Fernandez, M., Benito, M.J., Martin, A., Casquete, R., Cordoba, J.J. and Cordoba, M.G. (2016). Influence of starter culture and a protease on the generation of ACE-inhibitory and antioxidant bioactive nitrogen compounds in Iberian dry-fermented sausage 'salchichon'. *Heliyon*, 2(3): e0093.

Fitzgerald, R. and Murray, B.A. (2006). Bioactive peptides and lactic fermentations. *International Journal of Dairy Technology*, 59(2): 118-125.

Fujita, H. and Yamagami, T. (2001). Fermented soybean-derived *Touchi*-extract with anti-diabetic effect via α-glucosidase inhibitory action in a long-term administration study with KKA^y mice, *Life Science*, 70(2): 219-227.

Gagnaire, V., Molle, D., Herrouin, M. and Leonil, J. (2001). Peptides identified during Emmental cheese ripening: Origin and proteolytic systems involved. *Journal of Agriculture and Food Chemistry*, 49(9): 4402-4413.

Galvez, A.F., Chen, N., Macasieb, J. and Lumen, B.O.D. (2001). Chemopreventive property of a soybean peptide (Lunasin) that binds to deacetylated histones and inhibits acetylation. *Cancer Research*, 61: 7473-7478.

Gibbs, B.F., Zougman, A., Masse, R. and Mulligan, C. (2004). Production and characterisation of bioactive peptides from soy hydrolysate and soy-fermented food. *Food Research International*, 37(2): 123-131.

Giri, A., Nasu, M. and Ohshima, T. (2012). Bioactive properties of Japanese fermented fish paste, fish *miso*, using *koji* inoculated with *Aspergillus oryzae*. *International Journal of Nutrition and Food Science*, 1(1): 13-22.

Gobbetti, M., Ferranti, P., Smacchi, E., Goffredi, F. and Addeo, F. (2000). Production of angiotensin-I-converting-enzyme-inhibitory peptides in fermented milks started by *Lactobacillus delbrueckii* subsp. *bulgaricus* SS1 and *Lactococcus. lactis* subsp. *cremoris* FT4. *Applied Environmental Microbiology*, 66: 3898-3904.

Gomez-Ruiz, J.A., Ramos, M. and Recio, I. (2002). Angiotensin-converting enzyme-inhibitory peptides in Manchego cheeses manufactured with different starter cultures. *International Dairy Journal*, 12(8): 697-706.

Gonzalez, E., Vasconez, M., Lumen, B.O.D. and Nelson, R. (2004). Lunasin concentration in different soybean genotypes, commercial soy protein, and isoflavone products. *Journal of Agriculture and Food Chemistry*, 52: 5882-5887.

Gupta, V. and Nagar, R. (2010). Effect of cooking, fermentation, dehulling and utensils on antioxidants present in pearl millet *rabadi* – A traditional fermented food. *Journal of Food Science and Technology*, 47(1): 73-76.

Han, B.Z., Rombouts, F.M. and Robert Nout, M.J. (2004). Amino acid profiles of *sufu*, a Chinese fermented soybean food. *Journal of Food Composition Analysis*, 17(6): 689-698.

Hernandez-Ledesma, B., Amigo, L., Ramos, M. and Recio, I. (2004). Angiotensin converting enzyme inhibitory activity in commercial fermented products. Formation of peptides under simulated gastrointestinal digestion. *Journal of Agriculture and Food Chemistry*, 52: 1504-1510.

Hernandez-Ledesma, B., Miralles, B., Amigo, L., Ramos, M. and Recio, I. (2005). Identification of antioxidant and ACE-inhibitory peptides in fermented milk. *Journal of Science, Food and Agriculture*, 85: 1041-1048.

Hernández-Ledesma, B., Hsieh, C. and de Lumen, B.O. (2009). Antioxidant and anti-inflammatory properties of cancer preventive peptidelunasin in RAW 264.7 macrophages. *Biochemistry and Biophysical Research Communications*, 390: 803-808.

Ibe, S., Yoshida, K. and Kumada, K. (2006). Angiotensin I converting enzyme inhibitory activity of *natto*. *Japanese Journal of Food Science and Technology*, 53(3): 189-192.

Ichimura, T., Hu, J., Aita, D.Q. and Maruyama, S. (2003). Angiotensin I-converting enzyme inhibitory activity and insulin secretion stimulative activity of fermented fish sauce. *Journal of Bioscience and Bioengineering*, 96(5): 496-499.

Inoue, K., Gotou, T., Kitajima, H., Mizuno, S., Nakazawa, T. and Yamamoto, N. (2009). Release of antihypertensive peptides in *miso* paste during its fermentation, by the addition of casein. *Journal of Bioscience and Bioengineering*, 108: 111-115.

Islam, S. and Choi, H. (2009). Antidiabetic effect of Korean traditional *Baechu* (Chinese cabbage) *kimchi* in a type 2 diabetes model of rats. *Journal of Medicinal Food*, 12(2): 192-197.

Je, J.Y., Park, P.J., Byun, H.G., Jung, W.K. and Kim, S.K. (2005). Angiotensin I converting enzyme (ACE) inhibitory peptide derived from the sauce of fermented blue mussel, *Mytilus edulis*. *Bioresource Technology*, 96: 1624-1629.

Je, J.Y., Park, J.Y., Jung, W.K., Park, P.J. and Kim, S.K. (2005a). Isolation of angiotensin I converting enzyme (ACE) inhibitor from fermented oyster sauce, *Crassostrea gigas*. *Food Chemistry*, 90: 809-814.

Jemil, I., Mora, L., Nasri, R., Abdelhedi, O., Aristoy, M.C., Hajji, M., Nasri, M. and Toldra, F. (2016). A peptidomic approach for the identification of antioxidant and ACE-inhibitory peptides in sardinelle protein hydrolysates fermented by *Bacillus subtilis* A26 and *Bacillus amyloliquefaciens* An6. *Food Research International*, 89(1): 347-358.

Kabelova, I., Dvorakova, M., Cizkova, H., Dostalek, P. and Melzoch, K. (2008). Determination of free amino acids in beers: A comparison of Czech and foreign brands. *Journal of Food Composition and Analysis*, 21(8): 736-741.

Kim, S.H., Lee, Y.J. and Kwon, D.Y. (1999). Isolation of angiotensin-converting enzyme inhibitor from *Doenjang*. *Korean Journal of Food Science and Technology*, 31(3): 848-854.

Kim, H.J., Kang, S.G., Jaiswal, L., Li, J., Choi, J.H., Moon, S.M., Cho, J.Y. and Ham, K.S. (2016). Identification of four new angiotensin I-converting enzyme inhibitory peptides from fermented anchovy sauce. *Applied Biological Chemistry*, 59: 25-31.

Kim, S.H., Lee, J.Y., Balolong, M.P., Kim, J.E., Paik, H.D. and Kang, D.K. (2017). Identification and characterisation of a novel antioxidant peptide from bovine skim milk fermented by *Lactococcus lactis* SL6. *Korean Journal of Food Science and Animal Research*, 37: 402-409.

Kinariwala, D., Panchal, G., Sakure, A. and Hati, S. (2019). Exploring the potentiality of *Lactobacillus* cultures on the production of milk-derived bioactive peptides with antidiabetic activity. *International Journal of Peptide Research and Therapeutics*. https://doi.org/10.1007/s10989-019-09958-5

Kitagawa, M., Shiraishi, T., Yamamoto, S., Kutomi, R., Ohkoshi, Y., Sato, T., Wakui, H., Itoh, H., Miyamoto, A. and Yokota, S.I. (2017). Novel antimicrobial activities of a peptide derived from a Japanese soybean fermented food, *natto*, against *Streptococcus pneumoniae* and *Bacillus subtilis* group strains. *Applied and Industrial Microbiology Express*, 7(1): 127.

Kleekayai, T., Harnedy, P.A., O'Keeffe, M.B., Poyarkov, A.A., CunhaNeves, A., Suntornsuk, W. and FitzGerald, R.J. (2015). Extraction of antioxidant and ACE inhibitory peptides from Thai traditional fermented shrimp pastes. *Food Chemistry*, 176: 441-447.

Korhonen, H. and Pihlanto, A. (2003). Food-derived bioactive peptides – Opportunities for designing future foods. *Current Pharmaceutical Design*, 9: 1297-1308.

Korhonen, H. and Pihlanto, A. (2006). Bioactive peptides: Production and functionality. *International Dairy Journal*, 16: 945-960.

Kwon, D.Y., Hong, S.M., Ahn, I.S., Kim, M.J., Yang, H.J. and Park, S. (2011). Isoflavonoids and peptides from *meju*, long-term fermented soybeans, increase insulin sensitivity and exert insulinotropic effects *in vitro*. *Nutrition*, 27: 244-252.

Lam, Y., Galvez, A. and Lumen, B.O.D. (2003). Lunasin™ suppresses E1A-mediated transformation of mammalian cells but does not inhibit growth of immortalised and established cancer cell lines. *Nutrition Cancer*, 47(1): 88-94.

LeBlanc, J.G., Matar, C., Valdez, J.C., LeBlanc, J. and Perdigo, G. (2002). Immunomodulating effects of peptidic fractions issued from milk fermented with *Lactobacillus helveticus*. *Journal of Dairy Science*, 85: 2733-2742.

LeBlanc, J., Fliss, I. and Matar, C. (2004). Induction of a humoral immune response following an *Escherichia coli* O157:H7 infection with an immunomodulatory peptidic fraction derived from *Lactobacillus helveticus*-fermented milk. *Clinical and Diagnostic Laboratory Immunology*, 1171-1181.

LeBlanc, A.M., Matar, C., LeBlanc, N. and Perdigon, G. (2005). Effects of milk fermented by *Lactobacillus helveticus* R389 on a murine breast cancer model. *Breast Cancer Research*, 7: 477-486.

Lee, Y.G, Kim, J.Y., Lee, K.W., Kim, K.H. and Lee, H.J (2003). Peptides from anchovy sauce induce apoptosis in human lymphoma cells (U937) through the increase of Caspase 3 and 8 activities. *Annals of the New York Academy Sciences*, 1010: 399-404.

Lee, J.H., Nam, S.H., Seo, W.T., Yun, H.D., Hong, S.Y. and Kim, M.K. (2012). The production of surfactin during the fermentation of *cheonggukjang* by potential probiotic *Bacillus subtilis* CSY191 and the resultant growth suppression of MCF-7 human breast cancer cells. *Food Chemistry*, 131(4): 1347-1354.

Lee, M.H., Lee, J., Nam, Y.D., Lee, J.S., Seo, M.J. and Yi, S.H. (2016). Characterisation of antimicrobial lipopeptides produced by Bacillus sp. LM7 isolated from *chungkookjang*, a Korean traditional fermented soybean food. *International Journal of Food Microbiology*, 221: 12-18.

Li, Y. and Yu, J. (2015). Research progress in structure-activity relationship of bioactive peptides. *Journal of Medicinal Food*, 18(2): 147-156.

Li, C., Kwok, L.Y., Mi, Z., Bala, J., Xue, J., Yang, J., Ma, Y., Zhang, H. and Chen, Y. (2017). Characterisation of the angiotensin-converting enzyme inhibitory activity of fermented milks produced with *Lactobacillus casei*. *Journal of Dairy Science*, 100: 9495- 9507.

Liu, S.N., Han, Y. and Zhou, Z.J. (2011). Lactic acid bacteria in traditional fermented Chinese foods. *Food Research International*, 44: 643-651.

Lule, V.K., Garg, S., Pophaly, S.D., Hitesh and Tomar, S.K. (2015). Potential health benefits of lunasin: A multifaceted soy-derived bioactive peptide. *Journal of Food Science*, 80(3): 485-494.

Maeno, M., Yamamoto, N. and Takano, T. (1996). Identification of an antihypertensive peptide from casein hydrolysate produced by a proteinase from *Lactobacillus helveticus* CP790. *Journal of Dairy Science*, 79: 1316-1321.

Marshall, V.M. and Tamime, A.Y. (1997). Starter cultures employed in the manufacture of biofermented milks. *International Journal of Dairy Technology*, 50(1): 35-41.

Meyer, J., Butikofer, U., Walther, B., Wechsler, D. and Sieber, R. (2009). Hot topic: Changes in angiotensin-converting enzyme inhibition and concentration of the tripeptides Val-Pro-Pro and Ile-Pro-Pro during ripening of different Swiss cheese varieties. *Journal of Dairy Science*, 92(3): 826-836.

Migliore-Samour, D. and Jolle, P. (1988). Casein, a pro-hormone with an immunomodulating role for the newborn. *Experientia*, 44: 188-193.

Mora, L., Escudero, E., Aristoy, M.C. and Toldra, F. (2015). A peptidomic approach to study the contribution of added casein proteins to the peptide profile in Spanish dry-fermented sausages. *International Journal of Food Microbiology*, 212(6): 41-48.

Moreno, M.R.F., Leisner, J.J., Tee, L.K., Ley, C., Radu, S., Rusul, G., Vancanneyt, M. and De Vuyst, L. (2002). Microbial analysis of Malaysian *tempeh*, and characterisation of two bacteriocins produced by isolates of *Enterococcus faecium*. *Journal of Applied Microbiology*, 92: 147-157.

Moslehishad, M., Ehsani, M.R., Salami, M., Mirdamadi. S., Ezzatpanah, H., Naslaji, A.N. and Moosavi-Movahedi, A.A. (2013). The comparative assessment of ACE-inhibitory and antioxidant activities of peptide fractions obtained from fermented camel and bovine milk by *Lactobacillus rhamnosus* PTCC 1637. *International Dairy Journal*, 29: 82-87.

Muguerza, B., Ramos, M., Sanchez, E., Manso Miguel, M., Aleixandre, A., Degalo, M.A. and Recio, I. (2006). Antihypertensive activity of milk fermented by *Enterococcus faecalis* strains isolated from raw milk. *International Dairy Research*, 16(1): 61-69.

Nagasawa, T., Yonekura, T., Nishizawa, N. and Kitts, D. (2001). *In vitro* and *in vivo* inhibition of muscle lipid and protein oxidation by carnosine. *Molecular and Cellular Biochemistry*, 225: 29-34.

Najafian, L. and Babji, A.S. (2018). Fractionation and identification of novel antioxidant peptides from fermented fish (*pekasam*). *Journal of Food Measurement and Characterisation*, 12(3): 2174-2183.

Najafian, L. and Babji, A.S. (2019). Purification and identification of antioxidant peptides from fermented fish sauce (*Budu*). *Journal of Aquatic Food Product Technology*. https://doi.org/10.1080/10498850.2018.1559903

Nakahara, T., Sano, A., Yamaguchi, H., Sugimoto, K., Chikata, H., Kinoshita, E. and Uchida, R. (2010). Antihypertensive effect of peptide-enriched soy sauce-like seasoning and identification of its angiotensin I-converting enzyme inhibitory substances. *Journal of Agriculture and Food Chemistry*, 58(2): 821-827.

Nakamura, Y., Yamamoto, N., Sakai, K., Takano, T., Okubo, A. and Yamazaki, S. (1995). Purification and characterisation of angiotensin I-converting enzyme inhibitors from sour milk. *Journal of Dairy Science*, 78: 777.

Nakamura, K., Naramoto, K. and Koyama, M. (2013). Blood-pressure-lowering effect of fermented buckwheat sprouts in spontaneously hypertensive rats. *Journal of Functional Foods*, 5: 406-415.

Nielsen, M.S., Martinussen, T., Flambard, B., Sorensen, K.I. and Ott, J. (2009). Peptide profiles and angiotensin-I-converting enzyme inhibitory activity of fermented milk products:

Effect of bacterial strain, fermentation pH, and storage time. *International Dairy Journal*, 19(3): 155-165.

Nout, M.J.R. and Kiers, J.L. (2005). *Tempe* fermentation, innovation and functionality: Update into the third millennium. *Journal of Applied Microbiology*, 98(4): 789-805.

Ohata, M., Uchida, S., Zhou, L. and Arihara, K. (2016). Antioxidant activity of fermented meat sauce and isolation of an associated antioxidant peptide. *Food Chemistry*, 194: 1034-1039.

Okamoto, A., Hanagata, H., Matsumoto, E., Kawamura, Y., Koizumi, Y. and Yanagida, F. (1995). Angiotensin I converting enzyme inhibitory activities of various fermented foods. *Biotechnology and Biochemistry*, 59: 1147-1149.

Parmar, H., Hati, S. and Sakure, A. (2017). *In vitro* and *In silico* analysis of novel ACE-inhibitory bioactive peptides derived from fermented goat milk. *International Journal of Peptide Research and Therapeutics*, 24: 441-453.

Peres, C.M., Alves, M., Hernandez-Mendoza, A., Moreira, L., Silva, S., Bronze, M.R., Vilas-Boas, L., Peres, C. and Malcata, F.X. (2014). Novel isolates of Lactobacilli from fermented Portuguese olive as potential probiotics. *LWT–Food Science Technology*, 59: 234-246.

Pihlanto-Leppala, A., Antila, P., Mantsala, P. and Hellman, J. (1994). Opioid peptides produced by *in vitro* proteolysis of bovine caseins. *International Diary Journal*, 4: 281-301.

Quiros, A., Ramos, M., Muguerza, B., Delgado, M., Miguel, M., Alexaindre, A. and Recio, I. (2007). Identification of novel antihypertensive peptides in milk fermented with *Enterococcus faecalis*. *International Diary Journal*, 17(1): 33-41.

Rai, A.K., Pandey, A. and Sahoo, D. (2019). Biotechnological potential of yeasts in functional food industry. *Trends in Food Science and Technology*, 83: 129-137.

Rai, A.K and Jeyaram, K. (2017). Role of yeast in food fermentation. pp. 83-113. *In*: Tulasi Satyanarayana and Gotthard Kunze (Eds.). *Yeast Diversity in Human Welfare*. Springer, Singapore.

Rai, A.K., Sanjukta, S. and Jeyaram, K. (2017a). Production of angiotensin I converting enzyme inhibitory (ACE-I) peptides during milk fermentation and its role in treatment of hypertension. *Critical Reviews in Food Science and Nutrition*, 57: 2789–2800.

Rai, A.K., Sanjukta, S., Chourasia, R., Bhat, I., Bhardwaj, P.K. and Sahoo, D. (2017b). Production of bioactive hydrolysate using protease, β-glucosidase and α-amylase of *Bacillus* spp. isolated from *kinema*. *Bioresource Technology*, 235: 358-365.

Rai, A.K. and Jeyaram, K. (2015). Health benefits of functional proteins in fermented foods. pp. 455-476. *In*: J.P. Tamang (Ed.). *Health Benefits of Fermented Foods and Beverages*. CRC Press, Taylor and Francis Group of USA.

Rajapakse, N., Mendis, E., Jung, W.K., Je, J.Y. and Kim, S.K. (2005). Purification of a radical scavenging peptide from fermented mussel sauce and its antioxidant properties. *Food Research International*, 38: 175-182.

Rajasekaran, A., Kalaivani, M. and Sabitha, R. (2009). Antidiabetic activity of aqueous extract of *Monascus perpureus* fermented rice in high cholesterol diet fed streptozotocin-induced diabetic rats. *Asian Journal of Scientific Research*, 2(4): 180-189.

Ray, R.C. and Joshi, V.K. (2014). Fermented Foods: Past, present and future scenario. pp. 1-36. *In*: R.C. Ray and D. Montet (Eds.). *Microorganisms and Fermentation of Traditional Foods*. CRC Press, Boca Raton, Florida, USA.

Rinto, Dewanti, R., Yasni, S. and Suhartono M.T. (2017). Novel HMG-CoA reductase inhibitor peptide from *Lactobacillus acidophilus* isolated from Indonesian fermented food *Bekasam*. *Journal of Pharmaceutical, Chemical and Biological Sciences*, 5: 195-204.

Rui, X., Wen, D., Li, W., Chen, X., Jiang, M. and Dong, M. (2015). Enrichment of ACE inhibitory peptides in navy bean (*Phaseolus vulgaris*) using lactic acid bacteria. *Food & Function*, 6(2): 622-629.

Ryhanen, E.L., Pihlanto-Leppala, A. and Pahkala, E. (2001). A new type of ripened, low-fat cheese with bioactive properties. *International Dairy Journal*, 11(4): 441-447.

Sah, B.N.P., Vasiljevic, T., McKechnie, S. and Donkor, O.N. (2014). Effect of probiotics on antioxidant and antimutagenic activities of crude peptide extract from yoghurt. *Food Chemistry*, 156: 264-270.

Sanjukta, S., Rai, A.K. and Sahoo, D. (2017). Bioactive molecules in fermented soybean products and their potential health benefits. *In*: *Fermented Foods, Part II: Technological Interventions*, CPC Press.

Sanjukta, S. and Rai, A.K. (2016). Production of bioactive peptides during soybean fermentation and their potential health benefits. *Trends in Food Science and Technology*, 50: 1-10.

Sanjukta, S., Rai, A.K., Muhammed, A., Jeyaram, K. and Talukdar, N.C. (2015). Enhancement of antioxidant properties of two soybean varieties of Sikkim Himalayan region by proteolytic *Bacillus subtilis* fermentation. *Journal of Functional Foods*, 14: 650-658.

Sarmadi, B. and Ismail, A. (2010). Antioxidative peptides from food proteins: A review. *Peptides*, 31: 1949-1956.

Sato, K., Miyasaka, S., Tsuji, A. and Tachi, H. (2018). Isolation and characterisation of peptides with dipeptidyl peptidase IV (DPPIV) inhibitory activity from *natto* using DPPIV from *Aspergillus oryzae*. *Food Chemistry*, 261: 51-56.

Saito, T., Nakamura, T., Kitazawa, H., Kawai, Y. and Itoh, T. (2000). Isolation and structural analysis of antihypertensive peptides that exist naturally in Gouda cheese. *Journal of Dairy Science*, 83: 1434-1440.

Seppo, L., Jauhiainen, T., Poussa, T. and Korpela, R. (2003). A fermented milk high in bioactive peptides has a blood pressure-lowering effect in hypertensive subjects. *American Journal of Clinical Nutrition*, 77(2): 326-330.

Seppo, L., Kerojoki, O., Suomalainen, T. and Korpela, R. (2002). The effect of a *Lactobacillus helveticus* LBK-16 H fermented milk on hypertension: A pilot study on humans. *Milchwissenschaft*, 57(3): 124-127.

Settanni, L. and Moschetti, G. (2010). Non-starter lactic acid bacteria used to improve cheese quality and provide health benefits. *Food Microbiology*, 27(6): 691-697.

Singh, A.K. and Ramesh, A. (2008). Succession of dominant and antagonistic lactic acid bacteria in fermented cucumber: Insights from a PCR based-approach. *Food Microbiology*, 25: 278-287.

Skeggs, L.T., Kahn, J.E. and Shumway, N.P. (1956). Preparation and function of the hypertensin converting enzyme. *Journal of Experimental Medicine*, 103: 295-299.

Smacchi, E. and Gobbetti, M. (1998). Peptides from several Italian cheeses inhibitory to proteolytic enzymes of lactic acid bacteria, *Pseudomonas fluorescens* ATCC 948 and to the angiotensin I-converting enzyme. *Enzyme and Microbial Technology*, 22(8): 687-694.

Sourabh, A., Rai, A.K., Chauhan, A., Jeyaram, K., Taweechotipatr, M., Panesar, P.S., Sharma, R., Panesar, R., Kanwar, S.S., Walia, S., Tanasupawat, S., Sood, S., Joshi, V.K., Bali, V., Chauhan, V. and Kumar, V. (2015). Health-related issues and indigenous fermented products. pp. 303-343. *In*: V.K. Joshi (Ed.). *Indigenous Fermented Foods of South Asia*. CRC Press.

Sun, J., He, H. and Xie. B.J. (2004). Novel antioxidant peptides from fermented mushroom *Ganoderma lucidum*. *Journal of Agriculture and Food Chemistry*, 52: 6646-6652.

Tamam, B., Syah, D., Suhartono, M.T., Kusuma, W.A., Tachibana, S. and Lioe, H.N. (2019). Proteomic study of bioactive peptides from *tempe*. *Journal of Bioscience and Bioengineering*, 128: 241-248.

Torino, M.I., Limon, R.I., Martinez-Villaluenga, C., Makinen, S., Pihlanto, A., Vidal-Valverde, C. and Frias, J. (2013). Antioxidant and antihypertensive properties of liquid and solid state fermented lentils. *Food Chemistry*, 136(2): 1030-1037.

Vastag, Z., Popovic, L., Popovic, S., Petrovic, L. and Pericin, D. (2010). Antioxidant and angiotensin-I converting enzyme inhibitory activity in the water-soluble protein extract from Petrovac sausage (*Petrovska Kolbasa*). *Food Control*, 21: 1298-1302.

Vermeirssen, V., Van Camp, J. and Verstraete, W. (2004). Bioavailability of angiotensin I converting enzyme inhibitory peptides. Review. *British Journal of Nutrition*, 92: 357-366.

Vinderola, G., Matar, C., Palacios, J. and Perdigon, G. (2007). Mucosal immunomodulation by the non-bacterial fraction of milk fermented by *Lactobacillus helveticus* R389. *International Journal of Food Microbiology*, 115: 180-186.

Wang, Y.K., He, H.L., Chen, X.L., Sun, C.Y., Zhang, Y.Z. and Zhou, B.C. (2008). Production of novel angiotensin I-converting enzyme inhibitory peptides by fermentation of marine shrimp *Acetes chinensis* with *Lactobacillus fermentum* SM 605. *Applied Microbiology and Biotechnology*, 79: 785-791.

Weng, T.M. and Chen, M.T. (2011). Effect of two-step fermentation by *Rhizopus oligosporus* and *Bacillus subtilis* on protein of fermented soybean. *Food Science and Technology Resource*, 17(5): 393-400.

Wu, N., Xu, W., Liu, K., Xia, Y. and Shuangquan. (2019). Angiotensin-converting enzyme inhibitory peptides from *Lactobacillus delbrueckii* QS306 fermented milk. *Journal of Dairy Science*, 102(7): 5913-5921.

Wu, H., Rui, X., Li, W., Chen, X., Jiang, M. and Dong, M. (2015). *Mung* bean (*Vigna radiata*) as probiotic food through fermentation with *Lactobacillus plantarum* B1-6. *LWT–Food Science and Technology*, 63(1): 445-451.

Yamamoto, N., Akino, A. and Takano, T. (1994). Antihypertensive effect of the peptides derived from casein by an extracellular proteinase from *Lactobacillus helveticus* CP790. *Journal of Dairy Science*, 77: 917-922.

Yamamoto, N., Alcino, A. and Takano, T. (1994a). Antihypertensive effects of different kinds of fermented milk in spontaneously hypertensive rats. *Bioscience, Biotechnology and Biochemistry*, 58: 776.

Yang, J.H., Mau, J.L., Ko, P.T. and Huang, L.C. (2000). Antioxidant properties of fermented soybean broth. *Food Chemistry*, 71(2): 249-254.

Yeap, S.K., Ali, N.M., Yusof, H.M., Alitheen, N.B., Beh, B.K., Ho, W.Y., Koh, S.P. and Long, K. (2012). Antihyperglycemic effects of fermented and non fermented *mung* bean extracts on Alloxan-induced-diabetic mice. *Journal of Biomedicine and Biotechnology*, 285430, 1-7.

Young, J.F., Therkildsen, M., Ekstrand, B., Che, B.N., Larsen, M.K., Oksbjerg, N. and Stagsted, J. (2013). Novel aspects of health promoting compounds in meat. *Meat Science*, 95: 904-911.

Fibrinolytic Enzymes in Fermented Food Products

Yogesh Devaraj[1] and Prakash M. Halami[2]*

[1] Department of Studies in Molecular Biology, University of Mysore,
   Mysuru, Karnataka, India
[2] Microbiology and Fermentation Technology Department,
   CFTRI – Central Food Technological Research Institute, Mysore - 570 020, India

1. Introduction

In any living organism, enzymes have very prominent roles to play in most of the biological processes. Enzymes in therapies have gained significant ground in medical science. Various enzymes have been used as mucolytic agents, anti-coagulants, anti-fibrinolytic agents and digestive aids. Among the diseases, CVD (Cardio Vascular Disease) is a cause for the greater number of deaths worldwide. According to American Heart Association (2015), it is estimated that on an average, one person dies every 40 seconds because of heart attack or any other related CVD condition. In the Indian context, there were about 6.4 crores cases in 2015, related to CVD (based on *National Commission on Macroeconomics and Health Report*, NCMH).

The process of fibrin formation and its removal is balanced under the normal haemostatic state of the body. However, conditions favouring continuous formation of fibrin lead to undesirable as well as inappropriate clots which are dangerous for the normal blood circulation. In case of balanced hemostatic condition, the key enzyme which is responsible for fibrin degradation in the body is plasmin. However, during an unbalanced haemostasis, the clots continue to remain undegraded, further leading to pathological condition of thrombosis. By definition, thrombosis is a condition in which formation of undesirable blood clot occurs in blood vessels (Lopez-Sendon *et al.*, 1995). At present, the diseases related to thrombosis are rising at an alarming pace and growing as per the total number of related deaths each year. Fibrin is the key protein, which is involved in blood clot or thrombus, and by nature it is an insoluble protein formed by polymerisation of soluble fibrinogen molecules. Various thrombolytic molecules, such as t-PA (tissue-type plasminogen activator) (Collen and Lijnen, 2004), urokinase (Duffy, 2002) and molecules of bacterial origin, such as

---

*Corresponding author: prakashalami@cftri.res.in

streptokinase and staphylokinase, are also being used in the treatment of thrombosis-related conditions. Conversely, these are found to have many serious limitations which pose risks, such as internal haemorrhage apart from their high prices. Hence, these have persuaded the researchers to explore new resources for safer thrombolytic agents (Nakajima *et al.*, 1993; Bode *et al.*, 1996). Fermented foods are known to be one such potent source of fibrinolytic enzymes or such enzymes producing microorganisms.

Various fibrinolytic enzymes were identified from diverse fermented foods, such as *natto, chunghukjang, tofu, kimchi,* fermented shrimp paste *douchi, skipjack shikar* (fermented fish), *meju*, and many fermented foods from different regions of Asia (Mine *et al.*, 2005). The most important fibrinolytic enzyme producers from fermented foods belong to bacteria and major members of genus *Bacillus* in particular (Mine *et al.*, 2005). With the discovery and the studies on well-known fibrinolytic enzyme 'Nattokinase' from *Bacillus natto* with its potential and proven health benefits (Sumi *et al.*, 1987), various researchers worldwide have started to look for a fibrinolytic enzyme which is better, efficient and also specific for thrombosis-related disorders. The focus towards microbes as a novel and better source for fibrinolytic enzymes with diverse properties has attracted investigators in recent years (Goldhaber and Bounameaux, 2001; Tough, 2005). Hence, the present chapter focuses on fibrinolytic enzymes in various fermented foods and deals with the available literature on various microorganisms responsible for the secretion of fibrinolytic enzymes and related research in our laboratory. Possible application of these microbial fibrinolytic enzymes and their potential use as formulations is also discussed.

The main focus of our study is on the fibrinolytic enzymes-producing microorganisms (*Bacillus* spp.) from fermented foods (Fig. 1a). Some of the results related to the fibrinolytic activity of the enzyme producing *Bacillus* isolates from food sources in our laboratory are also presented. The enzymes were primarily assessed for their degradation ability on fibrin protein and techniques from basic fibrin plate assay (Fig. 1b), zymogram analysis (Fig. 1c), and pattern of fibrin degradation using SDS-PAGE (sodium dodecyl sulphate-polyacrylamide gel electrophoresis) (Fig. 1d) have been employed for the study. The fibrinolytic enzymes produced by the native *Bacillus* isolates were purified and characterised for physicochemical properties with reference to the nature of the enzymes, molecular weight, effect of temperature and pH, and the influence of various inhibitors, metal ions on enzyme activity, action on other protein substrates and analysis of fibrin degradation products through Tandem mass spectrometry (Yogesh and Halami, 2015a, b; Yogesh, 2016, 2017). The nucleotide sequence of fibrinolytic enzyme genes as well as MS/MS data analysis of enzymes were also employed to study the relatedness of fibrinolytic enzymes (Yogesh and Halami, 2018).

2. CVDs – Number One Killer

The diseases of civilisation, which are commonly referred as 'lifestyle diseases' or non-communicable diseases and these include, CVDs, diabetes, chronic liver disease, cancer, asthma, osteoporosis and stroke, etc. These are associated with the way an individual lives and the key reasons for such diseases may be the

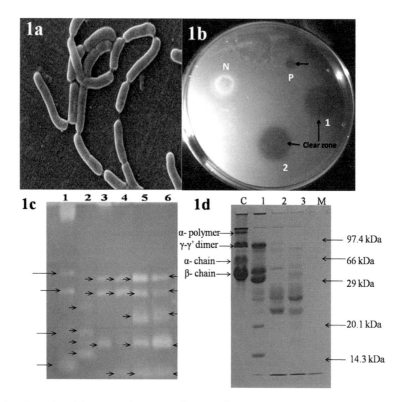

Fig. 1a: Scanning Electron Microscope image of native fibrinolytic enzyme producing *Bacillus* sp. isolated from fermented food. **1b:** Fibrin plate assay for detection of fibrinolytic protease in cell-free supernatant of various native *Bacillus* isolates, Isolate codes – (1) CFR15-protease, (2) CFR11-protease, P – Human plasmin (Positive control) and N – Phosphate buffer (negative control). The clear zone indicates the fibrinolytic activity. **1c:** Fibrin zymogram for fibrinolytic enzymes of native *Bacillus* isolates (cell-free supernatants of the *Bacillus* cultures were subjected to zymogram where fibrin is incorporated in gel). Position of the arrows indicates the clear zone due to fibrinolytic activity of the enzymes]. Isolate codes (1) Dchi15, (2) DCHI27, (3) DCHI26, (4) DCHI23, (5) DK63 and (6) BR-21. Multiple clear zones indicate the presence of multiple proteases with fibrinolytic activity. **1d:** Fibrin degradation assay using SDS-PAGE [Fibrin prepared from plasma clot was incubated (12 hours) for degradation using culture supernatant of different *Bacillus* isolates and the degradation products were analysed using SDS-PAGE. Degradation is confirmed with the disappearance of intact α-polymer, γ-γ' dimer, α-chain and β-chain (as seen in control-C). The bands lower to the β-chain represents fibrin degradation products]. Lane (C) – Control (no enzyme). Lane (1) degradation pattern for culture supernatant of M1, (2) Dchi15, (3) DK111 and M – Molecular weight marker

occupation, change in lifestyle, urbanisation, dietary habits along with habits of alcohol consumption and smoking, etc. (Gupta *et al.*, 2006; Hu, 2011). These diseases account for nearly 59 per cent of the 56.5 million deaths annually throughout the world. The situation is relatively alarming in India and according to the WHO (World Health Organisation, 2014), in near future, India is going to have a high number of cases of lifestyle disorders (Pappachan, 2011).

2.1. Burden of CVDs

CVDs are usually referred to diseases which are associated with the heart or blood vessels (WHO, 2015). These CVDs account for almost half of non-communicable diseases (NCDs) and have overtaken communicable diseases to become the world's most important disease burden as well as leading cause of deaths worldwide (Fig. 2). In a broader sense, CVDs comprise conditions of myocardial infarction, high blood pressure, peripheral vascular diseases, arrhythmias, valvular heart diseases and stroke (Mine *et al.*, 2005). At present, these diseases approximately account for about 17.3 million deaths annually and this figure is expected to increase to around 23.6 million by 2030. The population of low- and middle-income countries (LMIC) is likely to get affected as the medical and financial wealth to tackle CVDs are limited. Amongst all the non-communicable diseases, about 50 per cent loss account only for CVDs and is worth about \$7.28 trillion in LMIC from 2011 to 2025 (WHO, 2011).

Apart from deaths, CVDs also account for about 10 per cent of the DALYs (disability-adjusted life years) lost in LMIC (World Heart Federation, 2012). When the cost for CVDs is measured in both family and society, it conveys as loss in terms of income of the individual as well as productivity, and also their caregivers. The loss with respect to financial conditions is aggravated in many developing countries. Here the CVDs affect a larger fraction of working-age individuals and the estimated death rate is about eight per thousand in case of developed countries. In India, a similar trend can be assumed. It is estimated that 6.4 crores CVD cases and around one crore deaths occurred in India alone in the year 2015. This number is anticipated to increase significantly (NCMH, 2005). Hence, CVDs continue as the leading cause of deaths worldwide (WHO, 2011; American Heart Association, 2015).

The percentage of various diseases in the world and also in India is represented in Figs. 2a and 2b, respectively. In both the cases, CVDs account for a greater portion and the reason for the large number of deaths in these scenarios. Based on these statistics, CVDs take a greater share as per the World Heart Federation (WHF) and 'thrombosis' is one of the prime circulatory systems-related CVDs accountable

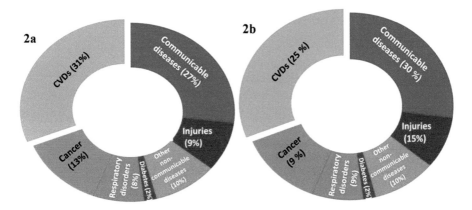

Fig. 2a: Causes of deaths worldwide (adapted from WHO, 2011), **2b:** Causes of deaths in India (adapted from Meruvu and Vangalapati, 2011)

for the greater number of deaths. It was only in recent years that the influence of thrombotic component in MI (myocardial infarction) and also the cases of sudden coronary death were recognised and acknowledged. A certain faction of individuals may have an anomalous inclination to develop thrombosis in veins or arteries and the incidence of thromboembolic events relatively early in life or suffer frequent events (Chan *et al.*, 2008).

2.2　Thrombosis, Pathophysiology of Thrombosis

Thrombosis is a condition in which the formation of undesirable blood clot or thrombus takes place inside the blood vessel (Lopez-Sendon *et al.*, 1995). This may be due to the activation of coagulation cascade leading to the generation of active thrombin and results in the formation of thrombus. The key component responsible for this clot is insoluble fibrin (Gorkum *et al.*, 1994; Lim *et al.*, 2008; Undas and Ariens, 2011). Under the normal physiological conditions of a healthy person, there is an optimal balance between the formation and degradation of clots. The deviation from this condition leads to abnormal homeostasis and results in either favouring thrombus formation or excessive bleeding, with the former representing the condition called 'thrombosis' (Lopez-Sendon *et al.*, 1995; Yin *et al.*, 2010). The thrombus can propagate into normal blood vessels and has chances to obstruct smooth bloodflow under certain pathological conditions. The formation of pathologic thrombus can take place in any blood vessel and at any site in the body. Based on the location of the thrombus, thrombosis is classified into coronary thrombosis, deep vein thrombosis, pulmonary thromboembolism, etc. The effects caused by the accumulated clots may be deleterious and may obstruct the smooth blood flow and circulation. Further, this may result in conditions like MI, Acute Ischemic Stroke (AIS) or related conditions, which may ultimately result in death (Mine *et al.*, 2005). There are different causes that predispose the body to the condition of thrombosis. These include atherosclerosis, changes in blood flow and metabolic disorders, such as hyperlipidemia, hypercoagulable states, diabetes mellitus, burns, trauma and smoking (Antoniades *et al.*, 2004; Pomp *et al.*, 2008).

2.3　Currently Used Drugs, Thrombolytic Agents and Their Limitations

Many drugs, which act as anti-coagulants and antiplatelets, have been employed for maintaining the fluidity of the blood. The drugs used for dissolution of the blood clots are referred as thrombolytic agents (Weitz, 2006; Schulman *et al.*, 2008) and these are employed for reducing the damage caused by occlusion or blockage inside the blood vessels. Presently, the anti-coagulants, thrombolytic agents and anti-platelet drugs get their important usage in management of stroke as well as heart diseases.

　　Till today, therapy using anti-coagulants remains the accepted and popular choice in most of the cases. The treatments, such as thrombolysis and inferior vena cava filtrations are used only in certain circumstances of thrombo-embolism. Several thrombolytic agents are employed in combination with anti-coagulants to achieve synergistic desirable anti-thrombotic results and secondary prevention (Schulman *et al.*, 2008). In the case of treatments, the anti-coagulants are recognised to increase the risk of bleeding; hence, treatment strategies should include factors like risk, etiology, benefit and patient preference, etc. and, more importantly, the cost.

There are some serious drawbacks linked with current thrombolytic therapies and these include hemorrhagic stroke and development of allergy in patients for the thrombolytics. Although a number of drugs are being used for the treatment of thrombosis and related disorders, heparin and antagonists of vitamin K remain as the common preferences and popular choice during the course of the therapy. However, the treatments, such as thrombolysis and inferior vena cava filtrations are employed for thromboembolism only in certain conditions. Due to the recognised risks of bleeding complications associated with therapy using anti-coagulants, factors of etiology need to be addressed. The commonly used thrombolytic agents in practice to treat CVDs, their chemical nature, mode of action and limitations, which include complications such as bleeding and allergic reaction, the drug as well as antigenicity effects are presented in Table 1.

Table 1: Currently Used Drugs/Molecules in Thrombosis Treatment,
Their Action and Limitations

Drug/Molecule and Chemical Nature	Mechanism of Action	Limitations	References
Aspirin (Acetylsalicylic acid)	Anti-platelet and inhibition of thrombin	Allergies and gastric bleeding	Schulman *et al.* (2008)
Streptokinase) (Protein (47 kDa))	Plasminogen activation	Antigenisity and bleeding complication	Banerjee et al. (2004)
Coumarin (Benzopyrones)	Vitamin K antagonist	Bleeding complication	Weitz (2006)
Heparin (Glycosamino glycan)	Anti-coagulant	Bleeding complication	Schulman *et al.* (2008)
Warfarin (Hydroxycoumarin derivative)	Anti-coagulant	Severe drug reaction, bleeding	Schulman *et al.* (2008)
Urokinase (Protein (55 kDa))	Plasminogen activation	Shorter half life and haemorrhage	Schulman *et al.* (2008)
Staphylokinase (Protein (16 kDa))	Plasminogen activation	Antigenisity and bleeding complication	Wu *et al.* (2003)
Human tissue plasminogen activator) (Protein (72 kDa))	Plasminogen activation	Shorter half-life and bleeding complication	Schulman *et al.* (2008)

2.4 Fibrinolytic System and Their Mechanism

In a healthy individual, under the normal condition of haemostasis, the formation of fibrin and its breakdown or lysis is critically controlled. This mechanism involves many substrates, activators, cofactors and inhibitors, etc. In this fibrinolytic system, fibrinogen, fibrin and inherent enzymes constitute the primary forms of the components. The key responsibility of this system is to ensure proper fluidity as well as control of blood loss in injuries. Fibrinogen is the soluble precursor for the insoluble fibrin as a general response to injury, which is a positive aspect in healing and health (Clark, 2001; Laurens *et al.*, 2006). However, in parallel, fibrin may also be accountable for undesirable and inappropriate blood clots in blood vessels which may lead to a greater risk factor, like myocardial infarction, stroke and related diseases of cardiovascular system.

2.4.1 Fibrinogen

Fibrinogen is a blood-borne soluble glycoprotein precursor which is synthesised by hepatocytes, and has a relative molecular weight of 340 kDa. It is a prime molecule responsible for process of blood clot at the site of injury (Herrick *et al.*, 1999). Structurally, fibrinogen is composed of three pairs of non-identical polypeptides referred as Aα, Bβ and γ chain. By the action of an enzyme called thrombin, this forms fibrin, which plugs the site of injury to prevent the blood loss. This process takes place by the action of a series of enzymes and molecules, ultimately resulting in formation of insoluble fibrin network, which can trap blood cells and platelets (Mosesson, 2005). Fibrinogen is also known to act as the protein responsible for adhesion, platelet aggregation and ultimately forms an insoluble clot at the final stage of the cascade (Herrick *et al.*, 1999; Undas and Ariens, 2011).

2.4.2: Fibrin

Fibrinogen, present in blood at the concentration of about 3 mg/ml, is one of the prominent and abundant proteins. Fibrin, an insoluble protein, acts as a hemostatic plug and is the key polymer formed in response to injury by the activation of fibrinogen. By the action of thrombin on fibrinogen, fibrinopeptide A and B are released and lead to the polymerisation of fibrinogen molecules (Mosesson 2002, 2005). In the case of normal haemostatic status, balance in the formation of fibrin and its degradation are controlled. However, deviation or its imbalance may lead to conditions of thrombosis or bleeding (Mine *et al.*, 2005). The vascular damage from injuries results in the activation of platelets and factors, which are involved in the coagulation cascade. This clotting machinery is essential in keeping the integrity of vascular system. Formation of fibrin occurs in several steps involving many molecules in succession. Limited proteolysis of many proenzymes involves the coagulation system activation (Yang *et al.*, 2000). Along with prevention of blood loss, fibrin also carries out numerous useful functions. These include binding of plasminogen and its activator, such as t-PA (tissue plasminogen activator) and form a complex which increases the conversion of plasminogen to plasmin (Tsurupa *et al.*, 2001). The accidental formation of fibrins or leftover fibrins after wound healing is degraded by inherent fibrinolytic system mediated by plasmin. A slight or acute variation from this route leads to conditions of thrombosis.

2.4.3 Fibrinolytic Enzymes

Fibrinolytic enzymes principally are proteases known to act on 'fibrin' protein and degrade it. The human fibrinolytic system comprises a few inherent molecules, such as t-PA, urokinase (Collen and Lijnen, 2004) and plasmin. These indirectly or directly are involved in fibrin degradation (Blasi and Sidenius, 2010). The main function of plasminogen activators is conversion of plasminogen (inactive) to plasmin (active), which further degrades fibrin. Streptokinase (SK) from *Streptococcus hemolyticus* (Arai *et al.*, 1995) or Staphylokinase from *Staphylococcus aureus* (Banerjee *et al.*, 2004) are a group of plasminogen activators (PAs) of bacterial origin. Currently, most of the plasminogen activators are under clinical usage for the treatment of thrombosis-related disorders. Even these are also associated with limitations, like antigenicity, bleeding complications, shorter half-life with respect to their biological activity after intravenous administration. In addition, complications associated with systemic activation of fibrinolytic system in blood, therapeutic doses and importantly their higher cost have limited their usage (Bode *et al.*, 1996; Killer *et al.*, 2010).

Inherent plasminogen activators, such as t-PA and urokinase, convert inactive plasminogen to active plasmin or fibrinolysin and result in degradation of fibrin clot (Cesarman-Maus and Hajjar, 2005). Currently, most of these molecules with the capability to activate plasminogen have been employed for clinical use in the treatment of thrombosis and related disorders and new sources have been explored for fibrinolytic enzymes with diverse and improved properties from microorganisms' various habitats.

The rising proportion of cardiovascular cases and fatalities due to thrombosis worldwide have raised concern in the researchers to identify substitute thrombolytic agents. Thus, it is essential to search for new sources of fibrinolytic enzymes and use of biotechnological approaches to develop the same for both prevention and management of heart diseases globally (Peng *et al.*, 2005; Mine *et al.*, 2005). Fibrinolytic agents have been identified from a variety of sources. Some of the snake venoms and plant proteases were extensively studied for their fibrinolytic and fibrinogenolytic properties (Rajesh *et al.*, 2007; Bernardes *et al.*, 2008) in a hope of their possible therapeutic applications and to reverse the effects of thrombosis. Similarly, several *in vitro* studies have been carried out for the characterisation of fibrinolytic enzymes from fungi (Kim *et al.*, 2008), algal sources (Matsubara *et al.*, 2000; Marsh and Williams, 2005; Costa *et al.*, 2010; Simkhada *et al.*, 2010), and bacteria (Wang *et al.*, 2008; Agrebi *et al.*, 2009). One of the major concerns with respect to fibrinolytic enzymes is their effectiveness and specificity towards fibrin. However, most of studied enzymes from various sources fail to answer this major limitation.

3. Functional Foods

Functional foods are those which have a significant beneficial effect on the health apart from providing basic nutrition to humans (Doyon and Labrecque, 2008). These are known to enhance the optimal health and also help in reducing the risk of diseases. Foods can be enriched with many components, such as enzymes, antimicrobial or antioxidant compounds, which are bioactive and contribute to their functionality. Many lactic acid bacteria (LAB) probiotics and bacteriocin-producing bacteria found in acidic fermented foods are known fermenters in the preparation of foods, such as *idli* and *dosa* (Devi and Halami, 2011, 2016, 2017). In case of foods, such as fermented soybean products and many alkaline fermented foods, *Bacillus* spp. are major fermenting bacteria and these foods are rich in natural bacterial enzymes which are capable of removing anti-nutritional components as well as undesirable blood clots when consumed and improve the health of an individual (Suzuki *et al.*, 2003; Mine *et al.*, 2005; Peng *et al.*, 2005). Hence, *Bacillus*-fermented soybean product is a functional food compared with normal soybean for thrombosis-related disorders. The research areas of functional foods and nutraceuticals are rapidly expanding worldwide. Many researchers are keenly working on the health benefits associated with foods, identifying and understanding their constituents which add to their functionality and mechanisms involved in physiological roles. These studies and findings contribute to a new nutritional paradigm, in which constituents of food go beyond their basic role as dietary essentials for sustaining life and growth, to one of preventing, managing, or delaying the premature commencement of chronic disease later in life (Fitzpatrick, 1999, 2000).

Fibrinolytic enzymes have been discovered from a variety of fermented foods and these have proven effective. They have been anticipated as strong fibrinolytic regimens. The food-grade microorganisms, with the potential to produce fibrinolytic enzymes, can be possibly applied in the food system which may add to its functionality. Hence, this can be used as functional food or as pharmaceutical preparation and applied for the prevention of thrombosis-related CVDs (Kim *et al.*, 1996; Hwang *et al.*, 2007), similar to *natto*.

3.1 Fermented Foods with Bacterial Fibrinolytic Enzymes

Along with the development of civilisation, fermented foods have been traditionally prepared and consumed for centuries. The characteristics of food are unique, depending on the climate and location. Raw materials for the preparation of fermented foods are obtained commonly from agricultural products available locally, such as cereals, legumes, vegetables, milk, fish, meat, and so on. Varieties of fermented foods have been known as the source for several fibrinolytic enzymes and have been the focus of research. Some of the well-studied fermented foods, as the source of fibrinolytic enzymes, include Japanese *natto* (Sumi *et al.*, 1987) and Korean *Chungkook-Jang* soy sauce (Kim *et al.*, 1996). Many fibrinolytic enzymes have been purified from these fermented foods and characterised for their properties. Greater diversity has been observed with respect to microorganisms and investigated for novel, better, effective and safer fibrinolytic enzymes.

Natto, a popular fermented food of Japan, prepared from boiled soybeans, has been consumed by oriental people for more than 1,000 years in Asia. The main fermenting microbe responsible in *natto* preparation is the *Bacillus subtilis natto*, an endospore-forming Gram-positive bacterium (formerly designated *Bacillus natto*). Nattokinase was first reported from '*natto*' and basically is an extracellular protease secreted by *B. subtilis* natto, which belongs to the family of alkaline serine protease. Nattokinase has been demonstrated to have greater thrombolytic activity to that of plasmin and increases the production of plasmin from plasminogen (Hsu *et al.*, 2009). It has been found to have proven health benefits, lower cost and safer than conventional thrombolytic agents (Sumi *et al.*, 1987, 1989; Fujita *et al.*, 1993;

Suzuki *et al.*, 2003; Peng *et al.*, 2005). With the popularity of nattokinase, various fibrinolytic enzymes from fermented foods, such as *kimchi, chunggukhang, tofu, skipjack shikara* (fermented fish), *douchi, meju,* fermented shrimp paste and many fermented foods from Asian countries have been explored (Mine *et al.*, 2005).

In fermented foods, *Bacillus* spp. are the major producers of these fibrinolytic enzymes among other microorganisms. Various *Bacillus* spp. have been subjected to extensive studies for different fibrinolytic enzymes and these enzymes were purified and characterised for their properties. The answer for the question why we need to consider *Bacillus* when several conventional thrombolytic agents are available is that in case of conventional thrombolytic agents (e.g. streptokinase, urokinase and tissue plasminogen activators, etc.) apart from their higher prices, these are found to have serious limitations with respect to the side effects (e.g. bleeding complications) (Bode *et al.*, 1996) and these problems have become the reasons for the search of better and safer alternatives.

Different fibrinolytic enzymes have been described from many species of bacteria which include *Streptococcus, Staphylococcus, Vibrio, Pseudomonas, Paenibacillus,* and *Chryseobacterium* (Kotb, 2013). However, members of genera *Bacillus* are the major groups which are well reported producers of fibrinolytic enzymes. History of health benefits of '*natto*' from centuries, a popular fermented soybean product and the presence of 'Nattokinase, the molecule of interest found in it and its promising outcome on safety evaluation directed the researchers to find many more fibrinolytic enzymes from various fermented foods (Mine *et al.*, 2005). Some of the popular fermented foods containing fibrinolytic enzymes from different sources have been enlisted in Table 2 and most of these enzymes are known to be serine and metallo-proteases in nature to act on fibrin.

3.2 *Bacillus* Associated in Indian Fermented Foods

Species of *Bacillus* are a less explored group among microorganisms with respect to fermented foods of Indian origin, with the exception of the northeastern region. Varieties of cereals and legumes are the staple food and important source of energy and nutrition in the world together. The types of cereal, legume and their products have regional importance with respect to cultivation and food habits (Sekar and Mariappan, 2007). Many cereals and legumes are the base for traditional foods and an essential part of the human diet since ages. With respect to Indian subcontinent, several of traditional fermented foods are popular, depending on geographical locations. Some of the fermented foods include *Maseura, Hawijar, Tungrymbai, Bekang, Kinema, Peruyaan, Tungtoh, Aakhone, Bemerthu, Idli, Dosa,* etc. (Soni *et al.*, 1986; Soni, 2007; Tamang *et al.*, 2012; Yogesh and Halami, 2015a). Generally, *Bacillus* species are soil bacteria; however, they have been found to be associated with a variety of foods, especially protein-rich foods. *Bacillus* spp. are known to produce a range of enzymes, including amylases, lipases and proteases. However, reports on these fermented foods as sources of fibrinolytic enzyme-producing microorganisms are rare (Mine *et al.*, 2005; Wang *et al.*, 2006).

Fibrinolytic enzymes in fermented foods impart a key attribute to the food and provide evidence with respect to the microorganism and its activity in foods.

Table 2: Reported Fibrinolytic Enzymes-producing Microorganisms Associated with Fermented Foods

Fermented Food	Key Ingredients	Fibrinolytic Enzyme- producing Microorganism associated	References
Natto	Soybean	*Bacillus subtilis natto*	Sumi *et al.* (1987)
Fermented *moong dal*	Moong dal	*B. circulans*CFR11	Yogesh and Halami (2015b)
Fermented shrimp paste	Shrimp	*Bacillus* sp.	Anh *et al.* (2015)
Vegetable cheese (*natto*)	Soybean	*Bacillus* sp.	Fujita *et al.* (1993)
Korean *Jeot gal*	Sea food/fish	*B. licheniformis* KJ-31	Hwang *et al.* (2007)
Meju	Soybean	*B. amyloliquefaciens* MJ5-41	Jo *et al.* (2011)
Cheonggukjang	Soybean	*B. subtilis* KCK-7	Paik *et al.* (2004)
Soybean grits	Soybean	*B. firmus* NA-1	Seo and Lee (2004)
Dosa batter	Rice, Urd dal, Methi	*B. subtilis* BR21	Yogesh and Halami (2015a)
Cheonggukjang	Soybean	*Bacilluis* sp. CK-11	Kim *et al.* (1996)
Fermented fish	Fish	*Bacillus* sp. KA 38	Kim *et al.* (1997)
Doenjang	Soybean	*Bacillus* sp.DJ-4	Kim and Choi (2000)
Tempeh	Soybean	*B. subtilis* TP-6	Kim *et al.* (2006)
Cheonggukjang	Soybean	*B. amyloliquefaciens* CH51	Kim *et al.* (2009)
Soybean paste	Soybean	*Bacillus* sp. KDO-13	Lee *et al.* (2001)
Dosa batter	Rice, Urd dal, Methi	*B. amyloliquefaciens* MCC2606	Yogesh *et al.* (2018)

Food product	Description	Substrate	Microorganism	Reference
Kishk	Egyptian fermented food	Parboiled wheat flour and milk	*B. megaterium* KSK07	Kotb (2015)
Fermented red bean		Red bean	*B. subtilis*	Chang *et al.* (2012)
Bekang, Hawaijar		Fermented Soya bean	*Bacillus* spp.	Yogesh (2016)
Fish sauce		Fish	*Virgibacillus halodenitrificans* SK 1-3-7	Montriwong *et al.* (2012)
Peru		Soybean	*Vagococcus lutrae*	Singh *et al.* (2014)
Tungrymbai		Soybean	*Vagococcus carniphilus*	Singh *et al.* (2014)
Peru		Soybean	*Pediococcus acidilactici*	Singh *et al.* (2014)
Perumb		Soybean	*Enterococcus gallinarum*	Singh *et al.* (2014)
Meso		Rice	*Aspergillus oryzae* KSK-3	Shirasaka *et al.* (2012)
Tempeh		Soybean	*Rhizopus* sp. and *Fusarium* sp	Sugimoto *et al.* (2007)
Starter for brewing rice wine, China		Rice	*Rhizopus chinensis* 12	Xiao-Lan *et al.* (2005)
Nagapi		Fish/shrimp	*B. subtilis*, *B. amyloliquefaciens*, *B. licheniformis*	Singh *et al.* (2014)
Hentak		Fish/shrimp	*Debaromyces fabryi*	Singh *et al.*, (2014)
Dang-pui-thu		Fish/shrimp	*Bacillus subtilis, Staphylococcus simulans, Streptomyces* sp.	Singh *et al.* (2014)

Fermented foods have been screened and established for the presence of fibrinolytic enzyme-producing microorganisms, such as fungi, LAB and *Bacillus* spp. (Peng *et al.*, 2005; Singh *et al.*, 2014; Yogesh, 2016; Yogesh *et al.*, 2018). Traditional fermented foods include those which are made at the domestic level and involve the simpler to sophisticated processes. The Indian subcontinent is rich in traditional fermented-food preparations practiced in each community and these in turn are dependent on availability of raw materials at particular geographical locations, seasons and dietary habits of people (Tamang *et al.*, 2012). These have paved way for the practice of preparing many traditional foods through fermentation. In India, most of the traditional food preparations pertain only at the household level and their popularity is making ground for commercial preparation of fermented food products using well-characterised fermenting microorganisms. Several of these fermented foods have been investigated from the scientific perspective for their health benefits.

In our study, several fibrinolytic enzyme-producing *Bacillus* spp. have been isolated from a variety of native fermented foods. These bacteria were identified and found to produce multiple proteases and have potent fibrin-hydrolysis ability assessed through fibrin plate and zymogram and specificity of the enzymes were studied (Yogesh and Halami, 2015a,b, 2017). The study was aimed at purifying and characterising the fibrinolytic enzymes from these strains. The bio-properties of these enzymes could provide useful indication or clues for their possible applications (Yogesh, 2016).

3.3 Nattokinase (EC 3.4.21.14) – A Well-studied Fibrinolytic Enzyme from *Bacillus* spp.

Large numbers of bacteria belonging to the genus *Bacillus* have been studied mainly for their proteases which possess fibrinolytic activity. The discovery of 'Nattokinase' and its beneficial attributes have drawn the attention of researchers to look for cheaper and orally administrable enzymes, which can act as thrombolytic agents (Sumi *et al.*, 1987; Fujita *et al.*, 1993). The enzyme 'Nattokinase' is basically a protease and belongs to the family of proteases known as subtilisins (Ghasemi *et al.*, 2012). Initially, Nattokinase was the only fibrinolytic enzyme from microorganisms whose *in vivo* thrombolytic effect has been best studied. The oral administration of Nattokinase, the fibrinolytic enzyme extracted from Japanese *natto*, was found to enhance fibrinolysis in experimentally-induced thrombosis in dogs (Sumi *et al.*, 1990) and angiography revealed the lysis of the thrombus (Kim *et al.*, 1996). More significantly, it was found that fibrinolytic activity, concentration of t-PA and products of fibrin degradation in the plasma doubled with the oral administration of Nattokinase to human subjects.

The mechanism of action with respect to fibrinolysis has been extensively studied for Nattokinase as compared to other microbial fibrinolytic enzymes. Nattokinase was found to get absorbed effectively across the intestinal tract in the rat model after intraduodenal administration and was found to induce fibrinolysis (Fujita *et al.*, 1995). The absorption enzyme across the intestinal tract led to the release of endogenous plasminogen activator which induced the fibrinolysis process in the occluded vessels (Sumi *et al.*, 1990). Nattokinase directly cleaves cross-linked fibrin and is also known to activate t-PA production, resulting in production of active plasmin from inactive plasminogen (Fujita *et al.*, 1995). In addition, Nattokinase enhances the process of fibrinolysis through cleaving and inactivation of PAI-1, a prime fibrinolysis inhibitor and regulates the total activity of fibrinolysis by its relative ratio with t-PA (Urano *et al.*, 2003).

In their study, Sumi et al. (1889) studied the efficiency of Nattokinase capsules in dissolving thrombi in dogs. The oral administration of four capsules of Nattokinase to the dogs with blood clots were induced experimentally in a major leg vein, indicating complete dissolution of clots (within five hours), and led to restoration of normal blood flow. However, in the case of a negative control, in which dogs that were administered a placebo, did not show any thrombolysis even after the duration of 18 hours (Sumi *et al.*, 1990). Moreover, it was also demonstrated that dietary supplementation of *natto* suppressed intimal thickening as well as modulated the lysis in rat models (Suzuki *et al.*, 2003b). In another study, thrombolytic effect of Nattokinase was examined on a thrombus with common carotid in the rat model. The endothelial cells of the blood vessel wall were subjected to injury by using acetic acid. Animals administered with Nattokinase were found to recover about 62 per cent of the blood flow, and those treated with plasmin regained about only 15.8 per cent, and those treated with elastases did not indicate any recovery. The conclusion of the study was that thrombolytic activity of Nattokinase *in vivo* was much potent as compared with plasmin or elastase (Fujita *et al.*, 1993).

A human trial was conducted that involved 12 healthy volunteers (six women and six men of age group between 21 and 55 years). Each individual was provided with 200 g of *natto* each day before breakfast and a series of blood plasma tests were carried out to assess their fibrinolytic activity. The results demonstrated that oral administration of *natto* improved the ability of participant individuals in dissolving blood clots. It was also found that the volunteers retained enhanced fibrinolytic activity post administration for about two to eight hours (Sumi *et al.*, 1989). Preference for fibrin over other protein substrates is one of the important properties of any thrombolytic agent to be effective. In our laboratory, *in vitro* studies using CFR15-protease produced from *B. amyloliqquefaciens* MCC2606 indicated that the enzyme was more fibrinolytic in nature than its preference toward other substrates. This unique property could be justified with the observed results that all the chains of fibrin were significantly degraded and in case of fibrinogen, γ-chain was resistant to degradative action of CFR15-protease (Yogesh *et al.*, 2018).

The presence and quantity of fibrinolytic enzymes in foods add to its functionality (Fujita *et al.*, 1993). This concept has led to the discovery and development of various enzymes with related as well as better kinds of fibrinolytic enzymes from diverse food systems (Mine *et al.*, 2005). In this present era, there is a high need for novel, cheaper, safer as well as superior fibrinolytic enzymes to address the alarming increase of CVD burden worldwide (WHO fact sheet, 2013). The various clusters of bacteria belonging to the genus *Bacillus* are a treasure for numerous potential and undiscovered fibrinolytic enzymes (Peng *et al.*, 2005; Venkatanagaraju and Divakar, 2015). In the present chapter, microbial fibrinolytic enzymes associated with fermented foods with their potential applications and future perspectives are discussed (Meruvu and Vangalapati, 2011).

3.4   Possible Application of Fibrinolytic Enzymes in Non-food Systems

Microorganisms produce a range of proteases and many have fibrinolytic activity. Several fibrinolytic enzymes have been studied and expressed in other host systems, such as *Bacillus* spp. and *Escherichia coli* (Xiao *et al.*, 2004; Kho *et al.*, 2005) for the ease of purification as well as better yield. Efforts were made to express genes of fibrinolytic enzymes in LAB system for their probable application as starter cultures (Liang *et al.*, 2007). Methods, like encapsulation of these enzymes in nano-capsules, can promise greater stability and oral applications. Attempts were made to encapsulate NKCP –a fibrinolytic enzyme preparation in Shellac by Law and Zhang (2006) and it helped in retention of about 60 per cent of the activity. Fibrinolytic enzyme from *cheonggukjang* (CGJ) was encapsulated by using alginate and this preparation had the stability for a broad range of pH and temperature compared with non-encapsulated form of enzyme (Ko *et al.*, 2008). Wei *et al.* (2012) have studied fibrinolytic enzyme Nattokinase from *Bacillus subtilis* LSSE-22. They employed chickpeas as substrate and used ethanol for the extraction and precipitation of the enzyme. methacrylic acid – Ethylacrylate was used for the encapsulation of the Nattokinase to enhance its stability at acidic pH.

Results of the experimental with respect to the effects of orally-administered Nattokinase on canine (Sumi *et al.*, 1990) and rat models (Suzuki *et al.*, 2003) along with the human trials (Sumi *et al.*, 1989; Omura *et al.*, 2004) signify their efficacy as well as safety. This has led to the commercial preparation of Nattokinase and some of the commercial Nattokinases are NSK-SD™, Cardiokinase™, Nattokinase NSK-SD, Natto-K, Orokinase, Best-Nattokinase, Nattozyme, Nattokinase-plus, Nattobiotic, Serracor-NK, etc. *Bacillus* probiotics are gaining grounds on par with LAB and with the ability of the *Bacillus* spp. to secrete fibrinolytic enzymes is another important property of the bacteria – this may add to its additional functionality. Hence, a *Bacillus* probiotic strain with the fibrinolytic enzyme-production ability is an additional beneficial attribute.

With the available information on fibrinolytic enzymes and looking towards the future perspective, there is a greater need for research to focus towards the exploration of novel, safer and better fibrinolytic enzymes. A lot of variations have been observed with respect to fibrinolytic enzymes from microorganisms. Much inclusive insight into sequences of both nucleotide and amino acid and the property of enzymes may add to the advancement in novel, safer and affordable thrombolytic agents, which can tackle the drawbacks of currently used conventional streptokinase, staphylokinase, tenecteplase, reteplase and other thrombolytics. The CVDs affect not only the well-being of an individual but also the quality of life as the cost of the treatment is high. Extensive studies on fibrinolytic enzymes may promise better molecules for preventing as well as strategies for the management of thrombosis-related diseases. The fibrinolytic enzymes from food-grade microorganisms can be used as functional food additives and may prove effective alternatives to currently-used conventional fibrinolytic agents.

4. Conclusion

Fibrin is a key protein responsible for both desirable and undesirable blood clots. The latter leads to thrombosis and related CVD conditions. Fibrinolytic agents are known to break down insoluble fibrin and dissolve blood clots; hence, they are used for treating thrombosis-related conditions, including pulmonary embolism and heart attack. Fermented foods are part of culture and civilisation. Fibrinolytic enzymes in fermented foods add to their functionality and help in improving cardiovascular health. Fibrinolytic enzymes from food-grade microorganisms will provide an adjunct to the expensive fibrinolytic enzymes that are currently in practice for the management of heart disease, as bulk quantities of enzymes can be conveniently and efficiently produced through biotechnological approaches. Additionally, these enzymes can be applied in the food system through food fortification or can be used as nutraceuticals, such that their application could prove effective in prevention of CVDs. Fermented foods, rich in fibrinolytic enzymes, may be used for therapy in thrombosis disorders and also can be used as strategic preventive measures for emerging heart diseases. New fibrinolytic enzyme needs to be purified and characterised with a thorough focus on its application as a better thrombolytic agent.

Acknowledgements

The authors thank Director, CSIR-CFTRI for the grant of major laboratory project on functional food formulation. YD acknowledges Head, Department of Studies in Molecular Biology, University of Mysore, Mysuru, for support and encouragement. The authors acknowledge kind approval of the Director, CSIR-CFTRI, Mysuru for publishing this work.

Abbreviations

AIS – Acute Ischemic Stroke
CVD – Cardio Vascular Disease
DALYs – Disability-adjusted life years
LMIC – Low- and middle-income countries
MI – Myocardial infarction
NCMH – National Commission on Macroeconomics and Health Report
SDS-PAGE – Sodium Dodecyl Sulphate – Polyacrylamide Gel Electrophoresis
SK – Streptokinase
t-PA – Tissue-type plasminogen activator
WHF – World Heart Federation
WHO – World Health Organisation

References

Agrebi, R., Haddar, A., Hajji, M., Frikha, F., Mani, L., Jellouli, K. and Nasri, M. (2009). Fibrinolytic enzymes from a newly-isolated marine bacterium *Bacillus subtilis* A26: Characterisation and statistical media optimisation. *Canadian Journal of Microbiology*, 55: 1049-1061.

American Heart Association Report (2015). (http://blog.heart.org/american-heart-association-statistical-report-tracks-global-figures-first-time/), accessed on 23rd August 2015.

Anh, D.B.Q., Mi, N.T.T., Huy, D.N.A. and Hung, P.V. (2015). Isolation and optimisation of growth condition of *Bacillus* sp. from fermented shrimp paste for high fibrinolytic enzyme production. *Arabian Journal for Science* and *Engineering*, 40: 23-28.

Antoniades, C., Tousoulis, D., Vasiliadou, C., Marinou, K., Tentolouris, C., Ntarladimas, I. and Stefanadis, C. (2004). Combined effects of smoking and hypercholesterolemia on inflammatory process, thrombosis/fibrinolysis system, and forearm hyperemic response. *The American Journal of Cardiology*, 94(9): 1181-1184.

Arai, K., Mimuro, J., Madoiwa, S., Matsuda, M., Sako, T. and Sakata, Y. (1995). Effect of staphylokinase concentration of plasminogen activation. *Biochimicae Biophysica Acta*, 1245(1): 69-75.

Banerjee, A., Chisti, Y. and Banerjee, U.C. (2004). Streptokinase – A clinically useful thrombolytic agent. *Biotechnology Advances*, 22: 287-307.

Bernardes, C.P., Santos-Filho, N.A., Costa, T.R., Gomes, M.S.R., Torres, F.S., Costa, J., Borges, M.H., Richardson, M., Dos-Santos, D.M., Pimenta, A.M.D.C., Homsi-Brandeburgo, M.I., Soares, A.M. and De-Oliveira, F. (2008). Isolation and structural characterisation of a new fibrino(geno)lytic metallo proteinase from *Bothropsmoojeni* snake venoms. *Toxicon*, 51: 574-584.

Blasi, F. and Sidenius, N. (2010). Theurokinase receptor: Focused cell surface proteolysis, cell adhesion and signalling. *FEBS Letters*, 584(9): 1923-1930.

Bode, C., Runge, M. and Smalling, R.W. (1996). The future of thrombolysis in the treatment of acute myocardial infarction. *European Heart Journal*, 17: 55-60.

Cesarman-Maus, G. and Hajjar, K.A. (2005). Molecular mechanisms of fibrinolysis. *British Journal of Haematology*, 129: 307-321.

Chan, M.Y., Andreotti, F. and Becker, R.C. (2008). Hypercoagulable states in cardiovascular disease. *Circulation*, 118: 2286-2297.

Chang, C.T., Wang, P.M., Hung, Y.F. and Chung, C.Y. (2012). Purification and biochemical properties of a fibrinolytic enzyme from *Bacillus subtilis*-fermented red bean. *Food Chemistry*, 133: 1611-1617.

Clark, R.A. (2001). Fibrin and wound healing. *Annals of the New York Academy of Science*, 936: 355-367.

Collen, D. and Lijnen, H.R. (2004). Tissue-type plasminogen activator: A historical perspective and personal account. *Journal of Thrombosis and Haemostasis*, 2: 541-546.

Costa, J.D.O., Fonseca, K.C., Garrote-Filho, M.S., Cunha, C.C., De-Freitas, M.V., Silva, H.S., Araújo, R.B., Penha-Silva, N. and De-Oliveira, F. (2010). Structural and functional comparison of proteolytic enzymes from plant latex and snake venoms. *Biochimie*, 92(12): 1760-1765.

Devi, M.S. and Halami, P.M. (2011). Detection and characterisation of pediocin PA-1/ACH like bacteriocin producing lactic acid bacteria. *Current Microbiology*, 63(2): 181-185.

Devi, S.M. and Halami, P.M. (2017). Probiotics from Fermented Foods. pp. 357-376. *In*: Kalia, V., Shouche, Y., Purohit, H., Rahi, P. (Eds.). *Mining of Microbial Wealth and Meta Genomics*. Springer, Singapore, ISBN 978-981-10-5707-6.

Doyon, M. and Labrecque, J.A. (2008). Functional foods: A conceptual definition. *British Food Journal*, 110: 1133-1149.

Duffy, M.J. (2002). Urokinase plasminogen activator and its inhibitor, pai-1, as prognostic markers in breast cancer: From pilot to level 1 evidence studies. *Clinical Chemistry*, 48(8): 1194-1197.

Fujita, M., Nomura, K., Hong, K., Ito, Y. and Asada, A. (1993). Purification and characterisation of a strong fibrinolytic enzyme (Nattokinase) in the vegetable cheese *natto*, a popular soybean fermented food in Japan. *Biochemistry and Biophysics Research Communication*, 197: 1340-1347.

Goldhaber, S.Z. and Bounameaux, H. (2001). Thrombolytic therapy in pulmonary embolism. *Seminars in Vascular Medicine*, 1(2): 213.

Gorkum, O.V., Veklich, Y.I., Medeved, L.V., Henschen, A.H. and Weisel, J.W. (1994). Role of the αC domain of fibrin in clot formation. *Biochemistry*, 33: 6986-6997.

Gupta, R., Misra, A., Pais, P., Rastogi, P. and Gupta, V.P. (2006). Correlation of regional cardiovascular disease mortality in India with lifestyle and nutritional factors. *International Journal of Cardiology*, 108(3): 291-300.

Herrick, S., Blanc, B.O., Gray, A. and Laurent, G. (1999). Fibrinogen. *International Journal of Biochemistry and Cell Biology*, 31: 741-746.

Hu, F.B. (2011). Globalisation of diabetes. *Diabetes Care*, 34(6): 1249-1257.

Hwang, K.J., Choi, K.H., Kim, M.J., Park, C.S. and Cha, J. (2007). Purification and characterisation of a new fibrinolytic enzyme of *Bacillus licheniformis* KJ-31 isolated from Korean traditional *jeot-gal*. *Journal of Microbiology and Biotechnology*, 17: 1469-1476.

Jo, H.D., Lee, H.A., Jeong, S.J. and Kim, J.H. (2011). Purification and characterisation of a major fibrinolytic enzyme from *Bacillus amyloliquefaciens* MJ5-41 isolated from Meju. *Journal of Microbiology and Biotechnology*, 21: 1166-1173.

Kho, C.W., Park, S.G., Cho, S., Lee, D.H. and Myung, P.K. (2005). Confirmation of VPR as a fibrinolytic enzyme present in extracellular proteins of *Bacillus subtilis*. *Protein Expression and Purification*, 39: 1-7.

Killer, M., Ladurner, G., Kunz, A.B. and Kraus, J. (2010). Current endovascular treatment of acute stroke and future aspects. *Drug Discovery Today*, 15: 640-647.

Kim, J.S., Kim, J.E., Choi, B.S., Park, S.E., Sapkota, K., Kim, S., Lee, H.-H., Kim, C.S., Park, Y., Kim, M.K., Kim, Y.S. and Kim, S.J. (2008). Purification and characterisation of fibrinolytic metalloprotease from *Perenniporia fraxinea* mycelia. *Mycological Research*, 112(8): 990-998.

Kim, S.H. and Choi, N.S. (2000). Purification and characterisation of subtilisin DJ-4 secreted by *Bacillus* sp. strain DJ-4 screened from Doen-Jang. *Bioscience, Biotechnology and Biochemistry*, 64: 1722-1725.

Kim, W., Choi, K. and Kim, Y. (1996). Purification and characterisation of a fibrinolytic enzyme produced from *Bacillus* sp. strain CK11-4 screened from Chungkook-Jang. *Applied and Environmental Microbiology*, 62: 2482-2488.

Ko, J.A., Koo, S.Y. and Park, H.J. (2008). Effects of alginate microencapsulation on the fibrinolytic activity of fermented soybean paste (Cheongguk-jang) extract. *Food Chemistry*, 111: 921-924.

Kotb, E. (2013). Activity assessment of microbial fibrinolytic enzymes. *Applied Microbiology and Biotechnology*, 97: 6647-6665.

Kotb, E. (2015). Purification and partial characterisation of serine fibrinolytic enzyme from *Bacillus megaterium* KSK-07 isolated from kishk, a traditional Egyptian fermented food. *Applied Biochemistry and Microbiology*, 51(1): 34-43.

Laurens, N., Koolwijk, P. and de Maat, M.P. (2006). Fibrin structure and wound healing. *Journal of Thrombosis and Haemostasis*, 4(5): 932-939.

Law, D. and Zhang, Z. (2006). Novel encapsulation of a fibrinolytic enzyme (Nattokinase) by Shellac. *XIVth International Workshop on Bioencapsulation*, CH. October 6-7, Lausanne, 1-4.

Lee, S.K., Bae, D.H., Kwon, T.J., Lee, S.B., Lee, H.H., Park, J.H., Heo, S. and Johnson, M.G. (2001). Purification and characterisation of a fibrinolytic enzyme from *Bacillus* sp. KDO-13 isolated from soybean paste. *Journal of Microbiology and Biotechnology*, 11(5): 845-852.

Lim, B.B., Lee, E.H., Sotomayor, M. and Schulten, K. (2008). Molecular basis of fibroin clot elasticity. *Structure*, 16: 449-459.

Liang, X., Zhang, L., Zhong, J and Huan, L. (2007). Secretory expression of a heterologous Nattokinase in *Lactococcus lactis*. *Applied Microbiology and Biotechnology*, 75: 95-101.

Lopez-Sendon, J., De Lopez, S.E., Bobadilla, J.F., Rubio, R., Bermejo, J. and Delcan, J.L. (1995). The efficacy of different thrombolytic drugs in the treatment of acute myocardial infarction. *Revista Española de Cardiología*, 48: 407-439.

Marsh, N. and Williams, V. (2005). Practical applications of snake venom toxins in haemostasis, *Toxicon*, 45(8): 1171-1181.

Matsubara, K., Hori, K., Matsuura, Y and Miyazawa, K. (2000). Purification and characterisation of a fibrinolytic enzyme and identification of fibrinogen-clotting enzyme in a marine green alga, *Codium divaricatum*. *Comparative Biochemistry and Physiology*, Part B: *Biochemistry and Molecular Biology*, 125(1): 137-143.

Meruvu, H. and Vangalapati, M. (2011). Nattokinase: A review on fibrinolytic enzyme. *International Journal of Chemical, Environmental and Pharmaceutical Research*, 2(1): 61-66.

Montriwong, A., Kaewphuak, S., Rodtong, S., Roytrakul, S. and Yongsawatdigul, J. (2012). Novel fibrinolytic enzymes from *Virgibacillus halodenitrificans* SK1-3-7 isolated from fish sauce fermentation. *Process Biochemistry*, 47(12): 2379-2387.

Mosesson, M.W. (2005). Fibrinogen and fibrin structure and functions. *Journal of Thrombosis and Haemostasis*, 3: 1894-1904.

Mosesson, M.W., Siebenlist, K.R., Hernandez, I., Wall, J.S. and Hainfeld, J.F. (2002). Fibrinogen assembly and crosslinking on a fibrin fragment E template. *Thrombosis and Haemostasis*, 87: 651-658.

Mine, Y., Wong, A.H.K. and Jiang, B. (2005). Fibrinolytic enzymes in Asian traditional fermented foods. *Food Research International*, 38: 243-250.

Nakajima, N., Mihara, H. and Sumi, H. (1993). Characterisation of potent fibrinolytic enzymes in earthworm, *Lumbricus rubellus*. *Bioscience, Biotechnology and Biochemistry*, 57(10): 1726-1730.

NCMH Report (2005). National Commission on Macroeconomics and Health, Estimations and Causal Analysis, Burden of Diseases. Background Paper in India, 1-5.

Omura, K., Kaketani, K., Maeda, H. and Hitosugi, M. (2004). Fibrinolytic and anti-thrombotic effect of the protein from *Bacillus subtilis (natto)* by the oral administration. *Journal of Japanese Society of Biorheology*, 18: 44-51.

Paik, H.D., Lee, S.K., Heo, S., Kim, S.Y., Lee, H. and Kwon, T.J. (2004). Purification and characterisation of the fibrinolytic enzyme produced by *Bacillus subtilis* KCK-7 from *Chungkook-jang*. *Journal of Microbiology and Biotechnology*, 14(4): 829-835.

Pappachan, M.J. (2011). Increasing prevalence of life style diseases: High time for action. *Indian Journal of Medical Research*, 134: 143-145.

Peng, Y., Yang, X.J. and Zhang, Y.Z. (2005). Microbial fibrinolytic enzymes: An overview of source, production, properties and thrombolytic activity *in vivo*. *Applied Microbiology and Biotechnology*, 69: 126-132.

Pomp, E.R., Rosendaal, F.R. and Doggen, C.J.M. (2008). Smoking increases the risk of venous thrombosis and acts synergistically with oral contraceptive use. *American Journal of Hematology*, 83(2): 97-102.

Rajesh, R., Shivaprasad, H.V., Raghavendragowda, C.D., Nataraju, A., Dhananjaya, B.L. and Vishwanath, B. (2007). Comparative study on plant latex proteases and their involvement in hemostasis: A special emphasis on clot inducing and dissolving properties. *Planta Medica*, 73(10): 1061-1067.

Schulman, S., Beyth, R.J., Kearon, C. and Levine, M.N. (2008). Hemorrhagic complications of anti-coagulant and thrombolytic treatment: American College of Chest Physicians evidence-based clinical practice guidelines. Anti-thrombotic and Thrombolytic Therapy, 8th ed., *ACCP Guidelines, Chest*, 133(6): 257S-298S.

Sekar, S. and Mariappan, S. (2007). Usage of traditional fermented products by Indian rural folks and IPR. *Indian Journal of Traditional Knowledge*, 6(1): 111-120.

Seo, J.H. and Lee, S.P. (2004). Production of fibrinolytic enzyme (KK) from soybean grits fermented by *Bacillus firmus* NA-1. *Journal of Medicinal Food*, 7(4): 442-449.

Simkhada, J.R., Mander, P., Cho, S.S. and Yoo, J.C. (2010). A novel fibrinolytic protease from *Streptomyces* sp. CS684. *Process Biochemistry*, 45: 88-93.

Singh, T.A., Devi, K.R., Ahmed, G. and Jeyaram, K. (2014). Microbial and endogenous origin of fibrinolytic activity in traditional fermented foods of northeast India. *Food Research International*, 55: 356-362.

Shirasaka, N., Naitou, M., Okamura, K., Kusuda, M., Fukuta, Y. and Terashita, T. (2012). Purification and characterisation of a fibrinolytic protease from *Aspergillus oryzae* KSK-3, *Mycoscience*. doi: 10.1007/s10267-011-0179-3

Sugimoto, S., Fujii, T., Morimiya, T., Johdo, O. and Nakamura, T. (2007). The fibrinolytic activity of a novel protease derived from a *tempeh* producing fungus, *Fusarium* sp. BLB. *Bioscience. Biotechnology and Biochemistry*, 71: 2184.

Soni, S.K. (2007). *Microbes: A Source of Energy for 21st Century*. New India Publishing Agency, New Delhi.

Soni, S., Sandhu, D., Vikhu, K. and Kamra, K. (1986). Microbiological studies on *dosa* fermentation. *Food Microbiology*, 3(1): 45-53.

Sumi, H., Hamada, H., Tsushima, H., Mihara, H. and Muraki, H. (1987). A novel fibrinolytic enzyme (nattokinase) in the vegetable cheese *Natto*: A typical and popular soybean food in the Japanese diet. *Experientia*, 43(10): 1110-1111.

Sumi, H., Hamada, H., Mihara, H., Nakanishi, K. and Hiratani, H. (1989). Fibrinolytic effect of the Japanese traditional food *natto* (Nattokinase). *Thrombosis and Haemostasis*, 62(1): 549.

Suzuki, Y., Kondo, K., Matsumoto, Y., Zhao, B.Q., Otsuguro, K., Maeda, T., Tsukamoto, Y., Urano, T. and Umemura, K. (2003). Dietary supplementation of fermented soybean, *natto*, suppresses intimal thickening and modulates the lysis of mural thrombi after endothelial injury in rat femoral artery. *Life Science*, 73: 1289-1298.

Tamang, J.P., Tamang, N., Thapa, S., Dewan, S., Tamang, B., Yonzan, H., Rai, A.M., Chettri, R., Chakrabarty, J. and Kharel, N. (2012). Microorganism and nutritional value of ethnic fermented foods and alcoholic beverages of north India. *Indian Journal of Traditional Knowledge*, 1: 7-25.

Tough, J. (2005). Thrombolytic therapy in acute myocardial infarction. *Nursing Standard*, 19(37): 55-64.

Tsurupa, G. and Medved, L. (2001). Identification and characterisation of novel tPA- and plasminogen-binding sites within fibrin (ogen) alpha C-domains. *Biochemistry*, 40: 801-808.

Undas, A. and Ariens, R.A.S. (2011). Fibrin clot structure and function, Arolein the pathophysiology of arterial and venous thromboembolic diseases. *Arteriosclerosis Thrombosis and Vascular Biology*, 31: e88-e99.

Venkatanagaraju, E. and Divakar, G. (2013). *Bacillus cereus* GD 55 strain improvement by physical and chemical mutagenesis for enhanced production of fibrinolytic protease. *International Journal of Pharmaceutical Sciences and Research*, 4(5): 81-93.

Wang, C., Ji, B., Li, B. and Ji, H. (2006). Enzymatic properties and identification of a fibrinolytic serine protease purified from *Bacillus subtilis* DC33. *World Journal of Microbiology and Biotechnology*, 22: 1365-1371.

Wang, S.H., Zhang, C., Yang, Y.L., Diao, M. and Bai, M.F. (2008). Screening of high fibrinolytic enzyme producing strain and characterisation of fibrinolytic enzyme produced from *B. subtilis* LD-8547. *World Journal of Microbiology and Biotechnology*, 24: 475-482.

Wei, X., Luo, M., Xie, Y., Yang, L., Li, H., Xu, L. and Liu, H. (2012). Strain screening, fermentation, separation, and encapsulation for production of *nattokinase* functional food. *Applied Biochemistry and Biotechnology*, 168(7): 1753-1764.

Weitz, J.I. (2006). Goodman and Gilman's: The Pharmacological Basis of Therapeutics. pp. 1325-1328. *In*: Blood Coagulation and Anti-coagulant, Thrombolytic and Anti-platelet Drugs, 11th ed.

World Health Statistics (2011). Geneva: Global Status Report on Non-communicable Diseases, 2010. http://www.who.int/ mediacentre/ factsheetts/fs317/en/- on 26 April 2014

World Heart Federation (2012). *Urbanisation and Cardiovascular Disease: Raising Heart-Healthy Children in Today's Cities*. Geneva, Switzerland.

World Health Organisation (2015). Fact Sheet on CVDs. http://www.wh o.int/cardiovascular_ diseases/en/- on 29 October

Wu, S., Castellino, F.J. and Wong, S. (2003). A fast-acting, modular structured Staphylokinase fusion with kringle-1 from human plasminogen as the fibrin-targeting domain offers improved clot lysis efficacy. *Journal of Biological Chemistry*, 278: 18199-18206.

Xiao, L., Zhang, R.H., Peng, Y. and Zhang, Y.Z. (2004). Highly efficient gene expression of a fibrinolytic enzyme (subtilisin DFE) in *Bacillus subtilis* mediated by the promoter of α-amylase gene from *Bacillus amyloliquefaciens*. *Biotechnology Letters*, 26: 1365-1369.

Yang, Z., Mochalkin, I. and Doolittle, R.F. (2000). A model of fibrin formation based on crystal structures of fibrinogen and fibrin fragments complexed with synthetic peptides. *Proceedings of National Academy of Science* (USA), 97(26): 14156-14161.

Yin, L.J., Lin, H.H. and Jiang, S.T. (2010). Bio-properties of potent Nattokinase from *Bacillus subtilis* YJ. *Journal of Agricultural and Food Chemistry*, 58: 5737-5742.

Yogesh, D. and Halami, P.M. (2015a). Evidence that multiple proteases of *Bacillus subtilis* can degrade fibrin and fibrinogen. *International Food Research Journal*, 22(4): 1662-1667.

Yogesh, D. and Halami, P.M. (2015b). A fibrin degrading serine-metalloprotease of *Bacillus circulans* with α-chain specificity. *Food Bioscience*, 11: 72-78.

Yogesh, D., Savitha Kumari, R. and Halami, P.M. (2018). Purification and characterisation of fibrinolytic protease from *Bacillus amyloliquefaciens* MCC2606 and analysis of fibrin degradation product by MS/MS. *Preparative Biochemistry and Biotechnology*, 48: 172-180.

Yogesh, D. and Halami, P.M. (2017). Fibrinolytic enzymes of *Bacillus* spp.: An overview. *International Food Research Journal*, 24(1): 35-47.

Yogesh, D. (2016). Characterisation of fibrinolytic enzyme of *Bacillus* species isolated from food source. Ph.D. thesis. University of Mysore, Mysuru, India.

Production of Antihypertensive Peptides during Food Fermentation

Brij Pal Singh[1]*, Harsh Panwar[2], Bharat Bhushan[3] and Vijay Kumar[1]

[1] Department of Microbiology, School of Science, RK University, Kasturbadham, Rajkot - 360020, Gujarat, India
[2] Department of Dairy Microbiology, College of Dairy Science and Technology, Guru Angad Dev Veterinary and Animal Sciences University, Ludhiana - 141004, Punjab, India
[3] Department of Basic and Applied Sciences, National Institute of Food Technology, Entrepreneurship and Management, Kundli, Sonepat - 131028, Haryana, India

1. Introduction

A basic understanding of the intriguing relationship between food and health has now been recognised as a key perspective to preventing diseases and attaining a better well-being. Being a source of amino acids, proteins hold an indispensable position in the food system and are a major nutrient for humans. Besides the nutritive value, many of the proteins that occur naturally in raw food material exert their physiological action either directly or upon hydrolysis *in vitro* or *in vivo*. Several proteins in the diet have specific biological functions, making them potential ingredients for functional foods (Korhonen *et al*., 1998). Bioactive peptides are latent within the sequence of proteins and can be released upon fermentation, enzymatic hydrolysis and food processing. These peptides exercise several biological functions, including antihypertensive, antioxidant, antithrombotic, antiadipogenic, antimicrobial, anti-inflammatory and immunomodulatory effects. These peptides generally have 2-20 amino acids and molecular masses of less than 6000 Da. Low molecular weight of peptides have greater ability to cross the intestinal barrier and get absorbed by the intestinal wall in order to exhibit their biofunctionality (Zhang *et al*., 2010; de Castro and Sato, 2015; Ahn *et al*., 2015).

Fermentation is the traditional method of preservation and value addition to various types of foods. It has long been used to change the physical and chemical nature of foods and make all ingredients more suitable for bioavailability in our system. Fermentation has now been recognised as an effective method for producing

*Corresponding author: bpsingh03@gmail.com

bioactive peptides as well as imparting extra benefit to foods other than their simple nutritive values. Microbial fermentation generates several biofunctional components with potential health benefits. Proteolytic microorganisms degrade proteins present in food during fermentation and release bioactive peptides, which execute many health-enhancing activities in the human body (Singh *et al.*, 2014; Singh and Vij, 2017). In Southeast Asian countries, such as China and Korea and in Japan in the Far East, food fermentation has long been used as a technique to preserve and improve the value of the food by producing acid and bioactive peptides, respectively (Sarmadia and Ismail, 2010). Milk and milk products are a keenly studied source of bioactive peptides, but many other food sources, such as egg, fish, oyster, rice, wheat, buckwheat, barley, and corn, soybean and radish seeds are now being explored as potent sources of bioactive peptides.

Hypertension is a major global health issue which elevates the risk of a large world population to chronic life-threatening diseases. More than 17 million people dying every year due to cardiovascular disease (CVD) worldwide and hypertension is one of the main causes for the development of CVD (WHO, 2003; Balti *et al.*, 2010; Chen *et al.*, 2014). The development of hypertension is influenced by many factors, such as genetic disposition, aging, overweight, lifestyle and nutrition. Angiotensin converting enzyme (ACE) is a non-specific dipeptidyl carboxypeptidase involved in the regulation of blood pressure by modulating the rennin-angiotensin system (RAS). ACE also catalyses degradation of bradykinin, which is a blood-pressure-lowering nonapeptide in the kallikrein–kinin system (Zhang *et al.*, 2006; Cat and Touyz, 2011). The inhibition of ACE is an effective target to manage essential hypertension. Antihypertensive peptides (AHPs) can block the interactions between angiotensin II (vasoconstrictor) and angiotensin receptors, which can directly contribute to reduced blood pressure (Aluko, 2015).

Recently focus on identifying dietary compounds instead of therapeutic drugs has increased in order to prevent hypertension. Bioactive peptides are one of the health-promoting components and are considered to promoting diverse activities, by modulating or improving several physiological functions (Erdmann *et al.*, 2008). Numerous studies have revealed antihypertensive effects resulting from fermented products containing bioactive peptides (Seppo *et al.*, 2003). Many bioactive peptides are multifunctional: they play regulatory roles as well as directly influence various developmental and metabolic processes. These peptides may have structural characteristics that allow them to interact with endogenous peptides and act as neurotransmitters, hormones and regulators (Hernandez-Ledesma *et al.*, 2014; de Castro and Sato, 2015). In contrast to synthetic substances, peptides are degraded into their basic amino acid units without leading to toxic metabolites (Fitzgerald and Murray, 2006). In light of the aforesaid benefits and intricate connection with human health, AHPs have been a subject of interest for decades. This chapter is an effort to catalogue the advancements in the production of AHPs through fermentation and to talk about mechanisms that are believed to be involved in antihypertensive activity.

2. Mechanism of Action of AHPs

The renin-angiotensin system (RAS) is an important hormone system that regulates blood pressure and fluid balance in the body. In this system, prorenin (a precursor of rennin) is converted into renin by the juxtaglomerular cells of the kidneys, when renal blood flow is reduced. Consequently, liver-derived angiotensinogen will be converted into angiotensin I by renin. Angiotensin I is subsequently converted to angiotensin II by the ACE found on the surface of vascular endothelial cells of the lungs. Angiotensin II is a potent vasoconstrictive that can increase blood pressure by constricting the blood vessels. It also induces the secretion of aldosterone, which causes increase in the reabsorption of sodium and water and excretion of potassium into the blood (Solomon *et al.*, 2005; Paul *et al.*, 2006; Yee *et al.*, 2010) (Fig. 1). This increases the volume of extracellular fluid in the body, thereby elevating blood pressure. If the RAS is abnormally active, blood pressure will be excessively high and will create the condition of high blood pressure, heart failure and many other clinical disorders (Jakubczyk *et al.*, 2013). Thus, the inhibition of ACE will cause a lowering in blood pressure through a vasodilator response. There are many drugs that are used to inhibit ACE and to lower blood pressure, i.e. captopril, enalaptril or ramipril. These synthetic pharmacological drugs cause several adverse effects on human beings like hypokalemia, joint pain, faintness, dry cough, skin rashes and angioneurotic edema. Therefore, scientific interest has increased in the search for natural antihypertensive

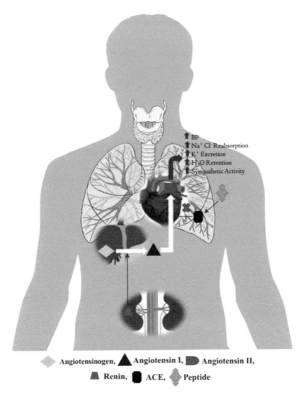

Fig. 1: Role of angiotensin converting enzyme (ACE) in blood pressure regulation via renin-angiotensin system (RAS). Inhibition of ACE by bioactive peptides lowers the blood pressure through a vasodilator response

compounds, such as peptides in food for use in the prevention and treatment of hypertension (Rai *et al.*, 2017; Daliri *et al.*, 2018; Chen *et al.*, 2018). Food-derived ACE-inhibitory peptides have obvious advantages over synthetic peptides. The first ACE-inhibitory peptide was reported long back from snake venom, but due to their lack of oral activity, they had partial pharmacological submissions (Ferreira *et al.*, 1970). The exact structural requirements of ACE-inhibitory peptides are the presence of hydrophobic amino acids, such as Trp, Tyr, Phe and Pro. These were found more favourable to interact with the active centre of ACE. In addition, the position of amino acid residues at C or N terminal region is of great importance in influencing the inhibitory action of ACE (Ondentti and Cushman, 1982).

In vitro methods are most common for determination of ACE-inhibitory activity of peptides in fermented food products. Commonly, the Cushman and Cheung (1971) spectrophotometric method is most widely used for assessment of ACE-inhibitory activity. This method uses HHL (hippuryl-His-Leu) as a substrate that can be hydrolyzsd by ACE into hippuric acid. Therefore, ACE-inhibitory peptides ensure that this reaction does not take place. This activity may be expressed as a percentage of ACE inhibition or as the minimum concentration of protein to inhibit 50 per cent of the enzymatic activity. Furthermore, Holmquist *et al.* (1979) developed a new improved assay, which is based on the shift of the absorption spectrum upon hydrolysis of a furanacryloyl tripeptide (FA-Phe-Gly-Gly, FAPGG) by ACE, leading to the production of corresponding amino acids (FA-Phe, FAP) and dipeptide (Gly-Gly, GG). In their study, Vermeirssen *et al.* (2002) further increased the sensitivity of the assay by adding an inhibitory compound in the reaction mixture by using the appropriate enzyme–substrate ratio and rabbit-lung-acetone extract as ACE.

Spontaneously hypertensive rats (SHR) are used as animal models to evaluate the antihypertensive effect of peptides (Beltrán-Barrientos *et al.*, 2016). Several human trials have been carried out to evaluate the antihypertensive effect of fermentation-generated peptides. For instance, a placebo-controlled study reported antihypertensive effect of sour milk in hypertensive patients (Hata *et al.*, 1996). Similarly, Aihara and co-workers (2005) found that daily ingestion of tablets containing *Lactobacillus helveticus*-fermented milk reduced the elevated blood pressure in hypertensive subjects. In a meta-analysis (12 controlled trials) of a total of 623 participants showed that VPP and IPP (milk-derived tripeptides) have hypotensive effect in hypertensive and pre-hypertensive subjects (Xu *et al.*, 2008).

2.1 Sources of AHPs

All food materials, especially those consisting of high capacity of protein have been exploited as promising sources of bioactive peptides. Milk and other dairy products are among the best precursors of bioactive peptides and have been extensively studied (Floris *et al.*, 2003). Bioactive peptides are encrypted in both major milk proteins, i.e. casein and whey proteins (α-lactalbumin, β-lactoglobulin, lactoferrin and immunoglobulins) (Belem *et al.*, 1999; Gobbetti *et al.*, 2002). Several studies indicated *in vitro* as well as *in vivo* generation of bioactive peptides from milk proteins. Bioactive peptides can be released during the manufacture of milk products, such as hydrolysed milk proteins used in baby formulas, clinical diets and as food ingredients consisting of peptides. In addition, proteolysis during milk fermentation and cheese ripening leads to the formation of various biofunctional peptides, including AHPs (Marshall *et al.*, 1994; Ashar *et al.*, 2004).

Besides milk proteins, bioactive peptides have also been isolated and characterised from other food protein sources, including egg, fish, meat products and marine waste or by-products (Li *et al.*, 2002; Yoshikawa *et al.*, 2003). Several bioactive peptides are also found in egg ovalbumin, chicken meat and fish products and comprise ACE-inhibitory and antioxidative properties (Yamamoto *et al.*, 2003). Insects, such as moth, fruit fly, ant, beetle and spider have been reported as a promising source of bioactive peptides. Royal jelly (RJ), a bee product rich in proteins, has also been found as a good source of ACE-inhibitory peptides. Marine processing wastes, such as substandard muscles, viscera, heads, skins, fins, frames, trimmings, shellfish and crustacean-shell waste have been demonstrated to contain high quality protein that could act as substrate for generation of novel bioactive peptides (Harnedy *et al.*, 2012). In fact, these by-products are relatively cheap and their utilisation in the production of bioactive peptides will not entirely cut the production cost but provide an effective waste disposal.

Soy, wheat, maize, rice, barley, buckwheat and sunflower are all plant sources of bioactive peptides. Based on a bioactive peptide sequence database analysis, rice prolamin is believed to be one of the best precursors of AHPs (Dziuba *et al.*, 2003). Immunomodulatory peptides derived from tryptic hydrolysates of rice and soybean proteins stimulate superoxide anions (reactive oxygen species – ROS), which triggers non-specific immune defence systems (Kitts and Weiler, 2003). Soybeans, an excellent source of dietary peptides, have antihypertensive, anticholesterol, and antioxidant activities and appear to prevent cancer (De Mejia and Ben, 2006; Singh and Vij, 2017; Singh and Vij, 2018). The search for naturally occurring bioactive peptides present in foods could be seen as potential nutraceutical due to the growing interest in their additional benefits beyond their nutritional values. Therefore, continuous efforts should be made into research and development of alternative sources of bioactive peptides, especially for those underutilised biological resources.

2.2   Production of AHPs

Although enzymatic hydrolysis is the primary method employed for obtaining pharmaceutical and research-grade bioactive peptides, we focus herein on generation of bioactive peptides during microbial fermentation. Recent studies indicate that fermentation is an efficient way to produce bioactive peptides and food-grade hydrolysed proteins (Singh *et al.*, 2014). Lactic acid bacteria (LAB), a large heterogeneous group of health-beneficial bacteria, majorly found in our digestive system and used in food fermentation since ancient eras, has now been explored for the production of bioactive peptides. Along with texture and flavour enhancement, LAB (lactic acid bacteria) also has a major role to play in the improvement of functional properties of fermented products (Singh and Vij, 2017).

The proteolytic system of LAB is complex and consists of three major components: (i) proteases bound to the cell wall that promote the initial hydrolysis of precursor protein into oligopeptides, (ii) specific transporters that transfer the oligopeptides to the cytoplasm, and (iii) intracellular peptidases that finish the hydrolysis process to convert oligopeptides into free amino acids or low molecular-weight peptides (Chaves-Lopez *et al.*, 2014). The ability of these microorganisms to produce proteolytic enzymes makes them potential producers of bioactive peptides. The proteolytic system of LAB has proteinases of broad specificity and is capable of releasing a large number of different oligopeptides. Oligopeptide transport systems are the main route of nitrogen entry into the cell and intracellular peptidases are required for complete degradation of accumulated peptides (Kunji *et al.*, 1996; Christensen *et al.*, 1999). Microbial fermentation is the cheapest process over enzymatic hydrolysis for bioactive peptide production because microorganisms are an inexpensive source of proteases and recognised as safe. Bacterial cultivation costs are comparatively low as they require minimal nutrition and short time of maturation. Moreover, proteases of LAB are expressed on the cell membrane, making purification protocols relatively easy and cheap (Agyei *et al.*, 2011).

Downstream processing of fermented hydrolysates to obtain purified peptides can be executed through a number of approaches. Chromatography techniques, including HPLC, ion exchange chromatography, reverse phase-HPLC, capillary liquids chromatography and a combination of the above methods are widely used to achieve fraction rich in bioactive peptides (Shahidi and Zhong, 2008; Pownall *et al.*, 2010). Membrane-based separation technology, microfiltration, ultrafiltration, nanofiltration and reverse osmosis are some other promising methods used for peptide separation through a porous membrane under applied pressure. Low-molecular-mass membrane cut-off is a common tool used for concentrating bioactive peptides (Singh *et al.*, 2014).

Moreover, several methods are available for identification of peptides, such as peptide/protein mass fingerprinting (PMF), peptide de novo sequencing and mass spectrometry (MALDI-TOF or ESI-TOF). In recent times, mass spectrometry is acknowledged as undisputed tool and the most sensitive technique to analyse peptide mixtures upon their proteolytic degradation. Two commonly used ionisation techniques, namely electrospray ionization (ESI) and matrix-assisted laser desorption/ionisation (MALDI), are well known as they offer very little fragmentation for measuring molecular mass. The mass of the resulting mixture of peptides is then compared to a database containing known proteins. A huge number of bioactive peptide databases, such as AHTPDB, Biopep, PeptideDB, APD and Peptidome are available online and have been projected to store information, including source, molecular weight, natural process, sequence, EC_{50}, reference to published work and more (Mamone *et al.*, 2009; Kumar *et al.*, 2014).

2.3   Current Status of AHPs

LABs are known to produce inhibitors of ACE in various amounts during milk fermentation. They hydrolyse milk proteins, mainly casein, into peptides, which can be used as nitrogen sources necessary for growth of these bacteria (FitzGerald and Murray, 2006). Several studies indicate that proteolytic system of *Lactobacillus helveticus, Lb. delbrueckii* ssp. *bulgaricus, Lactococcus lactis* ssp. *diacetylactis, Lc. lactis* ssp. *cremoris,* and *Streptococcus (Str.) salivarius* ssp. *thermophilu*s strains have the ability to hydrolyse milk proteins and release bioactive peptides with ACE-inhibitory activity (Table 1). On the same line, Tsai and co-workers (2008) identified ACE-inhibitory peptide (Tyr-Pro-Tyr-Tyr) from whey fraction of fermented milk and found blood pressure reduction in spontaneously hypertensive rats after oral administration of peptides. *Lb. helveticus* has been widely used as a potential dairy starter in cheese manufacturing and highly proteolytic strains producing bioactive peptides, ACE-inhibitory peptides in particular (Korhonen and Pihlanto, 2006). In two separate studies, *Lc. lactis* ssp. *cremoris* and *Lb. delbrueckii* ssp. *bulgaricus* were reported to produce ACE-inhibitory peptides from yoghurt-type fermented products (Gobbetti *et al.*, 2000). ACE-inhibitory peptides have also been identified in fermented donkey milk (Aspri *et al.*, 2018). A fermentation study done on caseins and cheese whey using commercial starter cultures has found that post-fermentative pepsin and trypsin digestion could specifically release ACE-inhibitory peptides from alpha-1-casein and beta-casein (Pihlanto-Leppala *et al.*, 1998). In another study, *Kluyveromyces marxianus* fermentation of bovine lactoferrin was found to produce six unique AHPs. Four out of six chemically-synthesised peptides (PYKLRP, DPYKLRP, GILRP and YKLRP) confirmed ACE-inhibitory effects in spontaneously hypertensive rats (Garcia-Tejedor *et al.*, 2014).

Two hypotensive tripeptides, VPP and IPP, produced by combined fermentation of *Lb. helveticus* and *Saccharomyces cerevisiae,* have been successfully commercialised in Japan in the form of brand name Calpis (Nakamura *et al.*, 1995). In a study, the levels of ACE-inhibitory peptides (VPP and IPP) increased during fermentation (by *Aspergillus oryzae)* of casein miso paste and consequently significant antihypertensive effects were seen in spontaneously hypertensive rats (Inoue *et al.*, 2009). Apart from LAB, *Bacillus subtilis* strain has also been used for the production of ACE-inhibitory peptides (DGVVYY) using tomato seeds by-products as a substrate (Moayedi *et al.*, 2018). Two ACE-inhibitory peptides (VQTL and LGYEN) were identified in walnut extract fermented by *B. subtilis* strain (Zheng *et al.*, 2017). Four ACE inhibitory peptides – IPP, LPP, VPP and KP – have been identified in cucumber fermented by LAB naturally present on the cucumbers (Fideler *et al.*, 2019). Similarly, *Pediococcus pentosaceus* strain has been used in a study to ferment soybean protein isolates and production of four potent ACE-inhibitory peptides, i.e. EDEVSFSP, SRPFNL, RSPFNL and ENPFNL (Daliri *et al.*, 2018). Three peptides – VAPFPEVFGK, LVYPFPGPLH, and FVAPEPFVFGKEK – have been identified in kombucha-fermented milk with high potency against ACE (Elkhtab *et al.*, 2017). In another study, Rho and co-workers (2009) observed that accelerating the process of fermented soybean extract at higher temperature (45°C) was beneficial for obtaining an effective ACE-inhibitory peptide from soybean. *Douchi*, a soybean product fermented by *Aspergillus egyptiacu*, also consists of ACE-inhibitory peptide, Phe-Ile-Gly (Zhang *et al.*, 2006).

Besides single starter, mixed fermentation using two or more strains/species/genus have also been performed for the development of bioactive peptides. In a similar study, pure strains of food grade *Lb. plantarum* LAT03, *Lb. plantarum* KLAT1, *Enterococcus faecalis* KE06, *Galactomyces geotrichum* KL20B, *Kluyveromyces marxianus* KL26A, *Pichia kudriavzevii* KL84A and *Torulaspora delbrueckii* KL66A were used in different combinations for milk fermentation and

Table 1: Antihypertensive Peptides (AHPs) Derived from Fermented Foods

Peptides Sequences*	Sources	References
MPFPKYPVQPF	Gouda cheeses	Saito et al. (2000
VPP, IPP	Swiss cheese, fermentation by *Lactobaillus helveticus* and *Saccharomyces cerevisiae*	Nakamura et al. (1995), Bütikofer et al. (2007), Bütikofer et al. (2008)
RPKHPIKHQ, RPKHPIK, RPKHPI, FVAPFPEVFGK, YQEPVLGPVRGPFPIIV	Cheddar cheeses (with probiotics)	Ong and Shah (2008)
LLR, EVLNENLLRF, FVAPFPEVFGK, YQEPVLGPVRGPF, YQEPVLGPVRGPFPI, RPKHPIKHQGLPQEV, YQEPVLGPVRGPFPIIV, RPKHPIKHQGLPQEVLNEN	Fresco cheese	Torres-Llanez et al. (2011)
SKVYPFPGPI, SKVYP	Fermented milk by *Lact. delbrueckii* ssp. *bulgaricus*	Ashar and Chand (2004)
LVYPFPGPIHNSLP, LVYPFPGPIH	Fermented milk by *Lb. jensenii*	Gobbetti et al. (2000)
GTW, GVW	Milk fermented by mixed lactic acid bacteria	Chen et al. (2007)
YPYY	Soy milk fermented by *Streptococcus thermophiles* and *Lactobacillus bulgaricus*	Tsai et al. (2008)
GY, AF, VP, AI, VG	Soy and wheat fermentation by *Aspergillus sojae*	Nakahara et al. (2010)

Peptide	Source	Reference
LIVTQ	Soy protein fermented by *Lactobacillus casei* spp. *pseudoplantarum*	Vallabha and Tiku (2014)
HHL	Korean fermented soybean paste	Shin et al. (2001)
DKIHPF, YQEPVL	Milk protein fermented by *Lb. rhamnosus* + digestion with pepsin and Corolase PP	Hernandez-Ledesma et al. (2004)
YLLF	Milk fermented by *Kluveromyces marxianus* var *marxianus*	Belem et al. (1999)
YPFP, AVPYPQR, TTMPLW	Milk protein by *Lactobacillus* GG enzymes+pepsin and trypsin	Rokka et al. (1997)
HY, VY, YGGY	Sake	Saito et al. (1994), Sarro et al. (1994)
LVYPFPGPIPNSLPQNIPP, LHLPLP	Milk fermented with *Enterococcus faecalis*	Quirós et al. (2007)
EVMAGNLYPG	Fermented sauce of blue mussel	Je et al. (2005)
DGVVYY	Fermentation of tomato seeds by-product using *Bacillus subtilis* strain	Moayedi et al. (2018)
EDEVSFSP, SRPFNL, RSPFNL, ENPFNL	Fermentation of soybean protein isolates by *Pediococcus pentosaceus*	Daliri et al. (2018)
LIVTQ	*Lactobacillus casei* spp. *pseudoplantrum* fermented soy protein concentrates	Vallabha and Tiku (2014)
SV, IF	Thai traditional fermented shrimp pastes	Kleekayai et al. (2015)
KP, IPP, LPP, VPP	Fermented cucumber	Fideler et al. (2019)
VAPFPEVFGK, LVYPFPGPLH, FVAPEPFVFGKEK	Kombucha-fermented milk	Elkhtab et al. (2017)

(Contd.)

Table 1: (*Contd.*)

Peptides Sequences*	Sources	References
LVQGS	Fermented soybean extract	Rho et al. (2009)
KEDDEEEQGEEE	Pea seeds fermented by *Lb.plantarum*	Jakubczyk et al. (2013)
FLG	Douchi, a soybean product fermented by *Aspergillus egyptiacus*	Zhang et al. (2006)
IFL, WL	Tofuyo, a fermented soybean food	Kuba et al. (2003)
VQTL, LGYEN	*Bacillus subtilis* fermented Walnut residue extract	Zheng et al. (2017)

* A – Alanine; R – Arginine; N – Asparagine; D – Aspartate; B – Aspartate or Asparagine; C – Cysteine; E – Glutamate; Q – Glutamine; Z – Glutamate or Glutamine; G – Glycine; H – Histidine; I – Isoleucine; L – Leucine; K – Lysine; M – Methionine; F – Phenylalanine; P – Proline; S – Serine; T – Threonine; W – Tryptophan; Y – Tyrosine; V – Valine

the co-culture with *G. geotrichum* KL20B, *Lb. plantarum* LAT03 and *E. faecalis* KE06 resulted in highest ACE inhibitory activity (70.6 per cent) (Chaves-Lopez *et al.*, 2014). In a molecular level study, X-prolyl dipeptidyl aminopeptidase (PepX) and general aminopeptidase (PepN) activity of the *Lb. helveticus* CNRZ32 strain was identified by using peptidase-negative variant of the same strain. Interestingly, ACE-inhibitory activities increased in mutant strain and it was suggested that peptidases in the wild type of strain were degrading the said peptides. Hence, genetic engineering approaches can also play a significant role in peptides synthesis of required effect (Kilpi *et al.*, 2007).

In a recent study, proteinases (immobilised sodium alginate) of *Lb. helveticus* strain were used to produce ACE-inhibitory peptides from whey protein. The peptides – KA, EN, DIS, EVD, LF, AIV, and VFK – were detected on the basolateral side when the generated hydrolysates transported across Caco-2 cell monolayers (Guo *et al.*, 2019). Jakubczyk *et al.* (2019) observed that the highest ACE-inhibitory activity of peptide fraction of faba bean seeds was obtained after fermented with *Lb. plantarum* strain for three days at 30°C. Differing to above-mentioned positive health benefits, some studies in pre-hypertensive and hypertensive subjects revealed the inefficient blood pressure-lowering outputs of *Lb. helveticus* fermented antihypertensive lactotripeptide-containing products in a clinical setting (Van Mierlo *et al.*, 2009; Usinger *et al.*, 2010). Similarly, a direct relationship between *in vitro* and *in vivo* ACE inhibition was hard to establish. For instance, a milk peptide (KVLPVPN) showed significant blood pressure-lowering activity *in vivo*, in spite of its low *in vitro* activity, indicating some other mechanism is also involved other than ACE inhibition (Wu *et al.*, 2017).

2.4   Bioavailability of AHPs

Bioavailability is the rate and extent to which a bioactive compound, such as a bioactive peptide or drug, reaches to the systemic circulation and is available at the site of action (Shargel and Yu, 1999). Bioactivity of peptides *in vitro* cannot be directly related to its *in vivo* effect because they may encounter degradation and modification in the intestine, vascular system and the liver. Therefore, they need to remain active during digestion by human proteases and be transported through the intestinal wall into the blood. Several studies, using *in vitro* digestion methods to analyse structural changes, bioavailability and digestibility of foods, observed that *in vitro* digestion systems are common and useful tools for analyses of foods and drugs (Singh and Vij, 2018). However, several differences are observed between *in vitro* models and *in vivo* studies (Hur *et al.*, 2011). It was seen that a small portion of bioactive peptides is sufficient to exert the specific function at tissue level once they pass through the intestine barrier (Gardner, 1988). Several studies indicated that a number of absorption mechanisms of peptides are available, such as paracellular route, passive diffusion, transport via carrier and endocytosis. In addition, lymphatic system is also a possible route of peptides absorption; here absorption of peptides is mainly effected by their permeability via the capillary of the portal circulation and lipid solubility (Deak and Csaky, 1984; Sarmadia and Ismail, 2010). It was observed by several workers that small peptides are absorbed more readily than large peptides.

They are able to cross the intestinal barrier and exert their biofunctionality at the tissue level (Grimble, 1994; Roberts *et al.*, 1999). However, the potency of bioactive peptides decreases as their molecular sizes increase and tripeptides containing proline residues at C-terminal region (i.e. Val-Pro-Pro or Ile-Pro-Pro) have been shown to be resistant to degradation by digestive enzymes and are able to reach the cardiovascular system in an active form (Vermeirssen *et al.*, 2004). Furthermore, different administration routes may influence the peptides' pharmacokinetics and their biological functionalities. There have been a few studies to evaluate the bioavailability of peptides following oral ingestion. This might be due to the complexity of the human gut, which is a difficult system to select the proper parameters for conducting gastrointestinal evaluation of a compound. Animal models are closely related to the human gut and are considered as the most preferred approach used for determination of gastric and intestinal digestive resistance of a compound. Besides that, several studies adopted *in vitro* simulatory proteolytic digestion as primary presumptive models for evaluating the gastrointestinal resistance of bioactive peptides (Hausch *et al.*, 2002; Picariello *et al.*, 2010).

3. Conclusion

Production of bioactive peptides and AHPs through fermentation of various food sources has gained increased recognition recently in the areas of food science and nutrition for their potential health-promoting benefits. Several products are already in the market, exploiting the potential application of these peptides in the form of ingredients in functional foods, novel foods, nutraceuticals, dietary supplements as well as pharmaceuticals with the purpose of delivering specific health benefits. In addition, improvement and development of sustainable technology for retaining the activity of peptides in fermented food should be of prime importance in order to deliver their health-promoting effects. Advanced biotechnology through proteomic, peptidomic, bioinformatics and chemometric approaches is expected to further develop a cost-effective production of pharmaceutical-grade AHPs as well as commercialisation of products carrying bioactive peptides. The mechanism of action of AHPs, absorption into the bloodstream, target sites and activities are other areas that require research. Bioavailability, toxicity and shelf-life are of great importance for fully defining these compounds as 'bioactive' before being labelled as a functional food or nutraceutical product. Several AHPs have shown very promising results in the clinical trial phase. Although peptides have lower ACE-inhibitory activity *in vitro* than the ACE-inhibitory drugs, yet they do not have the harmful side effects. They have been found of high specificity for their target tissues and no toxicity like synthetic chemical compounds; hence, they can typically be used as drugs with cumulative effects on the organism.

Abbreviations

AHPs – Antihypertensive peptides
CVD – Cardiovascular disease

ACE – Angiotensin converting enzyme
RAS – Rennin-angiotensin system
PMF – Protein mass fingerprinting
ROS – Reactive oxygen species
RJ – Royal jelly
LAB – Lactic acid bacteria

References

Agyei, D. and Danquah, K. (2011). Industrial-scale manufacturing of pharmaceutical-grade bioactive peptides. *Biotechnology Advances*, 29: 272-277.

Ahn, C.B., Cho, Y.S. and Je, J.Y. (2015). Purification and anti-inflammatory action of tripeptide from salmon pectoral fin byproduct protein hydrolysate. *Food Chemistry*, 168: 151-156.

Aihara, K., Kajimoto, O., Hirata, H., Takahashi, R. and Nakamura, Y. (2005). Effect of powdered fermented milk with *Lactobacillus helveticus* on subjects with high-normal blood pressure or mild hypertension. *Journal of American College of Nutrition*, 24: 257-265.

Aluko, R.E. (2015). Antihypertensive peptides from food proteins. *Annual Review of Food Science and Technology*, 6: 235-262.

Ashar, M.N. and Chand, R. (2004). Fermented milk containing ACE-inhibitory peptides reduces blood pressure in middle-aged hypertensive subjects. *Milchwissenshcaft*, 59: 363-366.

Aspri, M., Leni, G., Galaverna, G. and Papademas, P. (2018). Bioactive properties of fermented donkey milk, before and after in vitro simulated gastrointestinal digestion. *Food Chemistry*, 268: 476-484.

Balti, R., Nedjar-Arroume, N., Bougatef, A., Guillochon, D. and Nasri, M. (2010). Three novel angiotensin I-converting enzyme (ACE) inhibitory peptides from cuttlefish (*Sepia officinalis*) using digestive proteases. *Food Research International*, 43: 1136-1143.

Belem, M.A.F., Gibbs, B.F. and Lee, B.H. (1999). Proposing sequences for peptides derived from whey fermentation with potential bioactive sites. *Journal of Dairy Science*, 82: 486-493.

Beltrán-Barrientos, L.M., Hernández-Mendoza, A., Torres-Llanez, M.J., González-Córdova, A.F. and Vallejo-Córdoba, B. (2016). Invited review: Fermented milk as antihypertensive functional food. *Journal of Dairy Science*, 99(6): 4099-4110.

Bütikofer, U., Meyer, J., Sieber, R., Walther, B. and Wechsler, D. (2008). Occurrence of the angiotensin-converting enzyme-inhibiting tripeptides Val-Pro-Pro and Ile-Pro-Pro in different cheese varieties of Swiss origin. *Journal of Dairy Science*, 91(1): 29-38.

Bütikofer, U., Meyer, J., Sieber, R. and Wechsler, D. (2007). Quantification of the angiotensin-converting enzyme-inhibiting tripeptides Val-Pro-Pro and Ile-Pro-Pro in hard, semi-hard and soft cheeses. *International Dairy Journal*, 17(8): 968-975.

Cat, A.N. and Touyz, R.M. (2011). A new look at the renin-angiotensin system – focusing on the vascular system. *Peptides*, 32: 2141-2150.

Chaves-Lopez, C., Serio, A., Paparella, A., Martuscelli, M., Corsetti, A., Tofalo, R. and Suzzi, G. (2014). Impact of microbial cultures on proteolysis and release of bioactive peptides in fermented milk. *Food Microbiology*, 42: 117-121.

Chen, G.W., Tsai, J.S. and Pan, B.S. (2007). Purification of angiotensin I-converting enzyme inhibitory peptides and antihypertensive effect of milk produced by protease-facilitated lactic fermentation. *International Dairy Journal*, 17: 641-647.

Chen, L., Zhang, Q., Ji, Z., Shu, G. and Chen, H. (2018). Production and fermentation characteristics of angiotensin I-converting enzyme inhibitory peptides of goat milk fermented by a novel wild *Lactobacillus plantarum* 69. *LWT – Food Science and Technology*, 91: 532-540.

Chen, Y., Liu, W., Xue, J., Yang, J., Chen, X., Shao, Y., Kwok, L.Y., Bilige, M., Mang, L. and Zhang, H. (2014). Angiotensin-converting enzyme inhibitory activity of *Lactobacillus helveticus* strains from traditional fermented dairy foods and antihypertensive effect of fermented milk of strain H9. *Journal of Dairy Science*, 97(11): 6680-6692.

Christensen, J.E., Dudley, E.G., Pedersin, J.A. and Steele, J.L. (1999). Peptidases and amino acid catabolism in lactic acid bacteria. *Antonie Leeuwenhoek*, 76: 217-246.

Cushman, D.W. and Cheung, H.S. (1971). Spectrophotometric assay and properties of the angiotensin-converting enzyme of rabbit lung. *Biochemical Pharmacology*, 20(7): 1637-1648.

Daliri, E.B.M., Lee, B.H., Park, M.H., Kim, J.H. and Oh, D.H. (2018). Novel angiotensin I-converting enzyme inhibitory peptides from soybean protein isolates fermented by *Pediococcus pentosaceus* SDL1409. *LWT – Food Science and Technology*, 93: 88-93.

de Castro, R.J.S. and Sato, H.H. (2015). Biologically active peptides: Processes for their generation, purification and identification and applications as natural additives in the food and pharmaceutical industries. *Food Research International*, 74: 185-198.

De Mejia, E. and Ben, O. (2006). Soybean bioactive peptides: A new horizon in preventing chronic diseases. *Sexuality, Reproduction and Menopause*, 4(2): 91-95.

Deak, S.T. and Csáky, T.Z. (1984). Factors regulating the exchange of nutrients and drugs between lymph and blood in the small intestine. *Microcirc Endothelium Lymphatics*, 1: 569-588.

Dziuba, J., Iwaniak, A. and Minkiewicz, P. (2003). Computer-aided characteristics of proteins as potential precursors of bioactive peptides. *Polimery*, 48(1): 50-53.

Elkhtab, E., El-Alfy, M., Shenana, M., Mohamed, A. and Yousef, A.E. (2017). New potentially antihypertensive peptides liberated in milk during fermentation with selected lactic acid bacteria and Kombucha cultures. *Journal of Dairy Science*, 100(12): 9508-9520.

Erdman, K., Cheung, B.W.Y. and Schro der, H. (2008). The possible role of food-derived bioactive peptides in reducing the risk of cardiovascular disease. *Journal of Nutritional Biochemistry*, 19: 643-654.

Ferreira, S.H., Bartelt, D.C. and Greene, L.J. (1970). Isolation of bradykinin-potentiating peptides from *Bothrops jararaca* venom. *Biochemistry*, 9: 2583-2593.

Fideler, J., Johanningsmeier, S.D., Ekelöf, M. and Muddiman, D.C. (2019). Discovery and quantification of bioactive peptides in fermented cucumber by direct analysis IR-MALDESI mass spectrometry and LC-QQQ-MS. *Food Chemistry*, 271: 715-723.

Fitzgerald, R.J. and Murray, B.A. (2006). Bioactive peptides and lactic fermentations: Conference contribution. *International Journal of Dairy Technology*, 59(2): 118-125.

Floris, R., Recio, I., Berkhout, B. and Visser, S. (2003). Antibacterial and antiviral effects of milk proteins and derivatives thereof. *Current Pharmaceutical Design*, 9(16): 1257-1275.

Garcia-Tejedor, A., Sanchez-Rivera, L., Castello-Ruiz, M., Recio, I., Juan, B. and Salom Manzanares, P. (2014). Novel antihypertensive lactoferrin-derived peptides produced by *Kluyveromyces marxianus*: Gastrointestinal stability profile and *in vivo* angiotensin I-converting enzyme (ACE) inhibition. *Journal of Agriculture and Food Chemistry*, 62(7): 1609-1616.

Gardner, M.L.G. (1988). Gastrointestinal absorption of intact proteins. *Annual Review of Nutrition*, 8: 329-350.

Gobbetti, M., Ferranti, P., Smacchi, E., Goffredi, F. and Addeo, F. (2000). Production of angiotensin I-converting-enzyme-inhibitory peptides in fermented milks started by

Lactobacillus delbrueckii subsp. *bulgaricus* SS1 and *Lactococcus lactis* subsp. *cremoris* FT4. *Applied and Environmental Microbiology*, 66: 3898-3904.

Gobbetti, M., Stepaniak, L., De Angelis, M., Corsetti, A. and Di, Cagno. (2002). Latent bioactive peptides in milk proteins: Proteolytic activation and significance in dairy processing. *Critical Reviews in Food Science and Nutrition*, 42(3): 223-239.

Grimble, G.K. (1994). The significant of peptides in clinical nutrition. *Annual Review of Nutrition*, 14: 419-447.

Guo, Y., Jiang, X., Xiong, B., Zhang, T., Zeng, X., Wu, Z., Sun, Y. and Pan, D. (2019). Production and transepithelial transportation of angiotensin-I-converting enzyme (ACE)-inhibitory peptides from whey protein hydrolyzed by immobilised *Lactobacillus helveticus* proteinase. *Journal of Dairy Science*, 102(2): 961-975.

Harnedy, P.A. and Fitzgerald, R.J. (2012). Bioactive peptides from marine processing waste and shellfish: A review. *Journal of Functional Foods*, 4: 6-24.

Hata, Y., Yamamoto, M., Ohni, H., Nakajima, K., Nakamura, Y. and Takano, T. (1996). A placebo-controlled study of the effect of sour milk on blood pressure in hypertensive subjects. *American Journal of Clinical Nutrition*, 64: 767-771.

Hausch, F., Shan, L., Santiago, N.A., Gray, G.M. and Khosla, C. (2002). Intestinal digestive resistance of immunodominant gliadin peptides. *American Journal of Physiology-Gastrointestinal and Liver Physiology*. 283: 996-1003.

Hernández-Ledesma, B., Amigo, L., Ramos, M. and Recio, I. (2004). Angiotensin converting enzyme-inhibitory activity in commercial fermented products: Formation of peptides under simulated gastrointestinal digestion. *Journal of Agricultural and Food Chemistry*, 52(6): 1504-1510.

Hernandez-Ledesma, B., Garcia-Nebot, M.J., Fernandez-Tome, S., Amigo, L. and Recio, I. (2014). Dairy protein hydrolysates: Peptides for health benefits. *International Dairy Journal*, 38: 82-100.

Holmquist, B., Bünning, P. and Riordan, J.F. (1979). A continuous spectrophotometric assay for angiotensin converting enzyme. *Analytical Biochemistry*, 95(2): 540-548.

Hur, S.J., Lim, B.O., Decker, E.A. and McClements, D.J. (2011). *In vitro* human digestion models for food applications. *Food Chemistry*, 125(1): 1-12.

Inoue, K., Gotou, T., Kitajima, H., Mizuno, S., Nakazawa, T. and Yamamoto, N. (2009). Release of antihypertensive peptides in miso paste during its fermentation, by the addition of casein. *Journal of Bioscience and Bioengineering*, 108(2): 111-115.

Jakubczyk, A., Karaś, M., Baraniak, B. and Pietrzak, M. (2013). The impact of fermentation and *in vitro* digestion on formation angiotensin converting enzyme (ACE) inhibitory peptides from pea proteins. *Food Chemistry*, 141(4): 3774-3780.

Je, J.Y., Park, P.J., Byun, H.G., Jung, W.K. and Kim, S.K. (2005). Angiotensin I-converting enzyme (ACE) inhibitory peptide derived from the sauce of fermented blue mussel, *Mytilus edulis*. *Bioresource Technology*, 96(14): 1624-1629.

Kilpi, E.R., Kahala, M., Steele, J.L., Pihlanto, A.M. and Joutsjoki, V.V. (2007). Angiotensin I-converting enzyme inhibitory activity in milk fermented by wild-type and peptidase-deletion derivatives of *Lactobacillus helveticus* CNRZ32. *International Dairy Journal*, 17: 976-984.

Kitts, D.D. and Weiler, K. (2003). Bioactive proteins and peptides from food sources: Applications of bioprocesses used in isolation and recovery. *Current Pharmaceutical Design*, 9(16): 1309-1323.

Kleekayai, T., Harnedy, P.A., O'Keeffe, M.B., Poyarkov, A.A., CunhaNeves, A., Suntornsuk, W. and FitzGerald, R.J. (2015). Extraction of antioxidant and ACE inhibitory peptides from Thai traditional fermented shrimp pastes. *Food Chemistry*, 176: 441-447.

Korhonen, H. and Pihlanto, A. (2006). Bioactive peptides: Production and functionality. *International Dairy Journal*, 16: 945-960.

Korhonen, H., Pihlanto-Leppala, A., Rantamaki, P. and Tupasela, T. (1998). Impact of processing on bioactive proteins and peptides. *Trends in Food Science & Technology*, 9: 307-319.

Kuba, M., Tanaka, K., Tawata, S., Takeda, Y. and Yasuda, M. (2003). Angiotensin I-converting enzyme inhibitory peptides isolated from tofuyo-fermented soybean food. *Bioscience, Biotechnology and Biochemistry*, 67(6): 1278-1283.

Kumar, R., Chaudhary, K., Sharma, M., Nagpal, G., Chauhan, J.S., Singh, S., Gautam, A. and Raghava, G.P. (2014). AHTPDB: A comprehensive platform for analysis and presentation of antihypertensive peptides. *Nucleic Acids Research*, 43: 956-962.

Kunji, E.R., Mierau, I., Hagting, A., Poolman, B. and Konings, W.N. (1996). The proteolytic systems of lactic acid bacteria. *Antonie Van Leeuwenhoek*, 70: 187-221.

Li, C., Matsui, T., Matsumoto, K., Yamasaki, R. and Kawasaki, T. (2002). Latent production of angiotensin I-converting enzyme inhibitors from buckwheat protein. *Journal of Peptide Science*, 8(6): 267-274.

Mamone, G., Picariello, G., Caira, S., Addeo, F. and Ferranti, P. (2009). Analysis of food proteins and peptides by mass spectrometry-based techniques. *Journal of Chromatography A*, 1216: 7130-7142.

Marshall, W.E. (1994). Amino acids, peptides, and proteins. pp. 242-260. *In*: Goldberg, I. (Ed.). Functional Foods. Springer, Boston, MA. https://doi.org/10.1007/978-1-4615-2073-3_12

Moayedi, A., Mora, L., Aristoy, M.C., Safari, M., Hashemi, M. and Toldrá, F. (2018). Peptidomic analysis of antioxidant and ACE-inhibitory peptides obtained from tomato waste proteins fermented using *Bacillus subtilis*. *Food Chemistry*, 250: 180-187.

Nakahara, T., Sano, A., Yamaguchi, H., Sugimoto, K., Chikata, H., Kinoshita, E. and Uchida, R. (2010). Antihypertensive effect of peptide-enriched soy sauce-like seasoning and identification of its angiotensin I-converting enzyme inhibitory substances. *Journal of Agriculture and Food Chemistry*, 58: 821-827.

Nakamura, Y., Yamamoto, N., Sakai, K. and Takano, T. (1995). Antihypertensive effect of sour milk and peptides isolated from it that are inhibitors to angiotensin I-converting enzyme. *Journal of Dairy Science*, 78: 1253-1257.

Ondentti, M.A. and Cushman, D.W. (1982). Enzymes of the renin-angiotensin system and their inhibitors. *Annual Review of Biochemistry*, 51: 283-308.

Ong, L. and Shah, N.P. (2008). Influence of probiotic *Lactobacillus acidophilus* and *L. helveticus* on proteolysis, organic acid profiles, and ACE-inhibitory activity of cheddar cheeses ripened at 4, 8, and 12°C. *Journal of Food Science*, 73(3): 111-120.

Paul, M., Poyan Mehr, A. and Kreutz, R. (2006). Physiology of local renin-angiotensin systems. *Physiological Reviews*, 86(3): 747-803.

Picariello, G., Ferranti, P., Fierro, O., Mamone, G., Caira, S., di Luccia, A., Monica, S. and Addeo, F. (2010). Peptides surviving the simulated gastrointestinal digestion of milk proteins: Biological and toxicological implications. *Journal of Chromatography B*, 878: 295-308.

Pihlanto-Leppala, A., Rokka, T. and Korhonen, H. (1998). Angiotensin I-converting enzyme inhibitory peptides derived from bovine milk proteins. *International Dairy Journal*, 8: 325-331.

Pownall, T.L., Udenigwe, C.C. and Aluko, R.E. (2010). Amino acid composition and antioxidant properties of pea seed (*Pisum sativum* L.) enzymatic protein hydrolysate fractions. *Journal of Agriculture and Food Chemistry*, 58: 4712-4718.

Quirós, A., Ramos, M., Muguerza, B., Delgado, M.A., Miguel, M., Aleixandre, A. and Recio, I. (2007). Identification of novel antihypertensive peptides in milk fermented with *Enterococcus faecalis*. *International Dairy Journal*, 17(1): 33-41.

Rai, A.K., Sanjukta, S. and Jeyaram, K. (2017). Production of angiotensin I-converting enzyme inhibitory (ACE-I) peptides during milk fermentation and their role in reducing hypertension. *Critical Reviews in Food Science and Nutrition*, 57(13): 2789-2800.

Rho, S.J., Lee, J.S., Chung, Y.I., Kim, Y.W. and Lee, H.G. (2009). Purification and identification of an angiotensin I-converting enzyme inhibitory peptide from fermented soybean extract. *Process Biochemistry*, 44(4): 490-493.

Roberts, P.R., Burney, J.D., Black, K.W. and Zaloga, G.P. (1999). Effect of chain length on absorption of biologically active peptides from the gastrointestinal tract. *Digestion*, 60: 332-337.

Rokka, T., Syväoja, E.L., Tuominen, J. and Korhonen, H. (1997). Release of bioactive peptides by enzymatic proteolysis of *Lactobacillus* GG fermented UHT milk. *Milchwissenschaft*, 52: 675-678.

Saito, Y., Wanezaki, K., Kawato, A. and Imayasu, S. (1994). Structure and activity of angiotensin I-converting enzyme inhibitory peptides from sake and sake lees. *Bioscience, Biotechnology, Biochemistry*, 58: 1767-1771.

Sarmadia, B.H. and Ismail, A. (2010). Antioxidative peptides from food proteins: A review. *Peptides*, 31: 1949-1956.

Sarro, Y., Wanezaki, K., Kawato, A. and Imayasu, S. (1994). Antihypertensive effects of peptides in sake and its by-products on spontaneously hypertensive rats. *Bioscience, Biotechnology, Biochemistry*, 58: 812-816.

Seppo, L., Jauhiainen, T., Poussa, T. and Korpela, R. (2003). A fermented milk high in bioactive peptides has a blood pressure-lowering effect in hypertensive subjects. *American Journal of Clinical Nutrition*, 77: 326-330.

Shahidi, F. and Zhong, Y. (2008). Bioactive peptides. *Journal of AOAC International*, 91: 914-931.

Shargel, L. and Yu, A.B. (1999). *Applied Biophamaceutics & Pharmacokinetics*, 4th ed. New York: McGraw-Hill.

Shin, Z.I., Yu, R., Park, S.A., Chung, D.K., Ahn, C.W., Nam, H.S., Kim, K.S. and Lee, H.J. (2001). His-His-Leu: An angiotensin I converting enzyme inhibitory peptide derived from Korean soybean paste, exerts antihypertensive activity *in vivo*. *Journal of Agricultural and Food Chemistry*, 49(6): 3004-3009.

Singh, B.P. and Vij, S. (2017). Growth and bioactive peptides production potential of *Lactobacillus plantarum* strain C2 in soy milk: A LC-MS/MS-based revelation for peptides biofunctionality. *LWT – Food Science and Technology*, 86: 293-301.

Singh, B.P. and Vij, S. (2018). *In vitro* stability of bioactive peptides derived from fermented soy milk against heat treatment, pH and gastrointestinal enzymes. *LWT – Food Science and Technology*, 91: 303-307.

Singh, B.P., Vij, S. and Hati, S. (2014). Functional significance of bioactive peptides derived from soybean. *Peptides*, 54: 171-179.

Solomon, S.D. and Anavekar, N. (2005). A brief overview of inhibition of the renin-angiotensin system: Emphasis on blockade of the Angiotensin II Type-1 Receptor. *Medscape Cardiology*, 9(2): 1-6.

Torres-Llanez, M.J., González-Córdova, A.F., Hernandez-Mendoza, A., Garcia, H.S. and VallejoCordoba, B. (2011). Angiotensin-converting enzyme inhibitory activity in Mexican Fresco cheese. *Journal of Dairy Science*, 94: 3794-3800.

Tsai, J.S., Chen, T.J., Pan, B.S., Gong, S.D. and Chung, M.Y. (2008). Antihypertensive effect of bioactive peptides produced by protease-facilitated lactic acid fermentation of milk. *Food Chemistry*, 106: 552-558.

Usinger, L., Jensen, L.T., Flambard, B., Linneberg, A. and Ibsen, H. (2010). The antihypertensive effect of fermented milk in individuals with prehypertension or borderline hypertension. *Journal of Human Hypertension*, 24: 678-683.

Vallabha, V.S. and Tiku, P.K. (2014). Antihypertensive peptides derived from soy protein by fermentation. *International Journal of Peptide Research and Therapeutics*, 20(2): 161-168.

Van Mierlo, L.A.J., Koning, M.M.G., Van Zander, K.D. and Draijer, R. (2009). Lacto tripeptides do not lower ambulatory blood pressure in untreated whites: Results from 2 controlled multicenter crossover studies. *American Journal of Clinical Nutrition*, 89: 617-623.

Vermeirssen, V., Van Camp, J. and Verstraete, W. (2002). Optimisation and validation of an angiotensin-converting enzyme inhibition assay for the screening of bioactive peptides. *Journal of Biochemical and Biophysical Methods*, 51(1): 75-87.

Vermeirssen, V., Van Camp, J. and Verstraete, W. (2004). Bioavailability of angiotensin I converting enzyme inhibitory peptides. *British Journal of Nutrition*, 92: 357-366.

WHO. (2003). World Health Organisation (WHO)/International Society of Hypertension (ISH) statement on management of hypertension. *Journal of Hypertension*, 21: 1983-1992.

Wu, J., Liao, W. and Udenigwe, C.C. (2017). Revisiting the mechanisms of ACE inhibitory peptides from food proteins. *Trends in Food Science & Technology*, 69: 214-219.

Xu, J.Y., Qin, L.Q., Wang, P.Y., Li, W. and Chang, C. (2008). Effect of milk tripeptides on blood pressure: A meta-analysis of randomised controlled trials. *Nutrition*, 24: 933-940.

Yamamoto, N., Ejiri, M. and Mizuno, S. (2003). Biogenic peptides and their potential use. *Current Pharmaceutical Design*, 9(16): 1345-1355.

Yee, A.H., Burns, J.D. and Wijdicks, E.F. (2010). Cerebral salt wasting: Pathophysiology, diagnosis, and treatment. *Neurosurgery Clinics*, 21(2): 339-352.

Yoshikawa, M., Takahashi, M. and Yang, S. (2003). Delta opioid peptides derived from plant proteins. *Current Pharmaceutical Design*, 9(16): 1325-1330.

Zhang, J.H., Tatsumi, E., Ding, C.H. and Li, L.T. (2006). Angiotensin I-converting enzyme inhibitory peptides in douchi, a Chinese traditional fermented soybean product. *Food Chemistry*, 98(3): 551-557.

Zhang, L., Li, J. and Zhou, K. (2010). Chelating and radical scavenging activities of soy protein hydrolysates prepared from microbial proteases and their effect on meat lipid peroxidation. *Bioresource Technology*, 101(7): 2084-2089.

Zhang, Y., Lee, E.T., Devereux, R.B., Yeh, J., Best, L.G., Fabsitz, R.R. and Howard, B.V. (2006). Prehypertension, diabetes, and cardiovascular disease risk in a population-based sample: The strong heart study. *Hypertension*, 47: 410-414.

Zheng, X., Li, D.S. and Ding, K. (2017). Purification and identification of angiotensin I-converting enzyme inhibitory peptides from fermented walnut residues. *International Journal of Food Properties*, 20: 3326-3333.

Exopolysaccharides in Fermented Foods and Their Potential Health Benefits

Maria Antonia Pedrine Colabone Celligoi*, Gabrielly Terassi Bersanetti, Cristiani Baldo, Reginara Teixeira da Silva and Victoria Akemi Itakura Silveira

Department of Biochemistry and Biotechnology, Centre of Exact Science, State University of Londrina, 86057-970-Londrina, Paraná, Brazil

1. Introduction

The consumption of fermented foods and beverages has been reported since the development of human civilisation. This technique was based on the use of microorganisms and their metabolic activities in preserving foods (Wilburn and Ryan, 2017). Although, nowadays this method has gained much more attention due to the possibility of it enhancing shelf-life of foods naturally and also providing healthier food products with improved nutritional, functional and sensorial properties (Sanlier et al., 2017; Ripari, 2019; Sanjukta et al., 2017). Many scientific researchers state that fermented foods and beverages have beneficial microorganisms that improve bioavailability of nutrients and enrich the food with bioactive compounds (Patel and Prajapati, 2013; Khalil et al., 2018; Sanjukta and Rai, 2016).

Lactic acid bacteria (LAB) are part of a very important group of microorganisms in food processing due to their ability to promote positive effects on fermented foods, including improvement of sensory attributes and product quality (Lynch et al., 2018; Marco et al., 2017). Exopolysaccharides (EPSs) produced by LAB are a diverse group of polysaccharides that are produced by many species using substrates containing a high concentration of mono- and disaccharides (Marco et al., 2017; Lynch et al., 2018). The role of EPSs in food has been associated with several beneficial health functions, including prebiotic effect, antioxidant activity, antitumor activity and cholesterol reduction in the blood. EPSs have application in the food industry for several purposes: decreasing the addition of preservatives and improving the shelf-life; influencing viscosity and rheology; improving texture, sensory properties, mouth feel, freeze/thaw stability; use as softeners; as dietary fibres; as coating and

*Corresponding author: macelligoi@uel.br

moisturising agents (Ripari, 2019). These techno-functional properties of EPSs are related to their ability to bind water and retain moisture (Lynch *et al.*, 2018). This chapter discusses the current trends in the production and utilisation of EPSs in fermented foods while highlighting their potential health benefits.

2. Exopolysaccharides: Synthesis and Production of Microorganisms

EPSs are polysaccharides produced in the extracellular medium. They can be classified as homopolysaccharides (HoPS) or heteropolysaccharides (HePS), depending on the composition of the main chain and their mechanism of synthesis. Commonly, HoPS consist of only one type of sugar, glucose or fructose, and are named glucans or fructans, respectively. In contrast, HePS are formed by two or more types of monosaccharides, usually D-glucose, D-galactose, and L-rhamnose. However, other monosaccharides like fructose, fucose, mannose, N-acetylglucosides, and glucuronic acid can also be present. In addition, phosphate and acetyl groups are found in these molecules (Donot *et al.*, 2012; Lynch *et al.*, 2018).

EPSs biosynthesis can occur in the intra- and extracellular environment. In general, HoPS are produced externally and HePS inside the cells for later secretion (Lynch *et al.*, 2018). The biosynthesis of HoPS, such as levan and dextran, involves extracellular enzymes as levansucrase (EC 2.4.1.10) and dextranase (EC 2.4.1.5), which act on the hydrolysis of sucrose for polymerisation of the fructan and glucan chains (Gupta and Diwan, 2017). In the intracellular medium, the EPSs synthesis can be divided into three stages:

(I) assimilation of simple sugars and conversion into nucleotide derivatives;
(II) assembly of pentasaccharide subunits attached to a lipid carrier; and
(III) polymerisation of repeating units of pentasaccharides and EPSs secretion in the extracellular environment (Donot et al., 2012).

Among the EPSs-producing species in fermented foods, LAB are highly representative because they are a heterogeneous group of bacteria widely found in Nature. In addition, LAB have the status of Qualified Presumption of Safety (QPS) and are recognised as safe by the Scientific Committee of the European Food Safety Authority (EFSA) (Khalil *et al.*, 2018). The majority of LAB-producing EPSs are comprised by the genera *Streptococcus*, *Lactobacillus*, *Lactococcus*, *Leuconostoc*, and *Pediococcus* (Patel and Prajapati, 2013). Currently, the most studied microbial EPSs are levan, dextran, bacterial cellulose, pullulan, xanthan, gellan and hyaluronic acid (Table 1).

The production of microbial EPSs is closely related to the specific conditions of cell growth and the culture-medium composition. The main factors that influence EPSs production are the microorganism, carbon and nitrogen sources, presence of salts, temperature, pH, time of incubation, agitation and aeration (Zhang *et al.*, 2015a; Bilgi *et al.*, 2016; Du *et al.*, 2017). The most commonly used carbon sources for EPS production are sucrose and glucose in submerged fermentation with agitation (Bersaneti *et al.*, 2017; Kanimozhi *et al.*, 2017; Wang *et al.*, 2018b; Diana

Table 1: Main Microbial Exopolysaccharides and Their Producers

Type	EPSs	Monomers	Microorganisms	References
HoPS	Levan	Fructose	*Zymomonas mobilis; Gluconoacetobacter diazotrophicus; Bacillus subtilis; B.phenoliresistens*	Ernandes and Garcia-Cruz (2011), Idogawa et al. (2014), Bersaneti et al. (2017), Moussa et al. (2017), Gojgic-Cvijovic et al. (2019).
	Dextran	Glucose	*Weissella confusa; Leuconostoc mesenteroides; Lactobacillus sakei*	Sawale and Lele (2010), Kajala et al. (2016), Nácher-Vázquez et al. (2017), Diana et al. (2019).
	Bacterial celulose	Glucose	*Gluconacetobacter xylinus; Acetobacter xylinum; Komagataeibacter hansenii.*	Cheng et al. (2017), Qi et al. (2017), Uzyol and Saçan (2017), Ye et al. (2019), Dórame-Miranda et al. (2019).
	Pululan	Glucose	*Aureobasidium pullulans*	Jiang et al. (2018), Yang et al. (2018), Hilares et al. (2019), Badhwar et al. (2019).
HePS	Xanthan	D-glycosyl, D-mannosyl, D-glucuronic acid, acetyl and pyruvyl	*Xanthomonas* spp.	Li et al. (2017), Kumar et al. (2018), Silva et al. (2018), Demirci et al. (2019).
	Gelan	L-Ramnose, D-glucose and D-glucuronic acid	*Sphingomonas elodea*	Kang et al. (2017), Zia et al. (2018), Xu et al. (2019), Zhu et al. (2019).
	Hyaluronic acid	Glucuronic acid and N-acetylglucosamine	*Streptococcus zooepidemicus*	Pan et al. (2017), Westbrook et al. (2017), Chahuki et al. (2019), Huang et al. (2019).

et al., 2019). Low-cost substrate and industrial by-products, such as lignocellulosic materials, cheese whey, molasses and glycerol-rich products also can be used for the synthesis of EPSs in submerged fermentation (Roca *et al.*, 2015). In addition, the use of solid-state fermentation to produce EPSs, using organic residues, such as sugarcane bagasse, pecan peel, loaf residues, also was reported (Cheng *et al.*, 2017; Demirci *et al.*, 2019; Dórame-Miranda, 2019; Hilares *et al.*, 2019).

3. Fermented Foods

LAB are the most common microorganisms associated with fermented foods (Kook *et al.*, 2019). Many LAB species produce significant amounts of EPSs, a cell-wall component that acts as an important extracellular metabolite, providing health benefits to the host and also playing a main role in preservation and physicochemical characteristics of foods (Khalil *et al.*, 2018). The application of EPSs in foods can occur by two methods: *in situ* production or added as a purified ingredient. The utilisation as a bio-ingredient can be advantageous if applied in a known concentration and at a specific time as it could generate a more controllable system (Lynch *et al.*, 2018). On the other hand, the *ex situ* application of EPSs as a bio-ingredient needs its information as an additive in the product label, despite it being a natural source, it could have a negative impact as consumers are increasingly concerned about the use of such substances. Therefore, *in situ* production could be a viable method, since it does not necessitate labelling or any additional cost in purification (Waldherr and Vogel, 2009). The well-established recognition of the health benefits associated with the consumption of EPSs or food products containing these polymers has opened up an important market for consumers who seek food products with more natural appeal. This is showing up in many food sectors, some examples of which are as follows.

3.1   Dairy Products

Dairy products fall in the main food sector where EPSs have been utilised to contribute to techno-functional attributes. It can improve rheological characteristics, such as enhancing viscosity and firmness, reducing syneresis and raising sensory properties (Andhare *et al.*, 2015). EPSs can act as texturisers and stabilisers by interacting with water and other milk components, such as proteins and micelles, thus fortifying the casein network (Ripari, 2019).

In yoghurt, the presence of EPSs leads to a higher viscosity and less phase separation, without altering the flavour of the final product (London *et al.*, 2015; Caggianiello *et al.*, 2016; Ripari, 2019). In a study conducted by Ayyash *et al.* (2018), EPSs produced by *Streptococcus thermophilus* and *Lactobacillus delbrueckii* ssp. *bulgaricus* were incorporated in yoghurt, demonstrating great improvement in its texture. A mannose-rich EPSs producer-strain, *Lb. mucosae* DPC 6426, was used in yoghurt, conferring significantly higher techno-functional quality (London *et al.*, 2014b). Similar findings were observed with the use of *Streptococcus thermophilus.* EPSs-producing strain combined with *Lb. delbrueckii* to produce yoghurt, showed better texture and lower whey separation in comparison with yoghurt produced from commercial strains (Han *et al.*, 2016).

EPSs were also used in low-fat dairy products, thereby solving the weak texture and consistency problem due to their reduced milk-fat content (Patel and Prajapati, 2013). Low-fat cheese usually has low moisture-to-protein ratio, thereby generating a denser protein network that affects the cheese texture (Broadbent *et al.*, 2001). It was observed that the use of EPSs in cheese preparation resulted in a higher moisture level, similar to full-fat, which impacts positively in reduced-fat cheese structures. Awad *et al.* (2005) examined the influence of EPSs producer strain, *Lactococcus lactis* subsp. *Cremoris*, on a reduced-fat Cheddar cheese, concluding that the use of EPSs producer generated a cheese with more moisture than the control and with similar properties. Other studies corroborated with these findings (Dabour *et al.*, 2006; Costa *et al.*, 2010; Lynch *et al.* 2014; Lynch, 2018), indicating that the application of EPSs in reduced-fat dairy products acts as a bio-thickener, increasing the moisture levels and textural attributes significantly, without the use of chemical additives (Patel and Prajapati, 2013).

3.2   Bakery Products and Cereal-based Beverages

Recently the use of EPSs-producing strains was seen as a natural and cheaper alternative for the cereal and bakery industry, as an attempt to replace commercial hydrocolloids (Ryan *et al.*, 2015; Shi *et al.*, 2019). The presence of EPSs may positively affect the technological properties of dough, bread and beverages. EPS contributes to increase in water retention of the dough, better rheological properties, maintenance of bread structure, stability while frozen, increased loaf volume and crumb softness and delayed bread staling, thus increasing its shelf-life (Patel and Prajapati, 2013; Lynch *et al.*, 2017).

The EPS mechanisms involved in these properties are not quite understood, but it is proposed that dextran improves the dough stability and gas retention by building a polysaccharide structure that interacts with the gluten network, providing additional support (Lynch *et al.*, 2017). The enhanced crumb softness, delay in staling rate and increased shelf-life may be attributed to its ability to bind water and retain moisture, thereby retarding starch crystallisation (Galle *et al.*, 2012a). The well-studied EPS from LAB with potential application in the cereal industry is dextran (Patel and Prajapati, 2013; Lynch *et al.*, 2017). A study conducted by Katina *et al.* (2009) showed improved viscosity, greater volume and softness of sourdoughs with the presence of dextran produced by *Weissella confuse*. Hussein *et al.* (2010) reported increase in water absorption, dough development, stability and prolonged shelf-life in biscuit production with the implementation of *Lactobacillus helveticus* EPS producer in the formulation.

Bacterial EPS has also been applied to address the problem associated with gluten by contributing to production of gluten-free products. This is a promising application because usually these kinds of foods create problems in rheological properties, such as poor texture and viscosity. Furthermore, prebiotic effects of some EPSs could be seen as an additional advantage. EPSs produced by *Weissella cibaria* and *Lb. reuteri* resulted in improved dough rheology and quality of gluten-free sorghum (Galle *et al.*, 2012b). Other studies also demonstrated that EPSs can be successfully applied in gluten-free products to improve their functional and technological potentials (Schwab *et al.*, 2008; Rühmkorf *et al.*, 2012).

There is a crescent market in the development of non-dairy beverages due to lactose-intolerant population. The common substitutes for these are cereal-based fermented beverages. The incorporation of EPS-producing culture in these beverage preparations can improve the technical and functional quality, thereby generating drinks without phase separation and good stability during storage (Ripari, 2019). Lorusso *et al.* (2018) developed a quinoa-milk-fermented beverage employing *Lb. rhamnosus* SP1, *W. confusa* DSM 20194 EPSs-producing and *Lb. plantarum* T6B10 and observed that synthesised EPSs led to enhanced viscosity and water-holding capacity, with consequent increment in the beverage's textural properties. The presence of *Lb. fermentum* EPSs producer also proved beneficial in preparation of sorghum beers, Dolo and Pito, by improving the texture and sensorial features (Sawadogo-Lingani *et al.*, 2008).

3.3 Vegetable Products

EPSs-producing strains are also present in fermented vegetable products. In Hu *et al.* (2019) study, EPSs secreted by *Bacillus* sp. S-1 were isolated from sichuan pickles and the results obtained suggested its promising activity in quenching free radicals, indicating that it may be a natural source of antioxidant additives for nutraceuticals and functional foods.

Kimchi is a traditional fermented vegetable food in Korea and is also a source of LAB EPSs producer. Kook *et al.* (2019) revealed the immunomodulatory potential of EPSs derived from *Bacillus licheniformis* BioE-FL11 and *Leuconostoc mesenteroides* BioE-LMD18 isolated from *kimchi*, suggesting that these EPSs could exhibit dose-dependent anti-inflammatory effects.

Juvonen *et al.* (2015) highlighted the impact of EPSs-producing LAB on rheological and sensory characteristics of purred carrots. *Lactobacillus, Leuconostoc* and *Weissella* strains, capable of producing EPSs (dextran, levan and/or β-glucan), were used, indicating that dextran presence is correlated with thickness and β-glucan with elasticity. The presence of *W. confusa* and *Ln. lactis* also contributed to pureed carrots with thicker texture and pleasant odour and flavour. This presents a different approach to replace hydrocolloid additives as texturisers not only in carrot products, but also for other vegetables in general.

3.4 Meat Products

Consumption of meat products is often related to negative impacts on human health due to their high fat content, leading to the development of many diseases. EPSs application in meat products has led to an alternative to meet the consumers' demand for healthy products. Hilbig *et al.* (2019), in their studies, tried to reduce the fat content of the traditional German spreadable raw fermented sausages (Teewurst) through the application of *in situ* EPSs LAB. Their normal fat content is approximately 30-35 per cent and with the incorporation of EPSs, it was able to reduce to only 20 per cent, thus maintaining the acceptable range. The results of sensory evaluation also demonstrated that sausages containing HoPS were rated softer and better spreadable than the control.

Dertli *et al.* (2016) investigated the role of EPSs produced by *Lb. plantarum* 162 R and *Ln. mesenteroides* N6 on physicochemical, microbiological and textural aspects of the Turkish fermented sausage, sucuk. EPSs enhanced the final technological properties of sucuk sausages which became harder, less adhesive and tougher than the control samples. It is suggested that the use of EPSs reduces fat content due to its ability to cover fat-free areas on the protein particles, thus maintaining the spreadability (Hilbig *et al.*, 2019). Although there are not many researches on the role of EPSs in meat products, these studies clearly demonstrate their potential application in fermented meat products as an attempt to improve rheological properties and to create healthier products with reduced fat content. Besides all these positive techno-functional effects of EPSs in fermented foods, it can also bring various health benefits to the consumer, which has earned importance for these natural and non-toxic ingredients.

4. Exopolysaccharides and Health Benefits

EPSs are molecules that have innumerable health benefits for the host, including prebiotic activity which is responsible for modulating the growth of beneficial bacteria in the intestine, thus increasing the immune system response and improving the barrier against pathogenic bacteria that cause damage to the organism. Besides, the substances produced in the intestine by EPSs fermentation help reduce serum lipids, and also have antioxidant activity which is responsible for neutralising the reactive oxygen species that cause oxidative stress in the cells, thus preventing some diseases, such as cancer.

4.1 Prebiotic Activity

Prebiotic is defined as an undigestible food ingredient that may confer benefits on the host's health and associated with modulation of the intestinal microbiota. Prebiotics are not hydrolysed or absorbed in the stomach or small intestine and increase the colonisation of beneficial bacteria in the colon, such as *Bifidobacterium*, competing with pathogenic microorganisms by metabolic substrates. In addition, the metabolites formed on fermentation of the substrate induce beneficial effects on the host immune system, especially the lymphoid tissue associated with the intestine (Brownawell *et al.*, 2012; Parracho *et al.*, 2007). The intestinal microbiota ferments a wide range of biomolecules, mainly provided by the diet and which are not digested or absorbed in the superior gastrointestinal tract of the host, thus remaining available for fermentation by the colon microbiota. These include oligosaccharides (Bersaneti *et al.*, 2017), resistant starch, proteins, amino acids and non-polar polysaccharides (for example, celluloses, hemicelluloses, pectin and gums as EPSs) among others (Brownawell *et al.*, 2012; Gibson *et al.*, 2004).

The EPSs are considered unconventional sugars synthesised for cell protection against adverse environments. They are not degraded in the gastrointestinal tract and thus exert their prebiotic effect (Looijesteijn *et al.*, 2001; Salazar *et al.*, 2008; Paiva, 2013). The biological activity of this polymer depends on its chemical structure, which is associated with the type of the main chain, its ramification and molecular weight (Luna, 2016). The metabolism of these carbohydrates produces a variety of substances, such as short-chain fatty acids and organic acids, which can have varied effects on the organism, such as reduce the colonisation of potentially-pathogenic bacteria, such as *Citrobacter rodentium* (Wong *et al.*, 2006; Fanning *et al.*, 2012).

Olano-Martin *et al.* (2000) studied the prebiotic effects of dextran and oligodextran *in vitro* and observed that the oligodextran of low molecular weight caused a bifidogenic effect, similar to the effect described on fructooligosaccharide, a known prebiotic compound. This study also confirmed the reduction of undesirable bacteria, such as bacteroides and clostridia. Dal Bello *et al.* (2001) also studied the prebiotic properties of EPSs, testing the levan of *Lb. sanfranciscensis*, inulin and fructooligosaccharide (FOS) *in vitro*. The authors observed changes in the composition of intestinal bacteria during the fermentation of carbohydrates, with an increase in *Bifidobacterium* spp. using levan and inulin. The utilisation of levan and FOS resulted in the growth of *Eubacterium biforme* and *Clostridium perfringens*, respectively.

Prebiotic activity of EPSs was evaluated in the rat model (Hamdy *et al.*, 2018). The treatment of rats with 1 mL of EPSs for 60 days resulted in the increase of *Lactobacillus* count, with a gradual decrease in the *Escherichia coli*. Thus, it is possible to affirm that the increase of beneficial bacteria in the intestinal microbiota is related to the ingestion of the EPSs, attesting its prebiotic effect. On feeding EPSs an increase in the amount of *Bifidobacterium* and *Lactobacillus* was seen and reduction in the amount of *C. perfringens* in rats, with no change in the amount of *Enterobacteriaceae* and *Enterococcus* (Wu *et al.*, 2018). This indicates that EPSs obtained from *Bifidobacterium* is able to regulate the intestinal flora of normal rats (Wu *et al.*, 2018). The interaction between EPSs and human intestinal microbiota was studied *in vitro* by Liu *et al.* (2018). The results showed that the microbiota isolated from human faeces was able to degrade 50 per cent of the EPSs, thus promoting the growth of beneficial bacteria, mainly *Deinococcus thermus*. The authors suggest that the bacteria present in the human intestine are able to use the EPSs as a source of carbon and consequently induce benefits to the organism.

In addition, the prebiotic activity of the EPSs can have immunostimulant effect, which increases the immune response of the intestinal mucosa (Matsuzaki *et al.* 2018) and increases the elimination of free radicals due to antioxidant activity (Hu *et al.*, 2019). This also showed the anti-inflammatory or immunosuppressive effects (Nikolic *et al.*, 2012). By exhibiting these properties, EPSs can be incorporated in dietary products of patients with diseases associated with increased inflammatory condition, as allergy and intestinal diseases or other autoimmune disorders.

4.2   Immune System Modulation

Many food components have the capacity to modulate the immune system of the host. These compounds are classified as biological response-modifiers or immunomodulators. Many molecules have the ability to be biological response-modifiers, such as EPSs proteins, peptides, glycoproteins, lipopolysaccharides, lipid derivatives among others. Strategies have been developed to understand how cell signalling works and how its communication can influence the immune responses in the organism (Tzianabos, 2000).

Among the molecules that can induce modifications in the immune system, there are two types of EPSs with some physicochemical characteristics:

- The first is an acid EPS, which is characterised by the presence of a phosphate group (that is, negative charge) in its composition, being good inducers of the immune system response (Hidalgo-Cantabrana *et al.*, 2012). Authors have proved that phosphate is a molecule that induces the immune response, since the dephosphorylated EPSs reduce the stimulatory effect (Kitazawa *et al.*, 1998).
- The second type is the high molecular weight EPSs which can also operate as suppressors of the immune response. One example is the EPSs produced by *Lb. fermentum* Lf 2 that exhibits a high molecular weight and is composed of β-glucans and shows immunomodulatory activity in tests with mononuclear cells by reducing the expression levels of TLR-2 (peptidoglycans receptors), and by increasing TNF-α (tumor necrosis factor alpha) (Laws *et al.*, 2018).

In addition, EPSs with intermediate chains obtained after the hydrolysis process, are also capable of altering the immune system by affecting the degree of cytokine induction, suggesting that different cytokine signalling routes may be activated according to the molecular mass of the polysaccharide (Bleau *et al.*, 2010). This capacity to modulate the immune-system response in the host also improves the pathogen barrier (Patten *et al.*, 2014) and suppresses responses in inflammatory disorders (Bleau *et al.*, 2010). Some authors have studied the relation between EPSs and the immune response and have verified that the EPSs modulate the immune system.

The study by Paiva (2013) indicated a possible pro-inflammatory activity in the small intestine of rats, which were treated with dextran for seven days. The author also noted that this polysaccharide was able to modulate the immune response by stimulating the production of *Ig*A (considered first line of defence against infections) and TNF-α (tumor necrosis factor alpha). In addition to the reduction of IL-10 (interleukin 10), TGF-β (transforming growth factor beta) was also described. In their study, Balzaretti *et al.* (2016) also observed immunomodulatory properties with EPSs produced by *Lb. paracasei* DG strain. The authors reported an increased expression of the gene of the pro-inflammatory cytokines TNF-α and interleukin 6 (IL-6), particularly chemokines IL-8 in the human monocytic cell line THP-1. The results suggest that EPSs from LAB have the ability to stimulate the immune system.

The evaluation of mitogenic activity (capacity to stimulate the proliferation of immune cells) was studied by Kitazawa *et al.* (1998) by using EPSs produced by two strains of *Lb. delbrueckii*, which differed only in the presence of the phosphate group. According to the authors, only the phosphorylated polysaccharide was mitogenic for defence cells, whereas the dephosphorylated EPSs resulted in the loss of mitogenic activity of the lymphocytes. Another study, using the same EPSs, demonstrated a change in macrophage function, causing cytokine induction by increased mRNA expression, which was not observed with a neutral polysaccharide (absence of the phosphate group) (Nishimura-Uemura *et al.*, 2003). The stimulation of macrophage activity has also been demonstrated *in vitro* with EPSs produced by *Lb. paracasei* and *Lb. plantarum*. Both EPSs stimulated the proliferation of cytokines by macrophages, revealing that these molecules can be useful as immune modulators in pro-inflammatory responses (Liu *et al.*, 2011).

In recent research, the authors suggested that EPSs produced by *Lb. delbrueckii* have the potential to be used in protection against intestinal viruses, such as rotavirus. EPSs were also capable in decreasing the pro-inflammatory expression of chemokines IL-6 (Kanmani *et al.*, 2018). The EPSs produced by *Bacillus subtilis* also improved the immune response in rats fed for 49 days with one mL of EPSs, as observed by the increase of *Ig*A and *Ig*G antibodies. Besides, the results indicated protection against *Salmonella typhimurium* infection, which may cause liver inflammation (Hamdy *et al.*, 2018). Immunomodulatory activity was also reported in EPSs produced by *Lb. plantarum* JLK0142 isolated from tofu, a fermented food (Wang *et al.*, 2018a).

Hou *et al.* (2019) evaluated the *in vitro* immunoregulatory actions of an exopolysaccharide of the fungi genus, *Lachnum* (LEP) and its conjugation to a dipeptide (LEP-RH). The authors observed a similar behaviour in EPSs where treatment with the highest dose (200 μg/mL) of LEP and LEP-RH stimulated proliferation, secretion of cytokines (IL-2, IL-6 and TNF-α) of macrophages. Therefore, LEP and LEP-RH can potentially improve the immune response. In another study, *in vitro* and *in vivo* immune responses of EPSs were obtained from the probiotic strain, *Lb. casei* WXD030 which was evaluated. The authors observed that EPSs increased cell proliferation and phagocytic activity, as well as induced the production of nitric oxide, TNF-α, IL-1β and IL-6 in macrophage cells. Besides, EPSs also elevated the ovalbumin antibodies and T-cell proliferation. When used as an adjuvant *in vivo* in foot-and-mouth disease vaccine, EPSs greatly increased the production of antibodies, specific for the virus of that disease. Thus, the results suggest that EPSs is a safe and effective molecule for a broad spectrum of prophylactic and therapeutic vaccines (Xiu *et al.*, 2018).

4.3   Antioxidant Activity

Oxidative stress in living organisms is a consequence of increased or accumulation of reactive oxygen species (ROS), leading to tissue damage (Barbosa *et al.*, 2010). This process is initiated by free radicals causing oxidative damage of biological macromolecules, such as proteins, lipids and DNA (Nguyen and Nguyen, 2014). The chronicity of the process has important implications in many chronic diseases, including atherosclerosis, diabetes, obesity, neurodegenerative disorders and cancer (Zhang *et al.*, 2017b). The protection against free-radical damage is executed by superoxide dismutase, catalase, ascorbic acid, tocopherol and glutathione (Nguyen and Nguyen, 2014). Antioxidant supplements are necessary to combat oxidative stress if there is failure in the normal protective mechanism.

There are some techniques to estimate the antioxidant activity, such as DPPH (2,2-diphenyl-1-picrylhydrazyl), hydroxyl, and superoxide-anionscavenging activity and ferrous ion-chelating activity. The DPPH free radical is a radical with an unpaired valence electron at nitrogen bridge atom, and its free radical character can be neutralised upon exposure to a proton radical scavenger that is able to transfer a hydrogen atom or an electron to DPPH (Kodali, 2010; Liu *et al.*, 2011; Zhao *et al.*, 2018). The hydroxyl radical is a reactive oxygen species that can cause biological damage and lipid peroxidation by reacting with several biological molecules, such as proteins, lipids and carbohydrates (Cao *et al.*, 2007). In addition, the superoxide radical is a precursor of most reactive oxidative species, such as hydrogen peroxide and hydroxyl radical and singlet oxygen, resulting in tissue damage and cell death. Ferrous ion-chelating activity is based on the transition metal Fe^{2+} that can stimulate production of hydroxyl radicals, leading to lipid peroxidation (Duh, 1999). The dietary antioxidants commonly include bioactive peptides, microbial enzymes and polysaccharides. According to the literature, EPSs exhibit antioxidant, anti-inflammatory and angiogenic activities (Barbosa *et al.*, 2010) as well as the ability to modulate metabolism of specific microorganisms in the gut microbiota (Serafini *et al.*, 2014; Zhang *et al.*, 2017b).

In their study, Sun *et al.* (2013) showed the inhibition effects of EPSs from *Pleurotus eryngii* SI-02 at a dosage of 400 mg/L on hydroxyl, superoxide anion and DPPH radicals. According to the authors, the inhibition effect was higher than that observed in comparison to butylated hydroxytoluene (BHT), which is a known antioxidant (Sun *et al.*, 2013). EPSs produced by *Lb. plantarum* YW11 presented strong antioxidant activity *in vitro* and *in vivo*, modulating the gut microbiota and improving the antioxidant status of the host. Fermented milk with *Lb. plantarum* YW11 containing EPSs also showed favourable antioxidant and gut microbiota-regulating activities (Zhang *et al.*, 2017b).

EPSs from *Peanibacillus mucilaginosus* TKU032 showed strong antioxidant properties, especially DPPH free-radical-scavenging ability by donating hydrogen ions to DPPH radical (Liang *et al.*, 2016). It was also observed that the DPPH-scavenging activity increased in a dose-dependent (0-400 µg/mL) manner. The greatest scavenging rate of TKU032 EPSs was 80 per cent, which was higher than the other EPSs with the same activity. Besides, the half-maximal effective concentration of TKU032 EPSs (157.1 µg/mL) was lower than that observed in standard antioxidant ascorbic acid (191.7 µg/mL). According to authors, TKU032 EPSs was a potent and natural antioxidant that could be used as an alternative to synthetic antioxidants.

The EPSs extracted from yoghurt starter were found to possess antioxidant activities and the *in vitro* antioxidant assay showed that EPSs had the ability to scavenge DPPH free radicals (Ghalem, 2017). In another study, EPSs produced by *Bacillus amyloliquefaciens* GSBa-1 had the potential DPPH and hydroxyl-radical-scavenging activities and also exhibited moderate metal ion-chelating and superoxide anion-scavenging activities. The potential of the EPSs from *B. amyloliquefaciens* GSBa-1 to serve as a natural antioxidant for application in functional products was investigated (Zhao *et al.*, 2018). The authors believe that the DPPH radical-scavenging activity of the studied EPSs may be due to its lower molecular mass (45 KDa). It was also reported that the lower molecular weight of the polysaccharide might be a positive factor in its higher antioxidant activities (You *et al.*, 2013; Huang *et al.*, 2010).

Liu *et al.* (2010) also reported lower-molecular-weight EPSs with higher DPPH activity. In addition, Tsiapali *et al.* (2001) showed that the antioxidant activity of dextran is due to its radical scavenging ability. The dextran hydroxyl groups may possibly donate hydrogen for binding with hydroxyl radicals to achieve the scavenging activity. EPSs isolated from *Dixoniella grisea* had a stronger ability against superoxide radical anion than α-tocopherol (Sun *et al.*, 2009). *Brevibacterium otitidis* EPSs also exhibit a strong free radical scavenging effect, which can be comparable to vitamin C (Asker and Shawky, 2010). Wang *et al.* (2018b) also reported that EPSs from

Aerococcus uriaeequi HZ strains can scavenge free radicals as superoxide radical anion and hydroxyl radical and the activity of superoxide radical anion-free radical is comparable to vitamin C. There are varieties of mechanisms that can explain the hydroxyl radical scavenging activities of EPSs. One possibility was that EPSs could absorb radicals and terminate the radical reaction (Papageorgiou *et al.*, 2010).

The EPSs from *Lb. acidophilus* LA5 and *Bifidobacterium animalis* showed antioxidant activity by DPPH and hydroxyl free radicals scavenging activities and reducing power analysis, suggesting that EPSs have a high potential as natural antioxidants or bioactive additive in the food industry (Amiri *et al.*, 2019). Hu *et al.* (2019) showed that EPSs from *Bacillus* spp. possess strong quenching capacities on superoxide, hydroxyl and DPPH radicals in a dose-dependent manner. Moreover, the major EPSs component (BS-2) showed powerful protection against oxidative damage by inhibiting H_2O_2-induced apoptosis in RAW264.7 cells (Hu *et al.*, 2019). The BS-2 was able to scavenge DPPH, O_2 and OH in a concentration-dependent manner. In addition, BS-2 also exhibited notable protective effects in RAW264.7 cells treated with H_2O_2. BS-2 exhibited higher antioxidant activity in comparison to EPSs fraction with a low molecular weight (BS1). The BS-2 contained more hydroxyls than BS1 and this may be responsible for enhancing the ionic interaction and effective binding to the cell-membrane surface, promoting the interaction of free radicals.

Pan and co-workers (2019) studied two water-soluble EPSs (A14 EPSs-1 and A14 EPSs-2) isolated from the fungal endophytic strain A14. A14 EPSs-1 and A14 EPSs-2, both were able to scavenge DPPH radicals in concentration-dependent doses. At low concentrations (0.1 and 4.0 mg.mL^{-1}), A14 EPSs-1 and A14 EPSs-2 showed comparable scavenging activities, while the latter had a higher scavenging activity (47.22 and 2.30 per cent) than the earlier (23.74 and 4.15 per cent) at a concentration of 8 mg/mL (Pan *et al.*, 2019). Dietary antioxidants from fermented foods could avoid the manifestation of diseases associated with oxidative stress by blocking lipid peroxidation, scavenging ROS and promoting the regulation of intestinal environment (Coskun Cevher *et al.*, 2015). However, novel studies are still necessary to fully understand the antioxidant mechanism of microbial EPSs.

4.4 Antitumoral Activity

Cancer is responsible for 9.6 million deaths, in 2018, worldwide. According to the World Health Organisation, nearly 70 per cent of deaths occurred in low- and middle-income countries. High body mass index, lack of physical activity, low fruit and vegetable intake, consumption of tobacco and alcohol are the main traits responsible to one-third of deaths from cancer. The antitumor agents used currently in chemotherapy have strong activity against the cancer cells. However, the side effects, as hemopoietic suppression and immunotoxicity, drove the search for antitumor agents from natural sources. EPSs from LAB are safe natural sources that may be an alternative to synthetic antitumor agents (Wang *et al.*, 2014).

Studies showing antitumor activities of EPSs have increased in recent years. According to the literature, the antitumor effects of EPSs depend on the chemical composition, molecular weight, monosaccharide composition, structure of the polymeric backbone, side chains, and degree of branching (Jin *et al.*, 2003; Badel *et al.*, 2011; Jolly *et al.*, 2002; Sun *et al.*, 2018). Besides, the moderate sizes of polysaccharides, presence of mannose and glucose residues with repeating units are favourable for anticancer activity of EPSs (Li *et al.*, 2015). In the study by Li *et al.* (2015), three EPSs fractions (LHEPSs-1, LHEPSs-2 and LHEPSs-3) isolated from *Lb. helveticus* MB2-1 were used to assess the potential for antitumoral activity. The results showed that LHEPSs-1 significantly inhibited the proliferation of human colon cancer Caco-2 cells in time and concentration-dependent ways, suggesting that LHEPSs-1 may be suitable for use in functional foods and anti-colon cancer compound (Li *et al.*, 2015). In other study, EPSs-6 from *Bacillus megaterium* showed highest cytotoxicity against HepG2 cells. According to the authors, the presence of sulphur and uronic acids in the EPSs-6 structure is responsible for cytotoxicity (Abdelnasser *et al.*, 2018).

- *In vitro* antitumor activity was evaluated of three EPSs (EPSs-1a, EPSs-2a, and EPSs-3a) from *Streptococcus thermophilus* CH9. EPSs-3a exhibited higher antitumor activity against human liver cancer HepG2 cells, associated with cell apoptosis (Sun *et al.*, 2018). In contrast, EPSs-1a and EPSs 2a showed a lower antitumor activity. According to the authors, the differences in antitumor activity might be due to their differences in monosaccharide composition and contents of protein. EPSs-3a contained a higher ratio of glucose (63.93 per cent) and relatively higher contents of protein when compared with EPSs-1a and EPSs-2a. Zaidman and co-workers (2005) showed that EPSs from the mushrooms fruiting bodies that possess antitumor activity were mainly heteropolysaccharides presenting galactose, glucose, mannose and fructose (Zaidman *et al.*, 2005).

Using EPSs (c-EPSs) from *Lb. plantarum* 70810, the authors observed a significant inhibition in the proliferation of HepG-2, BGC-823 and HT-29 tumor cells. The authors suggest that c-EPSs might be suitable for use as functional foods and natural antitumor drugs (Wang *et al.*, 2014). In another study, the antitumor activity of EPSs from *Paecilomyces hepiali* HN1 (PHEPSs) was investigated, using three cell lines of human liver tumor – HepG2 cells, breast cancer MCF-7 cells and cervical cancer Hela cells. The results indicated that PHEPSs presented higher anti-proliferative effect against HepG2 cells than MCF-7 cells and Hela cells. At a concentration of 500 µg/mL, the inhibition rate of PHEPSs on HepG2 cells reached 62.58 after 72 hours of treatment (Wu *et al.*, 2014).

Xu *et al.* (2018) studied the antitumor activity of *Lachnum* YM130 (LEP-2a). This component presents mannose and galactose in its structure which could considerably enhance the inhibitory efficacy of 5-FU, a known antitumor agent, on Hela cells at the doses of 100, 200, 300, and 400 µg/mL. The results suggested that LEP-2a could be promising for developing as novel antitumor agents (Xu *et al.*, 2018). The antitumor effect of EPSs produced by *Fusarium* sp A14 on human hepatocellular carcinoma HepG2 cells was also demonstrated by Pan *et al.* (2019).

In addition to antitumor activity, EPSs may also be effective in wound healing, as demonstrated in the study by Insulkar *et al.* (2018). EPSs were produced by *Bacillus licheniformis* PASS26 and the authors observed an inhibition of MCF-7 cancer cells at concentrations of 0.8 and 1 mg/mL, reaching 46.2 and 67.8 per cent respectively. The effect of cell proliferation was assessed by a visual appearance of normal

HaCaT cells, with a migration from scratch to wound closure in the presence of EPSs. In conclusion, several studies reported that microbial EPSs have the potential for future applications in pharmacological formulations. Although many studies of EPSs antitumor activity have been published, the exact antitumor mechanism of these polysaccharides is not completely clarified and researchers are exploring the mechanism of action of EPSs.

4.5   Cholesterol-lowering Activity

Cardiovascular diseases are the main cause of death in the world (Mathers and Loncar, 2006). The mortality rate caused by vascular disorders, such as myocardial infarction, stroke, deep venous thrombosis, pulmonary embolism has significantly increased in recent years (Simkhada *et al.*, 2012; Mahajan *et al.*, 2012). The risk of heart attack is higher in individuals with hypercholesterolemia when compared to those who have normal blood lipid profiles. According to the American Heart Association, cardiovascular disease will remain the leading cause of death, affecting approximately 23.6 million people globally. Interestingly, the reduction of one per cent in serum low-density lipoprotein (LDL) decreases the risk of heart disease by around 3 per cent over a lifetime (Manson *et al.*, 1992). Moreover, the hypercholesterolemia is also a risk factor of metabolic syndrome and Type 2 diabetes mellitus (Gielen and Landmesser, 2014).

The ability of EPSs isolated from LAB in lowering the cholesterol has been investigated specially for β-glucans. It is known that β-glucans derived from oats can positively influence the cardiovascular health. The mechanism by which EPSs reduce cholesterol levels is not completely understood. According to the literature, EPSs could bind to cholesterol and promote its excretion, or indirectly increase conversion to bile through stimulation of increased numbers of microbes with bile-salt hydrolase activity, with a resultant decrease in bile levels (Welman, 2009; Korcz *et al.*, 2018).

In 2013, Duobin and co-workers described the potential of EPSs from *Pleurotus geesteranus* in preventing hyperglycemia in the experimental animals. Using streptozotocin-induced diabetic mice, EPSs were able to decrease plasma glucose, total cholesterol and triacylglycerol concentrations by 17.1 per cent, 18.8 per cent and 12.0 per cent, respectively (Duobin *et al.*, 2013). EPSs from *Lb. kefiranofaciens* reduced the serum cholesterol levels in rats fed with cholesterol-rich diet (Maeda and Omura, 2014). In another study, the supplementation of EPSs isolated from *kefir* grains significantly reduced high-fat diet-induced body gain, adipose tissue weight and very low-density lipoprotein cholesterol concentration (Lim *et al.*, 2017). Oat-based products fermented by using β-glucan producing *Pediococcus parvulus* were able to reduce total cholesterol levels in humans when compared to a control group (Martensson *et al.*, 2005).

The potential of *Lactobacillus* strains EPSs in hypocholesterolemic therapies was demonstrated by a study conducted by London *et al.* (2014a). The EPSs produced by *Lactobacillus* promote the decrease of serum triglycerides, total serum and liver cholesterol in mice that received a high-cholesterol diet (London *et al.*, 2014a). On the other hand, EPSs from *Pediococcus parvulus* did not cause any positive effect on blood lipids profile of hypercholesterolemia mice (Lindstrom *et al.*, 2012). In a study by Ghoneim *et al.* (2016), it was demonstrated that *B. subtilis* sp. EPSs reduced the total serum cholesterol, LDL, VLDL and triglycerides, and increased the HDL levels. In addition, EPSs were also able to decrease the atherogenic and coronary risks in mice (Ghoneim *et al.*, 2016). Using a different approach, Tok and Aslim (2010) showed the potential of *Lb. delbrueckii* subsp. *bulgaricus* to remove cholesterol from *in vitro* culture cells. All tested strains were able to remove the cholesterol from MRS broth. In addition, the presence of cholesterol also seemed to affect the EPSs level produced by the strains (Tok and Aslim, 2010).

In another study, EPSs from *Pleurotus eryngii* SI-04 showed a potent effect on mice treated with the high-fat diet-induced hyperlipidaemia. The results suggest that EPSs can be used as potential compounds and functional foods for the prevention of hyperlipidaemia (Zhang *et al.*, 2017b). EPSs from *Lb. plantarum* BR2 (Sasikumar *et al.*, 2017) and *Enterococcus faecium* K1 (Bhat and Bajaj, 2018) were found to significantly lower the cholesterol level in comparison to control. In summary, several experimental studies have shown that EPSs LAB may lower the blood cholesterol level and could be an alternative for hypercholesterolemia treatment. Nevertheless, the mechanism of cholesterol lowering is not totally understood and more data from human trials is needed to clarify such effect.

5. Conclusion

EPSs research has been receiving great attention since many years for allowing the development of several food products that have health benefits. The health benefits associated with functional foods with prebiotic effects have been partially attributed to EPSs. The increase in consumer demand for products containing EPSs encourages the industries to seek new polymers with appropriate characteristics. Thus, techniques to obtain high yields of EPSs, combined with analytical methods, metabolic engineering and *in vitro* models that better reflect physiology *in vivo* in the host organism, are valuable tools for establishing the correlation between the producing microorganism, EPSs structure and biological function. One of the challenges faced by industries is to get approval for new prebiotics or prebiotic fibres, even with available scientific evidence supporting pre-approval claims. Thus, continued research, particularly into the EPSs mechanism of action that promote modulation of health, will help to further validate the benefits associated with these polymers, reducing the risks of some diseases and improving human health.

Abbreviations

BHT – Butylated Hydroxytoluene
DPPH – 2, 2-diphenyl-1-picrylhydrazyl
EPSs – Exopolysaccharides
FOS – Fructooligosaccharide
HDL – High Density Lipoprotein
HePS – Heteropolysaccharides
HoPS – Homopolysaccharides

LAB – Lactic acid bacteria
LDL – Low-density lipoprotein
QPS – Qualified Presumption of Safety
ROS – Reactive Oxygen Species
TNF-α – Tumor Necrosis Factor Alpha
VLDL – Very Low Density Lipoprotein

References

Abdelnasser, S.M., Yahya, S.M.M., Mohamed, W.F., Asker, M.M.S., Shady, H.M.A., Mahmoud, M.G. and Gadallah, M.A. (2017). Antitumor exopolysaccharides derived from novel marine Bacillus: Isolation, characterisation aspect and biological activity. *Asian Pacific Journal Cancer Prevention*, 18: 1847-1854.

Amiri, S., Mokarrama, R.R., Khiabani, M.S., Bari, M.R. and Khaledabad, M.A. (2019). Exopolysaccharides production by *Lactobacillus acidophilus* LA5 and *Bifidobacterium animalis* subsp. lactis BB12: Optimisation of fermentation variables and characterisation of structure and bioactivities. *International Journal of Biological Macromolecules*, 123: 752-765.

Andhare, P., Chaukan, K., Dave, M. and Pathak, H. (2015). Microbial Exopolysaccharides: Advances in applications and future prospects, biotechnology, vol. 3. pp. 1-25. *In*: Rupinder Tewari (Ed.). *Microbial Biotechnology*. Studium Press LLC, Delhi, India.

Asker, M.M.S. and Shawky, B.T. (2010). Structural characterisation and antioxidant activity of an extracellular polysaccharide isolated from *Brevibacterium otitidis* BTS 44. *Food Chemistry*, 123: 315-320.

Awad, S., Hassan, A. and Muthukumarappan, K. (2005). Application of exopolysaccharide-producing cultures in reduced-fat Cheddar cheese: Texture and melting properties. *Journal of Dairy Science*, 88: 4204-4213.

Ayyash, M., Abu-Jdayil, B., Hamed, F. and Shaker, R. (2018). Rheological, textural, microstructural and sensory impact of exopolysaccharide-producing *Lactobacillus plantarum* isolated from camel milk on low-fat akawi cheese. *LWT – Food Science Technology*, 87: 423-431.

Badel, S., Bernardi, T. and Michaud, P. (2011). New perspectives for *Lactobacillus* exopolysaccharides. *Biotechnology Advance*, 29: 54-66.

Badhwar, P., Kumar, P. and Dubey, K.K. (2019). Extractive fermentation for process integration and amplified pullulan production by *A. pullulans* in aqueous two-phase systems. *Scientific Reports*, 9: 1-8.

Balzaretti, S., Taverniti, V., Guglielmetti, S., Fiore, W., Minuzzo, M., Ngo, H.N., Ngere, J.B., Sadiq, S., Humphreys, P.N. and Laws, A.P. (2016). A novel rhamnose-rich hetero-exopolysaccharide isolated from *Lactobacillus paracasei* DG activates THP-1 human monocytic cells. *Applied Environmental Microbiology*, 02702-02716.

Barbosa, K.B.F., Costa, N.M.B., Alfenas, R.C.G., Paula, S.O., Minim, V.P.R. and Bressan, J. (2010). Oxidative stress: Concept, implications and modulating factors. *Revista de Nutrição*, 23: 629-643.

Bersaneti, G.T., Pan, N.C., Baldo, C. and Celligoi, M.A.P.C. (2017). Co-production of fructooligosaccharides and levan by levansucrase from *Bacillus subtilis natto* with potential application in the food industry. *Applied Biochemistry and Biotechnology*, 184: 838-851.

Bhat, B. and Bajaj, B.K. (2018). Hypocholesterolemic and bioactive potential of exopolysaccharide from a probiotic *Enterococcus faecium* K1 isolated from kalarei. *Bioresource Technology*, 254: 264-267.

Bilgi, E., Bayir, E., Sendemir-Urkmez, A. and Hames, E.E. (2016). Optimisation of bacterial cellulose production by *Gluconaceto bacterxylinus* using carob and haricot bean. *International Journal of Biological Macromolecules*, 90: 2-10.

Bleau, C., Monges, A., Rashidan, K., Laverdure, J.P., Lacroix, M., Van Calsteren, M.R., Millette, M., Savard, R. and Lamontagne, L. (2010). Intermediate chains of exopolysaccharides from *Lactobacillus rhamnosus* RW-9595M increase IL-10 production by macrophages. *Journal of Applied Microbiology*, 108: 666-675.

Broadbent, J.R., McMahon, D.J., Oberg, C.J. and Welker, D.L. (2001). Use of exopolysaccharide-producing cultures to improve the functionality of low fat cheese. *International Dairy Journal*, 11: 433-439.

Brownawell, M.A., Caers, W., Gibson, G.R., Kendall, C.W.C., Lewis, K.D., Ringel, Y. and Slavins, J.L. (2012). Prebiotics and the health benefits of fibre: Current regulatory status, future research, and goals. *The Journal of Nutrition*, 142: 962-974.

Caggianiello, G., Kleerebezem, M. and Spano, G. (2016). Exopolysaccharides produced by lactic acid bacteria: From health-promoting benefits to stress tolerance mechanisms. *Applied Microbiology Biotechnology*, 100: 3877-3886.

Cao, S., Zhan, H., Fu, S.Y. and Chen, L. (2007). Regulation of superoxide anion radical during the oxygen delignification process. *Chinese Journal of Chemical Engineering*, 15: 132-137.

Chahuki, F.F., Aminzadeh, S., Jafarian, V., Tabandeh, F. and Khodabandeh, M. (2019). Hyaluronic acid production enhancement *via* genetically modification and culture medium optimisation in *Lactobacillus acidophilus*. *International Journal of Biological Macromolecules*, 121: 870-881.

Cheng, Z., Yang, R. and Liu, X. (2017). Production of bacterial cellulose by *Acetobacter xylinum* through utilizing acetic acid hydrolysate of bagasse as low-cost carbon source. *BioResources*, 12: 1190-1200.

Costa, N.E., Hannon, J.A., Guinee, T.P., Auty, M.A., McSweeney, P.L. and Beresford, T.P. (2010). Effect of exopolysaccharide produced by isogenic strains of *Lactococcus lactis* on half-fat Cheddar cheese. *Journal of Dairy Science*, 93: 3469-3486.

Coskun Cevher, S.C., Balabanli, B. and Aslim, B. (2015). Effects of probiotic supplementation on systemic and intestinal oxidant-antioxidant events in splenectomised rats. *Surgery Today*, 45: 1166-1172.

Dabour, N., Kheadr, E., Benhamou, N., Fliss, I. and LaPointe, G. (2006). Improvement of texture and structure of reduced-fat Cheddar cheese by exopolysaccharide-producing Lactococci. *Journal of Dairy Science*, 89: 95-110.

Dal Bello, F., Walter, J., Hertel, C. and Hammes, W.P. (2001). *In vitro* study of prebiotic properties of levan-type exopolysaccharides from Lactobacilli and non-digestible carbohydrates using denaturing gradient gel electrophoresis. *Systematic and Applied Microbiology*, 24: 232-237.

Demirci, A.S., Palabıyık, I., Apaydın, D., Mirik, M. and Gümüs, T. (2019). Xanthan gum biosynthesis using *Xanthomonas* isolates from waste bread: Process optimisation and fermentation kinetics. *LWT – Food Science and Technology*, 101: 40-47.

Dertli, E., Yilmaz, M.T., Tatlisu, N.B., Toker, O.S., Cankurt, H. and Sagdic, O. (2016). Effects of *in situ* exopolysaccharide production and fermentation conditions on physicochemical, microbiological, textural and microstructural properties of Turkish-type fermented sausage (sucuk). *Meat Science*, 121: 156-165.

Diana, C.R., Humberto, H.S. and Jorge, Y.F. (2019). Structural characterisation and rheological properties of dextran produced by native strains isolated of *Agave salmiana*. *Food Hydrocolloids*, 90: 1-8.

Donot, F., Fontana, A., Baccou, J.C. and Schorr-Galindo, S. (2012). Microbial exopolysaccharides: Main examples of synthesis, excretion, genetics and extraction. *Carbohydrate Polymers*, 87: 951-962.

Dórame-Miranda, R.F., Gámez-Meza, N., Medina-Juárez, L.Á., Ezquerra-Brauer, J.M., Ovando-Martínez, M. and Lizardi-Mendoza, J. (2019). Bacterial cellulose production by *Gluconacetobacter entanii* using pecan nutshell as carbon source and its chemical functionalisation. *Carbohydrate Polymers*, 207: 91-99.

Du, R., Xing, H., Yang, Y., Jiang, H., Zhou, Z. and Han, Y. (2017). Optimisation, purification and structural characterisation of a dextran produced by *Leuconostoc mesenteroides* isolated from Chinese sauerkraut. *Carbohydrate Polymers*, 174: 409-416.

Duh, P.D. (1999). Antioxidant activity of water extract of four Harng Jyur (*Chrysanthemum morifolium* Ramat) varieties in soybean oil emulsion. *Food Chemistry*, 66: 471-476.

Duobin, M., Yuping, M. Lujing, G. Aijing, Z., Jianqiang, Z. and Chunping, X. (2013). Fermentation characteristics in stirred-tank reactor of exopolysaccharides with hypolipidemic activity produced by *Pleurotus geesteranus* 5. *Anais da Academia Brasileira de Ciências*, 85: 1473-1481.

Ernandes, F.M.P.G. and Garcia-Cruz, C.H. (2011). Nutritional requirements of *Zymomonas mobilis* CCT 4494 for levan production. *Brazilian Archives of Biology and Technology*, 54: 589-600.

Fanning, S., Hall, L.J., Cronin, M., Zomer, A., MacSharry, J. and Goulding, D. (2012). Bifidobacterial surface-exopolysaccharide facilitates commensal-host interaction through immune modulation and pathogen protection. *Proceedings of the National Academy of Science*, 109: 2108-2113.

Galle, S., Schwab, C., Dal Bello, F., Coffey, A., Gänzle, M.G. and Arendt, E.K. (2012a). Comparison of the impact of dextran and reuteran on the quality of wheat sourdough bread. *Journal of Cereal Science*, 56: 531-537.

Galle, S., Schwab, C., Dal Bello, F., Coffey, A., Gänzle, M.G. and Arendt, E.K. (2012b). Influence of in-situ synthesized exopolysaccharides on the quality of gluten-free sorghum sourdough bread. *International Journal of Food Microbiology*, 155: 105-112.

Ghalem, B.R. (2017). Antioxidant and antimicrobial activities of exopolysaccharides from yoghurt starter. *Advances in Biochemistry*, 5: 97-101.

Ghoneim, M.A.M., Hassan, A.I., Mahmoud, M.G. and Asker. (2016). Effect of polysaccharide from *Bacillus subtilis* sp. on cardiovascular diseases and atherogenic indices in diabetic rats. *BMC Complementary and Alternative Medicine*, 16: 112-124.

Gibson, G.R., Probert, H.M., Van Loo, J., Rastall, R.A. and Roberfroid, M. (2004). Dietary modulation of the human colonic microbiota: Updating the concept of prebiotics. *Nutrition Research Reviews*, 17: 259-275.

Gielen, S. and Landmesser, U. (2014). The year in cardiology, 2013: Cardiovascular disease prevention. *European Heart Journal*, 35: 307-312.

Gojgic-Cvijovic, G.D., Jakovljevic, D.M., Loncarevic, B.D., Todorovic, N.M., Pergal, M.V., Ciric, J., Loos, V.P. and Vrvric, B.M.M. (2019). Production of levan by *Bacillus licheniformis* NS032 in sugar beet molasses-based medium. *International Journal of Biological Macromolecules*, 121: 142-151.

Gupta, P. and Diwan, B. (2017). Bacterial exopolysaccharide mediated heavy metal removal: A review on biosynthesis, mechanism and remediation strategies. *Biotechnology Reports*, 13: 58-71.

Hamdy, A.A., Elattal, N.A., Amin, M.A., Ali, A.E., Mansour, N.M., Awad, G.E., Farrag, A.R.H. and Esawy, M.A. (2018). *In vivo* assessment of possible probiotic properties of *Bacillus subtilis* and prebiotic properties of levan. *Biocatalysis and Agricultural Biotechnology*, 13: 190-197.

Han, X., Yang, Z., Jing, X., Yu, P., Zhang, Y., Yi, H. and Zhang, L. (2016). Improvement of the texture of yogurt by use of exopolysaccharide producing lactic acid bacteria. *BioMed Research International*, 1-6.

Hidalgo-Cantabrana, C., López, P., Gueimonde, M., de los Reyes-Gavilán, C.G., Suárez, A. and Margolles, A. (2012). Immune modulation capability of exopolysaccharides synthesised by lactic acid bacteria and bifidobacteria. *Probiotics and Antimicrobial Proteins*, 4: 227-237.

Hilares, R.T., Resende, J., Orsi, A.C., Ahmed, M.A., Lacerda, M.T., da Silva, S.S. and Santos, J.C. (2019). Exopolysaccharide (pullulan) production from sugarcane bagasse hydrolysate aiming to favor the development of biorefineries. *International Journal of Biological Macromolecules*, 127: 169-177.

Hilbig, J., Gisder, J., Loeffler, M., Prechtl, R.M., Herrmann, K. and Weiss, J. (2019). Influence of exopolysaccharide-producing lactic acid bacteria on the spreadability of fat-reduced raw fermented sausages (Teewurst). *Food Hydrocolloids*, 26: 734-743.

Hou, G., Chen, X., Li, J., Ye, Z., Zong, S. and Ye, M. (2019). Physicochemical properties, immunostimulatory activity of the *Lachnum* polysaccharide and polysaccharide-dipeptide conjugates. *Carbohydrate Polymers*, 206: 446-454.

Hu, X., Pang, X., Wang, P.G. and Chen, M. (2019). Isolation and characterisation of an antioxidant exopolysaccharide produced by *Bacillus* sp. S-1 from Sichuan pickles. *Carbohydrate Polymers*, 204: 9-16.

Huang, Q.L., Siu, K.C., Wang, W.Q., Heung, Y.C. and Wu, J.Y. (2010). Fractionation, characterisation and antioxidant activity of exopolysaccharides from fermentation broth of a *Cordyceps sinensis* fungus. *Process Biochemistry*, 48: 380-386.

Huang, T.L., Yang, S.H., Chen, Y.R., Liao, J.Y., Tang, Y. and Yang, K.C. (2019). The therapeutic effect of aucubin-supplemented hyaluronic acid on interleukin-1beta-stimulated human articular chondrocytes. *Phytomedicine*, 53: 1-8.

Hussein, A.S., Ibrahim, G.S., Asker, M.M.S. and Mahmoud, M.G. (2010). Exopolysaccharide from *Lactobacillus helveticus*: Identification of chemical structure and effect on biscuit quality. *Czech Journal of Food Sciences*, 28: 225-232.

Idogawa, N., Amamoto, R., Murata, K. and Kawai, S. (2014). Phosphate enhances levan production in the endophytic bacterium *Gluconacetobacter diazotrophicus* Pal 5. *Bioengineered*, 5: 173-179.

Insulkar, P., Kerkar, S. and Lele, S. (2018). Purification and structural-functional characterisation of an exopolysaccharide from *Bacillus licheniformis* PASS26 with *in vitro* antitumor and wound healing activities. *International Journal of Biological Macromolecules*, 120: 1441-1450.

Jiang, H., Xue, S.J., Li, Y.F., Liu, G.L., Chi, Z.M., Hu, Z. and Chi, Z. (2018). Efficient transformation of sucrose into high pullulan concentrations by *Aureobasidiumm elanogenum* TN1-2 isolated from a natural honey. *Food Chemistry*, 257: 29-35.

Jin, Y., Zhang, L., Zhang, M., Chen, L., Cheung, P.C.K., Oi, V.E.C. and Lin, Y. (2003). Antitumor activities of heteropolysaccharides of *Poriacocos* mycelia from different strains and culture media. *Carbohydrate Research*, 338: 1517-1521.

Juvonen, R., Honkapaa, K., Maina, N.H., Shi, Q., Viljanen, K., Maaheimo, H., Virkki, L., Tenkanen, M. and Lantto, R. (2015). The impact of fermentation with exopolysaccharide-producing lactic acid bacteria on rheological, chemical and sensory properties of pureed carrots (*Daucus carota* L.). *International Journal of Food Microbiology*, 207: 109-118.

Jolly, L., Vincent, S.J.F., Duboc, P. and Nesser, J. (2002). Exploiting exopolysaccharides from lactic acid bacteria. *Antonie Leeuwenhoek International Journal of General*, 82: 367-374.

Kajala, I., Mäkelä, J., Coda, R., Shukla, S., Shi, Q., Maina, N.H., Juvonen, R., Ekholm, P., Goyal, A., Tenkanen, M. and Katina, K. (2016). Rye bran as fermentation matrix boosts in situ dextran production by *Weissella confusa* compared to wheat bran. *Applied Microbiology and Biotechnology*, 100: 3499-3510.

Kanimozhi, J., Moorthy, I.G., Sivashankar, R. and Sivasubramanian, V. (2017). Optimisation of dextran production by *Weissella cibaria* NITCSK4 using response surface methodology-genetic algorithm based technology. *Carbohydrate Polymers*, 174: 103-110.

Kang, D., Cai, Z., Wei, Y. and Zhang, H. (2017). Structure and chain conformation characteristics of high acyl gellan gum polysaccharide in DMSO with sodium nitrate. *Polymer*, 128: 147-158.

Kanmani, P., Albarracin, L., Kobayashi, H., Iida, H., Komatsu, R., Akm, H.K., Ikedaohtsubo, W., Suda, Y., Aso, H. and Makino, S. (2018). Exopolysaccharides from *Lactobacillus delbrueckii* OLL1073R-1 modulate innate antiviral immune response in porcine intestinal epithelial cells. *Molecular Immunology*, 93: 253-265.

Katina, K., Maina, N.H., Juvonen, R., Flander, L., Johansson, L., Virkki, L., Tenkanen, M. and Laitila, A. (2009). *In situ* production and analysis of *Weissella confusa* dextran in wheat sourdough. *Food Microbiology*, 26: 734-743.

Khalil, E.S., Manap, M.Y., Mustafa, S., Amid, M., Alhelli, A.M. and Aljoubori, A. (2018). Probiotic characteristics of exopolysaccharides-producing *Lactobacillus* isolated from some traditional Malaysian fermented foods. *CyTA – Journal of Food*, 16(1): 287-298.

Kitazawa, H., Harata, T., Uemura, J., Saito, T., Kaneko, T. and Itoh, T. (1998). Phosphate group requirement for mitogenic activation of lymphocytes by an extracelular phosphopolysaccharide from *Lactobacillus delbrueckii* ssp. *Bulgaricus*. *International Journal Food Microbiology*, 40: 169-175.

Kodali, V.P. (2010). Antioxidant and free radical scavenging activities of an exopolysaccharide from a probiotic bacterium. *Biotechnology Journal*, 3: 245-251.

Kook, S.Y., Lee, Y., Jeong, E.C. and Kim, S. (2019). Immunomodulatory effects of exopolysaccharides produced by *Bacillus licheniformis* and *Leuconostoc mesenteroides* isolated from Korean *kimchi*. *Journal of Functional Foods*, 54: 211-219.

Korcz, E., Kerenyi, Z. and Varga, L. (2018). Dietary fibres, prebiotics and exopolysaccharides produced by lactic acid bacteria: Potential health benefits with special regard to cholesterol-lowering effects. *Food and Function*, 9: 3057-3068.

Kumar, A., Rao, K.M. and Han, S.S. (2018). Application of xanthan gum as polysaccharide in tissue engineering: A review. *Carbohydrate Polymers*, 180: 128-144.

Laws, A., Vitlic, A., Sadiq, S., Ahmed, H.I., Ale, E.C., Binetti, A.G., Collett, A. and Humpreys, P.N. (2018). Evidence for the modulation of the immune response in peripheral blood mononuclear cells after stimulation with a high molecular weight b-glucan isolated from *Lactobacillus fermentum* Lf2. doi: https://doi.org/10.1101/400 267.

Li, W., Xia, X., Tang, W., Ji, J., Rui, X., Chen, X., Jiang, M., Zhou, J., Zhang, Q. and Dong, M. (2015). Structural characterisation and anticancer activity of cell-bound exopolysaccharide from *Lactobacillus helveticus* MB2-1. *Journal of Agricultural Food Chemistry*, 8: 3454-3463.

Li, P., Zeng, Y., Xie, Y., Li, X., Kang, Y., Wang, Y., Xie, T. and Zhang, Y. (2017). Effect of pretreatment on the enzymatic hydrolysis of kitchen waste for xanthan production. *Bioresource Technology*, 223: 84-90.

Liang, T., Tseng, S. and Wang, S. (2016). Production and characterisation of antioxidant properties of exopolysaccharide(s) from *Peanibacillus mucilaginosus* TKU032. *Marine Drugs*, 40: 1-12.

Lim, J., Kale, M., Kim, D., Kim, H.S., Chon, J., Seo, K., Lee, H.G., Yokoyama, W. and Kim, H. (2017). Anti-obesity effect of exopolysaccharides isolated from *Kefir* grains. *Journal of Agricultural Food Chemistry*, 65: 10011-10019.

Lindstrom, C., Holst, O., Nilsson, L., Oste, R. and Andersson, K.E. (2012). Effects of *Pediococcus parvulus* 2.6 and its exopolysaccharide on plasma cholesterol levels and inflammatory markers in mice. *AMB Express*, 2: 1-9.

Liu, J., Luo, J., Ye, H., Sun, Y., Lu, Z. and Zeng, X. (2010). *In vitro* and *in vivo* antioxidant activity of exopolysaccharides from endophytic bacterium *Paenibacillus polymyxa* EJS-3. *Carbohydrate Polymers*, 82: 1278-1283.

Liu, C.F., Tseng, K.C., Chiang, S.S., Lee, B.H., Hsu, W.H. and Pan, T.M. (2011). Immunomodulatory and antioxidant potential of *Lactobacillus* exopolysaccharides. *Journal of Science Food Agricultural*, 91: 2284-2291.

Liu, G., Chen, H., Chen, J., Wang, X., Gu, Q. and Yin, Y. (2018). Effects of bifidobacteria-produced exopolysaccharides on human gut microbiota *in vitro*. *Applied Microbiology and Biotechnology*, 18: 9572-9576.

London, L.E., Chaurin, V., Auty, M.A., Fenelon, M.A., Fitzgerald, G.F., Ross, R.P. and Stanton, C. (2015). Use of *Lactobacillus mucosae* DPC 6426, an exopolysaccharide-producing strain, positively influences the techno-functional properties of yoghurt. *International Dairy Journal*, 40: 33-38.

London, L.E.E., Kumar, A.H.S., Wall, R., Casey, P.G., Sullivan, O.O., Shanahan, F., Hill, C., Cotter, P.D., Fitzgerald, G.F., Ross, R.P., Caplice, N.M. and Stanton, C. (2014a). Exopolysaccharide-producing probiotic lactobacilli reduce serum cholesterol and modify enteric microbiota in ApoE-deficient mice. *Journal of Nutrition*, 144: 1956-1962.

London, L.E., Price, N.P., Ryan, P., Wang, L., Auty, M.A., Fitzgerald, G.F., Stanton, C. and Ross, R.P. (2014b). Characterisation of a bovine isolate *Lactobacillus mucosae* DPC 6426 which produces an exopolysaccharide composed predominantly of mannose residues. Journal of Applied Microbiology, 117: 509-517.

Looijesteijn, P.J., Trapet, L., de Vries, E., Abee, T. and Hugenholtz, J. (2001). Physiological function of exopolysaccharides produced by *Lactococcus lactis*. *International Journal of Food Microbiology*, 64: 71-80.

Lorusso, A., Coda, R., Montemurro, M. and Rizzello, C.G. (2018). Use of selected lactic acid bacteria and quinoa flour for manufacturing novel yoghurt-like beverages. *Foods*, 7: 51.

Luna, W.N.S. (2016). *Acetilação do exopolissacarídeo (1®6)-β-d-glucana (lasiodiplodana): Derivatização química e caracterização, Dissertação apresentada ao Programa de Pós Graduação em Tecnologia de Processos Químicos e Bioquímicos, Universidade Tecnológica Federal do Paraná*, 1-79.

Lynch, K.M., Zannini, E., Coffey, A. and Arendt, E.K. (2018). Lactic acid bacteria exopolysaccharides in foods and beverages: Isolation, properties, characterisation, and health benefits. *Annual Review of Food Science and Technology*, 9: 155-176.

Lynch, K.M., Coffey, A. and Arendt, E.K. (2017). Exopolysaccharide-producing lactic acid bacteria: Their techno-functional role and potential application in gluten-free bread products. *Food Research International*, 110: 52-61.

Lynch, K.M., McSweeney, P.L., Arendt, E.K., Uniacke-Lowe, T., Galle, S. and Coffey, A. (2014). Isolation and characterisation of exopolysaccharide-producing *Weissella* and *Lactobacillus* and their application as adjunct cultures in Cheddar cheese. *International Dairy Journal*, 34: 125-134.

Maeda, H., Xia, Z. and Omura, K. (2004). Effects of *Pediococcus parvulus* 2.6 and its exopolysaccharide on plasma cholesterol levels and inflammatory markers in mice. *Biofactors*, 22: 197-201.

Marco, M.L., Heeney, D., Binda, S., Cifelli, C.J., Cotter, P.D., Foligné, B., Ganzle, M., Kort, R., Pasin, G., Pihlanto, A., Smid, E.J. and Hutkins, R. (2017). Health benefits of fermented foods: Microbiota and beyond. *Current Opinion in Biotechnology*, 44: 94-102.

Mahajan, P.M., Nayak, S. and Lele, S. (2012). Fibrinolytic enzyme from newly isolated marine bacterium *Bacillus subtilis* ICTF-1: Media optimisation, purification and characterisation. *Journal of Bioscience and Bioengineering*, 113: 307-314.

Manson, J.E., Tosteson, H., Ridker, P.M., Satterfield, S., Hebert, P., Connor, G.T.O., Buring, J.E. and Hennekens, C.H. (1992). The primary prevention of myocardial infarction. *The New England Journal of Medicine*, 326: 1406-1416.

Martensson, O., Biorklund, M., Lambo, A.M., Duenas-Chasco, M. and Irastorza, A. (2005). Fermented, ropy oat-based products reduce cholesterol levels and stimulate the bifidobacteria flora in humans. *Nutrition Research*, 25: 429-442.

Mathers, C.D. and Loncar, D. (2006). Projections of global mortality and burden of disease from 2002 to 2030. *PLoS Medicine*, 3: 442.

Matsuzaki, C., Takagaki, C., Higashimura, Y., Nakashima, Y., Hosomi, K., Kunisawa, J., Yamamoto, K. and Hisa, K. (2018). Immunostimulatory effect on dendritic cells of the adjuvant-active exopolysaccharide from *Leuconostoc mesenteroides* strain NTM048. *Bioscience Biotechnology Biochemistry*, 82: 1647-1651.

Moussa, T.A.A., Al-Qaysi, S.A.S., Thabit, Z.A. and Kadhem, S.B. (2017). Microbial levan from *Brachybacterium phenoliresistens*: Characterisation and enhancement of production. *Process Biochemistry*, 57: 9-15.

Nácher-Vázquez, M., Iturria, I., Zarour, K., Mohedano, M.L., Aznar, R., Pardo, M.Á. and López, P. (2017). Dextran production by *Lactobacillus sakei* MN1 coincides with reduced autoagglutination, biofilm formation and epithelial cell adhesion. *Carbohydrate Polymers*, 168: 22-31.

Nguyen, D.T. and Nguyen, T.H. (2014). Detection on antioxidant and cytotoxicity activities of exopolysaccharides isolated in plant-originated *Lactococcus lactis*. *Biomedical and Pharmacology Journal*, 7: 33-38.

Nishimura-Uemura, J., Kitazawa, H., Kawai, Y., Itoh, T., Oda, M. and Saito, T. (2003). Functional alteration of murine macrophages stimulated with extracellular polysaccharides from *Lactobacillus delbrueckii* ssp. *bulgaricus* OLL1073R-1. *Food Microbiology*, 20: 267-273.

Nikolic, M., López, P., Strahinic, I., Suárez, A., Kojic, M., Fernández-García, M., Topisirovic, L., Golic, N. and Ruas-Madiedo, P. (2012). Characterisation of the exopolysaccharide (EPS)-producing *Lactobacillus paraplantarum* BGCG11 and its non-EPS producing derivative strains as potential probiotics. *International Journal of Food Microbiology*, 158: 155-162.

Olano-Martin, E., Mountzouris, K.C., Gibson, G.R. and Rastall, R.A. (2000). *In vitro* fermentability of dextran, oligodextran and maltodextrin by human gut bacteria. *British Journal Nutrition*, 83: 247-255.

Paiva, I.M. (2013). *Caracterização estrutural e avaliação da capacidade imunomodulatória de exopolissacarídeos produzidos por Lactobacilos isolados de kefir, Dissertação apresentada ao Departamento de Biologia Geral do Instituto de Ciências Biológicas da Universidade Federal de Minas Gerais*, p. 1-99.

Pan, F., Hou, K., Li, D., Su, T. and Wu, W. (2019). Exopolysccharides from the fungal endophytic *Fusarium* sp. A14 isolated from *Fritillaria unibracteata* Hsiao et KC Hsia and their antioxidant and antiproliferation effects. *Journal of Bioscience and Bioengineering*, 127: 231-240.

Pan, N.C., Pereira, H.C.B., Silva, M.L.C., Vasconcelos, A.F.D. and Celligoi, M.A.P.C. (2017). Improvement production of hyaluronic acid by *Streptococcus zooepidemicus* in sugarcane molasses. *Applied Biochemistry and Biotechnology*, 182: 276-293.

Papageorgiou, S.K., Kouvelos, E.P., Favvas, E.P., Sapalidis, A.A., Romanos, G.E. and Katsaros, F.K. (2010). Metal-carboxylate interactions in metal-alginate complexes studied with FTIR spectroscopy. *Carbohydrate Research*, 345: 469-473.

Patel, A. and Prajapat, J.B. (2013). Food and health applications of exopolysaccharides produced by lactic acid bacteria. *Advances in Dairy Research*, 1: 1-8.

Patten, D.A., Leivers, S., Chadha, M.J., Maqsood, M., Humphreys, P.N., Laws, A.P. and Collett, A. (2014). The structure and immunomodulatory activity on intestinal epithelial cells of the EPSs isolated from *Lactobacillus helveticus* sp. Rosyjski and *Lactobacillus acidophilus* sp. *Carbohydrate Research*, 384: 119-127.

Parracho, H., Mccartney, A.L. and Gibson, G.R. (2007). Probiotics and prebiotics in infant nutrition. *Proceedings of Nutrition Society*, 66: 405-411.

Qi, G.X., Luo, M.T., Huang, C., Guo, H.J., Chen, X.F., Xiong, L., Wang, B., Lin, X.Q., Peng, F. and Chen, X.D. (2017). Comparison of bacterial cellulose production by *Gluconacetobacter xylinus* on bagasse acid and enzymatic hydrolysates. *Journal of Applied Polymer Science*, 134: 1-7.

Ripari, V. (2019). Techno-functional role of exopolysaccharides in cereal-based, yoghurt-like beverages. *Beverages*, 5: 16.

Roca, C., Alves, V.D., Freitas, F. and Reis, M.A.M. (2015). Exopolysaccharides enriched in rare sugars: Bacterial sources, production and applications. *Frontiers in Microbiology*, 6: 1-7.

Rühmkorf, C., Rübsam, H., Becker, T., Bork, C., Voiges, K., Mischnick, P. and Vogel, R.F. (2012). Effect of structurally different microbial homoexopolysaccharides on the quality of gluten-free bread. *European Food Research and Technology*, 235: 139-146.

Ryan, P.M., Ross, R.P., Fitzgerald, G.F., Caplice, N.M. and Stanton, C. (2015). Sugar coated: Exopolysaccharide producing lactic acid bacteria for food and human health applications. *Royal Society of Chemistry Food and Function*, 6: 679-693.

Sanjukta, S., Rai, A.K. and Sahoo, D. (2017). Bioactive molecules in fermented soybean products and their potential health benefits. pp. 97-121. *In*: R.C. Ray and D. Montet (Eds.). *Fermented Foods, Part II: Technological Interventions*. London, UK: CRC Press.

Sanjukta, S. and Rai, A.K. (2016). Production of bioactive peptides during soybean fermentation and their potential health benefits. *Trends in Food Science and Technology*, 50: 1-10.

Sanlier, N., Gokcen, B.B. and Sezgin, A.C. (2017). Health benefits of fermented foods. *Critical Reviews in Food Science and Nutrition*, 59: 506-527.

Salazar, N., Gueimonde, M., Hernandez-Barranco, A.M., Ruas-Madiedo, P. and de los Reyes-Gavil´na, C.G. (2008). Exopolysaccharides produced by intestinal *Bifidobacterium* strains act as fermentable substrates for human intestinal bacteria. *Applied and Environmental Microbiology*, 74: 4737-4745.

Sawale, S.D. and Lele, S.S. (2010). Statistical optimisation of media for dextran production by *Leuconostoc* sp., isolated from fermented idli batter. *Food Science and Biotechnology*, 19: 471-478.

Sasikumar, K., Vaikkath, D.K., Devendra, L. and Nampoothiri, K.M. (2017). An exopolysaccharide (EPS) from a *Lactobacillus plantarum* BR2 with potential benefits for making functional foods. *Bioresource Technology*, 241: 1152-1156.

Sawadogo-Lingani, H., Diawara, B., Traore, S. and Jakobsen, M. (2008). Technological properties of *Lactobacillus fermentum* involved in the processing of dolo and pito, West African sorghum beers, for the selection of starter cultures. *Journal of Applied Microbiology*, 104: 873-882.

Schwab, C., Mastrangelo, M., Corsetti, A. and Gänzle, M. (2008). Formation of oligosaccharides and polysaccharides by *Lactobacillus reuteri* LTH5448 and *Weissella cibaria* 10M in sorghum sourdoughs. *Cereal Chemistry Journal*, 85: 679-684.

Serafini, F., Turroni, F., Ruas-Madiedo, P., Lugli, G.A., Milani, C., Duranti, S., Zamboni, N., Bottacini, F., van Sinderen, D. and Margolles, A. (2014). *Kefir*-fermented milk and *kefiran* promote growth of *Bifidobacterium bifidum* PRL2010 and modulate its gene expression. *International Journal of Food Microbiology*, 178: 50-59.

Shi, Q., Hou, Y., Xu, Y., Kroghc, K.B.R.M. and Tenkanen, M. (2019). Enzymatic analysis of levan produced by lactic acid bacteria in fermented doughs. *Carbohydrate Polymers*, 208: 285-293.

Silva, J.A., Cardoso, L.G., Jesus Assis, D., Gomes, G.V.P., Oliveira, M.B.P.P., Souza, C.O. and Druzian, J.I. (2018). Xanthan gum production by *Xanthomonas campestris* pv. *campestris* IBSBF 1866 and 1867 from lignocellulosic agroindustrial wastes. *Applied Biochemistry and Biotechnology*, 186: 750-763.

Simkhada, J.R., Cho, S.S., Mander, P., Choi, Y.H. and Yoo, J.C. (2012). Purification, biochemical properties and antithrombotic effect of a novel *Streptomyces* enzyme on carrageenan-induced mice tail thrombosis model. *Thrombosis Research*, 129: 176-182.

Sun, L., Wang, C., Shi, Q. and Ma, C. (2009). Preparation of different molecular weight polysaccharides from *Porphyridium cruentum* and their antioxidant activities. *International Journal of Biological Macromolecules*, 45: 42-47.

Sun, N., Liu, H., Liu, S., Zhang, X., Chen, P., Li, W., Xu, X. and Tian, W. (2018). Purification, preliminary structure and antitumor activity of exopolysaccharide produced by *Streptococcus thermophiles* CH9. *Molecules*, 23: 2898-2910.

Sun, X., Hao, L., MaI, H., Li, T., Zheng, L., Zhao Ma, Z., Zhai, G., Wang, L., Gao, S., Liu, X., Jia, M. and Jia, L. (2013). Extraction and *in vitro* antioxidant activity of exopolysaccharide by *Pleurotus eryngii* SI-02. *Brazilian Journal of Microbiology*, 44: 1081-1088.

Tok, E. and Aslim, B. (2010). Cholesterol removal by some lactic acid bacteria that can be used as probiotic. *Microbiology Immunology*, 54: 257-264.

Tsiapali, E., Whaley, S., Kalbfleisch, J., Ensley, H.E., Browder, I.W. and Williams, D.L. (2001). Glucans exhibit weak antioxidant activity, but stimulate macrophage free radical activity. *Free Radical Biology and Medicine*, 30: 393-402.

Tzianabos, A.O. (2000). Polysaccharide immunomodulators as therapeutic agents: Structural aspects and biologic function polysaccharide immunomodulators as therapeutic agents. *Clinical Microbiology Reviews*, 13: 523-533.

Uzyol, H.K. and Saçan, M.T. (2017). Bacterial cellulose production by *Komagataeibacter hansenii* using algae-based glucose. *Environmental Science and Pollution Research*, 24: 11154-11162.

Waldherr, F.W. and Vogel, R.F. (2009). Commercial exploitation of homo-exopolysaccharides in non-dairy food systems. pp. 313-329. *In*: Ullrich, M. (Ed.). *Bacterial Polysaccharides: Current Innovations and Future Trends*. Academic Press, Norfolk, UK.

Wang, J., Wu, T., Fang, X., Min, W. and Yang, Z. (2018a). Characterisation and immunomodulatory activity of an exopolysaccharide produced by *Lactobacillus plantarum* JLK0142 isolated from fermented dairy tofu. *International Journal of Biological Macromolecules*, 115: 985-993.

Wang, C., Fan, O., Zhang, X., Lu, X., Xu, Y., Zhu, W., Zhang, J., Hao, W. and Hao, L. (2018b). Isolation, characterisation, and pharmaceutical applications of an exopolysaccharide from *Aerococcus Uriaeequi*. *Marine Drugs*, 16: 337-350.

Wang, K., Li, W., Rui, X., Chen, X., Mei Jiang, M. and Dong, M. (2014). Characterisation of a novel exopolysaccharide with antitumor activity from *Lactobacillus plantarum* 70810. *International Journal of Biological Macromolecules*, 63: 133-139.

Welman, A.D. (2009). Exploitation of exopolysaccharides from lactic acid bacteria: Nutritional and functional benefits. pp. 331–344. *In*: M. Ullrich (Ed.). *Bacterial Polysaccharides: Current Innovations and Future Trends*. Caister Academic Press, Norfolk, UK.

Westbrook, A.W., Ren, X., Moo-Young, M. and Chou, C.P. (2018). Engineering of cell membrane to enhance heterologous production of hyaluronic acid in *Bacillus subtilis*. *Biotechnology and Bioengineering*, 115: 216-231.

Wilburn, J.R. and Ryan, E.P. (2017). Fermented foods in health promotion and disease prevention: An overview A2 – Frias, Juana. pp. 3-19. *In*: C. Martinez-Villaluenga and E. Penas (Eds.). *Fermented Foods in Health and Disease Prevention*. Academic Press, Boston.

Wong, J.M., De Souza, R., Kendall, C.W., Emam, A. and Jenkins, D.J. (2006). Colonic health: Fermentation and short-chain fatty acids. *Journal of Clinical Gastroenterologyi*, 40: 235-243.

Wu, X., Li, P., Shen, Q., Liu, G.R. and Xu, R. (2018). The ecological regulating effect of exopolysaccharides produced by *Bifidobacterium* on intestinal microecological imbalance. *Ekoloji*, 27: 859-865.

Wu, Z., Lu, J., Wang, X., Hu, B., Ye, H., Fan, J., Abid, M. and Zeng, X. (2014). Optimisation for production of exopolysaccharides with antitumor activity *in vitro* from *Paecilomyces hepiali*. *Carbohydrate Polymers*, 99: 226-234.

Xiu, L., Zhang, H., Hu, Z., Liang, Y., Guo, S., Yang, M., Du, R. and Wang, X. (2018). Immunostimulatory activity of exopolysaccharides from probiotic *Lactobacillus casei* WXD030 strain as a novel adjuvant *in vitro* and *in vivo*. *Food and Agricultural Immunology*, 29: 1086-1105.

Xu, P., Yuan, R., Hou, G., Li, J. and Ye, M. (2018). Structural characterisation and *in vitro* antitumor activity of a novel exopolysaccharide from *Lachnum* YM130. *Applied Biochemistry and Biotechnology*, 185: 541-554.

Xu, X.-J., Fang, S., Li, Y.-H., Zhang, F., Shao, Z.-P., Zeng, Y.-T., Chen, J. and Meng, Y.-C. (2019). Effects of low acyl and high acyl gellan gum on the thermal stability of purple sweet potato anthocyanins in the presence of ascorbic acid. *Food Hydrocolloids*, 86: 116-123.

Yang, J., Zhang, Y., Zhao, S., Zhou, Q., Xin, X. and Chen, L. (2018). Statistical optimisation of medium for pullulan production by *Aureobasidium pullulans* NCPS2016 using fructose and soybean meal hydrolysates. *Molecules*, 23: 1-16.

Ye, J., Zheng, S., Zhang, Z., Yang, F., Ma, K., Feng, Y., Ke, M., Feng, Y., Zheng, J., Mao, D. and Yang, X. (2019). Bacterial cellulose production by *Acetobacter xylinum* ATCC 23767 using tobacco waste extract as culture medium. *Bioresource Technology*, 2174: 518-524.

You, L., Gao, Q., Feng, M., Yang, B., Ren, J. and Gu, L. (2013). Structural characterisation of polysaccharides from *Tricholoma matsutake* and their antioxidant and antitumor activities. *Food Chemistry*, 138: 2242-2249.

Zaidman, B.Z., Yassin, M., Mahajna, J. and Wasser, S.P. (2005). Medicinal mushroom modulators of molecular targets as cancer therapeutics. *Applied Microbiology and Biotechnology*, 67: 453-468.

Zia, K.M., Tabasum, S., Khan, M.F., Akram, N., Akhter, N., Noreen, A. and Zuber, M. (2018). Recent trends on gellan gum blends with natural and synthetic polymers: A review. *International Journal of Biological Macromolecules*, 109: 1068-1087.

Zhang, C., Li, J., Wang, J., Song, X., Zhang, J., Wu, Z., Hu, C., Gong, Z. and Jia, L. (2017a). Antihyperlipidaemic and hepatoprotective activities of acidic and enzymatic hydrolysis exopolysaccharides from *Pleurotus eryngii* SI-04. *Complementary and Alternative Medicine*, 17: 403-414.

Zhang, J., Zhao, X., Jiang, Y., Zhao, W., Guo, T., Cao, Y., Teng, J., Hao, X., Zhao, J. and Yang, Z. (2017b). Antioxidant status and gut microbiota change in an aging mouse model as influenced by exopolysaccharide produced by *Lactobacillus plantarum* YW11 isolated from Tibetan *kefir*. *Journal of Dairy Science*, 100: 6025-6041.

Zhang, J., Dong, Y., Fan, L., Jiao, Z. and Chen, Q. (2015). Optimisation of culture medium compositions for gellan gum production by a halobacterium *Sphingomonas paucimobilis*. *Carbohydrate Polymers*, 115: 694-700.

Zhao, W., Zhang, J., Jiang, Y., Zhao, X., Hao, X., Li, L. and Yang, Z. (2018). Characterisation and antioxidant activity of the exopolysaccharide produced by *Bacillus amyloliquefaciens* GSBa-1. *Microbiology and Biotechnology*, 28: 1282-1292.

Zhu, G., Guo, N., Yong, Y., Xiong, Y. and Tong, Q. (2019). Effect of 2-deoxy-d-glucose on gellan gum biosynthesis by *Sphingomonas paucimobilis*. *Bioprocess and Biosystems Engineering*, 1: 1-4.

Impact of Fermentation on Antinutritional Factors

Minna Kahala*, Sari Mäkinen and Anne Pihlanto

Natural Resources Institute Finland, Myllytie 1, 31600 Jokioinen, Finland

1. Introduction

Interest in plant-based diets is increasing continuously among consumers. Importance of cereals and grain legumes in diets is recognised and are recommended in the dietary guidelines. They are important constituents of the diet throughout the world and have traditionally been staple food and low-cost sources of proteins and nutrients for people, especially in developing countries (Abeshu and Kefale, 2017). However, the presence of antinutritional factors, having the potential to cause adverse effects on nutrition, or harmful compounds, in particular in pulses and cereals, reduces their nutritional value by decreasing the bioavailability of proteins, minerals and vitamins (Jain *et al.*, 2009; Khokhar and Owusu Apenten, 2009; Soetan and Oyewole, 2009). Antinutrients present in pulses and cereals include compounds, such as phytates, vicines, alkaloids, oligosaccharides, saponins, lectins and tannins, as well as amylase and protease inhibitors (Jain *et al.*, 2009; Khokhar and Owusu Apenten, 2009; Soetan and Oyewole, 2009). These factors are produced by plants to protect themselves but also restrict their use as food and feed. The knowledge of these compounds and processing of the raw materials is vital, especially in developing countries that depend mostly on plant-based diets (Soetan and Oyewole, 2009).

Fermentation is an important technique to improve the sensory characteristics and preserve fresh vegetables, fruits, cereals and legumes for feeding people and animals as the global population increases. Fermentation is an old technique which has been frequently utilised globally and has led to numerous traditional fermented products typical for specific areas. For example, *sauerkraut*, *kimchi*, sourdough bread, Indian *idli*, *dhokla*, Greek *tarhanas*, Indonesian *tempeh*, Japanese *natto*, Indian *kinema* and *bekang* (Sudarmadji and Markakis, 1977; Reddy *et al.*, 1982; Wei *et al.*, 2001; Omizu *et al.*, 2011) may serve as an essential part of the diet.

Fermentation has been shown to improve the nutritional quality of the legume and cereal products. Studies reveal the effect of fermentation on degradation of

---

*Corresponding author: minna.kahala@luke.fi

the antinutritional compounds and bioavailability of proteins and minerals. By optimisation of the process conditions and selecting well-known strains, adapted to the raw material, the nutritional value can be significantly enhanced. Studies on the diversity of the microbial community, as well as microbial dynamics during the manufacture of fermented products, can bring a better understanding of the fermentation-process characteristics. Information of metabolic differences of the strains is crucial for selecting suitable starters for obtaining the desired product in bioprocessing (Rizzello *et al.*, 2018).

This chapter summarises the present knowledge on the effects of fermentation processes on the antinutritional and harmful compounds of proteinaceous plants (Fig. 1). It also focuses especially on legumes, cereals and pseudocereals and the potential microbial processing possibilities for enhancing their safety and nutritional value.

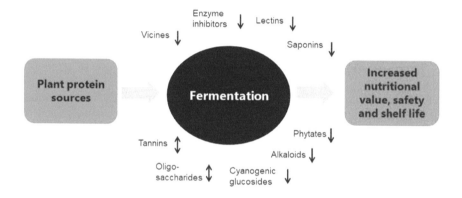

Fig. 1: Impact of fermentation on plant-based materials

2. Vicine and Convicine

Pyrimidine glycosides vicine (2,6-diamino-4,5-dihydroxypyrimidine 5-β-D-glucopyranoside) and convicine (2,4,5-trihydroxy-6-aminopyrimidine 5-β-D-glucopyranoside) are natural bioactive compounds in plants detected in various species of family Fabaceae (Pavlík *et al.*, 2002). The presence of pyrimidine glycosides in food is known to be involved in the formation of haemolytic anemia (favism). Ingestion of faba bean, which contains these glycosides, may result in hemolytic crisis in susceptible persons having genetic deficiency of glucose 6-phosphate dehydrogenase (G6PD) (Luzzatto and Arese, 2018). Vicines also have deleterious effects on monogastric animals, like chickens (Khamassi *et al.*, 2013). The hydrolysis of vicine and convicine generates the aglycones divicine and isouramil which are suggested to be the main factors in causing favism in humans (Pulkkinen *et al.*, 2016; Luzzatto and Arese, 2018). Degradation products, divicine and isouramil, have been shown to be fairly unstable and degrade in low pH; however, the decomposition pathways of the aglycones are still poorly known (Pulkkinen *et al.*, 2016).

Faba bean, belonging to the Fabaceae family, is an important and widely cultivated legume, which can be utilised both as food and feed. It has a valuable nutritional profile, yet faba bean contains two naturally-occurring toxic glycosides, vicine and convicine. Several studies have shown that the contents of vicines in faba beans can be reduced by fermentation (Coda *et al.*, 2015; Rizzello *et al.*, 2016). Certain lactic acid bacteria (LAB) strains, e.g. some *Lactobacillus* strains and some fungi, such as *Fusarium graminearum* and *Aspergillus oryzae* produce extracellular β-glucosidase, which participates in degradation of the vicines (McKay, 1992; Rizzello *et al.*, 2016). Especially, *Lactobacillus plantarum* strains, frequently used as starters in food fermentations, are reported to affect the degradation of vicines (McKay, 1992; Coda *et al.*, 2015; Rizzello *et al.*, 2016). Faba bean flour fermented with *Lb. plantarum* was shown to degrade the vicines completely after 48 hours of incubation (Rizzello *et al.*, 2016). *Streptococcus faeciens* was shown to convert convicine completely and vicine about 95 per cent (Donath and Kujawa, 1991). Detoxification of faba bean by fermentation using LAB was evidenced in the *ex vivo* hemolytic assay performed by Rizzello *et al.* (2016). The hemolysis of red blood cells in LAB-inoculated sample was considerably lower as compared to the spontaneously fermented control.

3. Saponins

Saponins are glycoside derivatives containing a carbohydrate moiety (mono/oligosaccharide) attached to an aglycone and can be found with varying contents in many plants, including legumes, root crops, tea, some cereals and medicinal herbs (Khokhar and Owusu Apenten, 2009; Qian *et al.*, 2018). Saponins are grouped according to their type of aglycone (Omizu *et al.*, 2011). They give a bitter taste, cause foaming in solutions, lower nutrient availability and can cause hemolysis in red blood cells (Khokhar and Owusu Apenten, 2009). On the other hand, saponins have also health-beneficial effects in humans, like reducing the cholesterol levels and risk of heart diseases (Jain *et al.*, 2009; Omizu *et al.*, 2011). They are not destroyed by cooking; however soaking, washing, abrasive process, germination and fermentation are known to reduce saponin content (Reddy and Pierson, 1994; Kumar *et al.*, 2010; Bolívar-Monsalve *et al.*, 2018). Pearling was shown to reduce the saponin content up to 79 per cent; however, also decrease in minerals was detected (Bolívar-Monsalve *et al.*, 2018).

β-glucuronidase produced by *Lactobacillus* and *Bacillus* strains hydrolyses saponins from glucosides to aglycones and can reduce hemolytic activity in tea seed saponins (Lv *et al.*, 2018; Qian *et al.*, 2018). Quinoa, native pseudocereal of South America, has a good nutritional value but contains triterpenic saponins which cause bitter taste and form complexes with some nutrients (Bolívar-Monsalve *et al.*, 2018). Fermentation with *Lb. plantarum* showed 71 per cent reduction of saponins in four days of quinoa dough process (Bolívar-Monsalve *et al.*, 2018). *Lactobacillus casei* subsp. *casei* produced β-glucuronidase in soybean fermentation, and thus hydrolysed soy saponins (Miyamoto *et al.*, 2012). *Bacillus*-fermented soybean foods of India, similar to Japanese *natto*, resulted in low reduction of saponins. Group A saponins were degraded, but showed only low decrease in total saponin content. Soybean seeds contain saponins (0.4-0.7 per cent, dry weight) while fermented foods contained 0.2-0.6 per cent, mainly consisting of group B saponins (Omizu *et al.*, 2011). Addition of *Streptococcus thermophilus* and *Bifidobacterium infantis* to soymilk reduced the saponin and phytate content and the fermented soymilk showed greater anti-proliferative effect against HT-29 and Caco-2 cells, which most probably is a sum of all antitumor-cell bioactive compounds present in fermented soymilk (Lai *et al.*, 2013). While processing of Lucerne for feed, saponins were reduced by *Lb. plantarum* strains, including a commercial starter, Lalmand (Tian *et al.*, 2018). Fermentation by *Lb. plantarum* and *Aspergillus niger* reduced the saponin levels in trembesi leaves, but the tolerable level, which is under 0.2 per cent saponin in feeds, wasn't reached (Sariri *et al.*, 2018).

4. Raffinose Family Oligosaccharides

Oligosaccharides are carbohydrates found in many plant tissues, particularly in seeds. Galacto-oligosaccharides generally include the (indigestible) β-(1,3 or 1,6) linked disaccharides and α-galactosyllactose (Gänzle and Follador, 2012). They are water soluble and thus concentrated with protein during the water extraction. Raffinose family oligosaccharides (RFO), of which the most commonly found are raffinose, stachyose and verbascose, are known as anti-nutrients present especially in cereals and legumes. They consist of one, two and three α-(1,6)-D-galactose units, respectively, linked to sucrose (Silvestroni *et al.*, 2002; Gänzle and Follador, 2012). They cannot be digested in the intestine because of the lack of alpha-1, 6-galactosidase enzyme in the humans and monogastric animals. This leads to passing of these saccharides along the small intestine without being degraded or absorbed. Consequently, they accumulate in the lower intestine following anaerobic fermentation by gas-producing intestinal bacteria in the large intestine, e.g. *Clostridium* spp. and *Bacteroides* spp. causing gastrointestinal symptoms (Silvestroni *et al.*, 2002; Adewumi and Odunfa, 2009; Teixeira *et al.*, 2012) including abdominal pain, nausea, diarrhoea and increased peristalsis (Silvestroni *et al.*, 2002). On the other hand, these indigestible fibres are also known to reduce the risk of intestinal cancer, fortify the immune system, increase excretion frequency and weight as well as HDL cholesterol level (Onder and Kahraman 2009). For example, wheat and rye contained 0.1-0.5 per cent, lupin 0.23 per cent and peeled buckwheat 0.73 per cent raffinose, while grain legumes were reported to contain 2-10 per cent RFO (Gänzle and Follador, 2012; Mattila *et al.*, 2018b). In the widely consumed soy, 40 per cent of the sugars present are α-galactosides (stachyose and raffinose) (Connes *et al.*, 2004)**.**

Certain microbes produce α-galactosidase and are capable of degrading the flatulence-causing indigestible oligosaccharides and converting them to digestible carbohydrates, which are further converted to organic acids (Adewumi and Odunfa, 2009; Teixeira *et al.*, 2012). α-galactosidase, encoded by *melA,* is found in Lactobacilli (Silvestroni *et al.*, 2002; Gänzle and Follador, 2012). In *Pediococcus pentosaceus* the fermentation of raffinose is known to be plasmid encoded and the presence of α-Gal is inducible (Gonzalez and Kunka, 1986). In addition to α-Gal activity, fructansucrases, such as glycoside hydrolase enzyme levansucrase, may participate in partial hydrolysis of raffinose oligosaccharides. This enzyme is also present in many Lactobacilli (Tieking *et al.*, 2003). The presence of enzymes hydrolysing raffinose family oligosaccharides is an important selection criterion when the starter mix is selected.

Several studies with different raw material have shown the effect of LAB on decreasing the RFO levels (Curiel *et al.*, 2015; Xu *et al.*, 2017; Rizzello *et al.*, 2018). Raffinose utilisation is not so widespread among different LAB species (Kostinek *et al.*, 2007; Rizzello *et al.*, 2018). A number of LAB strains, isolated from cassava fermentations and belonging to different groups (heterofermentative, homofermentative, rods, bacilli) showed raffinose fermenting ability. However, among the strains belonging to the group of the entero-, lacto- and streptococci, only 4 per cent of strains were positive for raffinose utilisation (Kostinek *et al.*, 2007). Furthermore, utilisation of stachyose was evidenced to be rare in all the LAB groups when compared to raffinose (Kostinek *et al.*, 2007). Studies indicate that many food-fermenting Lactobacilli and *Leuconostoc* species produce α-Gal and have been used to eliminate RFOs in food prepared from soy, pinto beans, cowpea or field pea flours, or pearl millet (Teixeira *et al.*, 2012). The level of locust bean oligosaccharides was shown to be reduced significantly by *Lb. plantarum, Lb. fermentum* and *Pediococcus acidilactici,* which were originally isolated from traditional African fermented foods (Adewumi and Odunfa, 2009).

Silvestroni *et al.* (2002) showed that in soy fermentations, *Lactobacillus* strains, i.e. *Lb. plantarum, Lb. fermentum, Lb. brevis,* and *Lb. buchneri,* can hydrolyse α-galactosides to digestible carbohydrates. These strains are commonly used in vegetable fermentations. *Lactobacillus reuteri* metabolised raffinose, stachyose and verbascose by levansucrase activity and accumulated RFOs as metabolic intermediates in faba bean and field pea fermentations (Teixeira *et al.*, 2012). In a study of Garro *et al.* (1998), *Lb. fermentum* reduced the content of stachyose by 73 per cent already after four hours of fermentation of soy. Raffinose and galactose were released but not consumed during the fermentation period. Presence of raffinose may also be the result of partial hydrolysis of longer chain RFOs (Teixeira *et al.*, 2012; Coda *et al.*, 2015). Fritsch *et al.* (2015) observed several strains, including *Lb. plantarum, P. pentosaceus* and *Lactococcus lactis* subsp. *lactis,* with raffinose and stachyose utilisation ability in fermentations of lupin. *Bifidobacterium lactis* degraded raffinose by 92-100 per cent and reduced stachyose content by approximately 23 per cent (Fritsch *et al.*, 2015). Several studies on sourdough processes have demonstrated an effect on the content of oligosaccharides (Rizzello *et al.*, 2010, 2016; Coda *et al.*, 2014, 2015). In faba bean sourdough process, raffinose, stachyose and verbascose amounts decreased significantly by *P. pentosaceus, Weissella cibaria* and *Weissella confusa* (Rizzello *et al.*, 2018). Raffinose was not in detectable levels in all the inoculated doughs at 48 hours. Raffinose content was shown to decrease significantly in several studied grains, wheat, barley, chickpea, lentil, quinoa, by fermentation with strains of *Lb. plantarum, Lb. rossiae* and *Lb. sanfranciscensis* (Montemurro *et al.*, 2019).

The enzymes from *Bacillus natto* has been shown to hydrolyse oligosaccharides and polysaccharides into reducing sugars during fermentation (Wei *et al.*, 2001). A significant reduction of total and particularly oligosaccharides of lupin was observed by yeasts *Candida utilis, Saccharomyces cerevisieae* and *Kluyveromyces lactis* reaching total reduction of oligosaccharides by all the strains after 72 hours of fermentation (Kasprowicz-Potocka *et al.*, 2018). Khattab and Arntfield (2009) demonstrated about 70 per cent reduction of RFOs in different legume seeds after 24-hour fermentation by *S. cerevisiae*. Complete degradation of oligosaccharides was obtained in lupine after 48 hours of fermentation by *Rhizopus oligosporus* (Jiménez-Martínez *et al.*, 2007).

Synthesis and activity of α-galactosidase, levansucrase or sucrose phosphorylase expression increased in LAB cultures where oligosaccharides were present (Garro *et al.*, 1998; Silvestroni *et al.*, 2002; Teixeira *et al.*, 2012) and growth of bifidobacteria was stimulated by the presence of non-digestible oligosaccharides (Fritsch *et al.*, 2015). Interestingly, exogenously supplied glucose inhibits the synthesis of the α-galactosidase enzyme (Silvestroni *et al.*, 2002). LAB with the RFO degrading activity have a role as starter strains in fermentations, but also as probiotics. They are supposed to be used as vehicles to deliver enzyme activity into the small intestine of mammals for improved digestion of antinutritional compounds (Connes *et al.*, 2004).

5. Phytic Acid

Phytic acid (myoinositol hexaphosphate), or phytate in a salt form is the primary storage form of both phosphate and inositol in plant seeds. It is widely present in legumes and cereals and represents 60-90 per cent of the total phosphate (Reddy and Pierson, 1994; Khokhar and Owusu Apenten, 2009; Kumar *et al.*, 2010). In the widely used soybean, phytate concentrations are up to 2.3 per cent of the grains dry weight (Karkle and Beleia, 2011). Content of phytic acid, condensed tannins and total phenolic compounds in some protein-rich plant materials are summarised in Table 1. Phytic acid is considered as an anti-nutrient since it acts as chelator of minerals, such as Ca, Mg, Fe and Zn, reducing their bioavailability, impeding their absorption and digestion in the small intestine (Reddy and Pierson, 1994; Khokhar and Owusu Apenten, 2009; Kumar *et al.*, 2010; Rizzello *et al.*, 2010). Phytate is also closely associated with proteins, reducing their solubility, functionality and digestibility (Reddy and Pierson, 1994; Khokhar and Owusu Apenten, 2009; Rizzello *et al.*, 2010;

Table 1: Content of Phytic Acid, Condensed Tannins and Total Phenolic Compounds in Protein-rich Plant Materials

	Phenolic Compounds GAE mg/100 g DM	Condensed Tannins mg/100 g DM	Phytic Acid g/100 g DM
Faba bean	364[a] - 6747[b]	470	0.6[c] - 1.6[a]
Quinoa	56[j] - 181[a]	-	1.1[k] - 1.2[a]
Lupin	163[a] - 696[d]	-	1.0[a] - 1.2[e]
Oil hemp seed	96[a] - 2224[h]	105[a]	3.5[a]
Oil hemp seed meal	ND	0.23[f]	7.5[i] - 22.5[f]
Rapeseed press cake	728[a]	119[a] - 1300[g]	2.8[g] - 3.9[a]

(a) Mattila *et al.* (2018a); (b) Chaieb *et al.* (2011); (c) Oomah *et al.* (2011); (d) Dalaram, 2017; (e) Trugo *et al.* (1993); (f) Pojić *et al.* (2014); (g) Khattab *et al.* (2010); (h) Vonapartis *et al.* (2015); (i) Russo and Reggiani (2015); (j) Diaz-Valencia *et al.* (2018); (k) Ruales and Nair (1993)

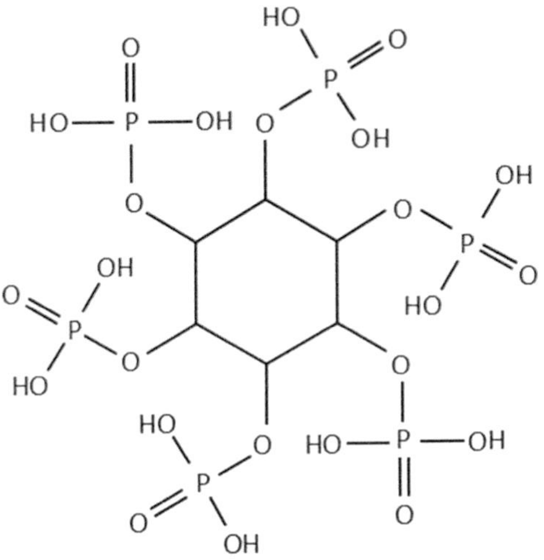

Fig. 2: Phytic acid

Gupta *et al.*, 2013). This is particularly important in developing countries where the diet is composed mainly of plants, thus causing deficiency of micronutrients (Gupta *et al.*, 2013). However, consumption of phytate may have also some beneficial effects for diabetics, for heart diseases and it may also provide protection against a variety of cancers mediated through antioxidative properties (Kumar *et al.*, 2010; Karkle and Beleia, 2011). Yet, the information on ingested levels of phytates, both for positive or negative effects, is still limited; thus optimal dosage cannot be recommended (Kumar *et al.*, 2010).

Nutritional value can be improved by dephosphorylation of phytate. By food processing techniques, such as germination, soaking, cooking and fermentation, phytate levels are reduced in foods (Ouwehand *et al.*, 2004; Kumar *et al.*, 2010). Phytases are enzymes, which may be naturally present in plant raw material or in microorganisms and are capable of making phosphate available by catalysing the hydrolysis of phytates to inorganic ortophosphates (Mullaney and Ullah, 2003). In addition, phytase can be generated by the small intestinal mucosa and gut-associated microbes (Kumar *et al.*, 2010). The most important sources of microbial phytases are bacteria and fungi, such as *Pseudomonas*, LAB, especially *Lactobacillus* spp., *A. niger*, yeasts *Pichia*, *Candida* and *Saccharomyces* (Kumar *et al.*, 2010; Fritsch *et al.*, 2015). The enzymatic degradation of phytic acid requires an optimum pH which activates the endogenous phytases (Reddy and Pierson, 1994; De Angelis *et al.*, 2003, Gupta *et al.*, 2013). Phytase activity is very sensitive to pH, and microbial phytases show a lower pH optimum than that determined for cereal phytases (Nielsen *et al.*, 2007). In bread making, the acidity of the dough has an important role in phytate degradation. In addition to natural phytases present in raw materials, sourdough LAB are also source of phytases and phytase activity is shown to increase during fermentation by strains like *P. pentosaceus*, *Lb. plantarum*, *Lb. rossiae*, *Leuconostoc kimchii*, *Lb. sanfranciscensis* in several legume and cereals sourdough processes

(Reddy and Pierson, 1994; Rizzello *et al.*, 2010; Montemurro *et al.*, 2019). The temperature and pH affected the phytic acid degradation, higher temperature and lower pH giving lower levels of phytic acid (Požrl *et al.*, 2009).

Food products, such as *tempeh, miso, koji* and soy sauce are produced by fermentation of soybeans with molds, *R. oligosporus* and *A. oryzae*, which are known to produce phytases (Kumar *et al.*, 2010). Phytic acid content of soybeans, fermented with *R. oligosporus* in tempe processing, was reduced by about 30 per cent during the preparation (Sudarmadji and Markakis, 1977). In whole-grain barley meal, iron absorption was shown to be improved while phytate content decreased via tempe fermentation. Pretreatment techniques or combination of different techniques, such as pearling, soaking and cooking the grains further improved the phytate hydrolysis in barley tempe process up to 90 per cent degradation (Eklund-Jonsson *et al.*, 2008). Also natural fermentation utilised in Indian foods, *dhokla, khama* and *idli*, lead to reduction of phytate (Reddy *et al.*, 1982); 88.3 per cent reduction in phytate content was obtained when pearl millet was fermented with mix pure cultures of yeast, *Saccharomyces diasticus, S. cerevisiae, Lb. brevis* and *Lb. fermentum*, at 30°C for 72 hours (Kaur *et al.*, 2014). Combined germination and lactic fermentation of sorghum and maize yielded more than 90 per cent degradation in phytate content (Svanberg *et al.*, 1993). The increase in soluble iron was also shown to strongly relate to enzymatic degradation of phytate. Kasprowicz-Potocka *et al.* (2015) observed reduction of phytate phosphorus concentrations by about 67 per cent for blue and by 29 per cent for yellow lupine seeds by *C. utilis*.

6. Lupine Alkaloids

Alkaloids are naturally-occurring toxic amine compounds that plants produce for protection against herbivores and microorganisms (Ortega-David and Rodriguez-Stouvenel, 2014; Fritsch *et al.*, 2015). Best-known leguminous plant containing alkaloids is lupine. Lupine (*Lupinus*) is a legume with a valuable nutritional composition but the use of this plant is limited since it is known to contain quinolizidine alkaloids, neurotoxins that cause bitter taste. By plant breeding the level of alkaloids has been reduced significantly; thus currently cultivated lupines are almost entirely sweet, containing alkaloids less than 0.05 per cent (Kasprowicz-Potocka *et al.*, 2018). Alkaloid content varies between lupine varieties, planting site and environmental conditions (Kasprowicz-Potocka *et al.*, 2018). Varieties cultivated, e.g. *Lupinus albus, L.mutabilis, L. campestris, L. angustifolius, L. luteus,* may contain alkaloids, such as lupanine, 13-hydroxylupanine, spartein, tetrahydroxyrombifolin and angustifoline in addition to other alkaloids with smaller amounts of isolupanine (Ortega-David and Rodriguez-Stouvenel, 2014; Fritsch *et al.*, 2015; Kasprowicz-Potocka *et al.*, 2018). Fritsch (2015) published alkaloid amounts of 0.5 per cent in bitter lupin and 0.02 per cent in sweet lupin, Kasprowicz-Potocka *et al.* (2018) reported 0.167 per cent alkaloid content in DM (dry matter) of seeds of narrow-leafed lupine, Ortega-David and Rodriguez-Stouvenel (2014) showed results with an average of 0.035 per cent of DM, whereas sweet variety *L. angustifolius* showed no detectable alkaloids (Signorini *et al.*, 2017). The maximum permitted level of quinolizidine alkaloid content in lupine flours and foods is set at 0.02 per cent (Fritsch *et al.*, 2015; Abeshu and Kefale, 2017).

The reduction of alkaloid content by fermentation depends on the preparation of media, microorganism type, pH and method of fermentation (Ortega-David and Rodríguez-Stouvenel, 2013; Abeshu and Kefale, 2017; Kasprowicz-Potocka *et al.*, 2018). During the tempe preparation by using *Rhizopus,* or by cofermentation with *Rhizopus* and *Propionibacterium*, 80 per cent or more of the alkaloid content was reduced in different lupine varieties in 72 hours (Signorini *et al.*, 2017). Similar results were reported in other traditional tempe processes, where depending on the lupine variety, 70-90 per cent lowered levels of alkaloids were obtained by *R. oligosporus* (Jiménez-Martínez *et al.*, 2007; Ortega-David and Rodriguez-Stouvenel, 2014). The significance of the particle size of the medium and suitable moisture content for degradation of quinolizidine alkaloids was shown (Ortega-David and Rodriguez-Stouvenel, 2014). Maximum degradation rate was obtained with the pH of 5.5 (Ortega-David and Rodríguez-Stouvenel, 2013; Ortega-David and Rodriguez-Stouvenel, 2014). Yeast strains *C. utilis, K. lactis*, and *S. cerevisiae* have been shown to yield a reduction of 5-16 per cent in the total alkaloid content in fermented products of the narrow-leafed lupine. Depending on the compound, the levels varied; angustifoline content decreased significantly in all the fermented products, while on the contrary, the content of isolupanine even increased (Kasprowicz-Potocka *et al.*, 2018). *C. utilis* was found to be the most effective because of the highest protein content, total reduction of RFOs and partial degradation of phytate and alkaloids (Kasprowicz-Potocka *et al.*, 2018).

7. Tannins

Tannins are a group of polyphenols formed as secondary metabolites in plants. Tannins are known for their antioxidant properties and   show chemoprotective potential. Berries and fruits are the best known sources of tannins, but they are also included in the chemical composition of proteinaceous plants, such as legumes. In contrast to the health-promoting effects of tannins, they can reduce the nutritive value of food products by forming tannin-protein complexes, resulting in inactivation of digestive enzymes and protein solubility. They can also reduce the dietary mineral bioavailability by binding, for example the ionisable iron (Brune *et al.*, 1989). Also, a lowered feed efficiency and growth depression of animals have been reported due to the tannins. The antinutritional activity of tannins can be reduced by processing, for example, by dehulling, soaking, cooking, germination and fermentation.

Tannins are known to be located mainly in the seed coats of legumes (Egounlety and Aworh, 2003). Díaz *et al.* (2010) characterised condensed tannins, tannin monomers and anthocyanins of common bean. The overall tannin content of common bean varieties ranged from 0-2 per cent of dry matter depending on the bean species and colour of the seed coat. Meanwhile, the overall average for tannins in seed coats was 20 per cent of dry matter. Condensed tannins of common bean varieties consisted mainly of catechin, gallocatechin and afzelechin monomeric units with respective proportions of 60, 25 and 15 per cent. Content of condensed tannins

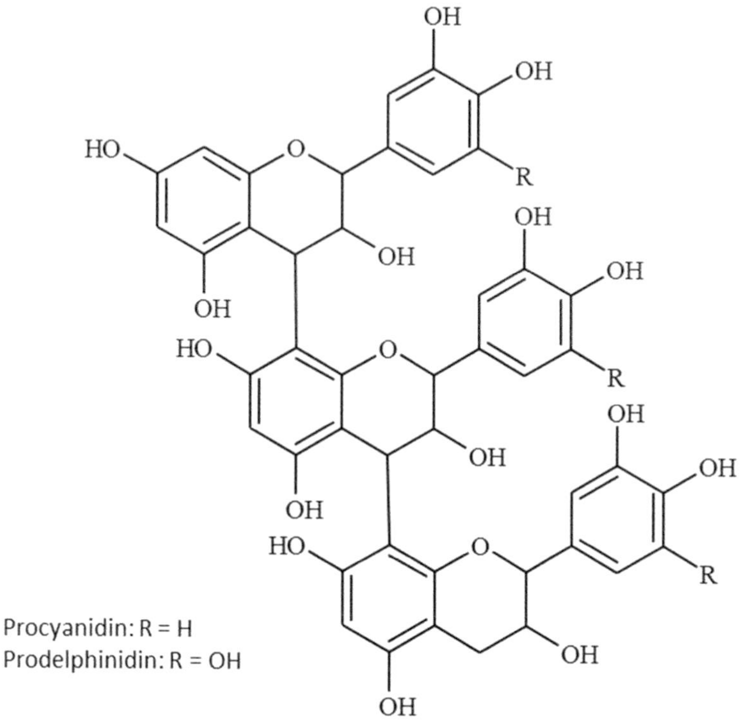

Procyanidin: R = H
Prodelphinidin: R = OH

Fig. 3: Condensed tannin (proanthocyanidin)

in some protein-rich plant materials are reviewed in Table 1. Dehulling has been found to be the most effective pre-treatment for removal of tannins from legumes in several studies (Egounlety and Aworh, 2003; Díaz *et al.*, 2010). Total removal of tannins with dehulling has been reported for example, for cowbean, soybean and groundbean (Egounlety and Aworh, 2003). Debranning has also been shown to reduce significantly tannin content of pseudocereals, such as pearl millet (Sharma and Kapoor, 1996). Soaking, germination, fermentation and heating are other options for decreasing tannin contents. Studies on the effects of fermentation on tannin content of plants have been conducted mainly in combination with different kinds of pre-treatments. Fermentation with *R. oligosporus* has been shown to remove residual tannins from dehulled and soaked legume materials (Egounlety and Aworh, 2003), while native fermentation of sorghum induced only a minor effect on the tannin content (Abdelhaleem *et al.*, 2008). The effects of fermentation on the tannin content depend mainly on the microbial strain applied. Fermentation with *Lb. plantarum* and *Rhodorotula* increased significantly the polyphenol content of pearl millet, while fermentation with *Lb. acidophilus* and *C. utilis* decreased the polyphenol content (Sharma and Kapoor, 1996). Fermentation with yeast strain *S. cerevisiae* and *Lactobacillus* strain, *Lb. brevis* also significantly reduced the polyphenol content of pearl millet (Khetarpaul and Chauhan, 1991). However, it should be taken into account that the polyphenols of these plants include also other compounds, except tannins. In general, the reduction of polyphenols after fermentation has been suggested to be due to activation of polyphenol oxidase (Dhankher and Chauhan, 1987).

8. Enzyme Inhibitors

Insecticidal proteins, like α-amylase inhibitors and protease inhibitors present in the seeds of legumes, have been suggested to play a major role in insect resistance. These are considered as most promising weapons that confer resistance against insects and are suggested to be eco-friendly alternative to synthetic pesticides. On the other hand, the same compounds, protease and α-amylase inhibitors, may have adverse effects on digestion and trigger an allergenic reaction in humans (Sagar and Dhall, 2018). Raw legumes and cereals are known to contain protease inhibitors with a serine residue, which interacts with trypsin and/or chymotrypsin proteases (Belitz and Weder, 1990). These protease inhibitors are proteins that form stoichiometric protease-inhibitor complexes with their respective enzymes and inhibit their activity in the gastrointestinal tract. They will bind to the digestive enzyme, either through competitive or allosteric mode of action to render the enzyme inactivate (Glencross, 2016). The presence of chymotrypsin inhibitors in animal diets will cause indigestion and abdominal pain, while pancreatic enlargement and growth depression can be caused by the presence of trypsin inhibitors (Kumar *et al.*, 2013). The amounts of protease inhibitors vary in different cereals, e.g. trypsin inhibitory activity of rye is 10-fold higher than that of wheat and oats (Mikola and Mikkonen, 1999). However, only a few reports have been published on their protease inhibitory activities, probably due to their lower inhibitor activity compared with legumes. Presence of α-amylase inhibitors is also evidenced in legumes and buckwheat. They inhibit human and porcine pancreatic amylases by non-competitive inhibition mechanisms (Santimone *et al.*, 2004). The complex formation of α-amylase inhibitors with α-amylases reduces starch digestion by inhibiting the hydrolysis of starch α-1,4-glycosidic bonds (Sagar and Dhall, 2018). Therefore, the presence of α-amylase inhibitor in the human diet can cause impaired carbohydrate digestion and associated digestive issues which could ultimately result in coeliac disease, leading to weight loss (Kumar *et al.*, 2013).

Kostekli and Karakaya (2017) found that trypsin inhibitory activity in mixed cereal flour decreased significantly during fermentation (dough) and baking. This decrease was about 3.6 times after fermentation and 8.6 times after baking. Moreover, chymotrypsin inhibitory activity of rye mix and mixed cereals flours was significantly less during fermentation, but baking had no effect. Interesting findings were a significant increase in chymotrypsin inhibitory activity after fermentation of whole wheat and a significant decrease in chymotrypsin inhibitory activity after fermentation of mixed cereals. This increase can be due to the formation of new protein fragments with chymotrypsin inhibitory activity during fermentation. Spontaneous fermentation of tef dough showed significant decrease of trypsin inhibitor activity, 68.9 per cent after 96 hours of fermentation (Urga *et al.*, 1997). Dehulling generally results in increase in the activity of the enzyme (trypsin, chymotrypsin and α-amylase) inhibitors found in legumes, though literature on this subject is not unambiguous. The extent of the dehulling effect on trypsin inhibitory activity appears to depend on variety. No effect was observed in two Egyptian varieties of chickpeas (Attia *et al.*, 1994), whereas Marquez and Alonso (1999) noted a 22 per cent reduction upon dehulling of Spanish chickpea cultivars. A slight increase (7.4 per cent) in α-amylase inhibitors was noted in lentils and faba beans (Shekib *et al.*, 1988; Alonso *et al.*, 2000). Deshpande *et al.* (1982) found that dehulling of whole dry beans increased the activities of all three enzyme inhibitors in the 10 dry bean cultivars investigated. Increases in the α-amylase inhibitory activity ranged from 27 per cent in the Sanilac cultivar to 188 per cent in the Great Northern cultivar; increases in trypsin inhibitory activity ranged from 1.6 per cent in the Viva pink cultivar to 36 per cent in the Sanilac cultivar and increases in chymotrypsin inhibitor activity ranged from 2.7 per cent in dark red kidney beans to 25 per cent in small red beans. The activity of chymotrypsin inhibitors in faba beans was reduced during cooking by 38-100 per cent (Vidal-Valverde *et al.*, 1997; Hernandez-Infante *et al.*, 1998) and completely absent in dry seeds (Jourdan *et al.*, 2007).

Germination reduces enzyme inhibitor activities but not to the same extent as thermal treatments. Reports for trypsin inhibitory activity indicate a 25 per cent reduction in trypsin inhibitory activity for faba beans after three days of germination in light (Alonso *et al.*, 2000) and a 31.8 per cent reduction when germinated for three days in the dark (Khalil, 1995). Greater inactivation of trypsin inhibitory activity (62.9 per cent), chymotrypsin inhibitory activity (73.4 per cent), and α-amylase inhibitory activity (67.1 per cent) were observed after five days of germination in the dark for Great Northern beans (Sathe *et al.*, 1983), and a 76 per cent reduction in trypsin inhibitory activity in white kidney beans was observed after seven days of germination in the light (Savelkoul *et al.*, 1994). Frias *et al.* (1995) reported a 15-18 per cent reduction in lentil trypsin inhibitory activity after six days of germination in the dark, whereas a 22 per cent reduction of trypsin inhibitory activity after six days of germination in the light was observed by Urbano *et al.* (1995). In dry beans, LAB fermentation of raw flour, flour soaked for 12 hours and flour soaked (12 hours) and then cooked (90 min.) resulted in 38, 50, and 95 per cent reductions in trypsin inhibitory activity, respectively. In a *tempeh* product made from chickpea flour, fermentation reduced trypsin inhibitory activity by 89.7 per cent (Abu-Salem and Abou-Arab, 2011).

9. Lectins (Hemagglutinins)

Plant lectins are natural carbohydrate reactive proteins that are frequently found in legumes, cereals and many plants used as farm feeds. In case of legume seeds, lectins represent approximately 2–10 per cent of total protein. Lectins are highly resistant to cooking and digestion and thus, they can reach the small intestine and blood circulation as biologically active compounds. They can prevent absorption of nutrients in the small intestine and enable the coagulation of red blood cells by affecting erythrocytes. Also, they can induce mild allergic or other subclinical effects in the higher animals and humans (Onder and Kahraman, 2009). Thus, to make these plant foods safe and to increase their nutritional value, lectins need to be inactivated or removed before consumption. Semi-quantitative haemagglutination assays and

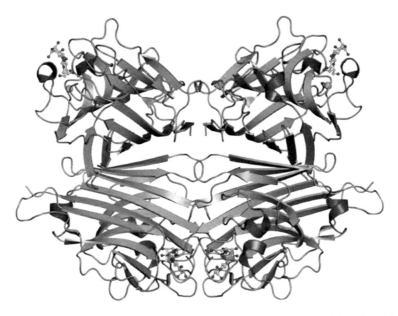

Fig. 4: Structure of legume lectin (http://www.ebi.ac.uk/https://commons.wikimedia.org/wiki/ File:PDB_1v6i_EBI.jpg). The figure is taken from the web page, Wikimedia Commons, the free media repository, and http://www.ebi.ac.uk/ should be appreciated.

more precise ELISA techniques are the most commonly used methods to determine the presence of lectins.

Intensive heat treatment is the most commonly used method for inactivating lectins. However, lectins are known to be more resistant to heat denaturation than other plant proteins and thus, heat processing can damage the other protein structures and reduce the nutritional value. Heat treatment is also rather expensive, especially when the product is used for animal feed. Combination of cooking, soaking and fermentation can be utilised to efficiently remove and/or inactivate lectins from various plants (Reddy and Pierson, 1994). Recently, also extrusion has been shown to inactivate 95 per cent of lentil lectins (Morales *et al.*, 2015). This far, information on the effects of fermentation on the lectin activity is rare; however, few reports have been published. Natural fermentation of lentil (*Lens culinaris*) with native microbial strains was shown to reduce lectin content significantly (Cuadrado *et al.*, 2002). The largest decrease in haemagglutination activity occurred after the first 24 hours of fermentation, and after 72 hours of fermentation the initial lectin content reduced by 98 per cent according to the ELISA assay. Results were confirmed by SDS-polyacrylamide gel electrophoresis and immunoblotting. Lentil flour concentration and fermentation temperature affected the removal of lectins; the best results were obtained with fermentation at 42°C with lentil flour concentration of 79 g/L. Solid-state fermentation using different fungal cultures has been shown to reduce anti-nutritional components of Jatropha seed cake, a by-product after biodiesel extraction. Fermentation was carried out for six, nine and 12 days with fungal strains, such as *Cunninghamella echinulata*. The unfermented Jatropha seed cake showed haemagglutination activity of 0.32 units, while after 12 days of fermentation, no haemagglutination activity was detected (Sharath *et al.*, 2014). However, according to Phengnuam and Suntornsuk (2013), solid-state fermentation of Jatropha seed cake with *Bacillus* strains, *B. subtilis* and *B. licheniformis* does not decrease the lectin contents. Fermentation in combination with soaking and cooking has been shown to reduce up to 95 per cent of the lectin contents of beans (Reddy and Pierson, 1994). Recently, it has also been proposed that certain dairy probiotic microorganisms, such as *P. acidipropionici* can bind to dietary lectins and thus reduce the antinutritional effects of them. According to the results, consumption of propionibacteria at the same time with plant lectins convalin A and jacalin could reduce the lectin-induced alterations in the gut and thus, may provide a tool to protect intestine from the adverse effects (Zárate *et al.*, 2017).

10.　Cyanogenic Glycosides

Cyanogenic glycosides are chemical compounds contained in foods that release hydrogen cyanide when chewed or digested. The act of chewing or digestion leads to hydrolysis of the substances, causing release of cyanide. There are approximately 25 known cyanogenic glycosides and these are generally found in the edible parts of plants, such as apples, apricots, cherries, peaches, plums, quinces, particularly in the seed of such fruits. The chemicals are also found in almonds, stone fruit, pome fruit, cassava, bamboo shoots, linseed/flaxseed, lima beans, coco yam, chick peas, cashews, and kirsch (Codex Committee on Contaminants in Foods, 2008).

　　Yeast and LAB are the most investigated microorganisms for the production of linamarase during cassava fermentation and the development of flavour. Natural β-glucosidase producing lactic acid bacteria strains can be applied for specific degradation of cyanogenic glycosides. Yeast, such as *S. cerevisiae*, is able to use cyanogenic glucosides and their metabolites during food processing, making it one of the microorganisms which is most involved in the cassava fermentation process (Lambri *et al.*, 2013). Oboh and Akindahunsi (2003), showed a marked decrease in the cyanide content of cassava products (cassava flour and gari) under solid-state fermentation with *S. cerevisiae* after 72 hours of fermentation. Another work by Gunawan *et al.* (2015) reported that after fermentation with *S. cerevisiae*, the cyanide content in cassava flour reduced by more than 88 per cent. The solid-state fermentation experiments showed that the cyanide content was reduced in line with an increase in fermentation time and after 60 hours of fermentation, the concentration of the cyanide in the leaves was below 10 ppm, bringing it to a safe level for human consumption (Hawashi *et al.*, 2019; WHO, 1995). Tefera *et al.* (2014) found that fermentation with *Lb. plantarum* lowers cassava cyanide content from 197.19 mg/g to 4.09 mg/g within 24 hours of inoculation. Several researchers isolated lactic acid bacteria strains belonging to *Lb. plantarum* and *Lb. paracasei* from African traditional fermented food, which are able to hydrolyze aryl-β-D-glucosidic bonds such as those in salicin, esculin and amygdalin. These observations indicate that most microbial β-D-glucosidases are acidic, belong to class 3 and are able to cleave aryl or alkyl glucosidic bonds.

11. Conclusion

In summary, the findings indicate that fermentation is a potential natural and feasible option for decreasing the level of antinutrients and improving the nutritional value of plant protein sources. Fermentation can be easily combined with pre-treatments, such as dehulling, soaking and cooking to reach adequate efficiency in antinutrient degradation in legumes and cereals. The biocatalysts produced by microbial starters involved in degradation of antinutritional factors need to be studied. Concerning some of the antinutrients, more research is needed to characterise food-grade microbial strains for their ability to degrade and/or inactivate these compounds.

Abbreviations

DM – Dry matter
ELISA – Enzyme linked immune sorbent assay
ex vivo – Experiment which takes place outside an organism
G6PD – 6-phosphate dehydrogenase
GAE – Gallic acid equivalent
HDL – High density lipoprotein
L – Litre
LAB – Lactic acid bacteria
RFO – Raffinose Family Oligosaccharides

Acknowledgements

The authors express their acknowledgements to Jarkko Hellström for making Figs. 2 and 3 and Dr Vesa Joutsjoki for critical reading of the chapter.

References

Abdelhaleem, W.H., El Tinay, A.H., Mustafa, A.I. and Babiker, E.E. (2008). Effect of fermentation, malt-pretreatment and cooking on antinutritional factors and protein digestibility of sorghum cultivars. *Pakistan Journal of Nutrition*, 7: 335-341.

Abeshu, Y. and Kefale, B. (2017). Effect of some traditional processing methods on nutritional composition and alkaloid content of lupin bean. *Int. J. Bioorg. Chem.*, 2: 174-179.

Abu-Salem, F.M. and Abou-Arab, E.A. (2011). Physico-chemical properties of *tempeh* produced from chickpea seeds. *Journal of American Science*, 7: 107-118.

Adewumi, G.A. and Odunfa, S.A. (2009). Effect of controlled fermentation on the oligosaccharides content of two common Nigerian Vigna unguiculata beans (drum and oloyin). *African Journal of Biotechnology*, 8: 2626-2630.

Alonso, R., Aguirre, A. and Marzo, F. (2000). Effects of extrusion and traditional processing methods on antinutrients and in vitro digestibility of protein and starch in faba and kidney beans. *Food Chemistry*, 68: 159-165.

Attia, R.S., Shehata, A.M.E.T., Aman, M.E. and Hamza, M.A. (1994). Effect of cooking and decortication on the physical properties, the chemical composition and the nutritive value of chickpea (*Cicer arietinum* L.). *Food Chemistry*, 50: 125-131.

Belitz, H.D. and Weder, J.K.P. (1990). Protein inhibitors of hydrolases in plant foodstuffs. *Food Reviews International*, 6: 151-211.

Bolívar-Monsalve, J., Ceballos-González, C., Ramírez-Toro, C. and Bolívar, G.A. (2018). Reduction in saponin content and production of gluten-free cream soup base using quinoa fermented with *Lactobacillus plantarum*. *Journal of Food Processing and Preservation*, 42.

Brune, M., Rossander, L. and Hallberg, L. (1989). Iron absorption and phenolic compounds: Importance of different phenolic structures. *European Journal of Clinical Nutrition*, 43: 547-557.

Chaieb, N., González, J.L., López-Mesas, M., Bouslama, M. and Valiente, M. (2011). Polyphenols content and antioxidant capacity of thirteen faba bean (*Vicia faba* L.) genotypes cultivated in Tunisia. *Food Research International*, 44: 970-977.

Coda, R., Cagno, R., Di, Gobbetti, M. and Rizzello, C.G. (2014). Sourdough lactic acid bacteria: Exploration of non-wheat cereal-based fermentation. *Food Microbiology*, 37: 51-58.

Coda, R., Melama, L., Rizzello, C.G., Curiel, J.A., Sibakov, J., Holopainen, U., Pulkkinen, M. and Sozer, N. (2015). Effect of air classification and fermentation by *Lactobacillus plantarum* VTT E-133328 on faba bean (*Vicia faba* L.) flour nutritional properties. *International Journal of Food Microbiology*, 193: 34-42.

Codex Committee on Contaminants in Foods (2008). Discussion paper on cyanogenic glycosides. FAO/WHO, Rome. CX/CF 09/3/11

Connes, C., Silvestroni, A., Leblanc, J.G., Juillard, V., de Giori, G.S., Sesma, F. and Piard, J.C. (2004). Towards probiotic lactic acid bacteria strains to remove raffinose-type sugars present in soy-derived products. *Lait*, 84: 207-214.

Cuadrado, C., Hajos, G., Burbano, C., Pedrosa, M.M., Ayet, G., Muzquiz, M., Pusztai, A. and Gelencser, E. (2002). Effect of natural fermentation on the lectin of lentils measured by immunological methods. *Food and Agricultural Immunology*, 14: 41-49.

Curiel, J.A., Coda, R., Centomani, I., Summo, C., Gobbetti, M. and Rizzello, C.G. (2015). Exploitation of the nutritional and functional characteristics of traditional Italian legumes: The potential of sourdough fermentation. *International Journal of Food Microbiology*, 196: 51-61.

Dalaram, I.S. (2017). Evaluation of total polyphenol content and antioxidant capacity of different verity lupin seeds. *Potravinarstvo*, 11: 26-34.

De Angelis, M., Gallo, G., Corbo, M.R., McSweeney, P.L.H., Faccia, M., Giovine, M. and Gobbetti, M. (2003). Phytase activity in sourdough lactic acid bacteria: Purification and characterisation of a phytase from *Lactobacillus sanfranciscensis* CB1. *International Journal of Food Microbiology*, 87: 259-270.

Deshpande, S.S., Sathe, S.K., Salunkhe, D.K. and Cornforth, D.P. (1982). Effects of dehulling on phytic acid, polyphenols, and enzyme inhibitors of dry beans (*Phaseolus vulgaris* L.). *Journal of Food Science*, 47: 1846-1850.

Dhankher, N. and Chauhan, B.M. (1987). Effect of temperature and fermentation time on phytic acid and polyphenol content of rabadi – A fermented pearl millet food. *Journal of Food Science*, 52: 828-829.

Díaz, A.M., Caldas, G.V. and Blair, M.W. (2010). Concentrations of condensed tannins and anthocyanins in common bean seed coats. *Food Research International*, 43: 595-601.

Diaz-Valencia, Y.K., Alca, J.J., Calori-Domingues, M.A., Zanabria-Galvez, S.J. and Da Cruz, S.H. (2018). Nutritional composition, total phenolic compounds and antioxidant activity of quinoa (*Chenopodium quinoa*, Willd.) of different colours. *Nova Biotechnologica et Chimica*, 77: 74-85.

Donath, R. and Kujawa, M. (1991). Untersuchungen zum Abbau von Vicin und Convicin in Ackerbohnenmehl durch ausgewählte Bakterienstämme. *Food/Nahrung*, 35: 449-453.

Egounlety, M. and Aworh, O.C. (2003). Effect of soaking, dehulling, cooking and fermentation with *Rhizopus oligosporus* on the oligosaccharides, trypsin inhibitor, phytic acid and tannins of soybean (*Glycine max* Merr.), cowpea (*Vigna unguiculata* L. Walp) and groundbean (*Macrotyloma geocarp*a Ha). *Journal of Food Engineering*, 56: 249-254.

Eklund-Jonsson, C., Sandberg, A.S., Hulthen, L. and Alminger, M.L. (2008). Tempe fermentation of whole grain barley increased human iron absorption and *in vitro* iron availability. *The Open Nutrition Journal*, 2: 42-47.

FAO/WHO (1995). Codex Standard for Edible Cassava Flour. Codex Standard 176-1989. Food and Agriculture Organisation and World Health Organisation of the United Nations, Rome, Italy.

Frias, J., Diaz-Pollan, C., Hedley, C.L. and Vidal-Valverde, C. (1995). Evolution of trypsin inhibitor activity during germination of lentils. *Journal of Agricultural and Food Chemistry*, 43: 2231-2234.

Fritsch, C., Vogel, R.F. and Toelstede, S. (2015). Fermentation performance of lactic acid bacteria in different lupin substrates—influence and degradation ability of antinutritives and secondary plant metabolites. *Journal of Applied Microbiology*, 119: 1075-1088.

Gänzle, M.G. and Follador, R. (2012). Metabolism of oligosaccharides and starch in Lactobacilli: A review. *Frontiers in Microbiology*, 3: 1-15.

Garro, M.S., de Valdez, G.F., Oliver, G. and de Giori1, G.S. (1998). Growth characteristics and fermentation products of *Streptococcus salivarius* subsp. *thermophilus*, *Lactobacillus casei* and *Lb. fermentum* in soymilk. *Zeitschrift fur Lebensmitteluntersuchung und Forschung A*, 206: 72-75.

Glencross, B. (2016). Understanding the nutritional and biological constraints of ingredients to optimise their application in aquaculture feeds. pp. 33–73. *In*: S.F. Nates (Ed.). *Aquafeed Formulation*. Academic Press, San Diego, CA, USA.

Gonzalez, C.F. and Kunka, B.S. (1986). Evidence for plasmid linkage of raffinose utilization and associated α-galactosidase and sucrose hydrolase activity in *Pediococcus pentosaceus*. *Applied and Environmental Microbiology*, 51: 105-109.

Gunawan, S., Widjaja, T., Zullaikah, S., Ernawati, L., Istianah, N., Aparamarta, H.W. and Prasetyoko, D. (2015). Effect of fermenting cassava with *Lactobacillus plantarum*, *Saccharomyces cereviseae*, and *Rhizopus oryzae* on the chemical composition of their flour. *International Food Research Journal*, 22: 1280-1287.

Gupta, R.K., Gangoliya, S.S. and Singh, N.K. (2013). Reduction of phytic acid and enhancement of bioavailable micronutrients in foodgrains. *Journal of Food Science and Technology*, 52: 676-684.

Hawashi, M., Aparamarta, H., Widjaja, T. and Gunawan, S. (2019). Optimisation of solid state fermentation conditions for cyanide content reduction in cassava leaves using response surface methodology. *International Journal of Technology*, 10: 624-633.

Hernandez-Infante, M., Sousa, V., Montalvo, I. and Tena, E. (1998). Impact of microwave heating on hemagglutinins, trypsin inhibitors and protein quality of selected legume seeds. *Plant Foods for Human Nutrition*, 52: 199-208.

Jain, A.K., Kumar, S. and Panwar, J.D.S. (2009). Antinutritional factors and their detoxification in pulses – A review. *Agricultural Reviews*, 30: 64-70.

Jiménez-Martínez, C., Hernández-Sánchez, H. and Dávila-Ortiz, G. (2007). Diminution of quinolizidine alkaloids, oligosaccharides and phenolic compounds from two species of Lupinus and soybean seeds by the effect of *Rhizopus oligosporus*. *Journal of the Science of Food and Agriculture*, 87: 1315-1322.

Jourdan, G.A., Norea, C.P.Z. and Brandelli, A. (2007). Inactivation of trypsin inhibitor activity from Brazilian varieties of beans (*Phaseolus vulgaris* L.). *Food Science and Technology International*, 13: 195-198.

Karkle, E.N.L. and Beleia, A. (2011). Effect of soaking and cooking on phytate concentration, minerals, and texture of food-type soybeans. *Ciência e Tecnologia de Alimentos*, 30: 1056-1060.

Kasprowicz-Potocka, M., Zaworska, A., Frankiewicz, A., Nowak, W., Gulewicz, P., Zduńczyk, Z. and Juśkiewicz, J. (2015). The nutritional value and physiological properties of diets with raw and *Candida utilis*-fermented lupin seeds in rats. *Food Technology and Biotechnology*, 53: 286-297.

Kasprowicz-Potocka, M., Zaworska, A., Gulewicz, P., Nowak, P. and Frankiewicz, A. (2018). The effect of fermentation of high alkaloid seeds of *Lupinus angustifolius* var. Karo by *Saccharomyces cerevisieae*, *Kluyveromyces lactis*, and *Candida utilis* on the chemical and microbial composition of products. *Journal of Food Processing and Preservation*, 42: e13487.

Kaur, K.D., Jha, A., Sabikhi, L. and Singh, A.K. (2014). Significance of coarse cereals in health and nutrition: A review. *Journal of Food Science and Technology*, 51: 1429-1441.

Khalil, A. (1995). The effect of cooking, autoclaving and germination on the nutritional quality of faba beans. *Food Chemistry*, 54: 177-182.

Khamassi, K., Ben Jeddi, F., Hobbs, D., Irigoyen, J., Stoddard, F., O'sullivan, D.M. and Jones, H. (2013). A baseline study of vicine-convicine levels in faba bean (*Vicia faba* L.) germplasm. *Plant Genetic Resources: Characterisation and Utilisation*, 11: 250-257.

Khattab, R.Y. and Arntfield, S.D. (2009). Nutritional quality of legume seeds as affected by some physical treatments 2. Antinutritional factors. *LWT–Food Science and Technology*, 42: 1113-1118.

Khattab, R., Goldberg, E., Lin, L. and Thiyam, U. (2010). Quantitative analysis and free-radical-scavenging activity of chlorophyll, phytic acid, and condensed tannins in canola. *Food Chemistry*, 122: 1266-1272.

Khetarpaul, N. and Chauhan, B.M. (1991). Sequential fermentation of pearl millet by yeasts and Lactobacilli-effect on the antinutrients and *in vitro* digestibility. *Plant Foods for Human Nutrition*, 41: 321-327.

Khokhar, S. and Apenten, R.K.O. (2003). Anti-nutritional factors in food legumes and effects of processing. pp. 82-116. *In*: V.R. Squires (Ed.). *The Role of Food, Agriculture, Forestry and Fisheries in Human Nutrition, Encyclopedia of Life Support Systems*. Vol. 4. EOLSS Publishers/UNESCO, Oxford, United Kingdom.

Kostekli, M. and Karakaya, S. (2017). Protease inhibitors in various flours and breads: Effect of fermentation, baking and in vitro digestion on trypsin and chymotrypsin inhibitory activities. *Food Chemistry*, 224: 62-68.

Kostinek, M., Specht, I., Edward, V.A., Pinto, C., Egounlety, M., Sossa, C., Mbugua, S., Dortu, C., Thonart, P., Taljaard, L., Mengu, M., Franz, C.M.A.P. and Holzapfel, W.H. (2007). Characterisation and biochemical properties of predominant lactic acid bacteria from fermenting cassava for selection as starter cultures. *International Journal of Food Microbiology*, 114: 342-351.

Kumar, S., Verma, A.K., Das, M., Jain, S.K. and Dwivedi, P.D. (2013). Clinical complications of kidney bean (*Phaseolus vulgaris* L.) consumption. *Nutrition*, 29: 821-827.

Kumar, V., Sinha, A.K., Makkar, H.P.S. and Becker, K. (2010). Dietary roles of phytate and phytase in human nutrition: A review. *Food Chemistry*, 120: 945-959.

Lai, L.R., Hsieh, S.C., Huang, H.Y. and Chou, C.C. (2013). Effect of lactic fermentation on the total phenolic, saponin and phytic acid contents as well as anti-colon cancer cell proliferation activity of soymilk. *Journal of Bioscience and Bioengineering*, 115: 552-556.

Lambri, M., Fumi, M.D., Roda, A. and De Faveri, D.M. (2013). Improved processing methods to reduce the total cyanide content of cassava roots from Burundi. *African Journal of Biotechnology*, 12(19): 2685-2691.

Luzzatto, L. and Arese, P. (2018). Favism and glucose-6-phosphate dehydrogenase deficiency. *New England Journal of Medicine*, 378: 60-71.

Lv, B., Sun, H., Huang, S., Feng, X., Jiang, T. and Li, C. (2018). Structure-guided engineering of the substrate specificity of a fungal β-glucuronidase toward triterpenoid saponins. *Journal of Biological Chemistry*, 293: 433-443.

Márquez, M.C. and Alonso, R. (1999). Inactivation of trypsin inhibitor in chickpea. *Journal of Food Composition and Analysis*, 12: 211-217.

Mattila, P.H., Pihlanto, A., Hellström, J., Pihlava, J.-M., Nurmi, M., Eurola, M., Mäkinen, S. and Jalava, T. (2018a). Contents of phytochemicals and antinutritional factors in commercial protein-rich plant products. *Food Quality and Safety*, 2: 213-219.

Mattila, P., Mäkinen, S., Eurola, M., Jalava, T., Pihlava, J.-M., Hellström, J. and Pihlanto, A. (2018b). Nutritional value of commercial protein-rich plant products. *Plant Foods for Human Nutrition*, 73: 108-115.

McKay, A.M. (1992). Hydrolysis of vicine and convicine from fababeans by microbial β-glucosidase enzymes. *Journal of Applied Bacteriology*, 72: 475-478.

Mikola, M. and Mikkonen, A. (1999). Occurrence and stabilities of oat trypsin and chymotrypsin inhibitors. *Journal of Cereal Science*, 30: 227-235.

Miyamoto, T., Kataoka, K., Miki, T., Yoshiki, Y., Okubo, K. and Yoneya, T. (2012). Fermentation effects of lactic acid bacteria on soybean saponins. *Food Preservation Science*, 25: 111-116.

Montemurro, M., Pontonio, E., Gobbetti, M. and Rizzello, C.G. (2019). Investigation of the nutritional, functional and technological effects of the sourdough fermentation of sprouted flours. *International Journal of Food Microbiology*, 302: 47-58.

Morales, P., Berrios, J.D.J., Varela, A., Burbano, C., Cuadrado, C., Muzquiz, M. and Pedrosa, M.M. (2015). Novel fiber-rich lentil flours as snack-type functional foods: An extrusion cooking effect on bioactive compounds. *Food and Function*, 6: 3135-3143.

Mullaney, E.J. and Ullah, A.H. (2003). The term phytase comprises several different classes of enzymes. *Biochemical and Biophysical Research Communications*, 312: 179-184.

Nielsen, M.M., Damstrup, M.L., Thomsen, A.D., Rasmussen, S.K. and Hansen, Å. (2007). Phytase activity and degradation of phytic acid during rye bread making. *European Food Research and Technology*, 225: 173-181.

Oboh, G. and Akindahunsi, A.A. (2003). Biochemical changes in cassava products (flour and gari) subjected to *Saccharomyces cerevisae* solid media fermentation. *Food Chemistry*, 82(4): 599-602

Omizu, Y., Tsukamoto, C., Chettri, R. and Tamang, J.P. (2011). Determination of saponin contents in raw soybean and fermented soybean foods of India. *Journal of Scientific and Industrial Research*, 70: 533-538.

Onder, M. and Kahraman, A. (2009). Antinutritional factors in food grain Legumes. *1st International Symposium on Sustainable Development*, pp. 40-44. June 9-10, 2009.

Oomah, B.D., Luc, G., Leprelle, C., Drover, J.C.G., Harrison, J.E. and Olson, M. (2011). Phenolics, phytic acid, and phytase in Canadian-grown low-tannin faba bean (*Vicia faba* L.) genotypes. *Journal of Agriculture and Food Chemistry*, 59: 3763-3771.

Ortega-David, E. and Rodriguez-Stouvenel, A. (2014). Bioprocessing of lupin cotyledons (*Lupinus mutabilis)* with *Rhizopus oligosporus* for reduction of quinolizidine alkaloids. *Journal of Food Processing & Technology*, 05: 1000323.

Ortega-David, E. and Rodríguez-Stouvenel, A. (2013). Degradation of quinolizidine alkaloids of lupin by *Rhizopus oligosporus*. *Applied Microbiology and Biotechnology*, 97: 4799-4810.

Ouwehand, A., Salminen, S. and Von Wright, A. (2004). *Lactic Acid Bacteria: Microbiological and Functional Aspects*, 3rd rev. and exp. ed. Ouwehand, A., Salminen, S. and Von Wright, A. (Eds.). Marcel Dekker, New York, USA.

Pavlík, M., Váňová, M., Laudová, V. and Harmatha, J. (2002). Fungitoxicity of natural heterocycle glucoside vicine obtained from *Vicia faba* L. against selected microscopic filamentous fungi. *Plant, Soil and Environment*, 48: 543-547.

Phengnuam, T. and Suntornsuk, W. (2013). Detoxification and anti-nutrients reduction of Jatropha curcas seed cake by *Bacillus* fermentation. *Journal of Bioscience and Bioengineering*, 115: 168-172.

Pojić, M., Mišan, A., Sakač, M., Hadnađev, T.D., Šarić, B., Milovanović, I. and Hadnađev, M. (2014). Characterisation of by-products originating from hemp oil processing. *Journal of Agriculture and Food Chemistry*, 62: 12436-12442.

Požrl, T., Kopjar, M., Kurent, I., Hribar, J., Janeš, A. and Simčič, M. (2009). Phytate degradation during breadmaking: The influence of flour type and breadmaking procedures. *Czech Journal of Food Sciences*, 27: 29-38.

Pulkkinen, M., Zhou, X., Lampi, A.M. and Piironen, V. (2016). Determination and stability of divicine and isouramil produced by enzymatic hydrolysis of vicine and convicine of faba bean. *Food Chemistry*, 212: 10-19.

Qian, B., Yin, L., Yao, X., Zhong, Y., Gui, J., Lu, F., Zhang, F. and Zhang, J. (2018). Effects of fermentation on the hemolytic activity and degradation of *Camellia oleifera* saponins by *Lactobacillus crustorum* and *Bacillus subtilis*. *FEMS Microbiology Letters*, 365.

Reddy, N.R. and Pierson, M.D. (1994). Reduction in antinutritional and toxic components in plant foods by fermentation. *Food Research International*, 27: 281-290.

Reddy, N.R., Sathe, S.K., Pierson, M.D. and Salunkhe, D.K. (1982). Idli, an Indian fermented food: A review. *Journal of Food Quality*, 5: 89-101.

Rizzello, C.G., Coda, R., Wang, Y., Verni, M., Kajala, I., Katina, K. and Laitila, A. (2018). Characterisation of indigenous *Pediococcus pentosaceus, Leuconostoc kimchii, Weissella cibaria* and *Weissella confusa* for faba bean bioprocessing. *International Journal of Food Microbiology*. https://doi.org/10.1016/j.ijfoodmicro.2018.08.014

Rizzello, C.G., Losito, I., Facchini, L., Katina, K., Palmisano, F., Gobbetti, M. and Coda, R. (2016). Degradation of vicine, convicine and their aglycones during fermentation of faba bean flour. *Scientific Reports*, 6: 1-11.

Rizzello, C.G., Nionelli, L., Coda, R., De Angelis, M. and Gobbetti, M. (2010). Effect of sourdough fermentation on stabilisation, and chemical and nutritional characteristics of wheat germ. *Food Chemistry*, 119: 1079-1089.

Ruales, J. and Nair, B.M. (1993). Saponins, phytic acid, tannins and protease inhibitors in quinoa (*Chenopodium quinoa*, Willd) seeds. *Food Chemistry*, 48: 137-143.

Russo, R. and Reggiani, R. (2015). Evaluation of protein concentration, amino acid profile and antinutritional compounds in hempseed meal from dioecious and monoecious varieties. *American Journal of Plant Sciences*, 6: 14-22.

Sagar, D. and Dhall, H. (2018). Legumes: Potential source of entomotoxic proteins – A review. *Legume Research: An International Journal*, 41: 639-646.

Santimone, M., Koukiekolo, R., Moreau, Y., Le Berre, V., Rougé, P., Marchis-Mouren, G., and Desseaux, V. (2004). Porcine pancreatic α-amylase inhibition by the kidney bean (*Phaseolus vulgaris)* inhibitor (α-AI1) and structural changes in the α-amylase inhibitor complex. *Biochimica et Biophysica Acta - Proteins and Proteomics*, 1696: 181-190.

Sariri, A.K., Mulyono, A.M.W. and Tari, A.I.N. (2018). The utilisation of microbes as a fermentation agent to reduce saponin in Trembesi leaves (*Sammanea saman*). *IOP Conference Series: Earth and Environmental Science*, 142.

Sathe, S.K., Deshpande, S.S., Reddy, N.R., Goll, D.E. and Salunkhe, D.K. (1983). Effects of germination on proteins, raffinose oligosaccharides, and antinutritional factors in the great northern beans (*Phaseolus vulgaris* L.). *Journal of Food Science*, 48: 1796-1800.

Savelkoul, F., Tamminga, S., Leenaars, P., Schering, J. and Ter Maat, D.W. (1994). The degradation of lectins, phaseolin and trypsin inhibitors during germination of white kidney beans. *Phaseolus vulgaris* L. *Plant Foods for Human Nutrition*, 45: 213-222.

Sharath, B.S., Mohankumar, B.V. and Somashekar, D. (2014). Bio-detoxification of phorbol esters and other anti-nutrients of *Jatropha curcas* seed cake by fungal cultures using solid-state fermentation. *Applied Biochemistry and Biotechnology*, 172: 2747-2757.

Sharma, A. and Kapoor, A.C. (1996). Levels of antinutritional factors in pearl millet as affected by processing treatments and various types of fermentation. *Plant Foods for Human Nutrition*, 49: 241-252.

Shekib, L.A., El-Iraqui, S.M. and Abo-Bakr, T.M. (1988). Studies on amylase inhibitors in some Egyptian legume seeds. *Plant Foods for Human Nutrition*, 38: 325-332.

Signorini, C., Carpen, A., Coletto, L., Borgonovo, G., Galanti, E., Capraro, J., Magni, C., Abate, A., Johnson, S.K., Duranti, M. and Scarafoni, A. (2017). Enhanced vitamin B12 production in an innovative lupin *tempeh* is due to synergic effects of *Rhizopus* and *Propionibacterium* in cofermentation. *International Journal of Food Sciences and Nutrition*, 69: 451-457.

Silvestroni, A., Connes, C., Sesma, F., De Giori, G.S. and Piard, J.C. (2002). Characterisation of the melA locus for α-galactosidase in *Lactobacillus plantarum*. *Applied and Environmental Microbiology*, 68: 5464-5471.

Soetan, K. and Oyewole, O. (2009). The need for adequate processing to reduce the anti-nutritional factors in plants used as human foods and animal feeds: A review. *African Journal of Food Science*, 3: 223-232.

Sudarmadji, S. and Markakis, P. (1977). The phytate and phytase of soybean *tempeh*. *Journal of the Science of Food and Agriculture*, 28: 381-383.

Svanberg, U., Lorri, W. and Sandberg, A.S. (1993). Lactic fermentation of non-tannin and high-tannin cereals: Effects on in vitro estimation of iron availability and phytate hydrolysis. *Journal of Food Science*, 58: 408-412.

Tefera, T., Ameha, K. and Biruhtesfa, A. (2014). Cassava based foods: Microbial fermentation by single starter culture towards cyanide reduction, protein enhancement and palatability. *International Food Research Journal*, 21(5): 1751.

Teixeira, J.S., McNeill, V. and Gänzle, M.G. (2012). Levansucrase and sucrose phoshorylase contribute to raffinose, stachyose, and verbascose metabolism by Lactobacilli. *Food Microbiology*, 31: 278-284.

Tian, J., Na, R., Yu, Z., Liu, Z., Liu, Z. and Yu, Y. (2018). Inoculant effects on the fermentation quality, chemical composition and saponin content of lucerne silage in a mixture with wheat bran or corn husk. *Animal Production Science*, 58: 2249.

Tieking, M., Ehrmann, M.A., Vogel, R.F., Korakli, M. and Ganzle, M.G. (2003). *In situ* production of exopolysaccharides during sourdough fermentation by cereal and intestinal isolates of lactic acid bacteria. *Applied and Environmental Microbiology*, 69: 945-952.

Trugo, L.C., Donangelo, C.M., Duarte, Y.A. and Tavares, C.L. (1993). Phytic acid and selected mineral composition of seed from wild species and cultivated varieties of lupin. *Food Chemistry*, 47: 391-394.

Urbano, G., Lopez-Jurado, M., Hernandez, J., Fernandez, M., Moreu, M.C., Frias, J., Diaz-Pollan, C., Prodanov, M. and Vidal-Valverde, C. (1995). Nutritional assessment of raw, heated, and germinated lentils. *Journal of Agricultural and Food Chemistry*, 43: 1871-1877.

Urga, K., Fite, A. and Biratu, E. (1997). Effect of natural fermentation on nutritional and antinutritional factors of tef (*Eragrostis tef*). *Ethiopian Journal of Health Development*, 11: 61-66.

Vidal-Valverde, C., Frias, J., Diaz-Pollan, C., Fernandez, M., Lopez-Jurado, M. and Urbano, G. (1997). Influence of processing on trypsin inhibitor activity of faba beans and its physiological effect. *Journal of Agricultural and Food Chemistry*, 45: 3559-3564.

Vonapartis, E., Aubin, M., Seguin, P., Mustafa, A.F. and Charron, J. (2015). Seed composition of ten industrial hemp cultivars approved for production in Canada. *Journal of Food Composition and Analysis*, 39: 8-12.

Wei, Q., Wolf-Hall, C. and Chang, K.C. (2001). *Natto* characteristics as affected by steaming time, Bacillus strain, and fermentation time. *Journal of Food Science*, 66: 167-173.

Xu, Y., Coda, R., Shi, Q., Tuomainen, P., Katina, K. and Tenkanen, M. (2017). Exopolysaccharides production during the fermentation of soybean and fava bean flours by *Leuconostoc mesenteroides* DSM 20343. *Journal of Agricultural and Food Chemistry*, 65: 2805-2815.

Zárate, G., Sáez, G.D. and Pérez Chaia, A. (2017). Dairy propionibacteria prevent the proliferative effect of plant lectins on SW480 cells and protect the metabolic activity of the intestinal microbiota *in vitro*. *Anaerobe*, 44: 58-65.

Prebiotics, Probiotics and Synbiotics: A New Integrated Approach in Functional Foods

Yogita Lugani[1], Balwinder Singh Sooch[1]* and Sachin Kumar[2]*

[1] Enzyme Biotechnology Laboratory, Department of Biotechnology,
   Punjabi University, Patiala, India
[2] Biochemical Conversion Division, Sardar Swaran Singh National Institute of
   Bio-Energy, Jalandhar-Kapurthala Road, Wadala Kalan, Kapurthala, India

1. Introduction

Over the past few centuries, there has been a tremendous improvement in the quality of life due to technological advancement, improved infrastructure and transportation facilities. However, the present lifestyle of people is leading to several health hazards due to physical inactivity and sedentary lifestyle. This affects the basic metabolism and results in major non-communicable and chronic diseases. One of the viable solutions to overcome these health hazards is to increase the intake of functional foods which possess physiological benefits, and reduce the risk of several diseases, besides providing nutritional benefits. Functional foods, also called as food supplements and nutraceuticals, have attracted worldwide attention due to their involvement in nutritional, physiology and immunological functions. Amongst these foods, synbiotics are key functional foods which are produced by the association of prebiotics and probiotics (Fig. 1). Prebiotics are non-digestible fibres, sugars, sugar alcohols and oligosaccharides which stimulate the growth of beneficial intestinal microbiota (Flesch et al., 2014). Probiotics are live microorganisms which confer health benefits on the host when administered in adequate doses (Saavedra, 2001). Recently, a term 'postbiotics' has been introduced which includes metabolic byproducts like acetaldehyde, bacteriocins, ethanol and organic acids from both live and heat killed probiotics (Kerry et al., 2018). Intestinal microbiota performs many metabolic activities, facilitates digestion and contributes to defence mechanism against pathogens (Patel and DuPont, 2015). Currently, more than 25 infectious diseases ranging from gastrointestinal diseases like irritable bowel syndrome,

*Corresponding authors: soochb@yahoo.com; sachin.kumar20@gov.in inflammatory bowel diseases and colorectal cancer to metabolic and neurological diseases, including autistic spectrum disorders, Alzheimer's disease, Parkinson's disease, chronic fatigue syndrome and autoimmune diseases (multiple sclerosis and rheumatoid arthritis) are linked with the alteration of intestinal microorganisms (Scott *et al.*, 2015). Prebiotics, probiotics and synbiotics help to alter the microenvironment of intestinal gut, resulting in favourable effects on many such disorders (McFarland, 2014; Mishra *et al.*, 2018; Farrokh *et al.*, 2019). The use of prebiotics, probiotics and synbiotics for prevention and treatment of many diseases is gaining acceptance in medical community; however, clinical evidence regarding their safety and efficacy is sparse. The objective of the present chapter is to understand the integrated approach used in functional foods using prebiotics, probiotics and synbiotics in a single product, their synergetic effect, mode of action in therapeutic and other applications, health and safety issues and global market status of these products.

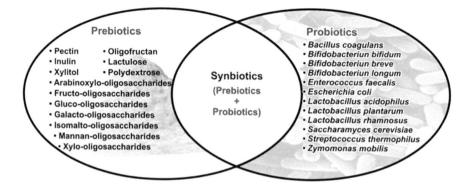

Fig. 1: Concept of synbiotics

2. Prebiotics, Probiotics and Postbiotics

The term prebiotics was employed during the mid nineties of the twentieth century by Gibson and Roberfroid, and later redefined by various researchers. The most recent definition of prebiotic is illustrated as a selectively fermented ingredient that allows specific changes, both in composition and activity in the gastrointestinal microbiota that confers benefits upon the host's well-being and health (Gibson *et al.*, 2004; Macfarlane *et al.*, 2006). Established prebiotics include inulin, fructo-oligosaccharide (FOS), galacto-oligosaccharide (GOS), lactulose and polydextrose, whereas isomalto-oligosaccharide (IMO), xylo-oligosaccharide (XOS), xylitol, and lactitol are emerging prebiotics (Femia *et al.*, 2010; Molina *et al.*, 2009; Lugani and Sooch, 2017; Lugani *et al.*, 2017; Lugani and Sooch, 2018; Farrokh *et al.*, 2019). These short-chain oligosaccharides are specifically utilised by the intestinal microflora and termed as bifidus factor because they promote the growth and activities of probiotic microorganisms in the gastrointestinal tract. A substance is classified as 'prebiotic' when it fulfils some criteria, like stability, safety, organoleptic properties, resistance to digestion in upper bowel, fermentability in colon and ability to promote the growth of beneficial bacteria in the gut (Gibson *et al.*, 2004). They are commonly used in cheese, fermented milk, cereals, salad dressing, beverages, edible coating, thermo-protectant, infant formulas, poultry, fishery, pig and cattle feed. Prebiotics have shown many therapeutic applications in cancer prevention, cholesterol removal, immunopotentiation, bone mineralisation, prevention of gastroenteritis, bacteriocin production, prevention and treatment of allergy, treatment of inflammatory bowel disease, improving gut microbiota and renal health (Bhadoria and Mahapatra, 2011; Anila and Bhalla, 2016; Yoo and Kim, 2016; Tsai *et al.*, 2019).

The term 'probiotic' has emerged from Greek words '*pro*' and '*bios*' meaning 'for life' (Gismondo *et al.*, 1999). This term was first used by Lilly and Stillwell in 1965 to describe the substances secreted by one microorganism to stimulate the growth of the other and thus is referred as support for life by naturally improving the overall health of the host organism. Probiotics are defined as live microorganisms that, when administered in adequate amounts, confer a health benefit to the host (FAO/WHO, 2001). The live microorganisms improve the host health due to its unique properties, like adherence to host epithelial tissue, acid resistance, bile resistance, elimination of pathogens, production of acids, hydrogen peroxide and bacteriocins to inhibit the growth of pathogens, safety, non-pathogenic and non-carcinogenic nature along with improving the intestinal microflora. The beneficial effects of probiotic strains are listed in Box 1. There are many factors which affect the functionality of probiotics, including strain characteristics, stability, viability, daily dosage, fermentation technology, storage conditions and interaction with some pharmaceuticals, microencapsulation, drying technology and target prebiotic. The microorganisms commonly used as probiotics include species of genera

---

Box 1: Beneficial Effects of Probiotic Strains

- GRAS (Generally Recognised as Safe) status
- Do not show any side effects even in immuno-compromised persons
- Show positive health effects with minimum effective dosage in products
- Possess immune-modulation effects with enhanced healing of damaged mucosa
- Down regulate over stimulated immune and inflammatory response
- Antagonistic effect against pathogenic microorganisms due to secretion of antibiotic like factors
- Degradation of toxin receptor
- Inhibition of auto-aggressive and allergic reactions
- Expansion of mucosal regulatory cells
- Reduce the amount of food required by the body due to complete digestion and metabolism of food
- Reduction of pro-inflammatory cytokines on lymphocytes and plasma
- Reduction of serum cholesterol
- Improvement of glucose and lipid metabolism
- Regulation of carcinogenesis at all stages of initiation, promotion and progression
- Change the composition of mucus secreted by colonocytes by altering their gene expression

Lactobacillus, Lactococcus, Enterococcus, Streptococcus, Pediococcus, Bacillis, Bifidobacterium, Saccharomyces and a virulent *Escherichia coli*. Different strains of same species of probiotics are always unique and they may have different areas of adherence (site-specificity), specific immunological effects and different modes of action on healthy *vs.* inflamed mucosal milieu (Soccol *et al.*, 2010; Tsai *et al.*, 2019). The most commonly used bacterial strains used as probiotics are *Lactobacillus,* and *Bifidobacterium* (Ranadheera *et al.*, 2010; Saez-Lara *et al.*, 2015; Mallappa *et al.*, 2019). Viable probiotics with $\geq 10^6$ cfu/mL in the small bowel and $\geq 10^8$ cfu/mL in the colon are recommended to show effective clinical results (Minelli and Benini, 2008). Nowadays, probiotics are used in many fermented and non-fermented foods and dairy products.

Postbiotics are defined as non-viable bacterial products or metabolic byproducts from probiotic microbes which show similar biological functions in the host cell as of probiotics (Patel and Denning, 2013). These metabolic byproducts include ethanol, hydrogen peroxide, bacteriocins, organic acids, and acetaldehyde and they are non-pathogenic, non-toxic and resistant to hydrolysis by mammalian enzymes (Islam, 2016). Postbiotics are used as an alternative to antibiotics due to their inhibitory property against pathogenic microbes (Ooi *et al.*, 2015) and to enhance *in vivo* and *in vitro* angiogenesis of epithelial cells by activation of $\alpha 2\beta 1$ integrin collagen receptors (Giorgetti *et al.*, 2015).

3. Synbiotics

Synbiotics (*syn* means 'together' and *bios* means 'life') are defined as a mixture of a prebiotics and probiotics that beneficially affect the host by enhancing the survival and implantation of live microbial-based dietary supplement machinery in the gut by selectively stimulating the growth and/or activating the metabolism of a specific or a few health-promoting bacteria (Gibson and Roberfroid, 1995; Roberfroid, 2002). Synbiotic is a fusion of 'prebiotics' and 'probiotics' and their symbiotic relationship results in a significant improvement of health (Mousavi *et al.*, 2015). On the one hand, synbiotics improve the survival of beneficial microbes added to food supplements, on the other hand, they also stimulate the proliferation of specific microbial strains which are native to the gastrointestinal tract (Gourbeyre *et al.*, 2011). Synbiotics are emerging as a new approach in functional foods due to their positive results on gut health, disease prevention and therapy; therefore, currently most of the research is focused on development of novel health-promoting foods with selected probiotic strains which have high efficiency in colonising the human gut and the ability to digest many forms of prebiotics. The health benefits of synbiotics are linked to synergetic effects of prebiotics and postbiotics.

4. Mechanism of Action

The major mechanisms which form the basis for beneficial effects of synbiotics are based on molecular and genetic studies, including immunomodulation of the host (Isolauri *et al.*, 2001), inhibition of bacterial toxin production (Brandao *et al.*, 1998), antagonism through production of antimicrobial substances (Vandenbergh,

1993) and competition with pathogens for adherence to epithelium and for nutrients (Guillot, 2003). The mucosal immunity is altered by probiotics due to a change in the local and systemic immune responses, such as T-cells, regulatory T-cells, natural killer cells, dendridic cells, immunoglobulin A producing B cells, epithelial cells (Zhang *et al.*, 2007). Probiotic microorganisms can stimulate the production of cytokines, including interleukin-10 (IL-10), transforming growth factor β (TGF) by expressing different toll like receptors (TLRs) like TLR2 and TLR4 which lead to regeneration of epithelial cells and inhibition of apoptosis (Nahoum *et al.*, 2004). Probiotic bacteria help to recognise the antigens on the surface of intestinal dendridic cells, resulting in development of both B- and T-cell responses (Williamson *et al.*, 2002). The antimicrobial effects of probiotics are also caused by alteration in the luminal environment, its antitoxin potential and competitive inhibition of adhesion and cellular invasion of pathogens which result in decrease of luminal pH along with production of antimicrobial substances (bacteriocins, organic acids, hydrogen peroxide). The prevention of toxin expression by probiotics is its another significant characteristic. Probiotic microorganisms possess the ability to metabolise complex carbohydrates, leading to production of organic acids and short-chain fatty acids like butyrate (Wong and Jenkins, 2007). The organic acids help in maintaining a lower pH which inhibits the growth and colonisation of pathogens (Asahara *et al.*, 2004). Butyrate helps to improve the production of mucin (glycoprotein) which helps in maintaining the integrity of epithelium (Paassen *et al.*, 2009), reduces bacterial translocation and maintain the organisation of tight junctions (Lewis *et al.*, 2010). Probiotics possess the ability to adhere to the surface of intestinal epithelium and thus prevent the invasion of pathogens through epithelium (Servin and Coconnier, 2003). Probiotic microorganisms competitively inhibit the growth of intestinal microbes and pathogens by the requirement of limiting resources, like iron. The enhancement of mucosal barrier is another mechanism for preventing many infectious diseases like Type 1 diabetes mellitus (Watts *et al.*, 2005), inflammatory bowel disease (Schmitz *et al.*, 1999) and diarrhoea (Sakaguchi *et al.*, 2002). The detoxifying property of probiotics is due to adsorption of toxins on the surface of probiotics and then metabolism of mycotoxins (McCormick, 2013) and by stimulation of metabolic pathways that lead to the production of antimicrobial substances, hydrolytic enzymes and vitamins (Oelschlaeger, 2010). Probiotics specifically enhance the immune response by enhancing anti-inflammatory short-chain fatty acids, allergen specific IgG, IL-10, TGF-β and diminishing response of IL-4, IL-5, IL-13, allergen-specific IgE (Virk and Wiersinga, 2019).

5. Health Benefits

The effectiveness of synbiotics for the treatment of many diseases, such as diabetes, obesity, cancer, cardiovascular disorders, hypercholestremia, common cold infections, allergic disorders, urinary tract infections, nosocomial infections, diarrheal illness, Crohn's disease, pouchitis, ulcerative colitis, necrotising enterocolitis, eczema, lactose intolerance, gastroenteritis, vaginal candidiasis, peptic ulcers, neurological disorders, mucositis and hepatic encephalopathy has already been proved in multiple previous studies (Zukiewicz-Sobczak *et al.*, 2014; Amara and Shibl, 2015; Kassaian *et al.*, 2017; Diaz *et al.*, 2018). The treatment of different diseases along with synbiotics and their dose are listed in Table 1. The most beneficial property of synbiotics is their anti-pathogenic activity due to release of a wide variety of anti-microbial compounds, such as peptides, ethanol, organic acids, hydrogen peroxide, diacetyl, acetaldehydes and bacteriocins (Simova *et al.*, 2009). Many neurochemicals which have been isolated from probiotic microorganisms including serotonin, γ-amino benzoic acid, dopamine, norepinephrine, acetylcholine (Kapoor *et al.*, 2016). Multiple aspects of synbiotics are illustrated in Fig. 2.

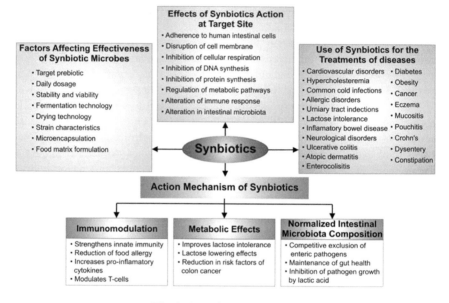

Fig. 2: Overview of synbiotics

Clostridium difficile infection can also be prevented by probiotics, either by degrading *C. difficile* A and B toxins or by increasing the expression of mucins with reduction of pathogen adherence (Moayyedi and Ford, 2011). Probiotics possess cholesterol-reducing ability by production of bile salt hydrolase (Lye *et al.*, 2010), by binding of cholesterol to surface of probiotic results in incorporation of cholesterol molecules into cellular membrane of probiotic microorganism, by synthesis of SCFAs, co-precipitation of cholesterol with deconjugated bile (Kumar *et al.*, 2012), and by conversion of cholesterol to coprostanolin (Gerald, 2014). Non-digestible prebiotics, such as oligo-fructose and inulin, help to improve the absorption of minerals, like Fe, Ca, Mg and Zn (Ahrans *et al.*, 2002). Probiotics maintain the vaginal and urogenital health by signalling the production of anti-inflamatory cytokines (Pessi *et al.*, 2000) and producing bio-surfactants and collagen-binding proteins which inhibit the binding of pathogens on the mucosal surface (Waigankar and Patel, 2011). Probiotic microorganisms maintain the oral health by producing bacteriocins against *Streptococcus pneumonia* and *S. pyogenes* which lead to prevention of tonsillitis, pharyngitis and otitis media (Pierro *et al.*, 2014) with reduction of gum bleeding

Table 1: Use of Synbiotics for Treatment of Different Diseases

S.No.	Disease	Treatment	Dose	Major Outcome	References
1.	Crohn's disease	*Bifidobacterium longum* plus Synergy 1	183 d/2×10¹¹ viable CFU twice daily	Improvement of clinical symptoms	Steed *et al.* (2010)
2.	Enterocolisitis	*Lactobacillus* alone or in combination with *Bifidobacterium*		Significant reduced incidence of disease	AlFaleh *et al.* (2011)
3.	Ulcerative colitis	*Lactobacillus reuteris* ATCC 55730	61 D/1×10¹⁰ CFU per day	Decreased expression of proinflamatory cytokines and increased expression of IL-10 in children	Oliva *et al.* (2012)
4.	Atopic dermatitis	Probiotc mixture (*Lactobacillus acidophilus, L. casei, L. salivarius, Bifidobacterium bifidum*)	Two bags of 2×10⁹ probiotic mixture	Effective in reducing serum cytokines and Ig E levels	Yesilova *et al.* (2012)
5.	Mucositis	*Lactobacillus brevis* CD2	2×10⁹ viable CFU/day	Effective in significantly reduced incidence of disease	Sharma *et al.* (2012)
6.	Dental caries	Pro-t-action containing *Lactobacillus paracasei* DSMZ16671	1-2 mg per candy piece	Reduced mutans streptococci in mouth	Holz *et al.* (2013)
7.	Bacillary dysentery	*Lactobacillus* GG plus fructooligosaccharides (FOS)	Lactol tablet: 1 tablet/d containing bacillus coagulant (150 million spores) and FOS (100 mg) for 3-5 d	Reduced the duration of fever, dysentery and rate of weight loss	Kahbazi *et al.* (2016)
8.	Type 2 diabetes	Probiotc mixture (*Lactobacillus acidophilus, Bifidobacteriumbifidum, B. lactis, B. longum*) with inulin plus maltodextrin	10⁹ viable CFU with 6 g/d of each culture and prebiotic for 6 months	Effective in management of prediabetes and preventing diabetes	Kassaian *et al.* (2017)

(Contd.)

Table 1: *(Contd.)*

S.No.	Disease	Treatment	Dose	Major Outcome	References
9.	Colon cancer	*Lactobacillus plantarum* HII11 plus inulin	10^8-10^9 CFU/g of cells and inulin (10%, w/w)	Reduced incidence of colon cancer by enhancing probiotic population and organic acids with reduction in pathogen load, microbial enzyme and putrefactive compounds	Chaiyasut *et al.* (2017)
10.	Ventilator-associated pneumonia	*Bifidobacterium breve* strain Yakult, *Lactobacillus casei* strain Shirota, galacto-oligosaccharides	-	Induction of host cell antimicrobial peptides, release of antimicrobial factors, stimulation of IgA production, antioxidant activity, suppression of immune cell proliferation, inhibition of epithelial NFκB activation	Virk and Wiersinga, 2019

and gingivitis (Vivekananda *et al.*, 2010). Probiotic bacteria and bacterial products, such as enzymes, proteins and immunotoxin secondary metabolites specifically target cancerous cells by inhibiting the growth of tumor cells, arresting cell cycle at different stages and through induction of apoptosis. Hence, microbial-based treatment is one of the emerging and promising remedies for cancer therapy (Sunkata *et al.*, 2014). The consumption of fermented beverages containing *Lactobacillus casei* was observed to improve the mood of healthy persons with less depression symptoms (Rao *et al.*, 2009). Dietary probiotic supplements help to prevent memory loss by enhanced expression of c-Fos in the hypothalamus and restoration of hypocampal brain-derived neurotropic factor (Collins *et al.*, 2012), thus helping to improve human brain-development functions (Tillisch, 2014; Umbrello and Espositi, 2016). Probiotics also help to reduce chronic kidney diseases by reducing blood urea, ammonia, p-cresol, indoxyl sulphate and p-cresyl sulphate (Fagunges *et al.*, 2018). The anti-obesity property of probiotics is facilitated by stimulation of sympathetic nervous system which results in lipolytic and thermogenic responses (Karimi *et al.*, 2015; Kerry *et al.*, 2018; Chang *et al.*, 2019). Recently, the preventive and therapeutic effects of *Faecalibacterium prausnitzii* ATCC 27766 have been reported on neuropsychiatric disorders, like anxiety and depression (Hao *et al.*, 2019). Similarly, the administration of prebiotics, such as arabinoxylans (ARXn) and synbiotics, is a promising treatment for dysbiosis by modulating the microbiome, regulation of neurotransmitters and alleviation of neurological manifestations (Salvucci, 2019). In another study, the role of probiotics and metabolites, like short-chain fatty acids (SCFAs), bile acids (BAs) and polysaccharide A (PSA) for the treatment of human autoimmunity has also been reported (Meng *et al.*, 2019).

6. Safety Aspects

The safety of probiotic strains is one of the major aspects affecting their applications and hence development of novel probiotic strains is becoming the essence of near future (Tsai *et al.*, 2019). Various regulatory aspects, such as the type of microbial strain, its safety and efficacy, stability, quality control of manufacturing, dosage and health claims need to be regulated before their implementation in treatment, therapy and diagnosis of a disease. The important parameter which needs to be controlled from the safety perspective is prevention of contamination during manufacturing and packaging. Another parameter for the effectiveness of probiotic product is the total number of live bacteria expressed as CFU (Colony Forming Units) and based on current regulations, the labelling of CFU per dose is mandatory on probiotic products (Sanders *et al.*, 2016). The probiotic microbes from synbiotics, ranging from 10^8-10^9 CFU per day, are recommended for consumption by Technical Regulations, 2005 (Stefe *et al.*, 2008). Probiotics are also considered safe in safety analysis conducted by Hempel *et al.* (2011). Probiotics are safe to be used in healthy individuals; however, more studies need to be carried out with premature neonates and immune-deficiency patients due to their strain specificity. The National Institute of Health has published a detailed assessment of safety aspects of probiotics (Ewe *et al.*, 2010), but there are no universal regulations for probiotics and these vary with the country. The clinical guidelines issued by the World Gastroenterology Organisation have recommended the use of specific probiotics for the treatment of pouchitis and ulcerative colitis (Bai and Ciacci, 2017). The benefits of probiotics for the treatment of acute gastroenteritis and antibiotic-associated diarrhoea have been recommended by different societies, like the European Society of Paediatric Gastroenterology, Hepatology and Nutrition (ESPGHAN) (Szajewska *et al.*, 2014), the American Academy of Paediatrics (Thomas and Greer, 2010), the World Gastroenterology Organisation (Guarner *et al.*, 2012) and the Canadian Paediatric Society (Marchand *et al.*, 2012). However, some of the previous studies showed the association of synbiotics with alteration of the immune response along with development of different serious infections, such as abscesses, bacteremia, endocarditis, meningitis, peritonitis and sepsis (Cannon *et al.*, 2005; Luong *et al.*, 2010). The other side effects of probiotics include infection in the immuno-compromised patients who are taking immune-suppressants, like azathioprine, cyclosporine, tacrolimus and chemotherapeutic agents. They also interact with some antifungal drugs, resulting in reduction of efficacy of probiotics (Sheehan *et al.*, 2007). Probiotic microorganisms can change the local production of vitamin K from the intestinal bacteria which affect the sensitivity to warfarin and different vitamin K antagonists (Marteau *et al.*, 2009). Hence, there is a dire need for further research on safety issues of probiotics and their dosage in drug formulations because microbial strains easily undergo genetic drift and shift, as also mutations.

7. Fermented Foods as a Source of Probiotics

Microorganisms associated with food fermentation have been reported to be good probiotics (Swain *et al.*, 2014; Chourasia *et al.*, 2020). The major groups of probiotic microorganisms include lactic acid bacteria, *Bacillus* spp. and yeasts (Nuraida, 2015; Kim *et al.*, 2014; Rai *et al.*, 2019). Globally, several studies are coming up with a wide range of food products for delivery of probiotics, including cheese, desserts, whey dairy products, soy products, ice cream, confectionary and breads as well as fermented vegetables and meat products (Franz *et al.*, 2014). LAB from several traditional Indonesian fermented foods possess probiotic properties (Nuraida, 2015). *Bacillus subtilis* strains from Korean fermented foods have shown to possess antibacterial properties apart from having probiotic attributes (Kim *et al.*, 2014). Similarly, African traditional fermented food products have shown to possess probiotic attributes (Mokoena *et al.*, 2016).

LAB isolated from fermented foods of Ladakh have probiotic effect (Angmo *et al.*, 2016; Chourasia *et al.*, 2020). In another study, exopolysaccharide-producing LAB isolated from Indian traditional fermented foods were tested for probiotic potential and antimicrobial properties (Patel *et al.*, 2014; Swain *et al.*, 2014). Similarly, probiotic potential of LAB isolated from *panchamirtham*, a South Indian ethnic fermented fruit mix was studied (Maheshwari *et al.*, 2019). Cheese products have been developed using probiotic bacteria, *Lactobacillus pentosus*, *Lb. plantarum*, *Lb. rhamnosus* and *Lb. brevis* (Blaiotta *et al.*, 2017; Ruiz-Moyano *et al.*, 2019). Even though there are reports on probiotic LAB from different traditional fermented foods (Angmo *et al.*, 2016), many traditional fermented foods across the globe need to be explored for their potential probiotic strains, which can be applied to add value to fermented foods with probiotic properties.

8. Current Market Status

The change in lifestyle, health-care costs, geriatric population and consumer's consciousness towards their well-being have enhanced the market growth of probiotic foods. Synbiotics have been used in different foods and food supplements for the management of different diseases and also to improve the health. The choice of probiotic strains in food products depends on many factors, including its safety, compatibility with the microbial strain, pH of product, strain viability through food processing, packaging and storage conditions. The list of major companies involved in manufacturing of synbiotics/probiotics-based products is given in Table 2. Many dairy-based (milk powder, ice cream, fermented milk, yoghurt) and non-dairy-based (cereals, soy based products, juices, nutrition bars) food products containing probiotics are available in the market.

The global market of probiotics was approximately USD 40.09 billion in 2017 and is expected to reach USD 65.87 billion by the end of 2024 with a Compound Annual Growth Rate (CAGR) of 7.35 per cent between 2018 and 2024. There is a significant growth in the global market for North American probiotics due to digestive problems caused by high consumption of processed foods, poor consumption habits and enhanced antibiotic use in this region (Zion Market Research, 2018). In Asia-Pacific, the food market of synbiotics is projected to grow at CAGR of 10 per cent, and a higher growth rate is expected between 2017-2023 (Synbiotics Food Market Analysis, 2018).

9. Conclusion

The present study concludes that there is a wide scope for using prebiotics, probiotics and synbiotics for preventive and therapeutic actions in various diseases. The global market for probiotic products in the form of fermented and non-fermented food supplements is increasing day by day. Although a lot of research has already been done on the action mechanism of probiotics and their safety studies, still many of these products are not approved by regulatory agencies. There is also a fact that many side effects have been observed with immunocompromised patients. Hence, there is further need for more detailed mechanistic studies to test the effectiveness and safety of different synbiotic products and their potential beneficial effects. Presently, *Lactobacillus* and *Bifidobacterium* are widely-used microbial strains for the production of probiotics; however the other valuable strains need to be studied extensively for their use as probiotics with well-designed clinical trials. The majority of currently-available probiotics have been designed with the target on the gastrointestinal tract and the focus of researchers is towards the development of cocktails of microbes with the target aimed on other sites of the body. Future studies on synbiotics may explain the beneficial effects of pre- and post-biotics on human health for developing bio-theraputic formulations. Genetically-modified microorganisms may express specific epitopes for upregulating the immune response against pathogens, downregulating the immune response against allergy reactions, delivery of oral vaccines and restoring antigen specific tolerance. Moreover, a broad

Table 2: Market Products of Probiotics

Company	Brand Name	Probiotic Strain
Ardeypharm	Mutaflor	*Escherichia coli* Nissle1917
Biocodex	Florastor	*Saccharomyces cerevisiae boulardii*
Biogaia	Biogaia Prebiotic	*Lactobacillus reuteri* ATCC 55,730
Bifodan	EcoVag	*Lactobacillus rhamnosus* PBO1, *L.gasseri* EB01
BioGaia	Prolectis	*Lactobacillus reuteri* DSM17938
Bio K+ Int.	Bio K+	*Lactobacillus acidophilus* CL1285 and *L. casei* LBC80R
CerbiosPharma	Bioflorin	*Enterococcus* LAB SF68
Chr. Hansen	Fem Dophilus	*Lactobacillus rhamnosus* GR1 and *L. reuteri* RC14
Danone	Activia	*Bifidobacterium animalis* DN173010
Danone	DanActive fermented milk	*Lactobacillus casei* DN-114001
Ganeden Biotech Inc.	Sustenex	*Bacillus coagulans* BC30
Ganaden Biotech Inc.	Sustenex	*Bacillus coagulans* BC30
Morinaga Milk Ind.	Bioclinic Naturals	*Bifidobacterium longum* BB536
Nestle	LC1	*Lactobacillus johnsanii* L1A
Next Foods, Inc.	Good Belly Proviva	*Lactobacillus plantarum* 299V
Oragenics Inc.	EvoraPlus	*Streptococcus uberis* KJ2
Procter and Gamble	Align	*Bifidobacterium infantis* 35,264

(Contd.)

Unique Biotech Ltd.	Florafix	*Saccharomyces boulardii*
Unique Biotech Ltd.	Bacipro	*Bacillus clausii*
Unique Biotech Ltd.	Provinorm	*Bifidobacterium bifidum, Lactobacillus acidophilus, L. rhamnosus, L. reuteri, L. plantarum, L. casei, L. fermentum*
Unique Biotech Ltd.	Flora IB	*Bifidobacterium bifidum, Lactobacillus acidophilus, L. rhamnosus, L. plantarum, L. casei, Saccharomyces boulardii*
Urex Biotech	Femdophilus	*Lactobacillus rhamnosus* GR-1
Valio Dairy	Culturelle	*Lactobacillus rhamnosus* GG
Wren Laboratories, Biocodex	Diar safe	*Saccharomyces cerevisiae*
Yakult	Yakult	*Lactobacillus casei* Shirota
Yakult	Bifine	*Bifidobacterium breve* Yakult

comprehensive relationship between genetic, microbial and environmental factors within humans is required to be understood for developing biomarkers against deadly diseases, like cancer and antibiotic-associated diseases. There is a further need for designing *in vivo* and *in vitro* studies to evaluate specific biomarkers of systemic and intestinal immunity. Further, clinical studies with large patient cohorts are also required for the selection of correct combination of probiotic strain, prebiotics and their dosage which may result in great efficacy alongwith their safety and limitations. Advancement in research in the field of metagenomics, system biology and bioinformatics further helps to modulate the intestinal health and immune response by supplementing postbiotics and synbiotics. The labelling of all the critical parameters, including the type of probiotic strain, time of administration, warnings, shipping and storage conditions greatly help researchers, clinicians and consumers to make the choice of probiotics for supporting gastrointestinal health.

Acknowledgements

The authors are thankful to the Department of Biotechnology, Punjabi University, Patiala and Bhai Kahn Singh Nabha Library, Punjabi University, Patiala for providing access to technical and scientific literature. The authors also acknowledge the financial support from UGC, MHRD, New Delhi in the form of Major Research Project.

Abbreviations

ARXn – Arabinoxylans
BAs – Bile acids
CFU – Colony Forming Units
ESPGHAN – European Society of Pediatric Gastroenterology, Hepatology and
                     Nutrition
FOS – Fructo-oligosaccharide
GOS – Galacto-oligosaccharide
IMO – Isomalto-oligosaccharide
LAB – Lactic acid bacteria
SCFAs – Short-chain fatty acids
TGF – Transforming growth factor β
TLRs – Toll like receptors
XOS – Xylo-oligosaccharide

References

Ahrans, K.E.S., Acil, Y. and Schrezenmeir, J. (2002). Effect of oligofructose or dietary calcium on repeated calcium and phosphorous balances, bone mineralisation and trabecular structure in oariectomised rats. *British Journal of Nutrition*, 88(4): 365-367.

AlFaleh, K., Anabrees, J., Bassler, D. and Kharfi, T.A. (2011). Probiotics for prevention of necrotising enterocolitis in preterm infants. *Cochrane Database of Systematic Reviews*, 16(3): 1-85.

Amara, A.A. and Shibl, A. (2015). Role of probiotics in health improvement, infection control and disease treatment and management. *Saudi Pharmaceutical Journal*, 23(2): 107-114.

Angmo, K., Kumari, A. and Bhalla, T.C. (2016). Probiotic characterisation of lactic acid bacteria isolated from fermented foods and beverage of Ladakh. *LWT–Food Science and Technology*, 66: 428-435.

Anila, K.K. and Bhalla, T.C. (2016). *In vitro* cholesterol assimilation and functional enzymatic activities of putative probiotic lactobacillus sp. isolated from fermented foods/beverages of Northwest India. *Journal of Nutrition and Food Sciences*, 6(02). https://doi.org/10.4172/2155-9600.1000467

Asahara, T., Shimizu, K., Nomoto, K., Hamabata, T., Ozawa, A. and Takeda, Y. (2004). Probiotic bifidobacteria protect mice from lethal infection with Shiga toxin-producing *Escherichia coli* O157:H7. *Infection and Immunity*, 72(4): 2240-2247.

Bai, J.C. and Ciacci, C. (2017). World gastroenterology organisation global guidelines: Celiac disease February 2017. *Journal of Clinical Gastroenterology*, 51(9): 755-768.

Bhadoria, P.B.S. and Mahapatra, S.C. (2011). Prospects, technological aspects and limitations of probiotics – A worldwide review. *European Journal of Food Research and Reviews*, 1: 23-42.

Blaiotta, G., Murru, N., Di Cerbo, A., Succi, M., Coppola, R. and Aponte, M. (2017). Commercially standardised process for probiotic 'Italico' cheese production. *LWT–Food Science and Technology*, 79: 601-608.

Brandao, R.L., Castro, I.M., Bambirra, E.A., Amaral, S.C., Fietto, L.G., Tropia, M.J., Neves, M.J., Santos, R.G.D., Gomes, N.C. and Nicoli, J.R. (1998). Intracellular signal triggered by cholera toxin in *Saccharomyces boulardii* and *Saccharomyces cerevisiae*. *Applied and Environmental Microbiology*, 64(2): 564-568.

Cannon, J.P., Lee, T.A., Bolanos, J.T. and Danziger, L.H. (2005). Pathogenic relevance of *Lactobacillus*: A retrospective review of over 200 cases. *European Journal of Clinical Microbiology and Infectious Diseases*, 24(1): 31-40.

Chaiyasut, C., Pattananandecha, T., Sililun, S., Suwannalert, P., Peerajan, S. and Sivamaruthi, B.S. (2017). Synbiotic preparation with lactic acid bacteria and inulin as a functional food: *In vivo* evaluation of microbial activities and preneoplastic abberant crypy foci. *Food Science and Technology*, 37(2): 328-336.

Chang, C.S., Ruan, J.W. and Kao, C.Y. (2019). An overview of microbiome-based strategies on anti-obesity. *Kaohsiung Journal of Medical Sciences*, 35(1): 7-16.

Chourasia, R., Phukon, L.C., Abedin, M., Sahoo, D., Singh, S.P. and Rai, A.K. (2020). Biotechnological approaches for the production of designer cheese with improved functionality. *Comprehensive Reviews in Food Science and Food Safety.*, doi: 10.1111/1541-4337.12680

Collins, S.M., Surette, M. and Bercik, P. (2012). The interplay between the intestinal microbiota and the brain. *Nature Reviews Microbiology*, 10(11): 735-742.

Diaz, J.P., Ojeda, F.J.R., Campos, M.G. and Gil, A. (2018). Immune-mediated mechanisms of action of probiotics and synbiotics in treating paediatric intestinal diseases. *Nutrients*, 10(1): 1-20.

Eguchi, K., Fujitani, N., Nakagawa, H. and Miyazaki, T. (2019). Prevention of respiratory syncytial virus infection with probiotic lactic acid bacterium *Lactobacillus gasseri* SBT2055. *Scientific Reports*, 9(1): 1-2.

Ewe, J.A., Abdullah, W.N.W. and Liong, M.T. (2010). Viability and growth characteristics of *Lactobacillus* in soymilk supplemented with B-vitamins. *International Journal of Food Sciences and Nutrition*, 61(1): 87-107.

Fagunges, R.A.B., Soder, T.F., Grokoski, K.C., Benetti, F. and Mendes, R.H. (2018). Probiotics in the treatment of chronic kidney disease: A systematic review. *Journal Brasileiro de Nefrologia*, 40(3): 278-286.

FAO/WHO (2001). A Report of a Joint FAO/WHO expert consultation on evaluation of health and nutritional properties of probiotics in food including powder milk with live lactic acid bacteria, 2 pp. www.fao.org/3/a-a0512e.pdf; accessed on October 3, 2018

Farrokh, A., Ehsani, M.R., Moayednia, N. and Nezhadd, A. (2019). The viability of probiotic bacteria and characteristics of uncultured cream-containing inulin. *Journal of Food Biosciences and Technology*, 9(1): 77-84.

Femia, A.P., Salvadori, M., Broekaert, W.F., Francois, I.E.J.A., Delcour, J.A., Courtin, C.M. and Caderni, G. (2010). Arabinoxylan-oligosaccharides (AXOS) reduce preneoplasstic lesions in the colon of rats treated with 1,2-dimethylhydrazine (DMH). *European Journal of Nutrition*, 49(2): 127-132.

Flesch, A.G.T., Poziomyck, A.K. and Damin, D.D.C. (2014). The therapeutic use of synbiotics. *Arquivos Brasileiros De Cirurgia Digestiva*, 27(3): 206-209.

Franz, C.M., Huch, M., Mathara, J.M., Abriouel, H., Benomar, N., Reid, G., Galvez, A. and Holzapfel, W.H. (2014). African fermented foods and probiotics. *International Journal of Food Microbiology*, 190: 84-96.

Gerald, P. (2014). Metabolism of cholesterol and bile acids by the gut microbiota. *Pathogens*, 3(1): 14-24.

Gibson, G.R. and Roberfroid, M.B. (1995). Dietary modulation of the human colonic microbiota: Introducing the concept of prebiotics. *Journal of Nutrition*, 125(6): 1401-1412.

Gibson, G.R., Probert, H.M., Loo, J.V., Rastall, R.A. and Roberfroid, M.B. (2004). Dietary modulation of the human colonic microbiota: Updating the concept of prebiotics. *Nutrition Research Reviews*, 17(2): 259-275.

Giorgetti, G.M., Brandimarte, G., Fabiocchi, F., Ricci, S., Flamini, P., Sandri, G., Trotta, M.C., Elisei, W., Penna, A., Lecca, P.G., Picchio, M. and Tursi, A. (2015). Interactions between innate immunity, microbiota and probiotics. *Journal of Immunology Research*, 2015: 1-20.

Gismondo, M.R., Drago, L. and Lombardi, A. (1999). Review of probiotics available to modify gastrointestinal flora. *International Journal of Antimicrobial Agents*, 12(4): 287-292.

Gourbeyre, P., Denery, S. and Bodinier, M. (2011). Probiotics, prebiotics and synbiotics: Impact on the gut immune system and allergic reactions. *Journal of Leukocyte Biology*, 89(5): 685-695.

Guarner, F., Khan, A.G., Garisch, J., Eliakim, R., Gangl, A., Thomson, A., Krabshuis, J., Lemair, T., Kaufmann, P., Paula, J.A., Fedorak, R., Shanahan, F., Sanders, M.E., Szajewska, H., Ramakrishna, B.S., Karakan, T. and Kim, N. (2012). World Gastroenterology Organisation global guidelines, probiotics and prebiotics, October 2011. *Journal of Clinical Gastroenterology*, 46(6): 468-481.

Guillot, J.F. (2003). Probiotic feed additives. *Journal of Veterinary Pharmacology and Therapeutics*, 26: 52-55.

Hao, Z., Wang, W., Guo, R. and Liu, H. (2019). *Faecalibacterium prausnitzii* (ATCC 27766) has preventive and therapeutic effects on chronic unpredictable mild stress-induced depression-like and anxiety-like behaviour in rats. *Psychoneuroendocrinology*, 104: 132-142.

Hempel, S., Newberry, S., Ruelaz, A., Wang, Z., Miles, J.N.V., Suttorp, M.J., Johnsen, B., Shanman, R., Slusser, W., Fu, N., Smith, A., Roth, E., Polak, J., Motala, A., Perry, T. and Shekelle, P.G. (2011). Safety of probiotics to reduce risk and prevent or treat disease. Evidence Report/Technology Assessment No. 200, Agency for Healthcare Research and Quality Rockville, MD, USA.

Holz, C., Alexander, C., Balcke, C., More, M.I., Nielson, A., Bauer, M., Junker, L., Gruenwald, J., Lang, C. and Pompejus, M. (2013). *Lactobacillus paracasei* DSMZ16671 reduces mutans *Streptococci*: A short-term pilot study. *Probiotics and Antimicrobial Proteins,* 5(4): 259-263.

Islam, S.U. (2016). Clinical uses of probiotics. *Medicine,* 95(5): 1-5.

Isolauri, E., Sutas, Y., Kankaanpaa, P., Arvilommi, H. and Salminen, S. (2001). Probiotics: Effects on immunity. *The American Journal of Clinical Nutrition,* 73(2): 444-450.

Kahbazi, M., Ebrahimi, M., Zarinfar, N., Arjomandzadegan, M., Fereydouni, T., Karimi, F. and Najmi, A.R. (2016). Efficacy of synbiotics for treatment of bacillary dysentery in children: A double-blind, randomised, placebo-controlled study. *Advances in Medicine,* 2016: 1-6.

Kapoor, D., Vyas, R.B., Lad, C. and Patel, M. (2016). Role of probiotics in management of alignment. *International Journal of Applied Pharmaceutical and Biological Research,* 1(3): 81-90.

Karimi, G., Sabran, M.R., Jamaluddin, R., Parvaneh, K., Mohterrudin, N., Ahmad, Z., Khazaai, H. and Khodavandi, A. (2015). The anti-obesity effects of *Lactobacillus casei* strain Shirota versus Orlistat on high fat diet-induced obese rats. *Food and Nutrition Research,* 59: 1-8.

Kassaian, N., Aminorroaya, A., Feizi, A., Jafari, P. and Amini, M. (2017). The effects of probiotic and synbiotic supplementation on metabolic syndrome indices in adults at risk of Type 2 diabetes: Study protocol for a randomised controlled trial. *Trials,* 18: 1-8.

Kerry, R.G., Patra, J.K., Gouda, S., Park, Y., Shin, H.S. and Das, G. (2018). Benefaction of probiotics for human health: A review. *Journal of Food and Drug Analysis,* 26(3): 927-939.

Kim, J.A., Park, M.S., Kang, S.A. and Ji, G.E. (2014). Production of γ-aminobutyric acid during fermentation of *Gastrodia Elata* Bl. by co-culture of *Lactobacillus brevis* GABA 100 with *Bifidobacterium bifidum* BGN4. *Food Science and Biotechnology,* 23: 459-466.

Kumar, M., Nagpal, R., Kumar, R., Hemalatha, R., Verma, V., Kumar, A., Chakraborty, C., Singh, B., Marotta, F., Jain, S. and Yadav, H. (2012). Cholesterol-lowering probiotics as potential biotherapeutics for metabolic diseases. *Experimental Diabetes Research,* 2012: 902-917.

Lewis, K., Lutgendorff, F., Phan, V., Soderholm, J.D., Sherman, P.M. and McKay, D.M. (2010). Enhanced translocation of bacteria across metabolically stressed epithelia is reduced by butyrate. *Inflammatory Bowel Diseases,* 16(7): 1138-1148.

Lilly, D.M. and Stillwell, R.H. (1965). Probiotics: Growth-promoting factors produced by microorganisms. *Science,* 147(3659): 747-748.

Lugani, Y. and Sooch, B.S. (2017). Xylitol, an emerging prebiotic: A review. *International Journal of Applied Pharmaceutical and Biological Research,* 2(2): 67-73.

Lugani, Y. and Sooch, B.S. (2018). Insights into fungal xylose reductases and its applications in xylitol production. pp. 121-144. *In*: S. Kumar, P. Dheeran, M. Taherzadeh and S. Khanal (Eds.). *Fungal Biorefineries,* Springer.

Lugani, Y., Oberoi, S. and Sooch, B.S. (2017). Xylitol: A sugar substitute for patients of diabetes mellitus. *World Journal of Pharmacy and Pharmaceutical Sciences,* 6(4): 741-749.

Luong, M.L., Sareyyupoglu, B., Nguyen, M.H., Silveira, F.P., Shields, R.K., Potoski, B.A., Pasculle, W.A., Clancy, C.J. and Toyoda, Y. (2010). *Lactobacillus* probiotic use in cardiothoracic transplant recipients: A link to invasive *Lactobacillus* infection. *Transplant Infectious Diseases,* 12(6): 561-564.

Lye, H.S., Ali, G.R.R. and Liong, M.T. (2010). Mechanisms of cholesterol removal by Lactobacilli under conditions that mimic the human gastrointestinal tract.. *International Dairy Journal,* 20(3): 169-175.

Macfarlane, S., Macfarlane, G.T. and Cummings, J.H. (2006). Review article: Prebiotics in the gastrointestinal tract. *Alimentary Pharmacology and Therapeutics*, 24(5): 701-714.

Maheshwari, S.U., Amutha, S., Anandham, R., Hemalatha, G., Senthil, N., Kwon, S.W. and Sivakumar, N. (2019). Characterisation of potential probiotic bacteria from '*panchamirtham*': A South Indian ethnic fermented fruit mix. *LWT–Food Science and Technology*, 116: 108540.

Mallappa, R.H., Singh, D.K., Rokana, N., Pradhan, D., Batish, V.K. and Grover, S. (2019). Screening and selection of probiotic *Lactobacillus* strains of Indian gut origin based on assessment of desired probiotic attributes combined with principal component and heat-map analysis. *LWT –Food Science and Technology*, 105: 272-281. (not in text)

Marchand, V. (2012). Canadian Paediatric Society, Nutrition and Gastroenterology Committee: Using probiotics in the paediatric population. *Paediatric and Child Health*, 17(10): 575-576.

Marteau, P., Seksik, P. and Jian, R. (2009). Probiotics and intestinal health effects: A clinical perspective. *British Journal of Nutrition*, 88(1): S51-S57.

McCormick, S.P. (2013). Microbial detoxification of mycotoxins. *Journal of Chemical Ecology*, 39(7): 907-918.

McFarland, L.V. (2014). Use of probiotics to correct dysbiosis of normal microbiota following disease or disruptive events: A systematic review. *BMJ Open*, 4: 1-18.

Meng, X., Zhou, H.Y., Shen, H.H., Lufumpa, E., Li, X.M., Guo, B. and Li, B.Z. (2019). Microbe-metabolite-host axis, two-way action in the pathogenesis and treatment of human autoimmunity. *Autoimmunity Reviews*, 18(5): 455-475.

Minelli, P.E.B. and Benini, A. (2008). Relationship between number of bacteria and their probiotic effects. *Microbial Ecology in Health and Disease*, 20(4): 180-183.

Mishra, S.S., Behera, P.K., Kar, B. and Ray, R.C. (2018). Advances in probiotics, prebiotics and nutraceuticals. pp. 121-142. *In*: S.K. Panda and P.H. Shetty (Eds.). *Innovations in Technologies for Fermented Foods and Beverage Industries*. Springer, Switzerland.

Moayyedi, P. and Ford, A.C. (2011). Symptom-based diagnostic criteria for irritable bowel syndrome: The more things change, the more they stay the same. *Gastroenterology Clinics of North America*, 40(1): 87-103.

Mokoena, M.P., Mutanda, T. and Olaniran, A.O. (2016). Perspectives on the probiotic potential of lactic acid bacteria from African traditional fermented foods and beverages. *Food Nutr. Res.*, 2016: 60.

Molina, J.P.A., Santamaría-Miranda, A., Luna-González, A., Martínez-Díaz, S.F. and Rojas-Contreras, M. (2009). Effect of potential probiotic bacteria on growth and survival of tilapia *Oreochromis niloticus* L., cultured in the laboratory under high density and suboptimum temperature. *Aquaculture Research*, 40(8): 887-894.

Mousavi, S.M., Seidavi, A., Dadashbeiki, M., Nthenge, A.K., Nahashon, S.N., Laudadio, V. and Tufarelli, V. (2015). Effect of a synbiotic (Biomin[R] IMBO) on growth performance traits of broiler chickens. *European Poultry Science*, 79: 1-15.

Nahoum, S.R., Paglino, J., Varzaneh, F.E., Edberg, S. and Medzhitov, R. (2004). Recognition of commensal microflora by toll-like receptors is required for intestinal homeostasis. *Cell*, 118(2): 229-241.

Nuraida, L. (2015). A review: Health-promoting lactic acid bacteria in traditional Indonesian fermented foods. *Food Science and Human Wellness*, 4(2): 47-55.

Oelschlaeger, T.A. (2010). Mechanisms of probiotic actions – A review. *International Journal of Medical Microbiology*, 300(1): 57-62.

Oliva, S., Narddo, G.D., Ferrari, F., Mallardo, S., Rossi, P., Patrizi, G., Cucchiara, S. and Stronati, L. (2012). Randomised clinical trial: The effectiveness of *Lactobacillus reuteri* ATCC 55730 rectal enema in children with active distal ulcerative colitis. *Alimentary Pharmacology and Therapeutics*, 35(3): 327-334.

Ooi, M.F., Mazlan, N., Foo, H.L., Loh, T.C., Mohamad, R., Rahim, R.A. and Ariff, A. (2015). Effects of carbon and nitrogen sources on bacteriocin-inhibitory activity of postbiotic metabolites produced by *Lactobacillus plantarum* I-UL4. *Malaysian Journal of Microbiology*, 11(2): 176-184.

Paassen, N.B., Vincent, A., Puiman, P.J., Sluis, M., Bouma, J., Boehm, G., Goudoever, J.B., Seuningen, I. and Renes, I.B. (2009). The regulation of intestinal mucin MUC2 expression by short-chain fatty acids: Implications for epithelial protection. *Biochemical Journal*, 420(2): 211-219.

Parvez, S., Malik, K., Ah Kang, S. and Kim, H.Y. (2006). Probiotics and their fermented food products are beneficial for health. *Journal of Applied Microbiology*, 100: 1171-1185.

Patel, A., Prajapati, J.B., Holst, O. and Ljungh, A. (2014). Determining probiotic potential of exopolysaccharide-producing LAB isolated from vegetables and traditional Indian fermented food products. *Food Bioscience*, 5: 27-33.

Patel, R.M. and Denning, P.W. (2013). Therapeutic use of prebiotics, probiotics and postbiotics to prevent necrotising enterocolitis: What is the current evidence? *Clinics in Perinatology*, 40(1): 11-25.

Patel, R.M. and DuPont, H.L. (2015). New approaches for bacteriotherapy: Prebiotics, new-generation probiotics and synbiotics. *Clinical Infectious Diseases*, 60(2): 108-121.

Pessi, T., Sutas, Y., Hurme, M. and Isolauri, E. (2000). Interleukin-10 generation in atopic children following oral *Lactobacillus rhamnosus* GG. *Clinical and Experimental Allergy*, 30(12): 1804-1808.

Pierro, F.D., Colombo, M., Zanvit, A., Risso, P. and Rottoli, A.S. (2014). Use of *Streptococcus salivarius* K12 in the prevention of streptococcal and viral pharyngotonsillitis in children. *Drug, Healthcare and Patient Safety*, 6: 15-20.

Rai, A. K., Pandey, A. and Sahoo, D. (2019). Biotechnological potential of yeasts in functional food industry. *Trends in Food Science & Technology*, 83: 129-137.

Ranadheera, R.D.C.S., Baines, S.K. and Adams, M.C. (2010). Importance of food in probiotic efficacy. *Food Research International*, 43(1): 1-7.

Rao, A.V., Bested, A.C., Beaulne, T.M., Katzman, M.A., Iorio, C., Berardi, J.M. and Logan, A.C. (2009). A randomised double-blind placebo-controlled pilot study of a probiotic in emotional symptoms of chronic fatigue syndrome. *Gut Pathogens*, 1(1): 6-12.

Roberfroid, M. (2002). Functional food concept and its application to prebiotics. *Digestive and Liver Disease*, 34(2): 105-110.

Ruiz-Moyano, S., Gonçalves dos Santos, M.T.P., Galván, A.I., Merchán, A.V., González, E., Córdoba, M. de G. and Benito, M.J. (2019). Screening of autochthonous lactic acid bacteria strains from artisanal soft cheese: Probiotic characteristics and prebiotic metabolism. *LWT –Food Science and Technology*, 114: 108388.

Saavedra, J.M. (2001). Clinical applications of probiotic agents. *The American Journal of Clinical Nutrition*, 73(6): 1147-1151.

Saez-Lara, M.J., Gomez-Llorente, C., Plaza-Diaz, J. and Gil, A. (2015). The role of probiotic lactic acid bacteria and bifidobacteria in the prevention and treatment of inflammatory bowel disease and other related diseases: A systematic review of randomised human clinical trials. *BioMed Research International*. https://doi.org/10.1155/2015/505878

Sakaguchi, T., Kohler, H., Gu, X., McCormick, B.A. and Reinecker, H.C. (2002). *Shigella flexneri* regulates tight junction associated proteins in human intestinal epithelial cells. *Cellular Microbiology*, 4(6): 367-381.

Salvucci, E. (2019). The human-microbiome superorganism and its modulation to restore health. *International Journal of Food Sciences and Nutrition*, 7: 1-15.

Sanders, M.E. (2016). Probiotics and microbioata composition. *BMC Medicine*, 14(1): 1-3.

Schmitz, H., Barmeyer, C., Fromm, M., Runkal, N., Foss, H.D., Bentzel, C.J., Riecken, E.O. and Schulzke, J.D. (1999). Altered tight junctions structure contributes to the impaired epithelial barrier function in ulcerative colitis. *Gastroenterology*, 116(2): 301-309.

Scott, K.P., Antoine, J.M., Midtvedt, T. and Hemert, S. (2015). Manipulating the gut microbiota to maintain health and treat disease. *Microbial Ecology in Health and Disease*, 26: 1-10.

Servin, A.L. and Coconnier, M.H. (2003). Adhesion of probiotic strains to the intestinal mucosa and interaction with pathogens. *Best Practice and Research: Clinical Gastroenterology*, 17(5): 741-754.

Sharma, A., Rath, G.K., Chaudhary, S.P., Thakar, A., Mohanti, B.K. and Bahadur, S. (2012). *Lactobacillus brevis* CD2 lozenges reduce radiation and chemotherapy induced mucositis in patients with head and neck cancer: A randomised double-blind placebo-controlled study. *European Journal of Cancer*, 48(6): 875-881.

Sheehan, V.M., Ross, P. and Fitzgerald, G.F. (2007). Assessing the acid tolerance and the technological robustness of probiotic cultures for fortification in fruit juices., *Innovative Food Science and Emerging Technologies*, 8(2): 279-284.

Simova, E.D., Beshkova, D.B. and Dimitrov, Z.P. (2009). Characterisation and antimicrobial spectrum of bacteriocins produced by lactic acid bacteria isolated from traditional Bulgarian dairy products., *Journal of Applied Microbiology*, 106(2): 692-701.

Soccol, C.R., Vandenberghe, L.P.S., Spier, M.R., Medeiros, A.B.P., Yamaguishi, C.T., Lindner, J.D.D., Pandey, A. and Soccol, V.T. (2010). The potential of probiotics: A review. *Food Technology and Biotechnology*, 48(4): 413-434.

Steed, H., Macfarlane, G.T., Blackett, K.L., Bahrami, B., Reynolds, N., Walsh, S.V., Cummings, J.H. and Macfarlane, S. (2010). Clinical trial: The microbiological and immunological effects of synbiotic consumption – A randomised double-blind placebo-controlled study in active Crohn's disease. *Alimentary Pharmacology and Therapeutics*, 32(7): 872-883.

Stefe, C.A., Alves, M.A.R. and Ribeiro, R.L. (2008). Probioticos, prebioticos e simbioticos-artigo de revisao. *Revista Saude e Meio Ambiente*, 1(3): 16-33.

Sunkata, R., Herring, J., Walker, L.T. and Verghese, M. (2014). Chemopreventive potential of probiotics and prebiotics. *Food and Nutrition Sciences*, 5(18): 1800-1809.

Swain, M.R., Anandharaj, M., Ray, R.C. and Parveen Rani, R. (2014). Fermented fruits and vegetables of Asia: A potential source of probiotics. Biotechnology Research International Article ID 250424, 19 pp.; http://dx.doi.org/10.1155/2014/2504

Synbiotics Foods Market Research Report (2018). Regional Trend, Industry Demand, Business Strategy, Current and Future Plans by Fast Forward Research. https://www.marketwatch.com/press-release/synbiotic-foods-market-analysis-2018-regional-trend-industry-demand-business-strategy-current-and-future-plans-by-fast-forward-research-2018-09-2-4 (accessed on September 27, 2018)

Szajewska, H., Guarino, A., Hojsak, I., Indrio, F., Kolacek, S., Shamir, R., Vandenplas, Y. and Weizman, Z. (2014). European Society for Pediatric Gastroenterology, Hepatology and Nutrition. Use of probiotics for management of acute gastroenteritis: A position paper by the ESPGHAN working group for probiotics and prebiotics. *Journal of Pediatric Gastroenterology and Nutrition*, 58(4): 531-539.

Thomas, D.W. and Greer, F.R. (2010). American Academy of Pediatrics Committee on Nutrition; American Academy of Pediatrics Section on Gastroenterology, Hepatology and Nutrition. Probiotics and prebiotics in pediatrics. *Paediatrics*, 126(6): 1217-1231.

Tillisch, K. (2014). The effects of gut microbiota on CNS function in humans. *Gut Microbes*, 5(3): 404-410.

Tsai, Y.L., Lin, T.L., Chang, C.J., Wu, T.R., Lai, W.F., Lu, C.C. and Lai, H.C. (2019). Probiotics, prebiotics and amelioration of diseases. *Journal of Biomedical Science*, 26(3): 1-8.

Umbrello, C. and Esposito, S. (2016). Microbiota and neurologic diseases: Potential effects of probiotics. *Journal of Translational Medicine*, 14: 1-11.

Vandenbergh, P.A. (1993). Lactic acid bacteria, their metabolic products and interference with microbial growth. *FEMS Microbiology Reviews*, 12(1-3): 221-237.

Virk, H.S. and Wiersinga, W.J. (2019). Current place of probiotics for VAP. *Critical Care*, 23: 1-3.

Vivekananda, M.R., Vandana, K.L. and Bhat, K.G. (2010). Effect of the probiotic *Lactobacilli peuteri* (Prodentis) in the management of periodontal disease: A preliminary randomised clinical trial. *Journal of Oral Microbiology*, 2: 1-9.

Waigankar, S.S. and Patel, V. (2011). Role of probiotics in urogenital healthcare. *Journal of Mid-life Health*, 2(1): 5-10.

Watts, T., Berti, I., Sapone, A., Geraduzzi, T., Not, T., Zielke, R. and Fasano, A. (2005). Role of the intestinal tight junction modulator zonulin in the pathogenesis of Type I diabetes in BB diabetic prone rats. *Proceedings of the National Academy of Sciences of the United States of America*, 102(8): 2916-2921.

Williamson, E., Bilsborough, J.M. and Viney, J.L. (2002). Regulation of mucosal dendridic cell function by receptor activator of NF-κB (RANK)/RANK ligand interactions: Impact on tolerance induction. *The Journal of Immunology*, 169(7): 3606-3612.

Wong, J.M. and Jenkins, D.J. (2007). Carbohydrate digestibility and metabolic effects. *The Journal of Nutrition*, 137(11): 2539-2546.

Yesilova, Y., Calka, O., Akdeniz, N. and Berktas, M. (2012). Effect of probiotics on the treatment of children with atopic dermatitis. *Annals of Dermatology*, 24(2): 189-193.

Yoo, J.Y. and Kim, S.S. (2016). Probiotics and prebiotics: Present status and future perspectives on metabolic disorders. *Nutrients*, 8(3): 1-20.

Zhang, Z., Hinrichs, D.J., Lu, H., Chen, H., Zhong, W. and Kolls, J.K. (2007). After interleukin-12p40, are interleukin-23 and interleukin-17 the next therapeutic targets for inflammatory bowel disease? *International Immunopharmacology*, 7(4): 409-416.

Zion™ Market Research (2018). A Report on Probiotics Market: Global Industry Perspective, Comprehensive Analysis and Forecast, 2018-2024. https://globenewswire.com/news-release/2018/06/21/1527822/0/en/Global-Probiotics-Market-Will-Reach-USD-65-87-Billion-by-2024-Zion-market-Research.html (accessed on September 27, 2018)

Zukiewicz-Sobczak, W., Wróblewska, P., Adamczuk, P. and Silny, W. (2014). Probiotic lactic acid bacteria and their potential in the prevention and treatment of allergic diseases. *Central European Journal of Immunology*, 39(1): 113-117.

Part III

Traditional Fermented Products and Health Benefits

Fermented Soybean Products and Their Health Benefits

Buddhiman Tamang*, Lalit Kumar Chaurasia, Kriti Ghatani and Ranjan Kaushal Tirwa

Department of Microbiology, Sikkim University, Samdur, Tadong - 737102, Sikkim, India

1. Introduction

Cultivation of soybean originated in northern China as early as 5000 BC and was considered as one of the sacred grains along with rice, wheat, barley and millet (Wang, 1997). Methods of soybean cultivation and soy-food preparation were gradually introduced in Japan, Korea and some other Far East countries about 1,100 years ago. It was officially introduced in the United States in 1900 and since 1954, the US became the leading producer of soybean. World soybean production in 2018-19 shows that US produced 123.66 million metric tonnes while China and India ranked 4th and 5th in the world with 15.9 and 11.0 million metric tonnes respectively (SOPA, 2019). The production of soybean in Japan and Southeast Asian countries in 2018-19 was less than 0.5 million metric tonnes (United States Department of Agriculture, 2019). Fermented soybean-based products are very popular and common in the markets all over the world. Because of high contents of amino acids, isoflavones, vitamins, minerals and proteins, soybean can serve as a great substitute for meat proteins (Wood, 1998; Kwak *et al.*, 2007).

Large numbers of studies have investigated bioactive molecules in soybean because it is uniquely rich in isoflavones and proteins (Hughes *et al.*, 2011) and other micronutrients. Due to the presence of flavonoids and other bioactive molecules, soybeans have beneficial effects on diabetes, hypertension and cardiovascular diseases and in osteoporosis-risk reduction (Ishida *et al.*, 1998), immunomodulations, breast and prostate cancer inhibition and other chronic diseases (Messina *et al.*, 1994; Li-Jun *et al.*, 2004). Bioactive compounds from soybean can also improve the renal functions, improve psychological disorders and skin functions (Anderson, 2008; Azadbakht and Esmaillzadeh, 2009; Atteritano *et al.*, 2015; Messina, 2016). Isoflavones in soybean are classified as both phytoestrogens and selective estrogen

---
*Corresponding author: bmtamang3@gmail.com

receptor modulators (Messina, 2016). All these facts indicate that consumption of fermented soy foods leads to a low incidence of some hormone-dependent diseases among the people living in Asian countries (Adlercreutz *et al.*, 1992).

Even though soybeans have a high nutritional value, raw soybeans with simple boiling preparation have low bioavailability. Fermentation of soybean is one of the methods widely practiced in Asian countries, mostly in Japan, northeast India, Nepal, Indonesia, southern China, Korea, Cambodia, Laos, Burma and Myanmar (Tamang, 2015; Nout *et al.*, 2007). Fermentation process breaks larger molecules, such as proteins into oligopeptides and amino acids, deconjugation of flavonoids into aglycones, destruction of undesirable components, such as trypsin inhibitors, hemaglutinin and oligosaccharides. This helps in the improvement of their digestibility and nutritional status, thereby increasing their bioavailability. With the improvement of sensory characteristics and certain nutritious components, fermentation has become a beneficial and safe method in product manufacture (Sharma *et al.*, 2015).

Two kinds of fermented soybean products are available – non-salted and salted products. Former groups include *kinema* and similar products (northeastern India, eastern Nepal and Bhutan), *Natto* (Japan), *tempeh* (Indonesia) which falls within the *natto* triangle (Tamang, 2015) or *kinema-natto-thua nao* triangle (Tamang, 2015) where Yunan province of China is in the centre of the triangle. Soybeans are fermented either by using natural microflora, such as in *kinema* or similar products of northeast India, Nepal, Bhutan and *thua nao* of Thailand or by use of the starter culture, as in *B. subtilis (natto)* for *natto* or molds for *tempeh* (Tamang, 2015). Salted fermented soybean, *doenjang* is fermented using *meju* as the starter which is a fermented soybean containing *Bacillus* and molds. Fermentation time of *Deonjang* is more than two months (Jeong *et al.*, 2014). While *chongkukjang* of Korea is of short duration, naturally fermented, salty *Bacillus*-fermented product (Nout *et al.*, 2007; Jeong *et al.*, 2014), *Douchi* of China can be salted or non-salted fermented by using molds (Tamang, 2015). Duration of fermentation varies from less than 30 days to 12 months, according to the type of *douche*. Cheese-like fermented soybean, *sufu* of China, is made from soybean curd (*tofu*) (Han *et al.*, 2001). *Miso* and *shoyu* of Japan are also mold fermented (Sugawara, 2020). The following sections describe the health benefits of different fermented soybean products.

2. Anti-thrombotic Activity of Fermented Soybean

Thrombosis is the formation of blood clots inside a blood vessel due to the aggregation of platelets around the site of injury. Subsequently, the blood changes into gel form in a multistep process with the help of clotting factors and finally transforms the soluble fibrinogen into a fibrin mesh. However, the clot formation is not the final solution. The clot must be dissolved when the wound is healed. Hence, fibrinolysis plays a crucial role since small clots form throughout the vessels and these need to be cleaned and without fibrinolysis, blood vessels would become gradually blocked over a period of time (Marieb and Hoehn, 2013). Normally, the process of coagulation and fibrinolysis is intricately regulated to balance healthy circulation within the body (Chapin and Hajjar, 2015). When the normal thrombosis-fibrinolysis system becomes unbalanced, blood clots will not dissolve and block the blood flow, leading to ischemic arterial syndrome, such as stroke and myocardial infarction, as well as venous syndromes, including deep vein thrombosis and resultant pulmonary emboli (McCarthy and Rinella, 2012).

As per the folk-medicine of Japan, *natto* was recommended for heart and vascular diseases to relieve fatigue and as an anti-beriberi agent (National Federation of Co-operatives on Natto, 1977). It's role in fibrinolytic activity was shown by Sumi *et al.* (1987) where a bean of *natto* fermented by *B. subtilis (natto),* when placed on a plate containing fibrin, developed a clear zone. Fibrinolytic enzyme was characterised and named as *nattokinase* (NK). NK was not only able to digest fibrin but also the plasmin substrate H-D-Val-Leu-Lys-pNA (Sumi *et al.,* 1987). One gram of *natto* extract had fibrinolytic activity of 40 caseinolytic unit (CU) plasmin or 1600 IU (international unit) urokinase. Shortening of euglobulin lysis time (ELT) and elevation of euglobulin fibrinolytic activity (EFA) was observed during two to eight hours, when the healthy adult volunteers were administered 200 g *natto* per day (Sumi *et al.*, 1990). *Natto* showed inhibitory effects on platelet aggregation induced by adenosine 5'diphosphate (ADP) and collagen (Park *et al.*, 2012a). Orally administrated *natto* also showed fibrinolytic activity with shortened euglobulin clot lysis time (ECLT) and prolonged partial thromboplastin time in hypercholesterolemia Spraque-Dawley (SD) male rats (Park *et al.*, 2012a).

Purification of the NK subsequently established its molecular weight to be 27.73 kDa with 275 amino acid residues (Fujita *et al.*, 1993). It is a serine protease sharing high degree of homology with subtilisins E (99.5 per cent) (Stahlt and Ferrari, 1984) and subtilisins J (Amylosacchariticus, 99.3 per cent) (Kurihara *et al.*, 1972). Using *in vitro* clot lysis assay, the cleavage of cross-linked fibrin by NK was six times more efficient than by plasmin as measured from k_{cat}/K_m (Fujita *et al.*, 1995). Further, consumption of 1.3g encapsulated NK thrice a day also increased the amount of fibrin or fibrinogen-degradation products (FDP) in serum, tissue plasminogen activator (TPA) and EFA (Sumi *et al.*, 1990). Angiographic examination of the femoral arteries of the dog after oral administration of NK showed complete dissolution of thrombi and also shortened the ELT. Similar results were also reported when 50-200g *natto* were administered to rat experimental pulmonary thrombosis and healthy human volunteers (Suzuki *et al.*, 2003). A decrease in thrombus count and plasma ELT, as well as an increase in tissue plasminogen activator (t-PA), indicate that *Bacillus natto* serves to activate plasma fibrinolysis *in vivo*. An open-label and self-controlled clinical study reported that daily oral administration of *nattokinase* capsules (800 mg/day) over a two-month period to healthy volunteers, patients with cardiovascular risk factors and patients undergoing dialysis led to decreased plasma levels of fibrinogen in all the three groups (Hsia *et al.*, 2009).

A double-blind and placebo-controlled cross-over intervention study involving 12 healthy volunteers who were orally administered a single-dose of 2,000 FU NK showed an increased D-dimer and other fibrin/fibrinogen degradation products, decreased blood coagulation factor VIII activity and increased blood antithrombin concentration (Kurosawa *et al.*, 2015). Ferric chloride ($FeCl_3$) induced arterial thrombosis was delayed in the rats in a dose-dependent manner by the administration of NK and doubling of occlusion time at the level of 160 mg/kg while at a very high dose of 500 mg/kg NK fully prevented the occlusion (Jang *et al.*, 2013). Similar results of NK effect on the dissolution of thrombosis was reported, using a carrageenan-induced model of rat thrombosis (Xu *et al.*, 2014).

Mechanistically, NK acts both by directly degrading the fibrin and increasing the release of t-PA, resulting in the release of plasmin (Hsia *et al.*, 2009; Kwon *et al.*, 2010; Dabbagh *et al.* 2014; Fujita *et al.*, 2016; Weng *et al.*, 2017). t-PA is under the control of plasminogen activator inhibitor 1 (PAI-1) and regulates fibrinolytic cascade (Anna *et al.*, 2012). NK has been found to cleave active recombinant prokaryotic PAI-1 into low-molecular-weight fragments at its active site Arg (346)-Met (347). NK was also found to increase the tissue-type plasminogen activator–induced fibrin clot lysis. Enhanced fibrinolytic activity observed in the absence of PAI-1 appeared to be induced through direct fibrin dissolution by NK (Urano *et al.*, 2001). NK also enhanced the production of clot-dissolving agents, such as urokinase through the conversion of prourokinase to urokinase (Fujita *et al.*, 1995; Milner and Makise, 2002). Furthermore, NK was shown to be capable of blocking thromboxane formation, resulting in an inhibition of platelet aggregation without producing the side effect of bleeding (Jang *et al.*, 2013). Thus, NK was found to be a potent antithrombotic agent, and, by reducing thrombus formation, was able to slow the progression of plaque formation and reverse the evolving atherosclerotic lesions (Suzuki *et al.*, 2003).

2.1 Benefits of Fermented Soybean in Cardiovascular Diseases

Cardiovascular disease (CVD) is one of the main causes of mortality and morbidity in the world (Benjamin *et al.*, 2017; Roth *et al.*, 2017). According to the Global Burden of Disease Study (GBD) 2015, in the United States, CVD is listed as the elementary cause of death, accounting for nearly 836,546 deaths, about one in every three deaths (Benjamin *et al.*, 2017) with an estimated 422.7 million of CVD in 2015 (Roth *et al.*, 2017). The rate of death seems to be increasing day by day due to major changes in lifestyle, like decreased physical activity along with increased cases of obesity. CVD deaths include deaths due to ischemic heart disease (IHD), stroke, hypertensive heart disease leading to heart failure and atrial fibrillation (Roth *et al.*, 2017). One major fundamental event in the development of a CVD is the presence of athermanous degeneration of the arteries where cholesterol is recognised as an environmental and genetic initiator of the disease (Steinberg, 2004). Besides elevated LDL cholesterol (LDL-C), other factors include chronic elevations in blood pressure (Kannel, 1996), prolonged hyperglycemia (Duckworth, 2001), inflammation and oxidative stress (Schulze and Lee, 2005).

Soybean products have gained much attention in recent past due to their properties, such as ability to reduce risk of CVD via lowering of LDL-C. The hypocholesterolemic properties of soy protein have been reported due to the presence bioactive peptides which exert their effects primarily through mechanisms involving the LDL-C receptor (LDLR) and regulation of bile acid (Torres *et al.*, 2006; Maki *et al.*, 2010). In the United States (US Food & Drug Administration, 2016) and Canada (Health Canada, 2016), soy protein (25g) claimed a reduced risk of Coronary Heart Diseases (CHD) (FDA, 2016).

Fermented soy milk has been reported to control hyperlipidemia in animal and human models by improving high-density lipoprotein (HDL) and low-density lipoprotein (LDL) ratios (Tsai *et al.*, 2014; Cavallini *et al.*, 2016). On testing soy milk prepared from genetically modified (GM) and non-GM soy for anti-hyperlipidemic effect on Syrian hamsters fed with a high-cholesterol diet, it was observed that isoflavone, aglycones (genistein and daidzein) did not contribute to anti-hyperlipidemic activity (Tsai *et al.*, 2014). However, a controlled clinical trial study conducted on hypercholesterolemic men, who were made to consume fermented and non-fermented soy milk for over a 42-day period, revealed isoflavones, especially aglycones, contributed to the anti-hypercholesterolemic effect of fermented soy milk. The soy milk was fermented with *E. faecium* CRL 183 and *L. jugurti* 416, and supplemented with isoflavones to a final isoflavone concentration of 51.26 mg/100g reduced serum total cholesterol, LDL-cholesterol and non-HDL-cholesterol compared to those consuming unfermented soy milk (total isoflavone concentration of 8.03 mg/100 g) (Cavallini *et al.*, 2016).

Several mechanisms of action have been proposed to explain how fermented soy milk may cause hypolipidemia. However, for cholesterol lowering, the bile acid deconjugation is considered to be the most accepted mechanism (Fig. 1). The deconjugation of bile acid by the gut microbes due to the presence of bile salt hydrolase (BSH) leads to the deconjugated bile acid in the colon, which passes in the faeces, thereby inducing more bile acid synthesis from cholesterol, leading to reduction in the level of serum cholesterol (Wang *et al.*, 2013; Long *et al.*, 2017). In a study, male Golden Syrian hamsters were fed a high-cholesterol diet supplemented with soy milk fermented with *B. breve* YIT 4065 where a reduction in plasma cholesterol level, triacylglycerol, VLDL, and LDL and increased faecal bile acid concentrations were observed. It was interesting to note the effects of fermented soy milk on lipid profiles and bile acid excretion when compared to unfermented soymilk milk. The fermented soy milk had bioactive components or LAB which were below the threshold required to affect lipid profiles (Kikuchi-Hayakawa *et al.*, 1998). In another study, the hypolipidemic effect of soy milk fermented with *Lb. plantarum* P-8 on hyperlipidemic rat model was studied and it was reported to lower the levels of serum cholesterol, increase the bile acid excretion in the faeces and modulate the gut microbiota with an increase in *Lactobacillus* spp., *Bifidobacterium* spp. and *Bacteroides* spp. and lesser numbers of *Clostridium* spp. (Wang *et al.*, 2013). A recent study has shown that the soy milk fermented by *Lactobacillus acidophilus, Lactobacillus casei* was effective against hypercholesterolemia on SD rats (Ahsan *et al.*, 2019).

Recent studies reported that *natto* consumption was associated with decreased risk of total CVD mortality, particularly ischaemic heart diseases (Nagata *et al.*, 2016). It was studied that the presence of a fibrinolytic enzyme called *nattokinase* (NK) is responsible for favourable effects on cardiovascular health being anti-atherosclerotic (Suzuki *et al.*, 2003; Ren *et al.*, 2017) as described in the previous section of the chapter. NK also has a lipid-lowering effect (Zhibian *et al.*, 2004; Ren *et al.*, 2017). Some workers believed that the prevention of arteriosclerosis was by its direct antioxidant effect, leading to reduced lipid peroxidation and inhibition of LDL oxidation (Iwai *et al.*, 2002). In a study, male white rabbits were fed a high

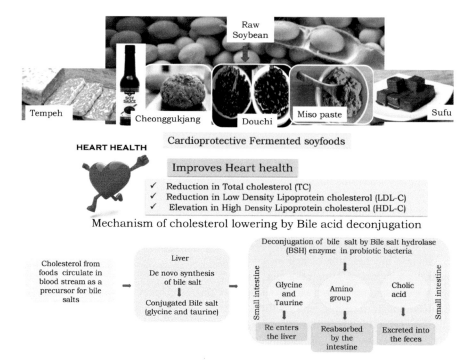

Fig. 1: Cholesterol-lowering effect of bile salt hydrolase-active probiotic bacteria in fermented soy foods

cholesterol diet supplemented with red ginseng (RG) or red ginseng with *nattokinase* (RGNK), revealing significant reduction on increased serum triglycerides levels and aortic plaque area suggesting RGNK as a potential therapeutic for atherogenesis (Kang *et al.*, 2014).

A recent clinical study was taken up in patients with atherosclerotic plaques divided into two control groups and fed NK and statin for 26 weeks (Ren *et al.*, 2017) where a significant reduction was observed in the common carotid artery intima-media thickness (CCA-IMT) on comparing with the condition before the treatment. There was a significant reduction in the carotid plaque size where the plaque reduction of NK-treated group were 36.6 per cent compared to 11.5 per cent in the statin-treated group. The data suggested that NK was a better alternative to statins and could be a better alternative therapy to CVD and patients with stroke. The mechanisms of the anti-atherosclerotic effect of NK is attributed to a combination of effects of antithrombotic, anticoagulant, antioxidant and lipid-lowering (Iwai *et al.*, 2002; Pais *et al.*, 2006; Kang *et al.*, 2014; Lin, 2017). NK was also reported to significantly reduce the increased serum triglycerides, total cholesterol, and LDL cholesterol (LDL-C) levels in animal models, showing hypolipidaemic effect (Wu *et al.*, 2009; Lin, 2017).

In a study, the evaluation of hypocholesterolemic effects of soybean and sweet lupine *tempeh* fermented with *Rhizopus oligosporus* was studied in 36 male hypercholesterolemic Sprague Dawley rats for four weeks. The rats were fed hypercholesterolemic diet + 3.5 per cent or 7 per cent protein either from soybean or sweet lupine *tempeh*; after four weeks there was a significant reduction in the levels of total cholesterol, LDL-C, VLDL-C, triglycerides, urea nitrogen, uric acid and increased HDL-C. The results indicated that *tempeh* obtained from soybean or sweet lupine with high protein level (7 per cent) significantly decreased the levels of lipid profiles, increased activity of liver enzymes and concentrations of uric acid and urea nitrogen exhibiting hypocholesterolemic, anti-atherogenic and hepatoprotective effects (Hassan *et al.*, 2014). In another recent study, the ameliorative effect of *Lactobacillus plantarum* and *R. oligosporus* fermented soy *tempeh* was studied in high-fat diet (HFD)-induced hyperglycemic rats. The combination of two was found to reduce insulin resistance, HbA1c, serum glucose, total cholesterol, triglyceride, free fatty acid, insulin and low-density lipoprotein contents and considerably increased the high-density lipoprotein content in experimental rats. The faeces of HFD rats indicated an increase in the LAB count, as well as in the bile acid, cholesterol, triglyceride, and short-chain fatty acid contents, thus indicating *tempeh* fermented with both *Lb. plantarum* and *Rhizopus oligosporus* to be a beneficial dietary supplement, showing hypocholesterolemic effect and abnormal carbohydrate metabolism (Huang *et al.*, 2018).

In a study, the effects of *doenjang* prepared from fermented soybean or brown rice on the body weight and lipid metabolism in C57BL/6N mice fed with high-fat diet were compared. After eight weeks of inducing the animals with the experimental diet, it was found that the diet supplemented with *doenjang* neutralised high fat-induced hyperlipidemia through modulation of lipogenesis and adipokine production. It was suggested that brown rice, in combination with rice bran or red ginseng marc *doenjang*, showed similar antiobesity effect and hypolipidemic action (Park *et al.*, 2012b). A study was conducted to investigate the cholesterol-lowering effect of *chongkukjang* fermented with *B. subtilis* DJI in male SD rats fed a high-fat/high-cholesterol diet for four weeks. The rats were divided into four groups and fed a normal diet, high-fat/high-cholesterol diet, high-fat/high-cholesterol diet with DJI *chongkukjang* and commercial *chongkukjang*. A significant decrease in body weight and liver and adipose tissue weights and serum total cholesterol was observed in rats fed *chongkukjang* powder in comparison with the high-fat diet group. The LDL-serum cholesterol levels, atherogenic index and cardiac risk factor-assessment indications decreased in *chongkukjang* powder-fed groups, thus suggesting the consumption of dietary *chongkukjang* may improve lipid metabolism and prevent obesity and hyperlipidemia (Kim *et al.*, 2010). Similarly in another study, *Chongkukjang,* fermented with poly γ-glutamic acid producing *Bacillus licheniformis*-67 was studied in diet-induced obese C57BL/6J mice to investigate the anti-obesity effects of soybean paste. It was observed that 30 per cent of *chongkukjang* supplemented group had lower serum and hepatic lipid profile, blood glucose, insulin and leptin along with the body weight and epididymal fat pad weight, thus indicating the prevention of obesity (Pichiah, 2016). The effect of *chongkukjang* on body composition, dyslipidemia and risk factors for atherosclerosis in a double-blind, randomised, controlled crossover trial that was conducted on 120 overweight/obese subjects (Byun *et al.*, 2016). They found that the percentage of body fat, lean body mass, waist circumference and waist-to-hip ratio of women in the *chongkukjang* group significantly improved as compared with the placebo group. Lipid profiles and high-sensitivity C-reactive protein of women in *chongkukjang* also improved significantly (Byun *et al.*, 2016).

2.2 Angiotensin Converting Enzyme (ACE) Inhibitors and Hypertension

Angiotensin I-converting enzyme (dipeptidyl peptidase A, EC 3.4.15.1 or ACE) is a monomeric, membrane-anchored carboxypeptidase that is activated by chloride and zinc that catalyse the conversion of angiotensin (Ang) I to angiotensin II. ACE has a broad specificity and hence may also act on non-vasoactive peptides. Being an exopeptidase, ACE cleaves dipeptides from the C-terminus of various oligopeptides. It can elevate blood pressure by converting the inactive decapeptide Ang I to potent vasoconstrictor Ang II (octapeptide) by removing the terminal dipeptide (Skeggs *et al.*, 1956). In addition, ACE can raise the blood pressure further by catalysing the degradation of a vasodialator peptide, a bradykinin, by sequential removal of carboxy terminal dipeptides (Li-Jun *et al.*, 2004). ACE exists in two distinct isoforms in humans – the somatic and smaller testicular types which are transcribed from a single gene located in the chromosome 17 having different initiation sites (Hubert *et al.*, 1991).

Renin-angiotensin system (RAS) is an important hormone system, which regulates blood pressure and fluid balance where ACE is the central component of this system. Angiotensinogen (Agt) is initially cleaved by rennin to Ang I, which in turn is cleaved to biologically active molecule Ang II by ACE. Ang II binds to Ang receptors (AT1 and AT2), which mediates a broad range of physiological functions that are linked to every system of the body, including the heart, kidney, vasculature, brain and immune system (Brenner *et al.*, 2001; Lu *et al.*, 2016). But a primary function of the RAS is in circulatory homeostasis, protecting body fluid volumes and abnormal activation of the RAS can contribute to the development of hypertension, cardiac hypertrophy and heart failure. In this regard, pharmacological inhibitors of the synthesis or activity of Ang II have proven immensely useful in cardio-vascular therapeutics. Studies in animal models have demonstrated that ACE inhibition reduces blood pressure and atherosclerosis (Lu *et al.*, 2012; Mann *et al.*, 2013; Chen *et al.*; 2013b). For example, ACE inhibitors are effective and widely used for the treatment of hypertension, congestive heart failure and kidney diseases (Brenner *et al.*, 2001; Svensson *et al.*, 2001; Mann *et al.*, 2013).

2.2.1 ACE Inhibitory Peptides Derived from Fermented Foods

The proteins present in soybean may be hydrolysed during processing or by microbial proteases during fermentation, and release small peptides possessing therapeutic properties, such as anti-hypertensive (Lu *et al.*, 2012; Lee and Hur, 2019a), antioxidant (Jin *et al.*, 2013), antimicrobial (Correa *et al.*, 2011), antidiabetic (Graham *et al.*, 2019), immunomodulatory activities (Fan *et al.*, 2019), etc. Such peptides having therapeutic properties are called bioactive peptides. Besides, these peptides can also be released during the digestion process in the gastrointestinal tract or by the starter culture during the fermentation process. Ferreira (Ferreira, 1965) first reported bradykinin potentiating peptides from the snake venom and subsequently isolated six such peptides. Amino acid sequence for one of the sequences was proposed to be Pyr-Lys-Trp-Ala-Pro (Ferreira *et al.*, 1970). Subsequently, these peptides were shown to possess the capability to inhibit ACE (Bakhle, 1968; Bakhle *et al.*, 1969). Later on, gelatine was used to obtain ACE inhibitory peptides by digesting gelatine with bacterial collagenase (Oshima *et al.*, 1979). At present such peptides have been reported from a variety of sources including protein-rich fermented foods, fermented goat milk (Shu *et al.*, 2019), yoghurt and similar products (Quirós *et al.*, 2007), cheese (Pripp *et al.*, 2006), fermented soybean foods (Kuba *et al.*, 2003; Zhang *et al.*, 2006b), fermented fish sauce (Ichimura *et al.*, 2003), fermented oyster sauce (Je *et al.*, 2005), grape wine (Pozo-Bayón *et al.*, 2007), and sake and sake lees (Saito *et al.*, 1994). Among non-fermented sources of ACE inhibitory peptides, beef meat (Lee and Hur, 2019b), egg white protein (Fan *et al.*, 2019), marine alga (Cao *et al.*, 2017; Sun *et al.*, 2019), *Agaricus biosporus* (Lau *et al.*, 2014), rice protein (Chen *et al.*, 2013a) and spent Brewer's yeast (Amorim *et al.*, 2019), etc. have been explored.

During soybean fermentation, the ACE inhibitory peptides are generated by proteolytic degradation of soybean protein fractions (glycinin and β-conglycinin) (Kuba *et al.*, 2005). Release of these inhibitory peptides is specific to the microbial strains used during fermentation (Sanjukta and Rai, 2016). Release of ACE inhibitory during fermentation has been shown in *douchi* fermented with *A. egypticus* revealing that the ACE inhibitory activity increased with fermentation time after gastrointestinal digestion of fermented soybean (Zhang *et al.*, 2006a). Chinese *sufu*, a fungus fermented soybean, cheese-like product, was also shown to exhibit ACE inhibitory activity (Ma *et al.*, 2013). ACE inhibitory activity increased with fermentation and ripening process by 11.55-37.61 per cent while the addition of salt decreased its production. A significant correlation was found between ACE inhibitory activity and peptide content, indicating the potential of *sufu* as a source of efficient ACE inhibitors (Ma *et al.*, 2013). Similarly, increase of ACE inhibitory peptides during soybean fermentation has been reported in *mao-tofu* fermented with *Mucor* spp. (Hang and Zhao, 2012) and *natto* fermented with *Bacillus subtilis* (*natto*) (Ibe *et al.*, 2006). Fermentation of soymilk with five different LAB (*Lb. casei, Lb. acidophilus, S. thermophilus, Lb. bulgaricus, B. longum*) followed by enzymatic treatment resulted in production of ACE inhibitory peptides (Tsai *et al.*, 2006). In another study, soymilk fermented with *E. faecium* strain isolated from raw soymilk resulted in production of peptides, exhibiting ACE inhibitory properties (Martinez-Villaluenga *et al.*, 2012).

Comparative study on the ACE inhibitory activities between the Japanese *tofuyo* and the Chinese *sufu* showed that Chinese *sufu* exhibited higher ACE inhibitory activity than *tofuyo* (Wang *et al.*, 2003). Presence of small peptides of molecular weight less than 10 kDa was found in the *sufu* extracts, though the nature of the peptides was not studied. In another study, among all of the fermented soy products, Chinese fermented soypaste showed lowest IC_{50} value of 0.012 mg/ml (Li *et al.*, 2010). It is known from this study that the reported IC_{50} value was lowest compared to that of other fermented soy products. In a comparative study of fermented soybean seasoning (FSS) and normal soy sauce, ACEI activity of FSS (IC_{50} = 454 mg/ml) was found lower than that of the regular soy sauce (IC_{50} = 1620 mg/ml) (Nakahara *et al.*, 2010). The ACE inhibitory peptides isolated from FSS and their IC_{50} values are presented in Table 1. After 30 hours of fermentation of soymilk by five different LAB strains, there was an increase in peptide content and the free amino acid content, which resulted in decrease in IC_{50} value of ACE inhibitory activity ranging between 9.28-0.66 mg powder/ml (Tsai *et al.*, 2006). Soy protein concentrate fermented with *Lactobacillus casei* sp. *pseudoplantarum* resulted in two ACE inhibitory peptide fractions, F2 and F3 with IC_{50} values of 17 and 30 mg/ml, respectively. The N-terminal sequence of the F2 peptide was found to be Leu-Ile-Val-Thr-Gln (Vallabha and Kaultiku, 2014).

Table 1: Some of the ACEI Peptides Derived from Fermented Soybean Products and Their IC_{50}

Source	Peptide Sequence	IC_{50} value (mg/ml)	References
Fermented soybean seasoning (FSS)	Ala-Trp	10	Nakahara *et al.* (2010)
	Ala-Tyr	48	
	Gly-Trp	30	
	Ser-Tyr	67	
	Gly-Tyr	97	
	Val-Pro	480	
	Ala-Phe	190	
	Ala-Ile	690	
	Val-Gly	1100	
Korean fermented soy paste	His-His-Leu	2.2	Shin *et al.* (2001b)
Fermented soy protein	Leu-Ile-Val-Thr-Gln	17	Vallabha and Kaultiku (2014)

ACEI peptides can be classified into three types (Fujita *et al.*, 2000): (1) true inhibitor type ACE inhibitory peptides, that are not affected by gastrointestinal digestion, (2) substrate type ACE inhibitory peptides that are converted peptides with weaker activity on gastrointestinal digestion, and (3) pro-drug type ACE inhibitory peptides that are converted to true inhibitors by gastrointestinal digestion. The presence of hydrophobic amino acids (Try, Phe, Trp, Ala, Ile, Val, and Met) or positively charged amino acids (Arg and Lys) as well as Pro at the C terminal position of the ACE inhibitory peptides show better affinity with ACE (Haque and Chand, 2008; Sanjukta *et al.*, 2015). ACE inhibitory peptides were also isolated and purified using chromatographic methods from the *tofuyo* (Kuba *et al.*, 2003). Peptides composed of the amino acid sequence, Ile-Phe-Leu and TrpeLeu, found in the α- and β-subunits of β-conglycinin and B-, B1A- and BX-subunits of glycinin respectively, showed good ACE inhibitory activity and were resistant to pepsin, chymotrypsin and trypsin treatments (Kuba *et al.*, 2003). Further studies on the inhibitor mechanism showed that both the peptides were non-competitive inhibitors.

A number of studies in animal models and in a few human volunteers showed that ACE inhibitory can reduce the blood pressure. Feeding of lyophilised fermented soybean curd, *tofuyo* also resulted in decrease of ACE activities in kidneys, decrease in blood pressure and decrease in serum total cholesterol (Kuba *et al.*, 2004). Oral administration of partially purified ACEI peptides from *natto*, fermented with *B. subtilis natto* in spontaneously hypertensive rats (SHR), showed decrease in systolic blood pressure within five hours of administration (Ibe *et al.*, 2009). *Chongkukjang* (or *chunggugjang*), fermented with *B. subtilis* CH-1023 was found to possess a potent antihypertensive peptide (Toshiro *et al.*, 2004). When 20 g *chongkukjang* was administered to humans, their systolic blood pressure reduced by 15mm Hg and diastolic blood pressure by 8 mm Hg after two hours of ingestion. Further, ACE inhibitory peptide Lys-Pro (0.083 mg/100 g sample) was purified and was shown to exhibit ACE-inhibitory activity of $IC_{50} = 32.1$ mM (Toshiro *et al.*, 2004). The results suggest that ACEI peptides formed during *douchi* fermentation were pro-drug-type or a mixture of pro-drug-type and true inhibitor-type peptides (Zhang *et al.*, 2006a). In another study, Shin *et al.* (2001) ACE inhibitory tri peptide, His-His-Leu from Korean fermented soy paste, was fed to SHR, which resulted in reduction of ACE activity in the aorta and led to lower systolic blood pressure in SHR (Shin *et al.*, 2001a). Similar reduction in the systolic blood pressure was observed in antihypertensive SHR upon long-term administration of fermented soy sauce (Nakahara *et al.*, 2010).

Though there are a number of potent ACE inhibitory peptides, such as accupril (quinapril), aceon (perindopril), altace (ramipril), capoten (captopril), etc., they possess side effects, such as dizziness, lightheadedness, tiredness, dry cough, nausea or vomiting. Hence ACEI derived from fermented foods can be preferred above synthetic drugs (e.g. Captopril) as they do not exhibit any side effects (Rho *et al.*, 2009). Moreover, fermented soybean foods with ACEI properties have the potential to be developed as functional foods.

2.3 Influence of Fermented Soybean on Immune System

There are many reports that show the influence of fermented soybean on immune system (Cao *et al.*, 2019; Lee *et al.*, 2017; Lefevre *et al.*, 2015; Sarkar and Nout, 2014; Singh *et al.*, 2014). *Doenjang* is a Korean food product prepared by fermentation of soybean with some *Bacillus* and fungal species for three to six months (Shin and Jeong, 2015), having anti-inflammatory property (Kim *et al.*, 2014). It also increases the CD4 and CD8 T cell proliferation in the stomach (Lee *et al.*, 2011), besides regulating the pro-inflammatory cytokinens, i.e. interleukin 6 (IL-6) and tumor necrosis factor α (Lee *et al.*, 2011).

2.3.1 *Immunomodulatory Peptides of Soybean*

The hydrolysis of soy proteins resulted in immunomodulatory peptides with the amino acid sequences of His-Cys-Gln-Arg-Pro-Arg and Gln-Arg-Pro-Arg (Singh *et al.*, 2014). These peptides carry phagocytosis activity (Singh *et al.*, 2014). Low molecular weight and positive charge of these peptides may provide the property to stimulate the lymphocyte proliferation at very low concentration (Mercier *et al.*, 2004). On hydrolyses of soy protein from trypsin, a derivative of α-subunit of β-conglycinin was obtained that stimulated phagocytosis (Maruyama *et al.*, 2003). Experiments on human monocyte model (U937) showed that the peptides from fermented soy milk influence the production of U937 cytokine (Masotti *et al.*, 2011). Soybean fermented products with *B. longum* R0175, *L. helveticus* R0052 and *S. thermophilus* ST5 were subjected to study the production of TGT-β1 by pro-inflammatory cytokine Tumor Necrosis Factor-α (TNF-α) challenged U937 cells and found tumor inhibiting property on mouse (Liu *et al.*, 2002). Soybean derived Lunasin and its similar peptides inhibit the inflammation by suppressing the NFkB pathway (de Mejia and Dia, 2009).

A lot of work has been done which focuses on the immunomodulating property of *natto* (Cao *et al.*, 2019). Pan *et al.* (2009) showing that direct administration of 16 mg *natto* per day to rat models can help in recovery of nerve injury (Pan *et al.*, 2009a). Bioactive proteins present in *natto* suppress the secretion of TNFα and IL-1β, which inhibit the apoptosis of Schwann cells (Pan *et al.*, 2009a). Bioactive peptides present in *natto* also suppress the migration of macrophages and reduce the cytokine levels in the nerve tissue of rat model in case of peripheral nerve injury (Pan *et al.*, 2009b). Functional peptides synthesised by *B. subtilis* (*natto*) reduce the LPS-induced IL-8 secretion (Azimirad *et al.*, 2017). Xu *et al.* (2012) reported that *B. subtilis* (*natto*) B4 in association with RAW264 murine macrophages increase the nitric oxide and inflammatory cytokine production (Xu *et al.*, 2012). *B. subtilis* produces cyclic lipopeptide, i.e. surfactin, that can efficiently reduce the expression of interferon γ (IFN-γ) and IL-6. It induces the nitric oxide synthase and nitric oxide (NO) in LPS-stimulated peritoneal macrophages of mouse by down-regulating toll-like receptor, which is induced by nuclear factor κB (NF-κB) (Zhang *et al.*, 2015). Evidences show that direct feeding of BALB/C mice with *B. subtilis* (*natto*) strains, BS02 and BS04, can enhance the production of IFNγ, resulting in increase of CD4 (T_H) cells in the spleen. It shows that *B. subtilis* strains, BS02 and BS04, stimulate the phagocytic activity of monocytes and cytotoxicity of natural killer cell (Gong *et al.*, 2017).

Studies of fermented soy milk also showed similar results as that of *natto*. Experiments conducted on BALB/c mice showed an increase in TNFα induced by splenocyte concanavalin A (Appukutty *et al.*, 2015). Human intestinal epithelial cell line HT 29 showed TNFα induced IL8 production on treatment with *B. longum* R0175 fermented soy milk (Wagar *et al.*, 2009). *L. helveticus* R0052 and *S. thermophilus* R0083 (SF-Lh) fermented soy milk treatment on IEC showed down-regulation of NFκB-regulated pro-inflammatory genes induced by TNFα (Lin *et al.*, 2016). Fermented soy milk with *L. helveticus* R0052 and *S. thermophilus* R0083 also produced isoflavonoid, known as aglycone, which decreases the pro-inflammatory cytokine production in IEC cell line (Champagne *et al.*, 2010). Like *natto,* soy milk fermented with lactic acid bacteria also carries the active compounds that down regulate the production of nitric oxide, which is an important factor to regulate the innate and adaptive immune system (Bogdan, 2015). Rat fed with *S. thermophilus, L. acidophilus* LA-5, and *B. bifidum* Bb-12 fermented soybean reduced the circulation of neutrophils and increased lymphocytes at the site of infection or injury, resulting in reduced inflammation (Niamah *et al.*, 2017).

Soy sauce is rich in polysaccharides which shows strong immunomodulatory potential (Cao *et al.*, 2019). A very well-studied example is that of acidic polysaccharides (APS1) from soy sauce induced the T_H1 based cytokines to activate the murine peritoneal macrophages (Matsushita *et al.*, 2006). APS1 also efficiently reduced the histamine secretion from basophilic leukaemia RBL-2H3 cells of rat and limited the passive cutaneous hypersensitivity type 1 reaction elicited in mouse model (Kobayashi *et al.*, 2004). Remarkable growth in immunoglobulin A (IgA) was reported in BALB/c mice intestine on transportation of APS1 across Caco-2 human IECs (Matsushita *et al.*, 2006). Direct administration of LAB strain *Tetragenococcus halophilus* Th221 induces IL-12 production, peritoneal macrophages of murine, stimulates T_H1 and suppresses serum IgE, which result in an ameliorating effect on allergic reaction (Matsushita *et al.*, 2009). However, a variety of bioactive proteins and other metabolites have been discovered, though more work is under progress to understand the detailed mechanism behind its immunomodulating abilities and its scope in futuristic pharmaceutical industry. There are a number of unexplored fermented soybean products in different regions of the world, including those from northeast India, Nepal and Bhutan that may carry highly functional metabolites with immunomodulating properties and can be used to deal with hypersensitivity and immune-system-related health disorders.

2.4 Benefits of Fermented Soybean in Type II Diabetes Mellitus

Type II diabetes is characterised by relative insulin deficiency caused by pancreatic β-cell dysfunction and insulin resistance by the target organs (DeFronzo, 2004). During the course of development of Type II diabetes, insulin resistance develops early but glucose tolerance remains normal because of a compensatory increase in insulin secretion (Polonsky *et al.*, 1996). One of the probable reasons for the development of insulin resistance is obesity. In comparison to the lean subjects, obese subjects have three to four times higher basal and total 24-hour rates of insulin secretion (Polonsky *et al.*, 1988). This is because the pancreatic β cell adapts to increased nutrient availability and insulin resistance by increasing its function and mass (Kaiser and Leibowitz, 2009). Thus, β-cell dysfunction and deficient pancreatic mass play crucial roles in the development of insulin resistance and subsequent β-cell compensation failure manifests as diabetes (Weir and Bonner-Weir, 2004).

Since the rate of endogenous glucose production is very high in diabetic people, the glucose released immediately after a meal further increases the post-prandial glucose (Firth *et al.*, 1986). This rise in post-prandial blood glucose can be avoided by retarding the action of α-amylase and α-glucosidase (Chen *et al.*, 2007); thus dietary glucose will not be immediately available for absorption to the intestinal brush border membrane Na^+/glucose transporters or to GLUT2 (Crane, 1965; Röder *et al.*, 2014). Many of the α-amylase and α-glucosidase inhibitors have been isolated from plants, such as *Salacia reticulata* (Yoshikawa *et al.*, 2002), the beans (Obiro *et al.*, 2009; Gupta *et al.*, 2014), *Punica granatum* flower (Li *et al.*, 2005), rye (Iulek *et al.*, 2000), wheat (Feng *et al.*, 1996) and from *Streptomyces* sp. (Geng *et al.*, 2008; 2009; Sun *et al.*, 2015). Three isoforms of α-amylase inhibitors are found in legumes:

α-amylase inhibitor isoform 1 (α- AI1), α-amylase inhibitor isoform 2 (Alpha-AI2), and α-amylase inhibitor isoform like (Alpha-AIL). The α-AI1 isoform displays activity in humans. α-AI1 is found in germs and seeds; and cannot be found in other plant parts (Iguti and Lajolo, 1991; Obiro *et al.*, 2008). Through *in vitro* research, it was proved that white beans bind α-amylase, forming a 1:1 complex blocking the enzyme. The research on rats showed that it decreases basic and postprandial glucose levels in blood. Moreover, it diminishes appetite, which, in turn, contributes to weight control (Tormo *et al.*, 2004). The water extract of *douchi* has been reported to have activity against rat intestinal α-glucosidase (Chen *et al.*, 2007). Similarly, the water-soluble extract of *touchi* also had strong anti-α-glucosidase activity and could significantly decrease the blood glucose level at 30 and 60 minutes of oral administration in rats and in humans (Fujita *et al.*, 2001a).

The administration of long-term naturally-fermented soybean paste (LFSP) using natural microflora, but not short-term fermented soybean paste (SFSP), protects high-fat diet (HFD)-fed obese mice against non-alcohol fatty liver disease (NAFLD) and insulin resistance (Kim *et al.*, 2018). Kim *et al.* (2018) reported that LFSP suppressed body weight gain in parallel with reduction in fat accumulation in mesenteric adipose tissue (MAT) and the liver *via* modulation of MAT lipolysis and hepatic lipid uptake. LFSP-treated mice had improved glucose tolerance and increased adiponectin levels concomitantly with enhanced AMP-activated protein kinase (AMPK) in skeletal muscle and suppressed expression of pro-inflammatory cytokines in skeletal muscle and the liver (Kim *et al.*, 2018). Extract prepared from *chongkukjang* that was previously fermented by *Bacillus licheniformis* showed decrease in triglyceride accumulation in 3T3-L1 adipocytes which was also confirmed by decrease in the mRNA expression of fatty acid synthetase and acetyl CoA carboxylase (Jeong *et al.*, 2018). Jeong *et al.* also observed that glucose-stimulated insulin secretion increased with high dosage of the ethanol and water soluble extract of *chongkukjang* in insulinoma cells compared to the control.

Soybean isoflavonoids and protein are effective in Type II diabetes; effective in overcoming insulin resistance and in controlling the blood glucose level (Ishihara *et al.*, 2003; Mezei *et al.*; 2003). The 6-o-malonylglucosides, principal isoflavones in the raw soybean, are converted into 6-o-acetylglucosides by heating and long-term fermentation (Barnes *et al.*, 1998; Murphy *et al.*, 1999). Besides, isoflavones are also deconjugated by the β-glucosidases of microbial origin into aglycones (genistein and daidzein) which seem to have greater activity than isoflavonoid glycones. Studies in male SD rats have shown that the serum total flavonoids were higher when fed with a diet containing isoflavone aglycone than in those rats fed with isoflavone glycoside-rich diet (Kawakami *et al.*, 2005). Besides, an isoflavone aglycone-rich diet reduced liver and serum total cholesterol levels and liver triglyceride levels in rats fed with cholesterol. Thus, fermented soybean with higher levels of isoflavonoid aglycones may be more effective in controlling glucose metabolism.

Genistein and daidzein at physiologically optimum concentrations exert beneficial effects on the functions of pancreatic β cells (Choi *et al.*, 2013). In their study, Choi *et al.* (2008) tested the influence of genistein and daidzein on glucose and insulin metabolism in non-fat mice, in which autoimmunological, insulin-dependent diabetes was developed. Isoflavonoids supplied at the dosage of 0.2 g/kg within nine weeks led to insulin production maintenance by pancreatic β cells, while with mice from the control group insulin production did not appear (Choi *et al.*, 2008). Another *in vivo* examination carried out on non-fat mice with streptozotocin (STZ)-induced diabetes, fed with fermented soya containing 0.222 g/kg genistin, had increased serum insulin and decreased serum glucose. Moreover, the death rate and incidences of cataract in the diabetic rats were markedly less among rats fed with soya containing higher dose of genistin (Lu *et al.*, 2008). Apart from increase in pancreatic insulin production, soya diet contributed to an improvement in peripheral tissue sensitivity to insulin, and decreased glycosylated haemoglobin (Kim *et al.*, 2008). Similar effect of *chongkukjang* on serum insulin and glucose levels was observed in 90 per cent pancreatectomised Type II diabetic animal model (Kwon *et al.*, 2007), C57BL/KsJ-db/db mice (Kim *et al.*, 2008). Therefore, *chongkukjang* delayed diabetic symptoms in Type II diabetic rats, and this was related to increased isoflavonoid aglycones, such as daidzein and genistein and small peptides. Similarly, water extract of *touchi*, decreased fasting and postprandial blood glucose levels in KKA^y diabetic rats decreased the amount of glycated haemoglobin in mild Type II diabetic patients (Fujita and Yamagami, 2001; Fujita *et al.*, 2001b).

2.5 Anticancer and Antitumor Property of Soybean Fermented Foods

A soybean cotyledon protein, 'lunasin', has been reported as anticancer and anti-inflammatory agent (Galvez *et al.*, 1997). Anticancer activity of synthetic lunasin has been studied on mammalian cells against chemical carcinogens and oncogenes RAS and E1A (Jeong *et al.*, 2007; Allen *et al.*, 2009). Several studies on animal models and epidemiologic analyses showed that soy protein can reduce the probability of prostrate, colon and mammary cancer cells (Badger *et al.*, 2005). Studies were also done on 7, 12-dimethylbenz(a)anthracene (DMBA) and methylcholanthrene treated (MCA) fibroblast NIH/3T3 cells line to analyse the anti-cancer property of soybean active peptides (Singh *et al.*, 2014). This study showed the significant effect of soy protein lunasin on cancer cells developed by chemical mutagens (Hsieh *et al.*, 2010). Studies on anticancer property of lunasin on mouse models with DMBA-induced mammary cancer showed that the soy protein like lunasin can be used as a regular component in our diet to reduce the risk of cancer (Lumen and O, 2005). Lunasin has structural homology to chromatin binding protein (CBP) (Lumen and O, 2005) and can penetrate into the mammalian cells within a few minutes and accumulate in nucleus in ~18 hours as a result, inhibiting the acetylation of histone proteins. On the other hand, it does not affect the normal cell lines (Badger *et al.*, 2005). Another group of researchers have reported that soybean meal contains certain bioactive peptides that can inhibit the cancer on liver (HepG-2), colon (HCT-116) and ling (NCL-H1299) cell lines at the rate of 70, 73 and 68 per cent, respectively (Rayaprolu *et al.*, 2013). One of the studies by Hwang and co-workers (2011) reported inhibition of breast cancer in MCF7 cell line by activating TGFβ pathway by the extract of fermented soybeans (Hwang *et al.*, 2011).

There are a number of reports that mention bioactive proteins isolated from fermented soy milk having remarkable effect on mutagenicity (Liu *et al.*, 2005;

Hsieh *et al.*, 2010), tumor development (Chen *et al.*, 2002; Lai *et al.*, 2013) and carcinogenesis (Ohta *et al.*, 2000). Toi *et al.* (2013) conducted a study that showed consumption of soy isoflavones and *L. casei* Shirota since adolescence can reduce the risk of breast cancer in women (Toi *et al.*, 2013). Soy milk fermented with *Bacillus infantis* and *S. thermophilis* possesses proteins that can reduce mutagenicity of carcinogens, like ether 3, 2-dimethyl 4-amino biphenyl or 4-nitroquinoline N-oxide in *Salmonella typhimurium* TA100 cells. However, unfermented soy milk does not show the antimutagenic property (Hsieh *et al.*, 2005). Mutagenesis induced by 4-nitroquinoline-N'-oxide and N-methyl-N-nitro-N-nitrosoguanidine can be reduced up to 68 and 45 per cent by soy milk *kefir* in *Salmonella nutagenicity* assay (Liu *et al.*, 2005). SD rats with induced mammary tumor by 2-amino-1-methyl-6-phenyliidazo [4,5-b] pyridine, were fed with lyophilised soy fermented milk with *B. breve* strain yakult for 20 weeks. The results showed that fermented milk reduced the growth and development of tumor cells in the SD rats (Ohta *et al.*, 2000). Fermented soy milk is the source of isoflavone aglycones that has been reported with a significant antitumor and anticancer property (Ohta *et al.*, 2000). Soy milk fermented with *Lactobacillus* strains has increased levels of isoflavone aglycones, like genistein and daidzein and their derivatives, like equol (Champagne *et al.*, 2009; Di Cagno *et al.*, 2010). A number of metagenomic studies have shown an inverse relation between the isoflavone aglycone intake with breast and prostate cancer development (Mahmoud *et al.*, 2008; Taylor *et al.*, 2009).

Mice fed with lyophilised soy milk *kefir* for one month were injected with sarcoma 180 mouse tumor cells under the abdominal skin. The results showed that the mouse supplemented with fermented soy milk had relatively smaller size of tumor than those in the control mice (Liu *et al.*, 2002). During soy milk fermentation, a novel compound, known as latifolicinin A, was produced, which had the potential to inhibit the MDA-MB-231 breast cancer cell proliferation (Chen *et al.*, 2002; Ke *et al.*, 2015). Bioactive proteins like Bowman-Birk and Lunasin were isolated from soy fermented foods like *natto, miso* and *tempeh*, and unfermented soy milk, with anticancer property (Hernández-Ledesma *et al.*, 2009; Cavazos *et al.*, 2012). Effect of fermentation time on anticancer and antimetastatic of *doenjang* was studied in the tumor-induced Balb/c mice (Jang *et al.*, 2006). *Doenjang* fermented for 24 months exhibited a two- to three-fold increase in antitumor effects on sarcoma-180-injected mice and antimetastatic effects in colon 26-M 3.1 cells in mice compared with three-or-six-month fermented *doenjang*. The 24-month fermentation was most effective in preventing cancer by decreasing tumor formation and increasing natural killer cell activity in spleen and glutathione *S*-transferase activity in liver of the mouse (Jung *et al.*, 2006). Traditional *doenjang* extracts demonstrated strong antimutagenic activities against various carcinogens/mutagens including aflatoxin (Park *et al.*, 1990, 2003). Traditionally fermented *doenjang* showed higher antimutagenic activity than raw soybeans, cooked soybeans, *meju* and other fermented soybeans in the Ames test (Park *et al.*, 2003). The active compounds that were identified are genistein, linoleic acid, β-sitosterol glucoside, soyasaponin, etc. The active compounds exhibited strong antimutagenic activities in the Ames test, SOS chromotest and *Drosophila* wing-spot test (Park *et al.*, 2003).

Zhao *et al.* (2013) demonstrated that three short-term fermented soybean foods (*shuidouchi, natto* and *chongkukjang*) were associated with a high degree of antimutagenic activities. At a concentration of 2 mg/mL, *chongkukjang* had highest growth inhibitory rate (78 per cent) on TCA8113 cells treated with *chongkukjang* as determined by a MTT. Similarly, *shuidouchi* and *natto*, the growth inhibition rate was 67 and 55 per cent respectively. When the expression of Bax, Bcl-2, and caspase-3 and -9 in TCA8113 cancer cells was analysed by RT-PCR and Western blot after 48 hours of incubation with 2 mg/mL sample solutions, it was found that *chongkukjang* treatment markedly altered the levels of pro-apoptotic Bax and anti-apoptotic Bcl-2. mRNA and protein expression of Bax was upregulated, while that of Bcl-2 significantly decreased. They demonstrated that *chongkukjang* had the most potent *in vitro* anticancer effect and can induce apoptosis (Zhao *et al.*, 2013). In their study, Seo *et al.* (2009) had similar observations where *K-chungkukjang* at the level of 1 mg/mL showed highest growth inhibitory effect of 87 per cent, followed by *H-chungkukjang* (8 5 per cent) and *MC-chungkukjang* (69 per cent). *K-chungkukjang* induced apoptosis as determined by 4,6-diamidino-2-phenylindole staining and exhibited increased bax and decreased bcl-2 mRNA expression (Seo *et al.*, 2009).

2.6 Benefits of Fermented Soybean in the Maintenance of Bone Density

There are many studies that establish a correlation between *natto* consumption and strengthening of bones and reducing the incidence of bone fracture (Kaneki *et al.*, 2001; Katsuyama *et al.*, 2002, 2004) due to the increase in bone mineralisation density (BMD) (Ikeda *et al.*, 2006). Regular intake of *natto* can reduce the case of hip fracture in post-menopausal women of Japan (Kaneki *et al.*, 2001). In a study on 550 post-menopausal and healthy women, regular consumption of *natto* increased BMD in hip bones (Ikeda *et al.*, 2006). Only fermented soybean had the property to increase the BMD of bones while the green soybeans, boiled beans, ground soybeans or *tofu* did not show such a property (Cao *et al.*, 2019).

The active constituent responsible for increase in BMD is vitamin K, which exists in two forms, as vitamin K_1 and K_2. The major source of vitamin K_1 is leafy and green vegetables, while vitamin K_2 is mostly produced by bacteria, which may be the reason for its abundance in fermented food and also some animal sources (Cao *et al.*, 2019). Vitamin K_2 promotes the osteoblastosis, osteocalcin carboxylation, and osteoclast differentiation and apoptosis (Villa *et al.*, 2017). *Natto* contains ~940 µg/100g menaquinone-7, which is a kind of vitamin K_2 (Ikeda *et al.*, 2006), and consumption of *natto* increases the level of menaquinone-7 in serum healthy humans (Tsukamoto *et al.*, 2000; Kaneki *et al.*, 2001). Many experiments on animal models showed a direct relationship between bone health and menquinone-7 content in *natto* (Cao *et al.*, 2019). When ovariectomised rats were supplemented with different concentrations of menaquinone-7 with *natto* in their diet for 77 days, it was found that rats fed on higher concentration of menaquinone-7 with *natto* had better bone strength and no bone loss (Yamaguchi *et al.*, 1999). On analyses of bones, a derivative of menaquinone-7, i.e. menaquinone-4 was found at an elevated concentration in femurs of experimental rats (Yamaguchi *et al.*, 1999). This experiment provides the direct evidence of direct role of *natto*-mediated bone strengthening. Further study on ovariectomised rats fed on menaquinone-7 supplemented diet for 150 days showed elevated dry weight, calcium content and BMD of femur, besides enhancing the γ-carboxylated osteocalcin of osteoblastic cells (Yamaguchi *et al.*, 2000).

Another possible mechanism behind bone strengthening may be the high calcium content in the fermented soybean (Cao *et al.*, 2019). Absorption of soluble calcium (Ca^{2+}) by the small intestine is essential for osteocyte development (Nordin, 1997; Guéguen and Pointillart, 2000). The primary reason of osteoporosis and lower bone mass is poor calcium absorption in intestine and deficiency of Ca^{2+} in diet (Cashman, 2002). *Natto* is not only rich in soluble Ca^{2+} but also contains *Bacillus* species that produces polyglutamic acids (PGA), which increases the Ca^{2+} absorption in the small intestine in post-menopausal women (Tanimoto *et al.*, 2007). Radioactive isotope of calcium (^{44}Ca) was supplied to post-menopausal women in organic juice in two doses in the form of $CaCO_3$ at the rate of 200 mg/200 g along with PGA at the rate of 60 mg/200 g. This was followed by injection of women volunteers with ^{42}Ca in the form of $CaCl_2$ solution for three or four weeks which resulted in higher concentration of Ca absorption by the intestine (Tanimoto *et al.*, 2007). Yang *et al.* (2008) showed that single dose of PGA remarkably increased the Ca^{2+} absorption in the small intestine of rats (Yang *et al.*, 2008). It lead to establish the hypothesis that *natto* and other similar fermented soybean products carry certain bioactive peptides that can bind to the Ca^{2+} and help in its absorption in the small intestine (Cao *et al.*, 2019).

In their study, Nirmala *et al.* (2019) investigated the effect of consumption of *doenjang* and non-fermented soybean on bone loss in senescent-accelerated mouse prone 6 (SAMP6). The octopaenic mice were fed with an equal dosage of non-fermented soybean (NS) or *doenjang* for 18 weeks. It was observed that supplementation of the diet with *doenjang* significantly improved the bone densities by 1.13-fold and bone length by 1.06-fold than in the control mice, while the NS-supplemented mice showed no significant improvement in either parameters. Further, they also observed that supplementation with *doenjang* effectively prevented bone loss in the osteopenic mice model by improvement in bone formation and reduction of osteoclastogenesis. The expression of BMP2, the key osteo-related signalling factor, Smd 1/5/8 and Runx2, a transcription factor for osteoblast differentiation were upregulated in both the NS and *doenjang* group. They suggested that DJ increased osteogenesis in SAMP6 mice via BMP2-Smad-Runx2 signalling. This study indicates that fermentation process can enhance soybean for maintaining the bone health (Nirmala *et al.*, 2019).

Not only fermented soybean but also the fermented soybean milk has been effective in elevation of bone density and reduction of osteoporosis. Lyophilised fermented soy milk with *Lb. plantarum* NTU 103 and *Lb. paracases* supsp. paracasei NTU 101, when fed to the ovariectomised C57BL/6J mice for eight weeks, were found with improved cell density, better microstructure of the femoral bone and thicker trabecular bone (Chiang and Pan, 2011). Similar types of experiments were conducted on rat and mice models to study the effect of fermented milk with 'Enterococcus faecium and L. jugurti' and 'Lactobacillus and Bifidobacterium' respectively on bone density. Both the above-mentioned experiments showed that the fermented soy milk containing isoflavone improves the BMD, microstructure of bone and thickness of trabecular bones (Shiguemoto *et al.*, 2007; Miura *et al.*, 2014). Isoflavones, like aglycone, genistein and daidzein, are rich in fermented soy milk products and improve the Ca^{2+} and vitamin D absorption. That is why fermented soy milk products are very effective in bone health improvement and prevention of osteoporosis (Champagne *et al.*, 2010; Chiang and Pan, 2011; Xiao *et al.*, 2018).

3. Conclusion

Fermented soybean products are consumed in many Asian countries and are a source of several bioactive compounds. The major bioactive compounds in fermented soybean products include bioactive peptides and polyphenols (particularly isoflavones). Specific starter can be explored for enhancement of selected bioactive compounds in fermented soybean products. Fermented soybean products have been shown to exhibit a wide range of health benefits including improvement of immune system, anti-thrombotic activity, bone density maintenance, anticancer property and prevention of Type II diabetes. There has been a detailed study on some of the fermented soybean products. Many fermented soybean products needs attention for metabolomics analysis and validation of specific health benefits. Molecular characterisation of bioactive compounds and validation of their health benefits can lead to commercialisation of some traditional fermented soybean products.

Abbreviations

ACE – Angiotensin converting enzyme
AMPK – AMP-activated protein kinase
APS1 – Acidic polysaccharides
BMD – Bone mineralisation density
BSH – Bile salt hydrolase
CBP – Chromatin binding protein
CCA-IMT – Common carotid artery intima-media thickness
CHD – Coronary Heart Diseases
CU – Caseinolytic unit
CVD – Cardiovascular disease
DMBA – 7,12-dimethylbenz (a) anthracene
EFA – Euglobulin fibrinolytic activity
FDP – Fibrinogen degradation products
FSS – Fermented soybean seasoning
HFD – High-fat diet
IU – International unit
IFN-γ – Interferon γ
LDL-C – LDL cholesterol
LDLR – LDL-C receptor
LFSP – Long-term naturally fermented soybean paste
MAT – Mesenteric adipose tissue
MCA – Methylcholanthrene treated

NAFLD – Non-alcohol fatty liver disease
NAFLD – Non-alcohol fatty liver disease
NK – *Nattokinase*
NO – Nitric oxide
PAI-1 – Plasminogen activator inhibitor 1
RAS – Renin-angiotensin system
PGA – Polyglutamic acids
RG – Red ginseng
RGNK – Red ginseng with *nattokinase*
RT-PCR – Real Time-Polymerase Chain Reaction
SD – Spraque-Dawley
SHR – Spontaneously hypertensive rats
STZ – Streptozotocin
TNF-α – Tumor Necrosis Factor-α
TPA – Tissue plasminogen activator

References

Adlercreutz, H., Mousavi, Y. and Höckerstedt, K. (1992). Diet and breast cancer. *Acta Oncologica*, 31: 175-181.

Ahsan, S., Zahoor, T., Shehzad, A. and Zia, M.A. (2019). Valuation of co-culture soymilk as a pragmatic approach on hyperglycemia and hypercholesterolemia in Sprague-Dawley rats. *Journal of Animal and Plant Sciences*, 29: 674-683.

Allen, C.M., Schwartz, S.J., Craft, N.E., Giovannucci, E.J., De Groff, V.L. and Clinton, S.K. (2009). Changes in plasma and oral mucosal lycopene isomer concentrations in healthy adults consuming standard servings of processed tomato products. *Nutrition and Cancer*, 47: 88-94.

Amorim, M., Pinheiro, H. and Pintado, M. (2019). Valorisation of spent brewer's yeast: Optimization of hydrolysis process towards the generation of stable ACE-inhibitory peptides. *LWT*, 111: 77-84.

Anderson, J.W. (2008). Beneficial effects of soy protein consumption for renal function. *Asia Pacific Journal of Clinical Nutrition*, 17: 324-328.

Anna, T.W., Brogran, H., Lo, E.H. and Wang, X. (2012). Plasminogen activator inhibitor-1 and thrombotic cerebrovascular diseases. *Stroke*, 43: 2833-2839.

Appukutty, M., Ramasamy, K., Rajan, S., Vellasamy, S., Ramasamy, R. and Radhakrishnan, A.K. (2015). Effect of orally administered soy milk fermented with *Lactobacillus plantarum* LAB12 and physical exercise on murine immune responses. *Beneficial Microbes*, 6: 491-496.

Atteritano, M., Mazzaferro, S., Mantuano, S., Bagnato, G.L. and Bagnato, G.F. (2015). Effects of infliximab on sister chromatid exchanges and chromosomal aberration in patients with rheumatoid arthritis. *Cytotechnology*, 68: 313-318.

Azadbakht, L. and Esmaillzadeh, A. (2009). Soy-protein consumption and kidney-related biomarkers among type 2 diabetics: a crossover, randomized clinical trial. *Journal of Renal Nutrition*, 19: 479-486.

Azimirad, M., Alebouyeh, M. and Naji, T. (2017). Inhibition of lipopolysaccharide-induced interleukin 8 in human adenocarcinoma cell line HT-29 by spore probiotics: *B. coagulans* and *B. subtilis* (natto). *Probiotics and Antimicrobial Proteins*, 9(1): 56-63.

Badger, T.M., Ronis, M.J.J., Simmen, R.C.M. and Simmen, F.A. (2005). Soy protein isolate and protection against cancer. *Journal of the American College of Nutrition*, 24: 146-149.

Bakhle, Y. (1968). Conversion of angiotensin I to angiotensin II by cell-free extracts of dog lung. *Nature* (London), 220: 919-921.

Bakhle, Y., Reynard, A. and Vane, J. (1969). Metabolism of the angiotensins in isolated perfused tissues. *Nature* (London), 222: 956.

Barnes, S., Coward, L., Kirk, M. and Sfakianos, J. (1998). HPLC-mass spectrometry analysis. *Experimental Biology and Medicine*, 217: 254-262.

Benjamin, E.J., Blaha, M.J., Chiuve, S.E., Carnethon, M., Dai, S., Simone, G.D., Ferguson, T.B., Ford, E., Furie, K., Gillespie, C., Go, A., Greenlund, K., Haase, N., Hailpern, S., Ho, P.M., Howard, V., Kissela, B., Kittner, S., Lackland, D., Lisabeth, L., Marelli, A., McDermott, M., Meigs, J., Mozaffarian, D., Mussolino, M., Nichol, G., Roger, V.L., Rosamond, W., Sacco, R., Sorlie, P., Roger, V.L., Thom, T., Wasserthiel-Smoller, S., Wong, N.D. and Wylie-Rosett, J. (2017). American Heart Association Statistics Committee and Stroke Statistics Subcommittee. Heart Disease and Stroke Statistics - 2017 Update: A report from the American Heart Association, Circulation, 135: e146-e603.

Bogdan, C. (2015). Nitric oxide synthase in innate and adaptive immunity: An update. *Trends in Immunology*, 36: 161-178.

Brenner, B.M., Cooper, M.E., de Zeeuw, D., Keane, W.F., Mitch, W.E., Parving, H.H., Remuzzi, G., Snapinn, S.M., Zhang, Z. and Shahnifar, S. (2001). Effects of losartan on renal and cardiovascular outcomes in patients with Type 2 diabetes and nephropathy. *The New England Journal of Medicine*, 345: 861-869.

Byun, M.S., Yu, O.K., Cha, Y.S. and Park, T.S. (2016). Korean traditional Chungkookjang improves body composition, lipid profiles and atherogenic indices in overweight/obese subjects: A double-blind, randomised, crossover, placebo-controlled clinical trial. *European Journal of Clinical Nutrition*, 70: 1116.

Cao, Z., Green-johnson, J.M., Buckley, N.D. and Lin, Q. (2019). Bioactivity of soy-based fermented foods: A review. *Biotechnology Advances*, 37: 223-238.

Cashman, K.D. (2002). Calcium intake, calcium bioavailability and bone health. *British Journal of Nutrition*, 87: 169-177.

Cavallini, D.C.U., Manzoni, M.S.J., Bedani, R., Roselino, M.N., Celiberto, L.S., Vendramini, R.C., de Valdez, G.F., Abdalla, D.S.P., Pinto, R.A., Rosetto, D., Valentini, S.R. and Rossi, E.A. (2016). Probiotic soy product supplemented with isoflavones improves the lipid profile of moderately hypercholesterolemic men: a randomised controlled trial. *Nutrients*, 8: 52.

Cavazos, A., Morales, E., Dia, V.P. and De Mejia, E.G. (2012). Analysis of lunasin in commercial and pilot plant produced soybean products and an improved method of lunasin purification. *Journal of Food Science*, 77: 1-7.

Champagne, C.P., Green-Johnson, J., Raymond, Y, Barrette, J. and Buckley, N. (2009). Selection of probiotic bacteria for the fermentation of a soy beverage in combination with *Streptococcus thermophilus*. *Food Research International*, 42: 612-621.

Champagne, C.P., Tompkins, T.A., Buckley, N.D. and Green-Johnson, J.M. (2010). Effect of fermentation by pure and mixed cultures of *Streptococcus thermophilus* and *Lactobacillus helveticus* on isoflavone and B-vitamin content of a fermented soy beverage. *Food Microbiology*, 27: 968-972.

Chapin, J.C. and Hajjar, K.A. (2015). Fibrinolysis and the control of blood coagulation. *Blood Review*, 29: 17-24.

Chen, J., Stavro, P.M. and Thompson, L.U. (2002). Dietary flaxseed inhibits human breast cancer growth and metastasis and downregulates expression of insulin-like growth factor and epidermal growth factor receptor. *Nutrition and Cancer*, 43(2): 214-226.

Chen, J., Cheng, Y., Yamaki, K. and Li, L. (2007). Anti-alpha-glucosidase activity of Chinese traditionally fermented soybean (douchi). *Food Chemistry*, 103: 1091-1096.

Chen, J., Liu, S., Ye, R., Cai, G., Ji, B. and Wu, Y. (2013a). Angiotensin-I converting enzyme (ACE) inhibitory tripeptides from rice protein hydrolysate: Purification and characterisation. *Journal of Functional Foods*, 5: 1684-1692.

Chen, X., Lu, H., Zhao, M., Tashiro, K., Cassis, L.A. and Daughtery, A. (2013b). Contributions of leukocyte angiotensin-converting enzyme to development of atherosclerosis. *Arteriosclerosis, Thrombosis, and Vascular Biology*, 33(9): 2075-2080.

Chiang, S.S. and Pan, T.M. (2011). Antiosteoporotic effects of *lactobacillus*-fermented soy skim milk on bone mineral density and the microstructure of femoral bone in ovariectomised mice. *Journal of Agricultural and Food Chemistry*, 59: 7734-7742.

Choi, M.S., Jung, U.J., Yeo, J., Kim, M.J. and Lee, M.K. (2008). Genistein and daidzein prevent diabetes onset by elevating insulin level and altering hepatic gluconeogenic and lipogenic enzyme activities in non-obese diabetic (NOD) mice. *Diabetes/Metabolism Research and Reviews*, 24: 74-81.

Choi, Y., Noh, J., Yun, S.-W., Kwon, Y.-I. and Kim, Y.-C. (2013). Effect of isoflavones from astragalus membranaceus on 3T3- L1 adipocyte differentiation and insulin sensitivity. *The FASEB Journal*, 27: 637.28.

Correa, A.P.F., Daroit, D.J., Coelho, J., Meria, S.M.M., Lopes, F.C., Segalin, J., Risso, P.H. and Brandelli, A. (2011). Antioxidant, antihypertensive and antimicrobial properties of ovine milk caseinate hydrolysed with a microbial protease. *Journal of the Science of Food and Agriculture*, 91: 2247-2254.

Crane, R. (1965). Na$^+$-dependent transport in the intestine and other animal tissues. *Federation Proceedings*, 24: 1000-1006.

Dabbagh, F., Negahdaripour, M., Berenjian, A., Behfar, A., Mohammadi, F., Zamani, M., Irajie, C. and Ghasemi, Y. (2014). Nattokinase: Production and application. *Applied Microbiology and Biotechnology*, 98: 9199-9206.

DeFronzo, R.A. (2004). Pathogenesis of Type 2 diabetes mellitus. *The Medical Clinics of North America*, 88: 787-835.

de Mejia, E.G. and Dia, V.P. (2009). Lunasin and lunasin-like peptides inhibit inflammation through suppression of NF-κB pathway in the macrophage. *Peptides*, 30: 2388-2398.

Di Cagno, R., Mazzacane, F., Rizzello, C.G., Vencintini, O., Silano, M., Giuliani, G., De Angelis, M. and Gobbeti, M. (2010). Synthesis of isoflavone aglycones and equol in soy milks fermented by food-related lactic acid bacteria and their effect on human intestinal caco-2 cells. *Journal of Agricultural and Food Chemistry*, 58: 10338-10346.

Duckworth, W.C. (2001). Hyperglycemia and cardiovascular disease. *Current Atherosclerosis Reports*, 3: 383-391.

Fan, H., Wang, J., Liao, W., Jiang, X. and Wu, J. (2019). Identification and characterisation of gastrointestinal-resistant angiotensin-converting enzyme inhibitory peptides from egg white proteins. *Journal of Agricultural and Food Chemistry*, 67: 7147-7156.

FDA (2016). Guidance Documents Regulatory Information.

Feng, G.H., Richardson, M., Chen, M.S., Kramer, K.J., Morgan, T.D. and Reeck, G.R. (1996). Alpha Amylase inhibitors from wheat: Amino acid sequences and patterns of inhibition of insect and Human α-amylases. *Biochemistry and Molecular Biology*, 26: 419-426.

Ferreira, S.H. (1965). A Bradykinin-potentiating factor (Bpf) present in the venom of *Bothrops jararaca*. *British Journal of Pharmacology*, 24: 163-169.

Ferreira, S.H., Bartelt, D.C. and Greene, L.J. (1970). Isolation of bradykinin-potentiating peptides from *Bothrops jararaca* venom. *Biochemistry*, 9: 2583-2593.

Firth, R.G., Bell, P.M., Marsh, H.M., Hansen, I. and Rizza, R.A. (1986). Postprandial hyperglycemia in patients with noninsulin-dependent diabetes mellitus: Role of hepatic and extrahepatic tissues. *Journal of Clinical Investigations*, 77(5): 1525-1532.

Fujita, H. and Yamagami, T. (2001). Fermented soybean-derived Touchi-extract with anti-diabetic effect via alpha-glucosidase inhibitory action in a long-term administration study with KKA y mice. *Journal of Nutrition Baltimor and Springfield then Betjesda*, 70: 219-227.

Fujita, H., Yamagami, T. and Ohshima, K. (2001a). Fermented soybean-derived water-soluble Touchi extract inhibits alpha-glucosidase and is antiglycemic in rats and humans after single oral. *Journal of Nutrition*, 131: 1211-1213.

Fujita, H., Yamagami, T. and Ohshima, K. (2001b). Long-term ingestion of a inhibitory activity is safe and effective in humans with borderline and mild Type-2 diabetes. *Journal of Nutrition*, 131: 2105-2108.

Fujita, H., Yokoyama, K. and Yoshikawa, M. (2000). Classification and antihypertensive activity of angiotensin I-converting enzyme inhibitory peptides derived from food proteins. *Food Chemistry and Toxicology*, 65: 564-569.

Fujita, M., Komura, K., Hong, K., Ito, Y. and Nishimuro, S. (1993). Purification and characterisation of a strong fibrinolytic enzyme (*nattokinase*) in the vegetable cheese *natto*, a popular soybean fermented food in Japan. *Biochemical and Biophysical Research Communications*, 197: 1340-1347.

Fujita, M., Ito, Y., Hong, K. and Nishimuro, S. (1995). Characterisation of *nattokinase*-degraded products from human fibrinogen or cross-linked fibrin. *Fibrinolysis*, 9: 157-164.

Fujita, M., Ohnishi, K., Takaoka, S., Ogasawara, K., Fukuyama, R. and Nakamuta, H. (2016). Antihypertensive effects of continuous oral administration of *nattokinase* and its fragments in spontaneously hypertensive rats. *Biological and Pharmaceutical Bulletin P*, 259: 715-716. doi: 10.1248/bpb.34.1696

Galvez, A., Revilleza, M. and Lumen, B. (1997). A novel methionine-rich protein from soybean cotyledon: Cloning and characterisation of cDNA. *Plant Physiology*, 57: 1567-1569.

Geng, P., Bai, G., Shi, Q., Zhang, L., Gao, Z. and Zhang, Q. (2009). Taxonomy of the Streptomyces strain ZG0656 that produces acarviostatin α-amylase inhibitors and analysis of their effects on blood glucose levels in mammalian systems. *Journal of Applied Microbiology*, 106: 525-533.

Geng, P., Qiu, F. and Bai, G. (2008). Four acarviosin-containing oligosaccharides identified from streptomyces coelicoflavus ZG0656 are potent inhibitors of α-amylase. *Carbohydrate Research*, 343(5): 882-892.

Gong, L., Huang, Q., Aikun, F., Wu, Y., Li, Y., Xu, X., Huang, Y., Yu, D. and Li, W. (2017). Spores of two probiotic *bacillus* spores enhance cellular immunity in BALB/C mice. *Canadian Journal of Microbiology*, 64: 799-808.

Graham, K., Rea, R., Simpson, P. and Stack, H. (2019). Enterococcus faecalis milk fermentates display antioxidant properties and inhibitory activity towards key enzymes linked to hypertension and hyperglycaemia. *Journal of Functional Foods*, 58: 292-300.

Guéguen, L. and Pointillart, A. (2000). The bioavailability of dietary calcium. *Journal of the American College of Nutrition*, 19: 119-136.

Gupta, M., Sharma, P. and Nath, A.K. (2014). Purification of a novel α-amylase inhibitor from local Himalayan bean (*Phaseolus vulgaris*) seeds with activity towards bruchid pests and human salivary amylase. *Journal of Food Science and Technology*, 51: 1286-1293.

Hang, B.Z., Beumer, R.R., Rombouts, F.M. and Robert Nout, M.J. (2001). Microbiological safety and quality of commercial sufu – A Chinese fermented soybean food. *Food Control*, 12: 541-547.

Haque, E. and Chand, R. (2008). Antihypertensive and antimicrobial bioactive peptides from milk proteins. *European Food Research and Technology*, 227: 7015.

Hassan, A.A., Rasmy, N.M., El-Gharably, A.M.A., El-Megied, A.A. and Gadalla, S.M.M. (2014). Hypocholesterolemic effects of soybean and sweet lupine *Tempeh* in hypercholesterolemic rats. *International Journal of Fermented Foods*, 3: 11.

Hernández-Ledesma, B., Hsieh, C.C. and de Lumen, B.O. (2009). Lunasin and Bowman-Birk protease inhibitor (BBI) in US commercial soy foods. *Food Chemistry*, 115: 574-580.

Hsia, C.H., Shen, M.C., Lin, J.S., Wen, Y.K., Hwang, K.L., Cham, T.M. and Yang, N.C. (2009). *Nattokinase* decreases plasma levels of fibrinogen, factor VII and factor VIII in human subjects. *Nutrition Research*, 29: 190-196.

Hsieh, C.C., Hernández-Ledesma, B. and De Lumen, B.O. (2010). Soybean peptide lunasin suppresses *in vitro* and *in vivo* 7,12-dimethylbenz[a]anthracene-induced tumorigenesis. *Journal of Food Science*, 75: 11-16.

Hsieh, M.L., Fang, S.W., Yu, R.C. and Chou, C.C. (2005). Possible mechanisms of antimutagenicity in fermented soymilk prepared with a coculture of *Streptococcus infantis* and *Bifidobacterium infantis*. *Journal of Food Protection*, 70: 1025-1028.

Huang, Y.C., Wu, B.H., Chu, Y.L., Chang, W.C. and Wu, M.C. (2018). Effects of *Tempeh* fermentation with *Lactobacillus plantarum* and *Rhizopus oligosporus* on streptozotocin-induced Type II diabetes mellitus in rats. *Nutrients*, 10: 1143.

Hubert, C., Houot, A., Corvol, P. and Soubrier, F. (1991). Structure of the angiotensin I-converting enzyme gene. Two alternate promoters correspond to evolutionary steps of a duplicated gene. *Journal of Biological Chemistry*, 266: 15377-15383.

Hughes, G.J., Ryan, D.J., Mukherjea, R. and Schasteen, C.S. (2011). Protein digestibility-corrected amino acid scores (PDCAAS) for soy protein isolates and concentrate: Criteria for evaluation. *Journal of Agricultural and Food Chemistry*, 59: 12707-12712.

Hwang, J.S., Yoo, H.J., Song, H.J., Kim, K.K., Chun, Y.J., Matsui, T. and Kim, H.B. (2011). Inflammation-related signaling pathways implicating TGFβ are revealed in the expression profiling of MCF7 cell treated with fermented soybean, Chungkookjang. *Nutrition and Cancer*, 63: 645-652.

Ibe, S., Yoshida, K. and Kumada, K. (2006). Angiotensin I converting enzyme inhibitory activity of *natto*, a traditional Japanese fermented food. *Nippon Shokuhin Kagaku Kogaku Kaishi* (in Japanese), 53: 189-192.

Ibe, S., Yoshida, K., Kumada, K., Tsurushin, S., Furusho, T. and Otobe, K. (2009). Antihypertensive effects of *natto*, a traditional japanese fermented food, in spontaneously hypertensive rats. *Food Science and Technology Research*, 15: 199-202.

Ichimura, T., Hu, J., Aita, D.Q. and Maruyama, S. (2003). Angiotensin I-converting enzyme inhibitory activity and insulin secretion stimulative activity of fermented fish sauce. *Journal of Bioscience and Bioengineering*, 96: 496-499.

Iguti, A. and Lajolo, F. (1991). Occurrence and purification of α-amylase isoinhibitors in bean (*Phaseolus vulgaris* L.) varieties. *Journal of Agricultural and Food Chemistry*, 39: 2131-2136.

Ikeda, Y., Iki, M., Morita, A., Kajita, E., Kagamimori, S., Kagawa, Y. and Yoneshima, H. (2006). Intake of fermented soybeans, *natto*, is associated with reduced bone loss in postmenopausal women: Japanese population-based osteoporosis (JPOS) study. *Nutritional Epidermiology*, 136: 1323-1328.

Ishida, M., Ishida, T., Thomas, S.M. and Berk, B.C. (1998). Activation of extracellular signal-regulated kinases (ERK1/2) by angiotensin II is dependent on c-Src in vascular smooth muscle cells. *Circulation Research*, 82: 7-12.

Ishihara, K., Oyaizu, S., Fukuchi, Y., Mizunoya, W., Segawa, K., Takahashi, M., Mita, Y., Fukuya, Y., Fushiki, T. and Yasumoto, K. (2003). A soybean peptide isolate diet promotes postprandial carbohydrate oxidation and energy expenditure in type II diabetic mice. *Journal of Nutrition*, 133: 752-757.

Iulek, J., Franco, O.L., Silva, M., Slivinski, C.T., Bloch, C.J. and Grossi de Sa, M.F. (2000). Purification, biochemical characterisation and partial primary structure of a new alpha-amylase inhibitor from *Secale cereale* (rye). *Int. J. Biochem. Cell Biol.*, 32: 1195-1204.

Iwai, K., Nakaya, N., Kawasaki and Matsue, H. (2002). Antioxidative functions of *natto*, a kind of fermented soybeans: Effect on LDL oxidation and lipid metabolism in cholesterol-fed rats. *Journal of Agricultural and Food Chemistry*, 50: 3597-3601.

Jang, J.Y., Kim, T.S., Cai, J., Kim, J., Kim, Y., Shin, K., Kim, K.S., Park, S.K., Lee, S.P., Choi, E.K., Rhee, M.H. and Kim, Y.B. (2013). *Nattokinase* improves blood flow by inhibiting platelet aggregation and thrombus formation. *Laboratory Animal Research*, 29: 221-225.

Je, J.Y.P., Jung, P.J., Kim, W.K. and Se, K. (2005). Amino acid changes in fermented oyster (*Crassostrea gigas*) sauce with different fermentation periods. *Food Chemistry*, 91: 15-18.

Jeong, H.J., Jeong, J.B., Kim, D.S., Park, J.H., Lee, J.B., Kweon, D.H., Chung, G.Y., Seo, E.W. and De Lumen, B. (2007). The cancer preventive peptide lunasin from wheat inhibits core histone acetylation. *Cancer Letters*, 255: 42-48.

Jeong, J.K., Chang, H.K. and Park, K.Y. (2014). Doenjang prepared with mixed starter cultures attenuates azoxymethane and dextran sulphate sodium-induced colitis-associated colon carcinogenesis in mice. *Journal of Carcinogenesis*, 13: 9.

Jeong, S., Jeong, D.Y., Kim, D.S. and Park, S. (2018). Chungkookjang with high contents of poly-γ-glutamic acid improves insulin sensitising activity in adipocytes and neuronal cells. *Nutrients*, 10: 1588-1595.

Jin, M.M., Zhang, L., Yu, H.X., Meng, Z. and Lu, S.RR. (2013). Protective effect of whey protein hydrolysates on H_2O_2-induced PC12 cells oxidative stress via a mitochondria-mediated pathway. *Food Chemistry*, 141: 847-852.

Jung, K.O., Park, S.Y. and Park, K.Y. (2006). Longer aging time increases the anticancer and antimetastatic properties of Deonjang. *Nutrition*, 22: 539-545. Burbank, Los Angeles County, Calif.

Kaiser, N. and Leibowitz, G. (2009). Failure of beta-cell adaptation in Type 2 diabetes: Lessons from animal models. *Front Biosci*, Landmark (Ed.). 14: 1099-1115.

Kaneki, M., Hodges, S., Hedges, S.J., Hosoi, T., Fujiwara, S., Lyons, A., Crean, S., Ishida, N., Nakagawa, M., Takechi, M., Sano, Y., Mizuno, Y., Hoshino, S., Miyao, M., Inoue, S., Horiki, K., Shiraki, M., Ouchi, Y. and Orimo, H. (2001). Japanese fermented soybean food as the major determinant of the large geographic difference in circulating levels of vitamin K2: Possible implications for hip-fracture risk. *Nutrition*, 17: 315-321.

Kang, S.J., Lim, Y. and Kim, A.J. (2014). Korean red ginseng combined with *nattokinase* ameliorates dyslipidemia and the area of aortic plaques in high cholesterol-diet fed rabbits. *Food Science and Biotechnology*, 23: 283-287.

Kannel, W.B. (1996). Blood pressure as a cardiovascular risk factor: Prevention and treatment. *JAMA*, 275: 1571-1576.

Katsuyama, H., Ideguchi, S., Fukunaga, M., Saijoh, K. and Sunami, S. (2002). Usual dietary intake of fermented soybeans (*Natto*) is associated with bone mineral density in premenopausal women departments of l. Public Health, 2. Health Care Medicine and with the rapidly increasing proportion of elderly in industrialized countries. *Jorunal of Nutrition Science and Vitaminology*, 48: 207-215.

Katsuyama, H., Ideguchi, S., Fukunaga, M., Fukunaga, T., Sajioh, K. and Sunami, S. (2004). Promotion of bone formation by fermented soybean (*Natto*) intake in premenopausal women. *Journal of Nutritional Science and Vitaminology*, 50: 114-120.

Kawakami, Y., Tsurugasaki, W., Nakamura, S. and Osada, K. (2005). Comparison of regulative functions between dietary soy isoflavones aglycone and glucoside on lipid metabolism in rats fed cholesterol. *Journal of Nutritional Biochemistry*, 16: 205-212.

Ke, Y.Y., Tsai, C.H., Yu, H.M., Jao, Y.C., Fang, J.M. and Wong, C.H. (2015). Latifolicinin A from a fermented soymilk product and the structure-activity relationship of synthetic analogues as inhibitors of breast cancer cell growth. *Journal of Agricultural and Food Chemistry*, 63: 9715-9721.

Kikuchi-Hayakawa, H., Onodera, N., Matsubara, S., Yasuda, Y., Shimakawa, Y. and Ishikawa, F. (1998). Effects of soya milk and bifidobacterium-fermented soya milk on plasma and liver lipids, and faecal steroids in hamsters fed on a cholesterol-free or cholesterol-enriched diet. *British Journal of Nutrition*, 79: 97-105.

Kim, A.R., Lee, J.J., Lee, H., Chang, H.C. and Lee, M.Y. (2010). Body-weight-loss and cholesterol-lowering effects of *Cheonggukjang* (a fermented soybean paste) given to rats fed a high-fat/high-cholesterol diet. *Korean Journal of Food Preservation*, 17(5).

Kim, D., Jeong, Y., Kwon, J., Moon, K.D., Kim, H.J., Jeon, S.M., Lee, M.K., Park, Y.B. and Choi, M.S. (2008). Beneficial effect of Chungkukjang on regulating blood glucose and pancreatic beta-cell functions in C75BL/KsJ-db/db mice. *Journal of Medicinal Food*, 11: 215-223.

Kim, K.A., Jang, S.E., Jeong, J.J., Yu, D.H., Han, M. and Kim, D. (2014). Doenjang, a Korean soybean paste, ameliorates TNBS-induced colitis in mice by suppressing gut microbial lipopolysaccharide production and NF-κB activation. *Journal of Functional Foods*, 11: 417-427.

Kim, M.S., Kim, B., Park, H., Ji, Y., Holzapfel, W., Kim, D.Y. and Hyun, C.K. (2018). Long-term fermented soybean paste improves metabolic parameters associated with non-alcoholic fatty liver disease and insulin resistance in high-fat diet-induced obese mice. *Biochemical and Biophysical Research Communications*, 495: 1744-1751.

Kobayashi, M., Matsushita, H., Yoshida, K., Tsukiyama, R.I., Sugimura, T. and Yamamoto, K. (2004). *In vitro* and *in vivo* anti-allergic activity of soy sauce. *International Journal of Molecular Medicine*, 14: 879-884.

Kuba, M., Tanaka, K., Tawata, S., Takeda, Y. and Yasuda, M. (2003). Angiotensin I-converting enzyme inhibitory peptides isolated from Tofuyo fermented soybean food. *Bioscience, Biotechnology, and Biochemistry*, 67: 1278-1283.

Kuba, M., Shinjo, S. and Yasuda, M. (2004). Antihypertensive and Hypocholesterolemic effects of Tofuyo in spontaneously hypertensive rats. *Journal of Health Science*, 50: 670-673.

Kuba, M., Tana, C., Tawata, S. and Yasuda, M. (2005). Production of angiotensin I-converting enzyme inhibitory peptides from soybean protein with Monascus purpureus acid proteinase. *Process Biochemistry*, 40: 2191-2196.

Kurihara, M., Markland, F.S. and Smith, E.L. (1972). Subtilisin Amylosacchariticus III. Isolation and sequence of the chymotryptic peptides and the complete amino acid sequence. *The Journal of Biological Chemistry*, 247: 5619-5631.

Kurosawa, Y., Nirengi, S., Homma, T., Esaki, K., Ohta, M., Clark, J.F. and Hamaoka, T. (2015). A single-dose of oral *nattokinase* potentiates thrombolysis and anti-coagulation profiles. *Scientific Reports*, 5: 11601.

Kwak, C.S., Lee, M.S. and Park, S.C. (2007). Higher antioxidant properties of Chungkookjang, a fermented soybean paste, may be due to increased aglycone and malonylglycoside isoflavone during fermentation. *Nutrition Research*, 27: 719-727.

Kwon, D.Y., Jang, J.S., Hong, S.M., Lee, J.E., Sung, S.R., Park, H.R. and Park, S. (2007). Long-term consumption of fermented soybean-derived Chungkookjang enhances insulinotropic action unlike soybeans in 90 per cent pancreatectomised diabetic rats. *European Journal of Nutrition*, 46: 44-52.

Kwon, D.Y., Daily, J.W., Kim, H.J. and Park, S. (2010). Antidiabetic effects of fermented soybean products on Type 2 diabetes. *Nutrition Research*, 30: 1-13.

Lai, L.R., Hsieh, S.C., Huang, H.Y. and Chou, C.C. (2013). Effect of lactic fermentation on the total phenolic, saponin and phytic acid contents as well as anti-colon cancer cell proliferation activity of soymilk. *Journal of Bioscience and Bioengineering*, 115: 552-556.

Lau, C.C., Abdullah, N., Shuib, A.S. and Aminudin, N. (2014). Novel angiotensin I-converting enzyme inhibitory peptides derived from edible mushroom *Agaricus bisporus* (J.E. Lange) Imbach identified by LC-MS/MS. *Food Chemistry*, 148: 396-401.

Lee, C., Youn, Y., Song, G. and Kim, Y. (2011). Immunostimulatory effects of traditional doenjang. *Journal of the Korean Society of Food Science and Nutrition*, 40: 1227-1234.

Lee, J.H., Paek, S.H., Shin, H.W., Lee, S.Y., Moon, B.S., Park, J.E., Lim, G.D., Kim, C.Y. and Heo, Y. (2017). Effect of fermented soyabean products intake on the overall immune safety and function in mice. *Journal of Veterinary Science*, 18: 25-32.

Lee, S.Y. and Hur, S.J. (2019a). Purification of novel angiotensin-converting enzyme inhibitory peptides from beef myofibrillar proteins and analysis of their effect in spontaneously hypertensive rat model. *Biomedicine & Pharmacotherapy*, 116: 109046.

Lee, S.Y. and Hur, S.J. (2019b). Biomedicine and pharmacotherapy purification of novel angiotensin converting enzyme inhibitory peptides from beef myofibrillar proteins and analysis of their effect in spontaneously hypertensive rat model. *Biomedicine & Pharmacotherapy*, 116: 109046.

Lefevre, M., Racedo, S.M., Ripert, G., Housez, B., Cazaubiel, M., Maudet, C., Justen, P., Marteau, P. and Urdaci, M.C. (2015). Probiotic strain *Bacillus subtilis* CU1 stimulates immune system of elderly during common infectious disease period: A randomised, double-blind placebo-controlled study. *Immunity and Ageing*, 1-11.

Li-Jun, Y., Li-Te, L., Zai-Gui, L., Tatsumi, E. and Saito, M. (2004). Changes in isoflavone contents and composition of sufu (fermented tofu) during manufacturing. *Food Chemistry*, 87: 587-592.

Li, F., Yin, L., Cheng, Y. and Saito, M. (2010). Angiotensin I-converting enzyme inhibitory activities of extracts from commercial Chinese style fermented soypaste. *Agricultural Research Quarterly*, 44(2): 167-172.

Li, Y., Wen, S., Kota, B.P., Peng, G., Li, G.Q., Yamahara, J. and Roufogalis, B.D. (2005). Punica granatum flower extract, a potent alpha-glucosidase inhibitor, improves postprandial hyperglycemia in Zucker diabetic fatty rats. *Journal of Ethnopharmacology*, 99: 239-244.

Lin, Q., Mathieu, O., Tompkins, T.A., Buckley, N.D. and Green-Johnson, J.M. (2016). Modulation of the TNFα-induced gene expression profile of intestinal epithelial cells by soy fermented with lactic acid bacteria. *Journal of Functional Foods*, 23: 400-411.

Lin, Y. (2017). A clinical study on the effect of *nattokinase* on carotid artery atherosclerosis and hyperlipidaemia. *Chinese Medical Journal*, 97: 2038.

Liu, J., Wang, S., Lin, Y. and Lin, C. (2002). Antitumor activity of milk *kefir* and soy milk *kefir* in tumor-bearing mice. *Nutrition and Cancer*, 44: 182-187.

Liu, J.R., Chen, M.J. and Lin, C.W. (2005). Antimutagenic and antioxidant properties of milk-*kefir* and soymilk-*kefir*. *Journal of Agricultural and Food Chemistry*, 53: 2467-2474.

Long, S.L., Gahan, C.G.M. and Joyce, S.A. (2017). Interactions between gut bacteria and bile in health and disease. *Molecular Aspects of Medicine*, 56: 54-65.

Lu, H., Balakrishnan, A., Howatt, D.A., Wu, C., Charnigo, R., Liau, G., Cassis, L.A. and Daughterty, A. (2012). Comparative effects of different modes of renin angiotensin system inhibition on hypercholesterolaemia-induced atherosclerosis. *British Journal of Pharmacology*, 165: 2000-2008.

Lu, H., Cassis, L.A., Koi, C.W., Vander Kooi, C.W. and Daugherty, A. (2016). Structure and functions of angiotensinogen. *Hypertension Research: Official Journal of the Japanese Society of Hypertension*, 39: 492-500.

Lu, M.P., Wang, R., Song, X., Chibbar, R., Wang, X., Wu, L. and Meng, Q.H. (2008). Dietary soy isoflavones increase insulin secretion and prevent the development of diabetic cataracts in streptozotocin-induced diabetic rats. *Nutrition Research*, 28: 464-471.

Lumen, B.O.D. (2005). Lunasin: A cancer-preventive soy peptide. *Nutrition Reviews*, 63: 16-21.

Ma, Y., Cheng, Y., Yin, L., Wang, J. and Li, L. (2013). Effects of processing and NaCl on angiotensin i-converting enzyme inhibitory activity and γ-aminobutyric acid content during sufu manufacturing. *Food Bioprocess Technology*, 6: 1782-1789.

Mahmoud, A., Yang, W. and Bosland, M. (2008). Soy isoflavones and prostate cancer: A review of molecular mechanisms. *Journal of Steroid Biochemistry and Molecular Biology*, 23: 1-7.

Maki, K.C., Butteiger, D.N., Rains, T.M., Lawless, A., Reeves, M.S., Schasteen, C. and Krul, E.S. (2010). Effects of soy protein on lipoprotein lipids and fecal bile acid excretion in men and women with moderate hypercholesterolemia. *Journal of Clinical Lipidology*, 4: 531-542.

Mann, J.F., Anderson, C., Gao, P., Gerstein, H.C., Boehm, M., Ryden, L., Sleight, P., Teo, K.K., Yusuf, S. and ONTARGET investigators (2013). Dual inhibition of the renin-angiotensin system in high-risk diabetes and risk for stroke and other outcomes: Results of the ONTARGET trial. *Journal of Hypertension*, 31: 414-421.

Marieb, E.N. and Hoehn, K.N. (2013). *Human Anatomy and Physiology*, 9th edn. Pearson. Boston.

Martinez-Villaluenga, C., Torino, M.I., Martín, V., Arroyo, R., Garcia-Mora, P., Padrola, I.E., Vidal-Valverde, C., Rodroguiz, J.M. and Friaz, J. (2012) . Multifunctional properties of soy milk fermented by *Enterococcus faecium* strains isolated from raw soy milk. *Journal of Agricultural and Food Chemistry*, 60: 10235-10244.

Maruyama, N., Maruyama, Y., Tsuruki, T., Okuda, E., Yoshikawa, M. and Utsumi, S. (2003). Creation of soybean β-conglycinin β with strong phagocytosis-stimulating activity. *Biochimica et Biophysica Acta (BBA) – Proteins and Proteomics*, 1648: 99-104.

Masotti, A.I., Buckley, N., Champagne, C.P. and Green-Johnson, J. (2011). Immunomodulatory bioactivity of soy and milk ferments on monocyte and macrophage models. *Food Research International*, 44: 2475-2481.

Matsushita, H., Kobayashi, M., Tsukiyama, R.I., Fujimoto, M., Suzuki, M., Tsuji, K. and Yamamoto, K. (2009). Stimulatory effect of Shoyu polysaccharides from soy sauce on the intestinal immune system. *International Journal of Molecular Medicine*, 23: 521-527.

Matsushita, H., Kobayashi, M., Tsukiyama, R. and Yamamoto, K. (2006). *In vitro* and *in vivo* immunomodulating activities of Shoyu polysaccharides from soy sauce. *International Journal of Molecular Medicine*, 17: 905-909.

McCarthy, E.M. and Rinella, M.E. (2012). The role of diet and nutrient composition in nonalcoholic fatty liver disease. *Journal of the Academy of Nutrition and Dietetics*, 112: 401-409.

Mercier, A., Gauthier, S.F. and Fliss, I. (2004). Immunomodulating effects of whey proteins and their enzymatic digests. *International Dairy Journal*, 14: 175-183.

Messina, M. (2016). Soy and health update: Evaluation of the clinical and epidemiologic literature. *Nutrients*, 8.

Messina, M., Persky, V., Setchell, K. and Barnes, S. (1994). Soy intake and cancer risk: A review of the *in vitro* and *in vivo* data. *Nutrition and Cancer*, 21: 113-131.

Mezei, O., Banz, W.J., Steger, R.W., Peluso, M.R., Winters, T.A. and Shay, N. (2003). Soy isoflavones exert antidiabetic and hypolipidemic effects through the PPAR pathways in obese Zucker rats and murine RAW 264.7 cells. *Journal of Nutrition*, 133: 1238-1243.

Milner, M. and Makise, K. (2002). *Natto* and its active A potent and safe thrombolytic agent. *Alternative and Complemenntary Therapies*, 57-64.

Miura, T., Kozai, Y., Kawamata, R., Wakao, H., Sakurai, T. and Kashima, I. (2014). Inhibitory effect of a fermented soy product from lactic acid bacteria (PS-B1) on deterioration of bone mass and quality in ovariectomised mice. *Oral Radiology*, 30: 45-52.

Murphy, P.A., Song, T., Buseman, G., Barua, K., Beecher, G.R., Trainer, D. and Holden, A. (1999). Isoflavones in retail and institutional soy foods. *Journal of Agriculture and Food Chemistry*, 47: 2697-2704.

Nagata, C., Wada, K., Tamura, T., Konishi, K., Goto, Y., Koda, S., Kawachi, T., Tsuji, M. and Nakamura, K. (2016). Dietary soy and *natto* intake and cardiovascular disease mortality in Japanese adults: The Takayama study. *The American Journal of Clinical Nutrition*, 105: 426-431.

Nakahara, T., Sano, A., Yamaguchi, H., Sugimoto, K., Chikata, H., Kinoshita, E. and Uchida, R. (2010). Antihypertensive effect of peptide-enriched soy sauce-like seasoning and identification of its angiotensin I-converting enzyme inhibitory substances. *Journal of Agricultural and Food Chemistry*, 58: 821-827.

National Federation of Co-operatives on Natto (1977). *A Historical Record of Natto*. Food Pionia, Natto Research Centre. Tokyo.

Niamah, A.K., Sahi, A.A. and Al-Sharifi A.S.N. (2017). Effect of feeding soy milk fermented by probiotic bacteria on some blood criteria and weight of experimental animals. *Probiotics and Antimicrobial Proteins*, 9: 284-291.

Nirmala, F.S., Lee, H., Kim, J.S., Jung, C.H., Ha, T.Y., Jang, Y.J. and Ahn, J. (2019). Fermentation improves the preventive effect of soybean against bone loss in senescence-accelerated mouse prone 6. *Journal of Food Science*, 84(2): 349-357.

Nordin, B.E.C. (1997). Calcium and osteoporosis. *Archivos Latinoamericanos de Nutricion*, 47: 13-16.

Nout, M.J.R., Sarkar, P.K. and Beuchat, L.R. (2007). Indigenous fermented foods. pp 817-835. *In*: *Food Microbiology: Fundamentals and Frontiers*, 3rd ed. American Society of Microbiology.

Obiro, W.C., Zhang, T. and Jiang, B. (2008). The nutraceutical role of the *Phaseolus vulgaris* α-amylase inhibitor. *British Journal of Nutrition*, 100: 1-12.

Obiro, W.C., Zhang, T. and Jiang, B. (2009). Alpha amylase inhibitor of Phaseolus_Obiro.pdf. *American Journal of Food Technology*, 2(4): 9-19.

Ohta, T., Nakatsugi, S., Watanabe, K., Kawamori, T., Ishikawa, F., Morotomi, M., Sugie, S., Toda, T., Sugimura, T. and Wakabayashi, K. (2000). Inhibitory effects of bifidobacterium-fermented soy milk on mammary carcinogenesis, with a partial contribution of its component isoflavones. *Carcinogenesis*, 21: 937-941.

Oshima, G., Shimabukuro, H. and Nagasawa, K. (1979). Peptide inhibitors of angiotensin I-converting enzyme in digests of gelatin by bacterial collagenase. *Biochimica et Biophysica Acta (BBA) – Enzymology*, 566: 128-137.

Pais, E., Alexy, T., Holsworth, J.R.J. and Meiselman, H. (2006). Effects of *nattokinase*, a pro-fibrinolytic enzyme, on red blood cell aggregation and whole blood viscosity. *Clinical Hemorheology and Microcirculation*, 35: 139-142.

Pan, H.C., Cheng, F.C., Chen, C.J., Lai, S.Z., Liu, M.J., Chang, M.H., Yang, Y.C., Yang, D.Y. and Ho, S.P. (2009a). Dietary supplement with fermented soybeans, *natto*, improved the neurobehavioral deficits after sciatic nerve injury in rats. *Neurological Research*, 31: 441-452.

Pan, H.C., Yang, D.Y., Ho, S.P., Sheu, M.L., Chen, C.J., Hwang, S.M., Chang, M.H. and Cheng, F.C. (2009b). Escalated regeneration in sciatic nerve crush injury by the combined therapy of human amniotic fluid mesenchymal stem cells and fermented soybean extracts, *natto*. *Journal of Biomedical Science*, 16: 1-12.

Park, K.Y., Jung, K.O., Rhee, S.H. and Choi, Y.H. (2003). Antimutagenic effects of doenjang (Korean fermented soypaste) and its active compounds. *Mutation Research/Fundamental and Molecular Mechanisms of Mutagenesis*, 523-524: 43-53.

Park, K., Kang, J.I., Kim, T. and Yeo, I. (2012a). The antithrombotic and fibrinolytic effect of natto in hypercholesterolemia rats. *Preventive Nutrition and Food Science*, 17: 78-82.

Park, K.Y., Moon, S.H., Baik, H.S. and Cheigh, H.S. (1990). Antimutagenic effect of doenjang (Korean fermented soy paste) toward aflatoxin. *Journal of the Korean Society of Food and Nutrition*, 19: 156-162.

Park, N.Y., Rico, C.W., Lee, S. and Kang, M.Y. (2012b). Comparative effects of doenjang prepared from soybean and brown rice on the body weight and lipid metabolism in high fat-fed mice. *Journal of Clinical Biochemistry and Nutrition*, 51(3): 235-240.

Pichiah, P.B.T. (2016). *Cheonggukjang*, a soybean paste fermented with *B. licheniformis*-67 prevents weight gain and improves glycemic control in high fat diet induced obese mice. *Journal of Clinical Biochemistry and Nutrition*, 58: 1-8.

Polonsky, K.S., Given, B.D., Hirsch, L., Shapiro, E.T., Tillil, H., Beebe, C., Galloway, J.A., Frank, B.H., Karrison, T. and Cauter, E.V. (1988). Quantitative study of insulin secretion and clearance in normal and obese subjects. *Journal of Clinical Investigations*, 81: 435-441.

Polonsky, K.S., Sturis, J. and Bell, G.I. (1996). Non-insulin-dependent diabetes mellitus: A genetically programmed failure of the beta cell to compensate for insulin resistance. *New England Journal of Medicine*, 334: 777-783.

Pozo-Bayón, M.Á., Alcaíde, J.M., Polo, M.C. and Pueyo, E. (2007). Angiotensin I-converting enzyme inhibitory compounds in white and red wines. *Food Chemistry*, 100: 43-47.

Pripp, A.H., Sørensen, R., Stepaniak, L. and Sørhaug, T. (2006). Relationship between proteolysis and angiotensin I-converting enzyme inhibition in different cheeses. *LWT–Food Science and Technology*, 39: 677-683.

Quirós, A., Ramos, M., Muguerza, B., Delgado, M.A., Miguel, M., Aleixandre, A. and Recio, I. (2007). Identification of novel antihypertensive peptides in milk fermented with *Enterococcus faecalis*. *International Dairy Journal*, 17: 33-41.

Rayaprolu, S.J., Mauromastakos, A., Kannan, A., Hettiarachchy, N.S. and Chen, P. (2013). Peptides derived from high oleic acid soybean meals inhibit colon, liver and lung cancer cell growth. *Food Research International*, 50: 282-288.

Ren, Y., Pan, X., Lyu, Q. and Liu, W. (2017). Biochemical characterisation of a fibrinolytic enzyme composed of multiple fragments. *Acta Biochimica et Biophysica Sinica*, 50: 227-229.

Rho, S.J., Lee, J.S., Chung, Y., Kim, Y.W. and Lee, H.G. (2009). Purification and identification of an angiotensin I-converting enzyme inhibitory peptide from fermented soybean extract. *Process Biochemistry*, 44: 490-493.

Röder, P.V., Geillinger, K.E., Zietek, T.S., Thorens, B., Koepsell, H. and Daniel, H. (2014). The role of SGLT1 and GLUT2 in intestinal glucose transport and sensing. *PLoS ONE*, 9: 1-10.

Roth, G.A., Johnson, C., Abajobir, A., Abd-Allah, F., Abera, S.F., Abyu, G., Ahmed, M., Aksut, B., Alam, T., Alam, K. and Alla, F. (2017). Global, regional, and national burden of cardiovascular diseases for 10 causes, 1990 to 2015. *Journal of the American College of Cardiology*, 70: 1-25.

Saito, Y., Wanezaki, (Nakamura) K., Kawato, A. and Imayasu, S. (1994). Structure and activity of angiotensin I-converting enzyme inhibitory peptides from sake and sake lees. *Bioscience, Biotechnology, and Biochemistry*, 58: 1767-1771.

Sanjukta, S. and Rai, A.K. (2016). Production of bioactive peptides during soybean fermentation and their potential health bene fits. *Trends in Food Science and Technology*, 50: 1-10.

Sanjukta, S., Rai, A.K., Muhammed, A., Jeyaram, K. and Talukdar, N.C. (2015) Enhancement of antioxidant properties of two soybean varieties of Sikkim Himalayan region by proteolytic *Bacillus subtilis*. *Journal of Functional Foods*, 14: 650-658.

Sarkar, P.K. and Nout, M.J.R. (2014). *Indigenous Foods Involving Alkaline Fermentation*. CRC Press, Taylor & Francis Group, Boca Raton.

Schulze, P.C. and Lee, R.T. (2005). Oxidative stress and atherosclerosis. *Current Atherosclerosis Reports*, 7: 242-248.

Seo, H.R., Kim, J.Y., Kim, J.H. and Park, K.Y. (2009). Identification of *Bacillus cereus* in a Chungkukjang that showed high anticancer effects against AGS human gastric adenocarcinoma cells. *Journal of Medicinal Food*, 12: 1274-1280.

Sharma, A., Kumari, S., Wongputtisin, P., Nout, M.J.R. and Sarkar, P.K. (2015). Optimisation of soybean processing into kinema, a *Bacillus*-fermented alkaline food, with respect to a minimum livel of antinutrients. *Journal of Applied Microbiology*, 119: 162-176.

Shiguemoto, G.E., Rossi, E.A., Baldissera, V., Gouveia, C.H., De Valdez Vargas, G.M.F. and de Andrade Perez, S.E. (2007). Isoflavone-supplemented soy yoghurt associated with resistive physical exercise increase bone mineral density of ovariectomized rats. *Maturitas*, 57: 261-270.

Shin, D. and Jeong, D. (2015). Korean traditional fermented soybean products: Jang. *Journal of Ethnic Foods*, 2: 2-7.

Shin, Z.I., Yu, R., Park, S.A., Chung, D.K., Ahn, C.W., Nam, H.S., Kim, K.S. and Lee, H.J. (2001a). His-His-Leu, an angiotensin I converting enzyme inhibitory peptide derived from Korean soybean paste, exerts antihypertensive activity *in vivo*. *Journal of Agricultural and Food Chemistry*, 49: 3004-3009.

Shin, Z.I., Yu, R., Park, S.A., Chung, D.K., Ahn, C.-W., Nam, H.-S., Kim, K.-S. and Lee, H.J. (2001). His-His-Leu, an angiotensin I converting enzyme inhibitory peptide derived from Korean soybean paste, exerts antihypertensive activity *in vivo*, *Journal of Agricultural and Food Chemistry*, 49(6): 3004-3009.

Shu, G., Shi, X., Chen, H., Ji, Z. and Meng, J. (2019). Optimisation of nutrient composition for producing ACE inhibitory peptides from goat milk fermented by *Lactobacillus bulgaricus* LB6. *Probiotics and Antimicrobial Proteins*, 11: 723-729.

Singh, B.P., Vij, S. and Hati, S. (2014). Functional significance of bioactive peptides derived from soybean. *Peptides*, 54: 171-179.

Skeggs, L., Kahn, J. and Shumway, N. (1956). Preparation and function of the hypertensin converting enzyme. *Journal of Experimental Medicine*, 103: 295-299.

SOPA (2019). World Soybean Production. *In*: The Soybean Processors Association of India. www.sopa.org/statistics/world-soybean-production/

Stahlt, M.L. and Ferrari, E. (1984). Replacement of the *Bacillus subtilis* subtilisin structural gene with an in vitro-derived deletion mutation. *Journal of Bacteriology*, 158: 411-418.

Steinberg, D. (2004). Thematic review series: The pathogenesis of atherosclerosis. An interpretive history of the cholesterol controversy: Part I. *Journal of Lipid Research*, 45: 1583-1593.

Sugawara, E. (2020). *Fermented Foods and Beverages of the World*. CRC Press, Taylor & Francis Group, New York.

Sumi, H., Hamada, H., Nakanishi, K. and Hiratani, H. (1990). Enhancement of the fibrinolytic activity in plasma by oral administration of *Nattokinase*, *Acta Haematol*. 84: 139-143.

Sumi, H., Hamada, H., Tsushima, H., Mihara, H. and Muraki, H. (1987). A novel fibrinolytic enzyme (*Nattokinase*) in the vegetable cheese *natto*, a typical and popular soybean food in the Japanese diet. *Experientia*, 43: 1110-1111.

Sun, S., Xu, X., Sun, X., Zhang, H., Chen, X. and Xu, N. (2019). Preparation and identification of ACE inhibitory peptides from the marine macroalga ulva intestinalis. *Marine Drugs*, 17: 179.

Sun, Z., Lu, W., Liu, P., Wang, H., Huang, Y., Zhao, Y. and Kong, Z.C. (2015). Isolation and characterisation of a proteinaceous α-amylase inhibitor AAI-CC5 from *Streptomyces* sp. CC5 from *Streptomyces* sp. CC5, and its gene cloning and expression. *Antonie van Leeuwenhoek*, 107: 345-356.

Suzuki, Y., Kondo, K., Matsumoto, Y., Zhao, B.Q., Otsuguro, K., Maeda, T., Tsukamoto, Y., Urano, T. and Umemura, K. (2003). Dietary supplementation of fermented soybean, *natto*, suppresses intimal thickening and modulates the lysis of mural thrombi after endothelial injury in rat femoral artery. *Life Sciences*, 73: 1289-1298.

Svensson, P., de Faire, U., Sleight, P., Yusuf, S. and Ostergren, J. (2001). Comparative effects of ramipril on ambulatory and office blood pressures: A HOPE substudy. *Hypertension*, (Dallas, Tex: 1979), 38: E28-32.

Tamang, J.P. (2015). Naturally fermented ethnic soybean foods of India. *Journal of Ethnic Foods*, 2: 8-17.

Tamang, J.P., Shin, D., Jung, S. and Chae, S. (2016). Functional properties of microorganisms in fermented foods. *Frontiers in Microbiology*, 7: 1-13.

Tanimoto, H., Fox, T., Eagles, J., Satoh, H., Nozawa, H., Okiyama, A., Morinaga, Y. and Fairweather-Tait, S.J. (2007). Acute effect of poly-γ-glutamic acid on calcium absorption in post-menopausal women. *Journal of the American College of Nutrition*, 26: 645-649.

Taylor, C.K., Levy, R.M., Elliott, J.C. and Burnett, B.P. (2009). The effect of genistein aglycone on cancer and cancer risk: A review of *in vitro*, preclinical, and clinical studies. *Nutrition Reviews*, 67: 398-415.

Toi, M., Hirota, S., Tomotaki, A., Sato, N., Hozumi, Y., Anan, K., Nagashima, T., Tokuda, Y., Masuda, N., Ohsumi, S., Ohno, S., Takahashi, M., Hayashi, H., Yamamoto, S. and Ohashi, Y. (2013). Probiotic beverage with soy isoflavone consumption for breast cancer prevention: A case-control study. *Current Nutrition and Food Science*, 9: 194-200.

Tormo, M., Gil-Exojo, I., de Tejada, A. and Campillo, J. (2004). Hypoglycaemic and anorexigenic activities of an α-amylase inhibitor from white kidney beans (*Phaseolus vulgaris*) in Wistar rats. *British Journal of Nutrition*, 92: 785-790.

Torres, N., Torre-Villalvazo, I. and Tovar, A.R. (2006). Regulation of lipid metabolism by soy protein and its implication in diseases mediated by lipid disorders. *The Journal of Nutritional Biochemistry*, 17: 365-373.

Toshiro, M., Jae, Y.H., Sung, H.J., Seok, L.D. and Bok, K.H. (2004). Isolation of angiotensin I converting enzyme inhibitory peptide from Chungkookjang. *Korean Journal of Microbiology*, 40: 355-358.

Tsai, J.S., Lin, T.C., Chen, J.L. and Pan, B.S. (2006). The inhibitory effects of freshwater clam (*Corbicula fluminea*, Muller) muscle protein hydrolysates on angiotensin I-converting enzyme. *Process Biochemistry*, 41: 2276-2281.

Tsai, T.Y., Chen, L.Y. and Pan, T.M. (2014). Effect of probiotic-fermented, genetically modified soy milk on hypercholesterolemia in hamsters. *Journal of Microbiology, Immunology and Infection*, 47: 1-8.

Tsukamoto, Y., Ichise, H., Kakuda, H. and Yamaguchi, M. (2000). Intake of fermented soybean (*natto*) increases circulating vitamin K2 (menaquinone-7) and γ-carboxylated osteocalcin concentration in normal individuals. *Journal of Bone and Mineral Metabolism*, 18: 216-222.

United States Department of Agriculture (2019). *World Agricultural Production.*

Urano, T., Ihara, H., Umemura, K., Suzuki, Y., Oike, M., Akita, S., Tsukamoto, Y., Suzuki, I. and Takada, A. (2001). The profibrinolytic enzyme subtilisin NAT purified from *Bacillus subtilis* cleaves and inactivates plasminogen activator inhibitor Type 1. *The Journal of Biological Chemistry*, 276: 24690-24696.

Vallabha, V. and Kaultiku, P. (2014). Antihypertensive peptides derived from soy protein by fermentation. *International Journal of Peptide Research and Therapeutics*, 20: 161.

Villa, J.K.D., Diaz, M.A.N., Pizziolo, V.R. and Martino, H.S.D. (2017). Effect of vitamin K in bone metabolism and vascular calcification: A review of mechanisms of action and evidences. *Critical Reviews in Food Science and Nutrition*, 57: 3959-3970.

Wagar, L.E., Champagne, C.P., Buckley, N.D., Raymond, Y. and Green-Johnson, J.M. (2009). Immunomodulatory properties of fermented soy and dairy milks prepared with lactic acid bacteria. *Journal of Food Science*, 74: 423-430.

Wang, L., Saito, M., Tatsumi, E. and Li, L. (2003). Antioxidative and angiotensin I-converting enzyme inhibitory activities of sufu (fermented tofu) extracts. *Japan Agricultural Research Quarterly: JARQ*, 37: 129-132.

Wang, X.L. (1997). *Zhong Guo Da Dou Zhi Ping* [Chinese Soybean Products], Zhong Guo Qing Gong Ye Chubanshe. China Light Industry Publisher, Beijing, China, pp. 8-30.

Wang, Z., Bao, Y., Zhang, Y., Zhang, J., Yao, G., Wang, S. and Zhang, H. (2013). Effect of soymilk fermented with *Lactobacillus plantarum* P-8 on lipid metabolism and fecal microbiota in experimental hyperlipidemic rats. *Food Biophysics*, 8: 43-49.

Weir, G.C. and Bonner-Weir, S. (2004). Five stages of evolving beta-cell dysfunction during progression to diabetes. *Diabetes*, 53: S16-S21.

Weng, Y., Yao, J., Sparks, S. and Wang, K.Y. (2017). *Nattokinase*: An oral antithrombotic agent for the prevention of cardiovascular disease. *International Journal of Molecular Sciences*, 18.

Wood, B.J.B. (1998). Protein-rich foods based on fermented vegetables. pp. 484-504. *In*: Wood, B.J.B. (Ed.). *Microbiology of Fermented Foods*. Springer US, Boston, MA.

Wu, D.J., Lin, C.S. and Lee, M.Y. (2009). Lipid-lowering effect of *Nattokinase* in patients with primary hypercholesterolemia. *Acta Cardiologica Sinica*, 25.

Xiao, Y., Zhang, S., Tong, H. and Shi, S. (2018). Comprehensive evaluation of the role of soy and isoflavone supplementation in humans and animals over the past two decades. *Phytotherapy Research*, 32: 384-394.

Xu, J., Du, M., Yang, X. and Chen, Q. (2014). Thrombolytic Effects *in vivo* of *Nattokinase* in a Carrageenan-Induced Rat Model of Thrombosis. pp. 247-253.

Xu, X., Huang, Q., Mao, Y., Cui, Z., Li, Y., Huang, Y., Rajput, I.R., Yu, D. and Li, W. (2012). Immunomodulatory effects of *Bacillus subtilis* (*natto*) B4 spores on murine macrophages. *Microbiology and Immunology*, 56: 817-824.

Yamaguchi, M., Kakuda, H., Gao, Y.H. and Tsukamoto, Y. (2000). Prolonged intake of fermented soybean (*natto*) diets containing vitamin K2 (menaquinone-7) prevents bone loss in ovariectomised rats. *Journal of Bone and Mineral Metabolism*, 18: 71-76.

Yamaguchi, M., Taguchi, H., Gao, Y.H., Igarashi, A. and Tsukamoto, Y. (1999). Effect of vitamin K2 (menaquinone-7) in fermented soybean (*natto*) on bone loss in ovariectomised rats. *Journal of Bone and Mineral Metabolism*, 17: 23-29.

Yang, L.C., Wu, J.B., Ho, G.H., Yang, S.C., Huang, Y.P. and Lin, W.C. (2008). Effects of poly-γ-glutamic acid on calcium absorption in rats. *Bioscience, Biotechnology, and Biochemistry*, 72: 3084-3090.

Yoshikawa, M., Morikawa, T., Matsuda, H., Tanabe, G. and Muraoka, O. (2002). Absolute stereostructure of potent alpha-glucosidase inhibitor, Salacinol, with unique thiosugar sulphonium sulphate inner salt structure from Salacia reticulata. *Bioorganic and Medicinal Chemistry*, 10: 1547-1554.

Zhang, J.H., Tatsumi, E., Ding, C.H. and Li, L.T. (2006a). Angiotensin I-converting enzyme inhibitory peptides in douchi, a Chinese traditional fermented soybean product. *Food Chemistry*, 98: 551-557.

Zhang, J.H., Tatsumi, E., Ding, C.-H. and Li, L.-T. (2006b). Angiotensin I-converting enzyme inhibitory peptides in douchi, a Chinese traditional fermented soybean product. *Food Chemistry*, 98(3): 551-557.

Zhang, X., Cao, D., Sun, X., Sun, S. and Xu, N. (2017). Purification and identification of a novel ACE inhibitory peptide from marine alga *Gracilariopsis lemaneiformis* protein hydrolysate. *European Food Fermentation and Technology*, 243: 1820-1837.

Zhang, Y., Liu, C., Dong, B., Ma, X., Hou, L., Cao, X. and Wang, C. (2015). Anti-inflammatory activity and mechanism of Surfactin in lipopolysaccharide-activated macrophages. *Inflammation*, 38: 756-764.

Zhao, X., Song, J.L., Wang, Q., Qian, Y., Li, G.J. and Pang, L. (2013). Comparisons of *shuidouchi*, *natto* and *cheonggukjang* in their physicochemical properties, and antimutagenic and anticancer effects. *Food Science and Biotechnology*, 22: 1077-1084.

Zhibian, D., Xiao, J., Hanhu, J., Shuxia, Z., Mingsheng, D. and Xiaoyan, Z. (2004). Study on the antioxidative activity and effects on experimental hyperlipidemia of *natto* extract [Ying yang xue bao]. *Acta Nutrimenta Sinica*, 26: 296-299.

Health Benefits of Fermented Alcoholic Beverages

Gabriela Rios-Corripio[1], Paola Hernández-Carranza[1], Irving Israel Ruiz-López[2], José Ángel Guerrero-Beltrán[3] and Carlos Enrique Ochoa-Velasco[1]*

[1] Facultad de Ciencias Químicas, Benemérita Universidad Autónoma de Puebla, Av. San Claudio y 18 Sur, Ciudad Universitaria, C.P. 72570, Puebla, Puebla, México
[2] Facultad de Ingeniería Química, Benemérita Universidad Autónoma de Puebla, Av. San Claudio y 18 Sur, Ciudad Universitaria, C.P. 72570, Puebla, Puebla, México
[3] Departamento de Ingeniería Química, Alimentos y Ambiental, Universidad de las Américas Puebla, Sta. Catarina Mártir, Cholula 72810, Puebla, México

1. Introduction

Nowadays consumers are demanding food products with high sensory and nutritional quality, as fresh as possible and free of any synthetic or chemical additives (Hernández-Carranza et al., 2019). However, the consumption of health-promoting compounds, such as fibre, vitamins, pigments, minerals, terpenoids, phenolic compounds among others has been gaining attention since they can help to maintain good health and reduce the appearance of different, chronic, degenerative diseases, such as cancer, diabetes, obesity, chronic heart diseases and others (Tapsell et al., 2016). Fruits and vegetables are a well-known source of bioactive compounds (González-Aguilar et al., 2008; Shashirekha et al., 2016); but their short shelf-life is one of the main constraints for their commercialisation (Sousa and Mahajan, 2011). Therefore, several processing technologies have been developed. Among them, the use of novel technologies (high hydrostatic pressure, pulsed electric field, ultraviolet-C light, microwave, ultrasound, etc.) is on the rise (Jermann et al., 2015; Priyadarshini et al., 2019). Nonetheless, the application of traditional technologies, such as drying, heating, cooling, freezing and fermentation is still in use (Awuah et al., 2007; Akeem et al., 2019).

Alcoholic fermentation is an anaerobic biochemical process in which sugars, such as glucose, fructose, and sucrose are consumed while ethanol and CO_2 are produced (Mani et al., 2018). Nevertheless, during the fermentation process, other interesting metabolic by-products are generated, contributing to their chemical composition

*Corresponding author: carlosenriqueov@hotmail.com

and sensorial properties (Buratti and Benedetti, 2016). For example, some volatile (citronellol, nerol, β-phenylethanol, etc.), non-volatile compounds (phenolic acids, flavonoids, tannins, etc.) and many other compounds develop during the fermentation process (Ugliano *et al.*, 2006; Vieira *et al.*, 2019). Among the fermented alcoholic beverages, wine and beer are the most widely consumed and studied. These drinks contain phenolic acids, flavonoids, stilbenes, tannins, carotenoids, etc. which present several human health benefits. However, other alcoholic beverages, such as cider, sake, mead, pulque, among others, are less studied but based on the raw material; they also may contain several health-promoting compounds.

This chapter aims to provide an overview of the fermentation process of the main alcoholic beverages and their health-promoting compounds and describe the beneficial health effects on consumption of fermented alcoholic beverages.

2. Fermentation Process and Alcoholic Fermented Beverages

In general, fermentation is an anaerobic biochemical process in which organic substrates change due to the action of enzymes present in the microorganisms (Rai *et al.*, 2019; Verardo *et al.*, 2020). Alcoholic fermentation (Fig. 1) is anaerobic transformation of sugars (glucose and fructose mainly) into ethanol and carbon dioxide, conducted generally by yeasts and some bacteria (Şanlier *et al.*, 2019).

$$C_6H_{12}O_6 \rightarrow 2C_3CH_2OH + 2CO_2$$

During the production of ethanol, many other interesting products are created, making it possible to turn the must into a more sensory and nutritionally-acceptable food product. In this aspect, compounds, such as esters, glycerol, succinic acid, diacetyl, acetoin, and 2,3-butanediol are also formed during the fermentation process, which increase the sensorial quality of fermented products. Furthermore,

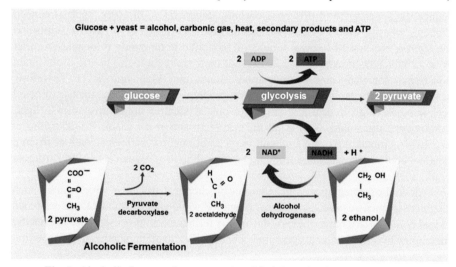

Fig. 1: Alcoholic fermentation process (modified from Garduño-García, 2014)

the fermentation process can increase not only the health-promoting compounds present in the raw materials from which they come, but also their bioavailability (Verardo *et al.*, 2020; Verni *et al.*, 2019). For example, in fruits and vegetables, the fermentation process may increase the antioxidant capacity of raw material due to the release of bioactive compounds from conjugated phytochemicals (Rai *et al.*, 2014; Septembre-Malaterre *et al.*, 2018), while proteins and peptides with antihypertensive, antithrombotic and antioxidants properties are also released through fermentation or hydrolysis processes (García *et al.*, 2013).

Although wine and beer are the most widely studied and consumed alcoholic fermented beverages, some other drinks are also important because they can impart beneficial effects to human health and which are associated with the raw material used in their fermentation process (Fig. 2). For example, wine contains anthocyanins, resveratrol, flavonols, tannins, catechins, hydroxycinnamic acids, tyrosol, hydroxytyrosol, etc. (Fernández-Mar *et al.*, 2012; Ciubara *et al.*, 2018); while, beer presents flavanones, flavanols, flavonols, catechins, and caffeic, gallic, and ferulic acids (Dabina-Bicka *et al.*, 2013). Cider, in turn, contains flavonols, dihydrochalcones, chlorogenic, dihydroshikimic, caffeic, *p*-coumaryl-quinic acids (Nogueira *et al.*, 2008); sake presents some phenolic acids (ferulic and *p*-coumaric acids) and pyroglutamyl (pGlu) peptides (Srinivasan *et al.*, 2007). On the other hand, pulque – a fermented beverage from the Agave sap – contains a consortium of microorganisms (*Lactobacillus, Streptococcus, Saccharomyces*, among others) that show several gastrointestinal benefits (Cervantes-Contreras and Pedroza-Rodríguez, 2007).

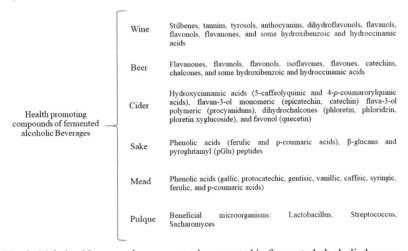

Fig. 2: Main health-promoting compounds presented in fermented alcoholic beverages

3. Health-promoting Compounds of Fermented Alcoholic Beverages

As mentioned before, fermented alcoholic beverages contain several health-promoting compounds, such as phenolic compounds, carotenoids, pigments and even some of them contain beneficial microorganisms. In the following lines, the main health-promoting compounds and beneficial microorganisms presented in them are described.

3.1 Phenolic Compounds

In plants, more than 8,000 phenolic compounds have been identified (Pandey and Rizvi, 2009). They are compounds produced from secondary metabolites of plants as a consequence of both normal development and in response to stress conditions (Naczk and Shahidi, 2006). Among phenolic compounds, phenolic acids, flavonoids, stilbenes, tannins, lignans and lignins are the most important.

3.1.1   Phenolic Acids

Phenolic acids are divided into hydroxycinnamic and hydroxybenzoic acids. In the former group, caffeic, ferulic, *p*-coumaric, and sinapic acids are most representative; while gallic, p-hydroxybenzoic, protocatechuic, vanillic, and syringic acids are most representative of the second group (Balansudram *et al.*, 2006). Phenolic acids show different health benefits, among them antioxidant capacity (increase with increasing degree of hydroxylation) for reducing oxidative stress is one of the most important (Rai *et al.*, 2010, 2014); however, anti-inflammatory (Oh *et al.*, 2005), anticancer (DeFuria *et al.*, 2009), antiobesity (cardiovascular diseases) (Lee *et al.*, 2011), and antidiabetes (Tourrel *et al.*, 2001) properties have also been associated with phenolic acids.

3.1.2 Flavonoids

Among phenolic compounds, flavonoids are the most abundant in the human diet (Luna-Guevara *et al.*, 2019). These include flavonols, flavones, flavanones, flavonols, isoflavones, flavanonols, and anthocyanins (Balansudram *et al.*, 2006). Flavonoids have been evaluated in several *in vitro* and *in vivo* studies related to their benefits on human health. In this aspect, flavonoids display beneficial effects against DNA-inducing damage by xenobiotics, inhibit reactive oxygen species (ROS) by producing antioxidant enzymes (Chien *et al.,* 2009), reduce atherosclerosis and coronary heart disease (Kodera *et al.*, 2009), and also display activity as anticarcinogenic and antitumoral in some kinds of cancers, such as ovarian (Liu *et al.*, 2010), prostate (Thakkar *et al.*, 2010) and skin (Wu *et al.*, 2016).

3.1.3   Stilbenes

Stilbenes are characterised by the presence of a 1,2-diphenylethylene nucleus with hydroxyl groups substituted in the aromatic rings (Ozcan *et al.*, 2014). Among stilbenes, resveratrol (3,4′,5-tri-hydroxystilbene) is the most studied. It showed anti-inflammatory, anticancer, immunomodulatory and hepatoprotective activities (Szkudelska and Szkudelski, 2010). Moreover, resveratrol showed a potent antidiabetic effect (Szkudelski and Szkudelska, 2015) because of increasing the insulin sensitivity and glucose uptake, improving the β-cells functions and decreasing insulin resistance (Ramadori *et al.*, 2009).

3.1.4 Lignans

Lignans are phenolic compounds that contain a 2,3-dibenzylbutane structure formed by the dimerisation of two cinnamic acid residues (Pandey and Rizvi, 2009). Lignans have been evaluated against ovarian cancer (Singh *et al.*, 2011), breast cancer (Raman and Lau, 1996) and diabetes (Attele *et al.*, 2002) with good results against these diseases.

3.1.5 Tannins

Tannins are a group of water-soluble polyphenols, which are divided into hydrolysable and condensed forms (Ozcan *et al.*, 2014). The former are esters of gallic acid (gallo- and ellagitannins), while the latter are the polymers of polyhidroxyflavan-3-ol monomers (Balansudram *et al.*, 2006). Some studies have shown that tannins could be used as a treatment for inflammation (Shukla *et al.*, 2008), atherosclerosis (Shukla *et al.*, 2008), LDL-cholesterol (Müller-Schwarze, 2009), cancer (Fu *et al.*, 2010), diabetes (Fu *et al.*, 2010), infections (Iso *et al.*, 2006) and other diseases.

3.2 Carotenoids

Carotenoids are pigments that play a major role in the protection of plants against photooxidative processes (Stahl and Sies, 2004). Carotenoids are classified as carotenes, xanthophylls and lycopene (Jomova and Valko, 2013). Like many other compounds, the health benefits of carotenoids are related to their antioxidant capacity (Krinsky, 2001). Among illnesses, carotenoids have shown beneficial effects against different disorders such as some forms of cancer (Nishino *et al.*, 2004; Mayne, 1996) and cardiovascular diseases (Riccioni, 2009).

3.3 Microorganisms

Although in alcoholic beverages several processes, such as filtration, clarification and pasteurisation are applied; in pulque, these processes are not employed. Thus, microorganisms responsible for the fermentation process are still present in the beverage when it is consumed. In this sense, the fermentation process of pulque is conducted by several microorganisms, such as *Lactobacillus* sp., *Leuconostoc mesenteroides*, *Lc. dextranicum*, *Saccharomyces cerevisiae*, and *Zymomonas mobilis*, which are naturally occurring in raw materials and in-process steps (Sánchez-Marroquín and Hope, 1953; Escalante *et al.*, 2004). The most important microorganisms related to the beneficial effects of pulque are the well-known probiotics, the lactic acid bacteria (LAB). The consumption of LAB is associated with beneficial effects on intestine (diarrhoea, colitis, flatulence, lactose intolerance, antibacterial, etc.) (Giles-Gómez *et al.*, 2016) and also because of their anti-inflammatory and hypercholesteremia (Torres-Maravilla *et al.*, 2016).

4. Alcoholic Beverages and Their Health Benefits

There are many scientific studies related to health benefits of low-to-moderate consumption of alcoholic beverages. In the following section, the health-promoting aspects of some alcoholic beverages are disused.

4.1 Wine

4.1.1 Overview

Wine is an alcoholic beverage [11 to 13% (v/v) of alcohol] produced by the fermentation of ripe and fresh grapes or fresh grape juice by *S. cerevisiae*. Some grape varieties most commonly used for winemaking are Chardonnay, Sauvignon Blanc, Viognier and Riesling, for white wines; while Tempranillo, Cabernet Sauvignon, Malbec, Mourvèdre, Pinot Noir, Grenache and Sangiovese are used for red wines (Patterson, 2010). The colour of red wine is produced by the skin addition during the fermentation process, providing not only colour to the wine but also antioxidant compounds, such as anthocyanins and tannins. The yellow, golden, greenish or pale color of white wine is produced by the colour of the grape pulp. On the other hand, rosé wines are made with the addition of skins for a shorter time as compared to red wines (Boulton, 2001). It is important to point out that several factors may affect the production of health-promoting compounds during the fermentation process of winemaking. Among these, temperature, pH, sugars concentration and acidity of must are the main factors (Swiegers *et al.*, 2005).

4.1.2 Wine Composition and Health Beneficial Effects

The main components of wine are water, carbohydrates, organic acids, minerals, alcohol, polyphenols and aromatic compounds. Although the harvest, fermentation and clarification processes can significantly modify the composition and concentration of phenolic compounds (Haseeb *et al.*, 2017), red wines show high amounts of phenolic compounds, such as stilbenes (resveratrol), flavonoids (catechin, epicatechin, quercetin, anthocyanins and procyanidins) and polymeric tannins (Castaldo *et al.*, 2019). Since the French paradox was revealed, the effect of wine consumption and its impact on cardiovascular diseases has been vastly studied. Several meta-analyses, *in vitro* and *in vivo* studies have reported that light-to-moderate consumption of wine reduces the oxidative stress, lipoproteins oxidation, blood viscosity, platelet adhesion to fibrinogen-coated surfaces, LDL-cholesterol and enhances cholesterol efflux from the vessel walls by increasing HDL-cholesterol levels (Svärdsudd, 1998; Rotondo *et al.*, 2001; Vogel, 2002; Vidavalur *et al.*, 2006; Lippi *et al.*, 2010).

Studies on the effect of wine consumption and cardiovascular diseases indicate that in human coronary endothelial cells and hypercholesterolemic mice, the wine treatment significantly reduced the activation of redox-sensitive genes (*ELK-1* and *p-JUN*) and increased the endothelial nitric oxide synthase expression (Napoli *et al.*, 2008). Similarly, red wine and resveratrol improve endothelial function by decreasing plasma ET-1 levels and increasing nitric oxide levels (Zou *et al.*, 2003), and by inhibiting ADP-induced platelet aggregation (Wang and Sang, 2002) in hypercholesterolemic rabbits. In humans, Ohta *et al.* (1994) indicated that wine consumption significantly increased serum HDL-cholesterol, Apo A-I, HDL3-cholesterol, LpA-I, and LpA-I:AII. They pointed out that the beneficial effect of wine was not only provided by the phenolic compounds of the wine, but also by the alcohol content. On the other hand, in a crossover study with high cardiovascular risk subjects, the four-week consumption of dealcoholised wine had shown a potential blood pressure lowering and nitric oxide raising (Chiva-Blanch *et al.*, 2012). Thus, moderate red wine consumption increases the total antioxidant capacity, vitamin E, and vitamin E/total cholesterol ration on blood serum (Apostolidou *et al.*, 2015), which reduces the red blood cell aggregation and increases the red blood cell deformability at higher shear stress, suggesting that wine consumption may reduce the risk of cardiovascular diseases (Toth *et al.*, 2014). Leighton and Urquiaga (2007) concluded that wine in moderation and as part of a healthy diet is directly responsible for changes that may help decrease the risk of cardiovascular disease.

As it is well mentioned by several authors, the effect of wine consumption on cardiovascular diseases is related to its content of phenolic compounds. Resveratrol is the most relevant phenolic compound of the wine; it exists as *cis* and *trans* isomers, ranging from 0.1 to 7 mg/L and 0.7 to 6.5 mg/L, respectively in red wine (Snopek *et al.*, 2018). Its cardio-protective effects include improving endothelial function and glucose metabolism, reducing inflammation and regulating blood lipids (Cavallini *et al.*, 2016). In this aspect, in a seven-day study, resveratrol greatly improved the post-ischemic recovery of heart rate and developed pressure recovery in a dose-dependent manner. The effect of resveratrol as a cardioprotector may be due to its antioxidant capacity because it reduced the levels of tissue malondialdehyde (MDA), without increasing the nitric oxide levels (Mokni *et al.*, 2007). Moreover, resveratrol also increased the SOD (superoxide dismutase) and POD (peroxidase) activities, reduced the myocardial free iron level probably due to resveratrol act on L-type voltage-dependent Ca^{2+} channels decreasing iron overload, or by the modulation of iron-regulatory hormone (hepcidin); this hormone is autocrine which causes heart dysfunction (Mokni *et al.*, 2013). It is important to mention that a high concentration of resveratrol (> 300 mg per day) may display adverse effects on the cardiovascular risk biomarkers (Mankowski *et al.*, 2020). Anthocyanins, highly presented in red grape skin and red wine, also display beneficial effects on cardiovascular diseases. Cyanidin-3-glucoside induced a two-fold increase in nitric oxide production and nitric oxide synthase by activating the Src kinase, and consequently, the ERK 1/2 kinase in bovine artery endothelial cells (Xu *et al.*, 2004). Similarly, malvidin-3-*O*-glucoside showed heart protection against ischemia/reperfusion injury by inducing Akt, ERK1/2 and GSK3-β phosphorylation in rats (Quintieri *et al.*, 2013).

On the other hand, wine and its phenolic compounds have been evaluated against different kinds of cancer, such as lung (Duan *et al.*, 2018), breast (Waffo-Téguo *et al.*, 2001), osteosarcoma (Peng and Jang, 2018), cervical (Ruíz *et al.*, 2018), prostate (Tenta *et al.*, 2017) and colorectal (Del Pino-García *et al.*, 2017) among others. Red and white wine extracts inhibited the proliferation of PC-3 cancer cell line in a dose-dependent manner by regulating the glutathione, hydrogen peroxide and nitrogen oxide in the cells (Tenta *et al.*, 2017). In breast cancer cell lines (MCF7, T47D, and MDA-MB-231) wine produced a time- and dose-dependent inhibitory effect probably caused by an agonistic effect of polyphenols on the estrogen receptor (Damianaki *et al.*, 2000).

As in other diseases, resveratrol is the main investigated compound of wine. In this aspect, results indicated that resveratrol affects the cancer cells in different ways. In a first approach, the effect of resveratrol on cell inhibition growth may be caused by the interruption of different signal transduction pathways or by inhibiting cell proliferation and increasing cells arrested at the G2/M phase of the cell cycle (Okabe *et al.*, 1997; Mgbonyebi *et al.*, 1998). However, current studies have delved into its effects and some of them indicated different mechanisms of cancer cells' inactivation. For example, in colorectal (DLD1), cervical (HeLa) and breast cancer (MCF-7) cells, resveratrol induced apoptosis by down-regulating the expression of PKM2-mediated endoplasmic reticulum stress and mitochondrial fission (Wu *et al.*, 2016). While in osteosarcoma cells, resveratrol inhibited cell viability, self-renewal ability and tumorigenesis by decreasing cytokines synthesis and inhibiting JAK/ STAT3 signalling (Peng and Jiang, 2018). In breast (MDA-MB-231 and MCF-7) and lung (H1299) cancer cell lines, resveratrol inhibited proliferation and induced senescence process by elevating ROS generation through mitochondrial dysfunction and up-regulated DLC1 expression with down-regulation of DNMT (Ji *et al.*, 2018). While in cervical cancer cells (HeLa), resveratrol significantly decreased both glycolysis and oxidative phosphorylation protein contents and fluxes. It also increased the ROS production (two to three times), diminished the super-oxide dismutase activity and inhibited the RAD51 gene expression, as a result of which the cancer cells were not able to repair by themselves (Ruíz *et al.*, 2018). Similar to resveratrol, other important compounds of wine, such as ellagic acid, trans-astringin and trans-piceatannol display significant anticancer effects (Waffo-Téguo *et al.*, 2001; Duan *et al.*, 2018).

Although resveratrol showed beneficial effects against inflammatory and neurodegenerative diseases (Panaro *et al.*, 2012; Rodríguez-Ramiro *et al.*, 2011; Navarro-Cruz *et al.*, 2017; Wang *et al.*, 2018; Silva and Vauzour 2018; Corpas *et al.*, 2019), no clinical effects of wine on these diseases was reported. Health benefits of wine and its metabolites are presented in Table 1.

4.2 Beer

4.2.1 Overview

Beer is a fermented alcoholic (3.5-10 percentage v/v) beverage made with malted barley, water, hops and yeast (*S. cerevisiae*) as the main raw materials. Beer has been included in the human diet since at least 5000 years BC. Its origin dates back to years ago. There is some evidence that beer originated in China and in Mesopotamia. Later it extended to Egypt and Europe; nowadays, beer is produced around the world (Poelmans and Swinnen, 2011). In general, the main steps during beer brewing are malting of barley, boiling with hops (fermentation process during the required time depends on the type of beer) and bottling (McKay *et al.*, 2011).

4.2.2 Beer Composition and Health Beneficial Effects

As a product of cereals, beer contains important nutrients, such as carbohydrates, proteins, amino acids (alanine, glutamic acid, proline, etc.), calcium, magnesium, phosphorus, potassium, fluoride and folate (De Gaetano *et al.*, 2016; USDA, 2015). Beer also contains health-promoting compounds like tannins, phenolic acids (gallic, caffeic, *p*-coumaric, and ferulic acids) and flavonoids (catechin, epicatechin, gallocatechin, kaempferol, myricetin, quercetin, proanthocyanidin monomers,

Table 1: Beneficial Effects of Wine and Its Compounds against Different Diseases

Compound	Concentration	Disease	Assay	Effect	References
Red wine	100 μL	Cardiovascular	*In vitro* (coronary endothelial cells) and *In vivo* (mice)	Both *in vitro* and *in vivo* results indicated that p-JUN and ELK-1 proteins expression were increased in a 100%, and endothelial nitric oxide synthase increased ~50% in mice.	Napoli *et al.* (2008)
Red wine and resveratrol	RW: 4 mL/kg/d R: 4 mg/kg/d	Cardiovascular	*In vivo* (rabbits)	Hypercholesterolemic rabbits showed platelet aggregation rate (PAR) increased from 39.5% in normal animals to 61.0% in the high cholesterol fed group. Resveratrol inhibited ADP-induced platelet aggregation *in vivo* by maintaining the PAR at 35.7% vs. 39.5% in control rabbits.	Wang and Sang (2002)
Polyphenols	30 g/d for 4 wk	Cardiovascular	*In vivo* (humans)	The consumption of alcohol-free red wine generated a decrease in the systolic (-5.8 mm Hg) and diastolic (-2.3 mm Hg) blood pressure, and an increase in the nitric oxide (4.1 μmol/L).	Chiva-Blanch *et al.* (2012)
Resveratrol and phenolic compounds	R: 0- 20 μmol/L PC: 0.25 g/100 mL	Stress oxidative	*In vivo* (rats)	Resveratrol (20 μmol/L) displayed *in vivo* antioxidant capacity. Polyphenols reduced lipid peroxidation (0.25 g/100 mL).	Cavallini *et al.* (2016)
Resveratrol	Acute test (1-100 μM) and long time (25 mg/kg body weight)	Cardiovascular	*In vivo* (rats)	In acute test, resveratrol did not display any beneficial effects. After 7 days of resveratrol treatment, a great improvement on post-ischemic recovery of heart rate in short-term or long-term was observed. Moreover, resveratrol increased pressure in drastic ischemic, showing an increase of 51 ± 12 mmHg compared to untreated rats.	Mokni *et al.* (2007)

(Contd.)

Table 1: *(Contd.)*

Compound	Concentration	Disease	Assay	Effect	References
Resveratrol	25 mg/kg body weight	Cardiovascular	*In vivo* (rats)	On reperfusion, resveratrol significantly increased heart rate (213%), dropped heart lipid peroxidation (72%), and free iron (21%). Besides, resveratrol increased total SOD and POD and reduced CAT activities in rats.	Mokni *et al.* (2013)
Resveratrol	1000 mg/d	Cardiovascular	*In vivo* (humans)	Biomarkers sVCAM-1 (86.4 ng/mL) and tPAI (1.4 ng/mL) were significantly increased by resveratrol treatment.	Mankowski *et al.* (2020)
Cyanidin-3-glucoside	0.1 μmol/L	Cardiovascular	*In vitro* (bovine endothelial cells)	Anthocyanin induced a 2-fold increase of nitric oxide production and nitric oxide synthase by activating the Src kinase and consequently the ERK 1/2 kinase	Xu *et al.* (2004)
Red wineand Malvidin	RW: 1-1000 ng/mL M:10^{-10} to 10^{-6} mol/L	Cardiovascular	*In vitro* (rat heart)	Red grape skins extract induced positive inotropic and negative lusitropic effects. Furthemore, malvidin elicited negative inotropism and lusitropism and coronary dilation. These effects dependent of pathways as PI3K/Akt/NOS-NO.	Quintieri *et al.* (2013)
Red wine	10 μL	Cancer	*In vitro* (breast cancer cell lines)	In cell lines (MCF7, T47D, and MDA-MB-231) wine produced a time- and dose-dependent inhibitory effect, especially after short incubation times IC_{50} 0.14, 0.09, and 1.3 pM at day 2, and 0.16, 0.9, and 0.23 pM at day 5, respectively.	Damianaki *et al.* (2000)

Resveratrol	25, 50, 100, 150, 200 μM	Cancer	*In vitro* (cancer cell lines)	The IC$_{50}$ values of resveratrol against colorectal (DLD1), cervical (HeLa), and breast cancer (MCF-7) cells were 75, 50, and 50 μM, respectively. P. PKM2-mediated mitochondrial fission and fusion participate in Resveratrol-induced cell apoptosis.	Wu *et al.* (2016)
Resveratrol	10, 20, 40 μM	Cancer	*In vitro* (osteosarcoma stem cells)	In hFOB 1.19, MG-63 and MNNG/HOS cell lines, resveratrol inhibited cell viability in a 95, 50, and 40%, respectively. Resveratrol decreased cytokines synthesis and inhibited JAK2 (1 to 1.15) and STAT3 (1 to 1.25) signaling.	Peng and Jiang (2018)
Resveratrol	25, 50, 75, 100, 150, 200 μM	Cancer	*In vitro* (human breast cancer cell lines)	The presence of 100 μM resveratrol induced more than 50% of growth inhibition (50.97% for MCF-7, 49.40% for MDA-MB-231, and 52.81% for H1299). Resveratrol decreased DNMT1 (0.6 to 0.3) and increased DLC1 (0.1 to 1.4) expression by ROS generation.	Ji *et al.* (2018)
Resveratrol	200 μM	Cancer	*In vitro* (cancer cell lines)	Resveratrol showed higher potency to decrease growth of metastatic HeLa and MDA-MB-231 (IC$_{50}$ = 200–250 μM) cells after 48 h exposure. Resveratrol decreased both glycolysis and oxidative phosphorylation (OxPhos) protein contents (30–90%) and fluxes (40–70%) vs. non-treated cells.	Rodriguez-Enriquez *et al.* (2019)

(Contd.)

Table 1: *(Contd.)*

Compound	Concentration	Disease	Assay	Effect	References
Resveratrol	137 µM	Cancer	*In vitro* (human cervical cancer cell line)	Cell viability (60.9%) and the levels of RAD51 (21.4%) of HeLa cells decreased after resveratrol treatment.	Ruiz *et al.* (2018)
Resveratrol, and ellagic acid.	R and EG: 25 and 50 µmol/L	Anti-lung cancer activity	*In vitro* (adenocarcinoma cell lines)	Compounds of wine such as resveratrol (30 to 68%) and ellagic acid (34 to 70%) displayed high anticancer effects in lung cancer cell line (HOP62).	Waffo-Téguo *et al.* (2001), Duan *et al.* (2018)

RW: Red wine, R: Resveratrol, M: Malvidin, PC: phenolic compounds proanthocyanidin dimers, proanthocyanidin trimmers and proanthocyanidin) (De Gaetano *et al.*, 2016; Marques *et al.*, 2017; Wannenmacher *et al.*, 2018).

Moderate consumption of beer could be related to important health outcomes. In this aspect, although the mechanisms have not been fully understood, beer intake has been linked to several beneficial health effects (Nagendra *et al.*, 2011; De Gaetano *et al.*, 2016). For example, the antioxidant compounds, folates, amino acids and protein content of beer may be related to the beneficial effects on vascular disorders (Arranz *et al.*, 2012). The effect of phenolic compounds of beer on vascular disorders has been evaluated by different approaches. Vinson *et al.* (2003) pointed out that both lager and dark beers significantly inhibited atherosclerosis; however, at high doses (1/2 diluted beer), lager beer significantly decreased cholesterol and triglycerides, acting as *in vivo* antioxidant by decreasing the oxidisability of lower-density lipoproteins. In this line, De Gaetano *et al.* (2016) indicated that drinking a pint of beer every day could reduce heart problems by 30-35 per cent. This beneficial effect is associated with the polyphenolic compounds of beer, which may reduce the LDL-cholesterol and increase the HDL-cholesterol levels on plasma. The effect of phenolic compounds of beer has been pointed out in different studies. For example, men who drink beer for 30 days showed a positive effect on plasma lipid levels, plasma antioxidants and anticoagulant activities (Gorinstein *et al.*, 2007). Padro *et al.* (2018) indicated that after a four-week consumption of traditional beer (15g alcohol and 640mg polyphenols/660 mL) or alcohol-free beer (0g alcohol/day) by overweight or obese class 1 individuals, significantly improved the antioxidant capacity of HDL-cholesterol and facilitated cell-cholesterol efflux, preventing lipid deposition in the vessel wall. Similarly, an increase in the circulating endothelial progenitor cells (cells with the ability to repair and maintain endothelial integrity and function) in peripheral blood were observed in men with high cardiovascular risk who consumed beer (30g alcohol/d) or its equivalent number of phenolic compounds (Chiva-Blanch *et al.*, 2014). Thus, phenolic compounds may be responsible for reducing leukocyte adhesion molecules and inflammatory biomarkers, while alcohol could improve the lipid profile and reduce some inflammatory plasma biomarkers related to atherosclerosis (Chiva-Blanch *et al.*, 2015).

On the other hand, the effect of beer consumption on vascular diseases could be due to other compounds. For example, Mayer *et al.* (2001) indicated that beer contains folate and vitamins B6 and B12, which influence the reduction of total homocysteine concentration in blood, thus contributing to the protective effect of beer consumption on cardiovascular disease. Gorinstein *et al.* (2002) concluded that beer with a high content of protein and amino acids reduces the total LDL-cholesterols and total triglycerides. And at last, some reports have indicated that β-glucans found in beer could reduce the level of bad cholesterol in the blood (De Gaetano *et al.*, 2016).

Beer has also displayed beneficial effects against the development of possibly prostate, breast, bowel, ovarian and blood cancers (Sohrabvandi *et al.*, 2012). In this aspect, the inhibition of angiogenesis by beer consumption could prevent tumor growth and metastasis (Boehm *et al.*, 1997). Different *in vivo* and *in vitro* studies have indicated that humulone (α-lupulic acid from hops in beer), a compound isolated from hops extract, is a strong inhibitor of angiogenesis (Shimamura *et al.*, 2001). On the other hand, six flavonoids (xanthohumol, 2',4',6',4-tetrahydroxy-3'-prenylchalcone, 2',4',6',4-tetrahydroxy-3'-geranylchalcone, dehydrocycloxanthohumol, dehydro-cycloxanthohumol hydrate and isoxanthohumol) present in beer inhibited the growth of human breast cancer (MCF-7), colon cancer (HT-29) and ovarian cancer (A-2780) cells *in vitro*. All flavonoids inhibited the growth of cancer cells in a dose- and time-response manner. However, xanthohumol and isoxanthohumol may have the potential chemoprotective activity against breast and ovarian cancer lines because of inhibited DNA synthesis (Miranda *et al.*, 1999).

Xanthohumol has an exceptionally broad spectrum of inhibitory mechanisms at the initiation, promotion and progression stage of carcinogenesis (Gerhäuser, 2005). It potently modulates the activity of enzymes involved in carcinogen metabolism and detoxification. It inhibits tumor growth in the early stages by inhibiting angiogenesis and inflammatory signals; in the progression phase, it may inhibit DNA synthesis, induce cell cycle arrest in S phase, apoptosis and cell differentiation (Gerhauser *et al.*, 2002). Moreover, xanthohumol can scavenge reactive oxygen species, including hydroxyl- and peroxyl radicals, and inhibit superoxide anion radical and nitric oxide production by suppressing the expression of inducible nitric oxide synthase (Zhao *et al.*, 2003). Besides, xanthohumol demonstrates anti-inflammatory properties by inhibiting cyclooxygenase-1 and cyclooxygenase-2 activity (Gerhauser *et al.*, 2002). Other compounds in beer with anticarcinogenic capacity are dehydrocycloxanthohumol, isoxanthohumol and other prenilflavonoids (Chung *et al.*, 2009).

In an analysis conducted by Jain *et al.* (1998), the relationship between various dietary factors, including alcohol and other beverages, and the risk of prostate cancer was evaluated. They reported that beer consumption (>10 g alcohol/d) significantly decreases the risk of prostate cancer in comparison with non-beer drinking people. On the other hand, Arimoto-Kobayashi *et al.* (1999) indicated that some alcoholic beverages (wines, sake and beer) have antimutagenic effects against heterocycles amines [preactivated Trp-P-1, Trp-P-2(NHOH) and Glu-P-1(NHOH)] in *Salmonella*. In this sense, beer also inhibited the O^6-MG formation by MNNG in the DNA of the bacteria in a dose-response manner. The antimutagenic effect of beer was also corroborated by *in vitro* assays on hamster cells (Nozawa *et al.*, 2004) and *in vivo* assays using mice (Arimoto-Kobayashi *et al.*, 1999). Antimutagenic effects of beer are attributed to the phenolic compounds of hops.

Other health-related benefits of beer include the stimulation of the immune system (immunomodulation). Winkler *et al.* (2006) investigated the ability of beer to stimulate and un-stimulate peripheral blood mononuclear cells (PBMC) of healthy donors to interfere with immune activation cascades by monitoring neopterin formation and tryptophan degradation as well as the release of IFN-g on the cells. Their data demonstrated that beer reduces the production of neopterin and degradation of tryptophan, both biochemical pathways being induced during the cell-mediated immune response. They concluded that the immunosuppressive capacity of beer could be related to its anti-inflammatory nature. Beer intake is also linked to reducing the risk of osteoporosis (Tobe *et al.*, 1997) and dementia (Ruitenberg *et al.*, 2002) diseases.

On the other hand, recent studies have also dispelled the misconception that beer drinkers are more obese than either nondrinkers or drinkers of wine or spirits (Kondo, 2004). A cross-sectional study of approximately 2,400 men and women indicated that beer intake is not associated with a largely increased waist-hip ratio or body mass index (BMI) (Bobak *et al.*, 2003). Beer consumption may stimulate gastrointestinal motility via the muscarinic M3 receptor (Fujii *et al.*, 2002). Health benefits of beer and its metabolites are presented in Table 2.

4.3 Sake

4.3.1 Overview

Sake is a traditional Japanese alcoholic drink (15 percentage v/v) with a tradition of more than 1,300 years (Kitagaki and Kitamoto, 2013). Sake is a beverage made mainly of rice. The fermentation process is more complex in comparison with other alcoholic beverages. It is fermented from steamed white rice using koji (*Aspergillus oryzae*) and the mold converts rice starch into sugars by amylases; then, the sake yeast *S. cerevisiae* converts the sugars to ethanol in fermentation vats. The fermentation process takes place for approximately 3-5 weeks (Davoodi *et al.*, 2019; Akaike *et al.*, 2020).

4.3.2 Sake Composition and Health Beneficial Effects

In Japan, sake production is regulated by the jurisdiction of the National Tax Agency. According to Japanese liquor tax law, sake must have a slightly higher alcohol content than that of wine but less than 20 per cent of alcohol (Akaike *et al.*, 2020). Sake has a mild taste with little acidity, bitterness or astringency. In terms of its chemical composition, sake contains more protein and carbohydrates, which provide it a higher taste than other alcoholic beverages (Kumar-Anal, 2019). Sake contains significant levels of nitrogenous components, such as amino acids and minor amounts of organic acids (Japan Sake and Shochu Makers Association, 2011). As is expected, sake presents phenolic compounds (ferulic and *p*-coumaric acids) derived from rice endosperm cell wall; these compounds displayed several potential health benefits (Srinivasan *et al.*, 2007).

Among the beneficial health effects of sake, some studies have indicated that topical application or the consumption of sake concentrate or its major compounds (α-ethyl glucoside and organic acids) reduced the effect of UVB (ultraviolet B) irradiation on the skin (epidermal barrier disruption) of mice. This effect may be due to sake or α-ethyl glucoside-induced replacement of normal cells by the acceleration of keratinocyte differentiation (Hirotsune *et al.*, 2005). The beneficial effect was even higher than that obtained with red wine and beer. In rats, sake concentrate and α-ethyl glucoside consumption displayed a hepatoprotective effect against D-galactosamine-induced liver injury. The results indicated that sake concentrate and α-ethyl glucoside significantly reduced the alanine aminotransferase and aspartate aminotransferase activities by inhibiting the induction of interleukin-6 (Izu *et al.*, 2019).

On the other hand, different studies about the beneficial effects of sake by-products have been reported. The oral administration of sake yeast as a supplement

Table 2: Beneficial Effects of Beer and Compounds Against Different Diseases

Beverage	Concentration	Disease	Assay	Effect	References
Beer	High dose: 1/2-diluted lager or dark beer Low dose:1/10-diluted lager or dark beer	Cardiovascular	*In vivo* (hamsters)	At the high dose, both lager and dark beer significantly reduced the plasma cholesterol (178-219 mg/dL) and triglycerides (619-912 mg/dL). The quality of antioxidant from both beers is much higher than that of vitamin C or vitamin E.	Vinson *et al.* (2003)
Beer	11 Pilsner beer (92.4 g alcohol per wk)	Atherosclerosis and cardiovascular diseases	*In vivo* (humans)	Consumption of eleven Pilsner beer significantly increased the levels of folate and total homocysteine concentrations in blood.	Mayer *et al.* (2001)
Beer	330 mL for 30 d	Cardiovascular	*In vivo* (humans)	The beer-supplemented diet decreased serum lipids levels. In the experimental group (EG) versus the control group (CG): total cholesterol by 19.3%, LDL-C by 29.6%, and triglycerides by 36.6%. The increase in the HDL-C in the EG vs. the CG group was not significant (p>0.05).	Gorinstein *et al.* (2007)
Beer and its phenolic compounds	Beer: 660 mL beer for 4 wk PC: 640 mg for 4 wk	Cardiovascular	*In vivo* (humans)	Levels of HDL-C and Total-C were significantly increased after intervention with traditional beer in the subgroup of subjects with LDL-C levels <130 mg/dL (moderate CVD risk), whereas this effect was not found in subjects with LDL-C > 130 mg/dL. The risk of CVD value was of 7.8, being the risk 8.7 in women and 7.3 in men.	Padro *et al.* (2018)
Beer	660 mL/d of beer with and without alcohol	Cardiovascular	*In vivo* (humans)	After the beer and non-alcoholic beer interventions, the number of circulating endothelial progenitor cells significantly increased by 8 and 5 units, respectively.	Chiva *et al.* (2014)

Beer	5.1-5.3 g alcohol/100 mL for 4 wk	Cardiovascular	*In vivo* (rats)	Beer with high amount of protein and amino acids presented significant reduction in the level of TC, LDL-C and TG.	Gorinstein *et al.* (2002)
Humulone	10-100 mM	Cancer	*In vitro* (vascular endothelial and tumor cells)	Humulone strongly inhibited angiogenesis, clearly producing an avascular zone in CAM. Humulone at 10 mM inhibited endothelial cell proliferation by 80%. Humulone had a strong inhibitory effect on the growth of endothelial cells. Humulone at 100 mM substantially inhibited the production of VEGF by KOP2.16 cells and Co26 cells. The inhibition was more significant in tumor cells than in endothelial cells at 10–100 mM of humulone.	Shimamura *et al.* (2001)
Hops	Hop flavonoids (0.01, 0.1, 1.0, 10, or 100 mM)	Cancer	*In vitro* (breast, colon, and ovarian cells)	MCF-7 cells were inhibited by 50% (IC) at a concentration of 13.3, 15.7 and 15.3 mM for XN, DX and IX, respectively. HT-29 cells were more resistant than MCF-7 cells to these flavonoids. In A-2780 cells, XN was highly antiproliferative with IC values of 0.52 and 5.2 mM after 2 and 4 days of treatment, respectively. At 100 mM, all the hop flavonoids were cytotoxic in the three cell lines.	Miranda *et al.* (1999)
Hops	12.5-100 µg/mL	Cancer	*In vitro* (rats)	Bioactive compounds obtained from hops extract significantly inhibited nitric oxide production by suppressing the expression of nitric oxide synthase.	Zhao *et al.* (2003)

(Contd.)

Table 2: *(Contd.)*

Beverage	Concentration	Disease	Assay	Effect	References
Hops	Hop flavonoids (0.01, 0.1, 1.0, 10, or 100 mM).	Cancer	*In vivo* (cells)	Both XN and IX were identified as potent inhibitors of Cyp1A activity in vitro, with IC_{50} of 0.022 mM for XN and 0.30 mM for IX. XN was found to induce the specific activity of QR in a dose-dependent manner with a CD value of 1.67 mM XN was characterized as an effective anti-inflammatory agent. It doses dependently inhibited the activity of the constitutive form of cyclooxygenase Cox-1 with an of 16.6 mM (n =2). XN also inhibited the activity of the inducible Cox-2, which is linked to carcinogenesis.	Gerhauser *et al.* (2002)
Beer	Three different types of beer. Alcohol content 2-4% (v/v)	Anti-inflammatory	*In vitro* (cells)	Compared to unstimulated cells, phytohaemagglutinin increased production of neopterin and also triggered the degradation of tryptophan. All types of beer (2-4% dilution) were found to counteract these stimulation-induced effects and significant reduction of neopterin formation and tryptophan degradation was observed.	Winkler *et al.* (2006)

PC: phenolic compounds, XN: xanthohumol, DX: dehydrocycloxanthohumol, IX: isoxanthohumol in the treatment of induced diabetic rats reduced the levels of malondialdehyde, glucose, cholesterol and triglycerides while increasing the levels of insulin, glutathione and total antioxidants in diabetic and non-diabetic rats. Moreover, sake yeast consumption showed a reduction in pro-inflammatory markers, such as IL-6, TNF-α and C-reactive protein in a dose-response manner (Davoodi *et al.*, 2019). On the other hand, koji glycosylceramide decreased the liver cholesterol of obese mice by increasing the bile acid production caused by the expression of *CYP7A1* and *ABCG8* genes (Hamajima *et al.*, 2019). Moreover, the authors pointed out that glycosylceramide may suppress obesity and metabolic disorders because it is an antigen recognised by natural killer T cells, which display this immunosuppressive cytokine IL-10 effect.

Sake cake, a by-product obtained during filtration of sake beverage, contains high levels of branched-chain amino acids. These compounds may reduce ROS production by increasing mitochondrial biogenesis and sirtuin 1 expression in cardiac and skeletal muscle of mice. An additional benefit of sake-cake consumption is the increase in vitamin B_6 and B_3 in plasma. These vitamins showed several beneficial effects on health in the metabolism of proteins, lipids, carbohydrates, neurotransmitters and nucleic acids. Thus, sake cake may be beneficial in maintaining brain and motor function in elderly people (Izu *et al.*, 2019). Similarly, the consumption of a mixture of sake cake and rice malt increased mucin levels and improved the intestinal microbiota of mice. These results suggest that sake cake and rice malt can be used against intestinal disorders because they can improve the gut barrier function by over-expressing the Muc3 mRNA. Moreover, after four weeks of sake cake and rice malt, the *Lactobacillus* population dramatically increased, indicating that this product may act as a prebiotic (sake yeast residue) for beneficial microorganisms. Health benefits of sake and its metabolites are presented in Table 3.

4.4 Cider

4.4.1 Overview

Cider is likely to be similar in age to beer and wine because there is mention of alcoholic beverages made from apples in Greek and Roman literature (McKay *et al.*, 2011). It can be defined as fermented alcoholic beverages made from apple juice (Bamforth, 2004). Ciders have been traditionally produced for years in England, France (Normandy and Brittany), northern Spain, Ireland and Germany. However recently, countries such as Argentina, Austria, Australia, Belgium, Canada, China, Finland, New Zealand, South Africa, Sweden, Switzerland, and the United States have also been incorporated in the production of ciders (Jarvis, 2014). Cider is the fermented juice of freshly crushed and pressed apples. Traditionally, apple juice fermentation is conducted by the natural microorganisms present in the fruits at harvest. Fermentation process is done at 15-25°C until all the fermentable sugars have been used (three to eight weeks, depending on temperature). After the fermentation and aging process, the blended cider can be carbonated and packaged in bottles, cans, barrels or kegs for distribution and sale, or it can be packaged as a non-carbonated product (McKay *et al.*, 2011; Jarvis, 2014).

Table 3: Beneficial Effects of Sake and Compounds Against Different Diseases

Beverage	Concentration	Disease	Assay	Effect	References
Sake yeast	15, 30 or 45 mg/kg	Diabetes mellitus type 2	*In vitro* (rats)	Sake yeast reduced the levels of malondialdehyde (4 to 2 nmol/mL), glucose (300 to 200 mg/dL), cholesterol (125 to 100 mg/dL), and triglycerides (140 to 120 mg/dL) and increased levels of insulin (30 to 50 mU/L), glutathione (1 to 3.5 nmol/mL), and total antioxidant status (1.2 to 3.5 nmol/mL).	Davoodi *et al.* (2019)
Koji	Koji glycosylceramide (1% of the total diet)	Cholesterol metabolism	*In vitro* (mice)	Koji glycosylceramide decreased the liver cholesterol (8.62 to 6.83 mg/liver).	Hamajima *et al.* (2019)
Sake cake	10% of the total diet composition	Metabolic changes	*In vitro* (mice)	Sake cake increased the final body weight (29.6 to 32.5 g) and decreased glucose (8.02 to 6.62 mmol/L) in blood. The activity of alanine aminotransferase (36.8 to 41.9U/L) and aspartate aminotransferase (185.8 to 209.1 U/L). The branched chain amino acids were increased on muscle and brain.	Izu *et al.* (2019)
Sake cake and rice malt	10% of the total diet composition	Changes in intestinal microbiota	*In vitro* (mice)	After 4-week feeding, Fecal mucin level was significantly increased (230 µg GalNAc/g feces) with the diet of a mixture of sake cake and rice malt. Proportion of *Lactobacillaceae, Porphyromonadaceae,* and *Prevotellaceae* were significantly higher in the group fed with sake cake and rice malt.	Kawakami *et al.* (2020)

4.4.2 Cider Composition and Health Beneficial Effects

The main compounds of cider are water, sugars, organic acids (lactic, acetic and malic), alcohol (1.2-8.5percentage v/v), minerals (Na, K, Ca, Mg, Cu, Fe, Mn, and Zn), vitamins (B2, B5, B6, B8, B9, and B12) and phenolic compounds, such as (+)-catechin, (−)-epicatechin, phloridzin, rutin, procyanidin B2, quercetin, quercitrin as also gallic, protocatechuic, chlorogenic, caffeic, *p*-coumaric, ferulic and cinnamic acids (Joshi *et al.*, 2017; Barreira *et al.*, 2019; Panesar *et al.*, 2017; Wei *et al.*, 2020).

It is well known that phenolic compounds of apples are responsible for bitterness, astringency, colour and may also partly contribute to the aroma of cider. For example, phenolic compounds, such as procyanidins are responsible for cider's bitterness and astringency; colour is provided by the oxidation of chlorogenic acid, procyanidins, catechin and phloridzin, while some hydroxycinnamic acids are precursors of volatile compounds responsible for the cider aroma (Herrero *et al.*, 1999; Verdu *et al.*, 2014). Phenolic compounds of apples not only provide sensory characteristics of cider but also their consumption is inversely correlated with the development of diseases, such as asthma, diabetes, cancer, inflammation, neural, pulmonary, liver and cardiovascular diseases (Hossen *et al.*, 2017). Moreover, according to Zuo *et al.* (2019), eight days of the fermentation process of apple juice to become cider increases the phenolic acids (110.0 per cent) and flavonoids (38.8 per cent). Thus, the health-beneficial effects of cider consumption may be similar or even higher than the consumption of apple. However, on the available literature, only one research about the effect of cider consumption on human health was conducted. In this study, the effect of apple-derived beverages against oxidative damage caused by H_2O_2 in SH-SY5Y and BV2 cells was evaluated. Results indicated that ciders protected the SH-SY5Y cells against oxidative injury, diminishing the ROS levels to 50-60 per cent, reducing the neuro-inflammation process in BV2 cells by increasing the levels of anti-inflammatory interleukin (IL-10) and reducing the levels of pro-inflammatory cytokine (TNF-α). During *in vivo* assays, mice were injected with LPS-induced oxidative stress and inflammation. The results indicated that cider intake increased the levels of antioxidant (SOD1, CAT, and GPx1) enzymes in the brain. Moreover, the malondialdehyde level and nitric oxide production were reduced by apple juice and ciders by reducing the expression of inducible nitric oxide synthase and nuclear factor kappa-light-chain-enhancer of activated B cells (Alvariño *et al.*, 2020). The results of this study provide evidence of the antioxidant and anti-inflammatory effects of apple-derived beverages, suggesting that their consumption may be a good approach for treating neurodegeneration diseases.

4.5 Pulque

4.5.1 Overview

Pulque is a traditional Mexican alcoholic beverage obtained by fermentation of aguamiel, a fresh sap extracted from several species of *Agave* (mainly *A. salmiana*, *A. americana* or *A. mapisaga*). Pulque was known by pre-Hispanic cultures, like *metoctli, iztacoctlli* or *poliuhquioctli;* however, the Spanish decided to call it *pulque* (Picazo *et al.*, 2019). Pulque is produced mainly from the spontaneous fermentation process of aguamiel conducted by its natural microbiota. The fermentation process of pulque is carried out in fed-bath fermentation at 25-50°C for 12 to 24 hours (Valadez-Blanco *et al.*, 2012).

In general, four steps are required for obtaining pulque: castration, pit scraping and aguamiel extraction, seed preparation and fermentation (Parsons and Darling, 2000). Mature agave plants of six to 15-years of age are used to obtain the aguamiel. The liquid is placed in cow-leather, glass-fibre, plastic or wood barrels, where inoculation (seed addition) and fermentation processes are conducted (Parsons and Darling, 2000; Ramírez-Rancaño, 2000). The fermentation process finishes when the desired alcohol grade (3-6 per cent), acetic notes and viscosity are reached (Escalante *et al.*, 2012). Generally, aguamiel fermentation is performed under non-aseptic conditions. Earlier studies about the microbiology of pulque indicated that homo- and hetero-fermentative LAB (*Lactobacillus* sp., *Leuconostoc mesenteroides*), yeast (*S. cerevisiae*) and α-proteobacteria (*Z. mobilis*) are the main organisms responsible for the fermentation process.

4.5.2 Pulque Composition and Health Beneficial Effects

Pulque is an attractive alcoholic beverage due to its chemical composition, which consists of carbohydrates (fructose, glucose, sucrose, fructooligosaccharides), free amino acids, proteins, vitamins (ascorbic acid, thiamin, riboflavin) and minerals (iron) (Backstrand *et al.*, 2002; AGARED, 2017). However, pulque has been studied due to its content of probiotic microorganisms and their effects against gastrointestinal disorders and intestinal infections (Cervantes-Contreras and Pedroza-Rodríguez, 2007; Giles-Gómez *et al.*, 2016).

Probiotic strains isolated from pulque were evaluated for their beneficial effects on human intestinal epithelial cells and immunomodulatory and anti-inflammatory properties on a mouse model (Torres-Maravilla *et al.*, 2016). Results indicated that Lactobacilli isolates displayed a higher binding capacity to intestinal epithelial (HT-29) cells; significantly suppressing the interleukin-8 secretion by the HT-29 cells, which indicate the anti-inflammatory effects of the bacteria. These results were *in vivo* corroborated with a murine colitis model induced by the dinitro-benzene sulphonic acid (DNBS). Results indicated that probiotic microorganisms decrease the intestinal hyperpermeability induced by DNBS. Moreover, levels of IFN-γ, IL-22, IL-17A, and IL-5 were significantly reduced in mice treated with isolated probiotics. On the other hand, *Leuconostoc mesenteroides* isolated from pulque displayed *in vitro* and *in vivo* gastric resistance (lysozyme, acid pH and bile acids) and antibacterial activity against *Listeria monocytogenes,* enteropathogenic *Escherichia coli, Salmonella enterica* serovar Typhi and *S. enterica* serovar Typhimurium by producing fermentative metabolites, such as lactic and acetic acids, fatty acids, hydrogen peroxide or diacetyl and antimicrobial proteins, such as bacteriocins (Giles-Gómez *et al.*, 2016). Similar results were obtained by Cervantes-Elizarrarás *et al.* (2019), who indicated that LAB isolated from pulque and aguamiel showed antimicrobial activity against *E. coli, Staphylococcus aureus* and *Helicobacter pylori*; moreover, strains can inhibit the urease activity of *H. pylori*. Health benefits of pulque and its metabolites are presented in Table 4.

Table 4: Beneficial Effects of Pulque and Compounds Against Different Diseases

Beverage	Concentration	Disease	Assay	Effect	References
Pulque	Isolated probiotic strains (1×10⁹ CFU/mL)	Anti-inflammatory activity	*In vivo* (mice)	*Lactobacillus sanfranciscensis* improved mice health by a reduction of weight loss (13%). *L. plantarum* LBH1062, *L. plantarum* LBH1063, and *L. paracasei* spp. *paracasei* LBH1065 isolated from pulque displayed a higher binding capacity to HT-29-MTX cells.	Torres-Maravilla *et al.* (2016)
Pulque	*Leuconostoc mesenteroides* (10⁹ CFU/mL)	Probiotic potential and Antibacterial activity	*In vitro*	Strain *L. mesenteroides* was resistant to lysozyme (90%/30 min), acid pH of 2.5 (75%) and bile salts (100% at concentration of 0.3 and 1 %). Isolated microorganism displayed antimicrobial activity against *Listeriamonocytogenes*, enteropathogenic *Escherichiacoli*, *Salmonellaenterica* serovar Typhi and *S. enterica* serovar Typhimurium	Giles-Gómez *et al.* (2016)
Pulque	LAB (~1.5 x 10⁸ CFU/mL)	Probiotic potential and antimicrobial effect	*In vitro*	LAB strains were sensitive to the majority of antibiotics (MIC ≤ 8 µg/mL), all LABs were resistant to vancomycin at highest concentration (MIC > 328 µg/mL), 60% of the LAB isolated (6 of 10) had inhibitory activity against *E. coli*, *S. aureus* and *H. pylori*.	Cervantes-Elizarrarás *et al.* (2019)
Pulque	1 mL of pulque	Probiotic potential	*In vitro*	*Lactobacillus casei* isolated from pulque survived to acidity, gastric juice, and bile salts (over 4 h). *L. casei* showed bile salt hydrolase activity (BSH) over primary and secondary bile salts principally to taurocholate (671.72 U/mg).	González-Vázquez *et al.* (2015)

LAB: lactic acid bacteria

4.6   Mead or Honey Wine

4.6.1   Overview

Mead or honey wine is a traditional alcoholic beverage (8-18% v/v) obtained from yeast fermentation of honey and water (Mendes-Ferreira *et al.*, 2010). This beverage is as old as other fermented drinks and probably has a venerable history as beer, cider and wine. The first mention of mead is probably found in the *Rigveda* books, from the ancient Vedic religion of India (17th century BC). Since then, mead has been known to be produced as a popular beverage in all the three continents of the Old World, especially in ancient Greece and the Celtic and Germanic kingdoms of northern and western Europe (McKay *et al.*, 2011). Mead is a popular beverage in many African countries, notably Ethiopia and South Africa; however, in recent years, the production of mead in Australia, New Zealand and the United States has increased along with the number of craft breweries and wineries. Like other alcoholic beverages, mead can be produced by natural yeasts (spontaneous fermentation) or using cultured wine yeasts; the latter method currently prevails (McKay *et al.*, 2011).

4.6.2   Mead Composition and Health Beneficial Effects

Mead contains ethanol and many other compounds, such as sugars (glucose and fructose), acids (gluconic, formic, malic, lactic acetic, citric, succinic, fumaric and propionic acids), vitamins, mineral and phenolic compounds (gallic acid, protocatechuic acid, gentisic acid, vanillic acid, caffeic acid, syringic acid, vanillin, ferulic acid and *p*-coumaric acid) (Švecová *et al.*, 2015). It is important to mention that the phenolic compounds of mead, as well as other elements of its composition, depend on the honey floral source and other compounds used in its preparation. (Gupta and Sharma, 2009).

According to the available literature, there is a lack of scientific research and epidemiological studies related to the beneficial health effects that the consumption of mead may have. However, some beneficial effects of this beverage are related to their bioactive compounds. For example, mead demonstrates a high antioxidant activity, which is usually associated with the protective effects as antithrombotic, antiischemic, antioxidant and vasorelaxant (Idris *et al.*, 2011; Khalil and Sulaiman, 2010; Cavaco and Figueira, 2016).

5.   Conclusion

Although fermented alcoholic beverages are widely consumed worldwide, their consumption is always under review because of several human health problems associated with their consumption. However, throughout the development of this chapter, it has been shown that fermented alcoholic beverages contain several health-promoting compounds that are associated with a decreased risk of different diseases. According to this review, a light to moderate consumption of fermented alcoholic beverages is recommended to obtain all the benefits from them. However, further clinical studies are needed to evaluate the real effects of fermented alcoholic beverages against different human diseases.

Abbreviations

BC – Before Christ
CAT – Catalase
DNBS – Dinitro-benzene sulfonic acid
HDL – High Density Lipoprotein
IFN-γ – Interferon gamma
IL – Interleukin
LAB – Lactic Acid Bacteria
LDL – Low Density Lipoprotein
MDA – Malondialdehyde
PBMC – Peripheral blood mononuclear cells
POD – Peroxidase
ROS – Reactive oxygen species
SOD – Superoxide dismutase
UVB – Ultraviolet B

Acknowledgement

The author Gabriela Rios-Corripio would like to thank the Secretaría de Educación Pública (SEP) for her fellowship grant (VD-C.U. 004/2020) and the Benemérita Universidad Autónoma de Puebla for the support provided during her postdoctoral studies.

References

AGARED (2017). Red temática mexicana aprovechamiento integral sustentable y biotecnología de los agaves, El pulque: bebida ancestral que perdura hasta la actualidad. *In*: *Panorama del aprovechamiento de los agaves en México*, pp. 131-157.

Akaike, M., Miyagawa, H., Kimura, Y., Terasaki, M., Kusaba, Y., Kitagaki, H. and Nishida, H. (2020). Chemical and bacterial components in sake and sake production process. *Current Microbiology*, 77: 632-637.

Akeem, S.A., Kolawole, F.L., Joseph, J.K., Kayode, R.M.O. and Akintayo, O.A. (2019). Traditional food processing techniques and micronutrients bioavailability of plant and plant-based foods: A review. *Annals of Food Science and Technology*, 20: 30-41.

Alvariño, R., Alonso, E., Alfonso, A. and Botana, L.M. (2019). Neuroprotective effects of apple-derived drinks in a mice model of inflammation. *Molecular Nutrition and Food Research*, 64: 1901017.

Apostolidou, C., Adamopoulos, K., Lymperaki, E., Iliadis, S., Papapreponis, P. and Kourtidou-Papadeli, C. (2015). Cardiovascular risk and benefits from antioxidant dietary intervention with red wine in asymptomatic hypercholesterolemics. *Clinical Nutrition ESPEN*, 10: e224-233.

Arimoto-Kobayashi, S., Inada, N., Anma, N., Shimada, H. and Hayatsu, H. (1999). Induced mutations in M13mp2 phage DNA exposed to N-nitrosopyrrolidine with UVA irradiation. *Environmental and Molecular Mutagenesis*, 34: 24-29.

Arranz, S., Chiva-Blanch, G., Valderas-Martínez, P., Medina-Remón, A., Lamuela-Raventós, R.M. and Estruch, R. (2012). Wine, beer, alcohol and polyphenols on cardiovascular disease and cancer. *Nutrients*, 4: 759-781.

Attele, A.S., Zhou, Y.-P., Xie, J.-T., Wu, J.A., Zhang, L., Dey, L., Pugh, W., Rue, P.A., Polonsky, K.S. and Yuan, C.-S. (2002). Antidiabetic effects of panax ginseng berry extract and the identification of an effective component. *Diabetes*, 51: 1851-1858.

Awuah, G., Ramaswamy, H.S. and Economides, A. (2007). Thermal processing and quality: Principles and overview. *Chemical Engineering and Processing*, 46: 584-602.

Backstrand, J.R., Allen, L.H., Black, A.K., de Mata, M. and Pelto, G.H. (2002). Diet and iron status of nonpregnant women in rural Central Mexico. *American Journal of Clinical Nutrition*, 76: 156-164.

Balansudram, N., Sundram, K. and Samman, S. (2006). Phenolic compounds in plants and agri-industrial by-products: Antioxidant activity, occurrence and potential uses. *Food Chemistry*, 99: 191-203.

Bamforth, C. (2004). *Fermented Beverage Production*. Kluwer Academic/Plenum Publishers, New York.

Barreira, J.C.M., Arraibi, A.A. and Ferreira, I.C.F.R. (2019). Bioactive and functional compounds in apple pomace from juice and cider manufacturing: Potential use in dermal formulations. *Trends in Food Science and Technology*, 90: 76-87.

Bobak, M., Skodova, Z. and Marmot, M. (2003). Beer and obesity: A cross-sectional study. *European Journal of Clinical Nutrition*, 57: 1250-1253.

Boehm, T., Folkman, J., Browder, T. and O'Reilly, M.S. (1997). Antiangiogenic therapy of experimental cancer does not induce acquired drug resistance. *Nature*, 390: 404-407.

Boulton, R. (2001). The co-pigmentation of anthocyanins and its role in the colour of red wine: A critical review. *American Journal of Enology and Viticulture*, 52: 67-87.

Buratti, S. and Benedetti, S. (2016). Alcoholic fermentation using electronic nose and electronic tongue. pp. 291-299. *In:* M.L.R. Mendez (Eds.). *Electronic Noses and Tongues in Food Science*. London: Academic Press LTD-Elsevier Science.

Castaldo, L., Narváez, A., Izzo, L., Graziani, G., Gaspari, A., Di Minno, G. and Ritieni, A. (2019). Red wine consumption and cardiovascular health. *Molecules*, 24: 3626.

Cavaco, T. and Figueira, A.C. (2016). Functional properties of honey and some traditional honey products from Portugal. pp. 339-352. *In*: K. Kristborgsson and S. Otles (Eds.). *Functional Properties of Traditional Foods*, Springer, NY.

Cavallini, G., Straniero, S., Donati, A. and Bergamini, E. (2016). Resveratrol requires red wine polyphenols for optimum antioxidant activity. *Journal of Nutrition and Health*, 20: 540-545.

Cervantes-Contreras, M. and Pedroza-Rodríguez, A.M. (2007). El pulque: Características microbiológicas y contenido alcohólico mediante espectroscopia Raman. *NOVA – Publicación Científica en Ciencias Biomédicas*, 5: 135-146.

Cervantes-Elizarrarás, A., Cruz-Cansino, N. del S., Ramírez-Moreno, E., Vega-Sánchez, V., Velázquez-Guadarrama, N., Zafra-Rojas, Q.Y. and Piloni-Martini, J. (2019). *In vitro* probiotic potential of lactic acid bacteria isolated from aguamiel and pulque and antibacterial activity against pathogens. *Applied Sciences*, 9: 601.

Chien, S.-C., Young, P.H., Hsu, Y.-J., Chen, C.H., Tien, Y.J., Shiu, S.-Y., Li, T.-H., Yang, C.-W., Marimuthu, P., Tsai, L.F.-L. and Yang, W.-C. (2009). Anti-diabetic properties of three common *Bidens pilosa* variants in Taiwan. *Phytochemistry*, 70: 1246-1254.

Chiva-Blanch, G., Condines, X., Magraner, E., Roth, I., Valderas-Martínez, P., Arranz, S., Casas, R., Martínez-Huélamo, M., Vallverdú-Queralt, A., Quifer-Rada, P., Lamuela-Raventos, R.M. and Estruch, R. (2014). The non-alcoholic fraction of beer increases stromal cell derived Factor 1 and the number of circulating endothelial progenitor cells in high cardiovascular risk subjects: A randomised clinical trial. *Atherosclerosis*, 233: 518-524.

Chiva-Blanch, G., Magraner, E., Condines, X., Valderas-Martínez, P., Roth, I., Arranza, S., Casas, R., Navarro, M., Hervas, A., Sisó, A., Martínez-Huélamo, M., Vallverdú-Queralt, A., Quifer-Rada, P., Lamuela-Raventos, R.M. and Estruch, R. (2015). Effects of alcohol and polyphenols from beer on atherosclerotic biomarkers in high cardiovascular risk men: A randomized feeding trial. *Nutrition, Metabolism and Cardiovascular Diseases*, 25: 36-45.

Chiva-Blanch, G., Urpi-Sarda, M., Ros, E., Arranz, S., Valderas-Martínez, P., Casas, R., Sacanella, E., Llorach, R., Lamuela-Raventos, R.M., Andres-Lacueva, C. and Estruch, R. (2012). Dealcoholised red wine decreases systolic and diastolic blood pressure and increases plasma nitric oxide. *Circulation Research*, 28: 1065-1068.

Chung, W.-G., Miranda, C.L., Stevens, J.F. and Maier, C.S. (2009). Hop proanthocyanidins induce apoptosis, protein carbonylation, and cytoskeleton disorganization in human colorectal adenocarcinoma cells via reactive oxygen species. *Food and Chemical Toxicology*, 47: 827-836.

Ciubara, A.B., Tudor, R.C., Nechita, L., Tita, O., Ciubara, A., Turliuc, S. and Raftu, G. (2018). The composition of bioactive compounds in wine and their possible influence on osteoporosis and on bone consolidation. *Revista de Chimie*, 69: 1247-1253.

Corpas, R., Griñán-Ferré, C., Rodríguez-Farré, E., Pallàs, M. and Sanfeliu, C. (2019). Resveratrol induces brain resilience against Alzheimer neurodegeneration through proteostasis enhancement. *Molecular Neurobiology*, 56: 1502-1516.

Dabina-Bicka, I., Karklina, D., Kruma, Z. and Dimins, F. (2013). Bioactive compounds in Latvian beer. *Proceedings of the Latvia University of Agriculture*, 30: 325.

Damianaki, A., Bakogeorgou, E., Kampa, M., Notas, G., Hatzoglou, A., Panagiotou, S., Gemetzi, C., Kouroumalis, E., Martin, P.-M. and Castanas, E. (2000). Potent inhibitory action of red wine polyphenols on human breast cancer cells. *Journal of Cellular Biochemistry*, 78: 429-441.

Davoodi, M., Karimooy, F.N., Budde, T., Ortega-Martinez, S. and Moradi-Kor, N. (2019). Beneficial effects of Japanese sake yeast supplement on biochemical, antioxidant, and anti-inflammatory factors in streptozotocin-induced diabetic rats. *Diabetes, Metabolic Syndrome and Obesity: Targets and Therapy*, 12: 1667-1673.

DeFuria, J., Bennett, G., Strissel, K.J., Perfield, J.W., Milbury, P.E., Greenberg, A.S. and Obin, M.S. (2009). Dietary blueberry attenuates whole-body insulin resistance in high fat-fed mice by reducing adipocyte death and its inflammatory sequelae. *The Journal of Nutrition*, 139: 1510-1516.

de Gaetano, G., Costanzo, S., Di Castelnuovo, A., Badimon, L., Bejko, D., Alkerwi, A., Chiva-Blanch, G., Estruch, R., La Vecchia, C., Panico, S., Pounis, G., Sofi, F., Stranges, S., Trevisan, M., Ursini, F., Cerletti, C., Donati, M.B. and Lacoviello, L. (2016). Effects of moderate beer consumption on health and disease: A consensus document. *Nutrition, Metabolism and Cardiovascular Diseases*, 26: 443-467.

Del Pino-García, R., González-SanJosé, M.L., Rivero-Pérez, M.D., García-Lomillo, J. and Muñiz, P. (2017). The effects of heat treatment on the phenolic composition and antioxidant capacity of red wine pomace seasonings. *Food Chemistry*, 15: 1723-1732.

Duan, J., Zhan, J.-C., Wang, G.-Z., Zhao, X.-C., Huang, W.-D. and Zhou, G.-B. (2018). The red wine component ellagic acid induces autophagy and exhibits anti-lung cancer activity *in vitro* and *in vivo*. *Journal of Cellular and Molecular Medicine*, 23: 143-154.

Escalante, A., Giles-Gómez, M., Esquivel-Flores, G., Matus-Acuña, V., Moreno-Terrazas, R. and López-Munguía, A. (2012). Pulquefermentation. pp. 691-706. *In*: Y.H. Hui (Eds.). *Handbook of Plant-based Fermented Food and Beverage Technology*. Boca Raton, FL, CRC Press.

Escalante, A., Rodríguez, M.E., Martínez, A., López-Munguía, A., Bolívar, F. and Gosset, G. (2004). Characterisation of bacterial diversity in pulque, a traditional Mexican alcoholic fermented beverage, as determined by 16S rDNA analysis. *FEMS Microbiology Letters*, 235: 273-279.

Fernández-Mar, M.I., Mateos, R., García-Parrilla, M.C., Puertas, B. and Cantos-Villar, E. (2012). Bioactive compounds in wine: Resveratrol, hydroxytyrosol and melatonin: A review. *Food Chemistry*, 130: 797-813.

Fu, Z., Zhang, W., Zhen, W., Lum, H., Nadler, J., Bassaganya-Riera, J., Zhenquan, J., Wang, Y., Misra, H. and Liu, D. (2010). Genistein induces pancreatic β-Cell proliferation through activation of multiple signaling pathways and prevents insulin-deficient diabetes in mice. *Endocrinology*, 151: 3026-3037.

Fujii, W., Hori, H., Yokoo, Y., Suwa, Y., Nukaya, H. and Taniyama, K. (2002). Beer congener stimulates gastrointestinal motility via the muscarinic acetylcholine receptors. *Alcoholism: Clinical and Experimental Research*, 26: 677-681.

Garduño-García, A., López-Cruz, I.L., Ruíz-García, A. and Martínez-Romero, S. (2014). Simulación del proceso de fermentación de cerveza artesanal. *Ingeniería, Investigación y Tecnología*, 15: 221-232.

García, M.C., Puchalska, P., Esteve, C. and Marina, M.L. (2013). Vegetable foods: A cheap source of proteins and peptides with antihypertensive, antioxidant, and other less occurrence bioactivities. *Talanta*, 106: 328-349.

Gerhäuser, C. (2005). Beer constituents as potential cancer chemopreventive agents. *European Journal of Cancer*, 41: 1941-1954.

Gerhauser, C., Alt, A., Heiss, E., Gamal-Eldeen, A., Klimo, K., Knauft, J., Neumann, I., Scherf, H.R., Frank, N., Bartsch, H. and Becker, H. (2002). Cancer chemopreventive activity of xanthohumol, a natural product derived from hop. *Molecular Cancer Therapeutics*, 1: 959-969.

Giles-Gómez, M., Sandoval-García, J.G., Matus, V., Campos-Quintana, I., Bolívar, F. and Escalante, A. (2016). *In vitro* and *in vivo* probiotic assessment of *Leuconostoc mesenteroides* P45 isolated from pulque, a Mexican traditional alcoholic beverage. *Springer Plus*, 5: 708.

González-Aguilar, G.A., Robles-Sánchez, R.M., Martínez-Téllez, M.A., Olivas, G.I., Alvarez-Parrilla, E. and de la Rosa, L.A. (2014). Bioactive compounds in fruits: Health benefits and effect of storage conditions. *Stewart Postharvest Review*, 3: 1-10.

González-Vázquez, R., Azaola-Espinosa, A., Mayorga-Reyes, L., Reyes-Nava, L.A., Shah, N.P. and Rivera-Espinoza, Y. (2015). Isolation, identification and partial characterisation of a *Lactobacillus casei* strain with bile salt hydrolase activity from pulque. *Probiotics and Antimicrobial Proteins*, 7: 242-248.

Gorinstein, S., Caspi, A., Libman, I., Leontowicz, H., Leontowicz, M., Tashma, Z., Katrich, E., Jastrzebski, Z. and Trakhtenberg, S. (2007). Bioactivity of beer and its influence on human metabolism. *International Journal of Food Sciences and Nutrition*, 58: 94-107.

Gorinstein, S., Leontowicz, H., Lojek, A., Leontowicz, M., Číž, M., Gonzalez Stager, M.A., Bastias, M.J.M., Toledo, F., Arancibia-Avila, P. and Trakhtenberg, S. (2002). Hypolipidemic effect of beer proteins in experiment on rats. *LWT–Food Science and Technology*, 35: 265-271.

Gupta, J.K. and Sharma, R. (2009). Review paper production technology and quality characteristics of mead and fruit-honey wines: A review technology of mead. *Natural Product Radiance*, 8: 345-355.

Hamajima, H., Tanaka, M., Miyagawa, M., Sakamoto, M., Nakamura, T., Yanagita, T., Nishimukai, M., Mitsutake, S., Nakayama, J., Nagao, K. and Kitagaki, H. (2019). Kojiglycosylceramide commonly contained in Japanese traditional fermented foods alters cholesterol metabolism in obese mice. *Bioscience, Biotechnology, and Biochemistry*, 83: 1514-1522.

Haseeb, S., Alexander, B. and Baranchuk, A. (2017). Wine and cardiovascular health: A comprehensive review. *Circulation*, 136: 1434-1448.

Hernández-Carranza, P., Jattar-Santiagoa, K.J., Avila-Sosa, R., Pérez-Xochipa, I., Guerrero-Beltrán, J.A., Ochoa-Velasco, C.E. and Ruiz-López, I.I. (2019). Antioxidant fortification of yogurt with red cactus pear peel and its mucilage. *CYTA – Journal of Food*, 17: 824-833.

Herrero, M., García, L.A. and Díaz, M. (2006). Volatile compounds in cider: Inoculation time and fermentation temperature effects. *Journal of the Institute of Brewing*, 112: 210-214.

Hirotsune, M., Haratake, A., Komiya, A., Sugita, J., Tachihara, T., Komai, T., Hizume, K., Ozeki, K. and Ikemoto, T. (2005). Effect of ingested concentrate and components of sake on epidermal permeability barrier disruption by UVB irradiation. *Journal of Agricultural and Food Chemistry*, 53: 948-952.

Hossen, M.S., Ali, M.Y., Jahurul, M.H.A., Abdel-Daim, M.M., Gan, S.H. and Khalil, M.I. (2017). Beneficial roles of honey polyphenols against some human degenerative diseases: A review. *Pharmacological Reports*, 69: 1194-1205.

Idris, Y.M.A., Mariod, A.A. and Hamad, S.I. (2011). Physicochemical properties, phenolic contents and antioxidant activity of Sudanese honey. *International Journal of Food Properties*, 14: 450-458.

Iso, H., Date, C., Wakai, K., Fukui, M., Tamakoshi, A. and the JACC Study Group. (2006). The relationship between green tea and total caffeine intake and risk for self-reported Type 2 diabetes among Japanese adults. *Annals of Internal Medicine*, 144: 554-562.

Izu, H., Shibata, S., Fujii, T. and Matsubara, K. (2019). Sake cake (sake-kasu) ingestion increases branched-chain amino acids in the plasma, muscles, and brains of senescence-accelerated mice prone 8. *Bioscience, Biotechnology, and Biochemistry*, 83: 1490-1497.

Jain, M.G., Hislop, G.T., Howe, G.R., Burch, J.D. and Ghadirian, P. (1998). Alcohol and other beverage use and prostate cancer risk among Canadian men. *International Journal of Cancer*, 78: 707-711.

Japan Sake and Shochu Makers Association. (2011). A Comprehensive Guide to Japanese Sake. nrib.go.jp/English/sake/pdf/guidesse01.pdf. accessed on March 17, 2020

Jarvis, B. (2014). Cider (Cyder; Hard Cider). *Encyclopedia of Food Microbiology*, 437-443.

Jermann, C., Koutchma, T., Margas, E., Leadley, C. and Ros-Polski, V. (2015). Mapping trends in novel and emerging food processing technologies around the world. *Innovative Food Science and Emerging Technologies*, 31: 14-27.

Ji, S., Zheng, Z., Liu, S., Ren, G., Gao, J., Zhang, Y. and Li, G. (2018). Resveratrol promotes oxidative stress to drive DLC1 mediated cellular senescence in cancer cells. *Experimental Cell Research*, 370: 292-302.

Jomova, K. and Valko, M. (2013). Health protective effects of carotenoids and their interactions with other biological antioxidants. *European Journal of Medicinal Chemistry*, 70: 102-110.

Joshi, V.K., Panesar, P.S., Rana, V.S. and Kaur, S. (2017). Science and technology of fruit wines: An overview. pp. 1-72. *In*: M. Kosseva, V.K. Joshi and P.S. Panesar (Eds.). *Science and Technology of Fruit Wine Production*. Academic Press, London.

Kawakami, S., Ito, R., Maruki-Uchida, H., Kamei, A., Yasuoka, A., Toyoda, T., Ishijima, T., Nishimura, E., Morita, M., Sai, M., Abe, K. and Okada, S. (2020). Intake of a mixture of sake cake and rice malt increases mucin levels and changes in intestinal microbiota in mice. *Nutrients*, 12: 449.

Khalil, M. and Sulaiman, S. (2010). The potential role of honey and its polyphenols in preventing heart disease: A review. *African Journal of Traditional, Complementary and Alternative Medicines*, 7: 315-321.

Kitagaki, H. and Kitamoto, K. (2013). Breeding research on sake yeasts in Japan: History, recent technological advances, and future perspectives. *Annual Review of Food Science and Technology*, 4: 215-235.

Kodera, T., Yamada, S., Yamamoto, Y., Hara, A., Tanaka, Y., Seno, M., Umezawa, K., Tekei, I. and Kojima, I. (2009). Administration of conophylline and betacellulin-δ4 increases the β-cell mass in neonatal sptreptozotocin-treated rats. *Endocrine Journal*, 56: 799-806.

Kondo, K. (2004). Beer and health: Preventive effects of beer components on lifestyle-related diseases. *BioFactors*, 22: 303-310.

Krinsky, N.I. (2001). Carotenoids as antioxidants. *Nutrition*, 17: 815-817.

Kumar-Anal, A. (2019). Quality ingredients and safety concerns for traditional fermented foods and beverages from Asia: A review. *Fermentation*, 5: 8.

Lee, D.C., Sui, X., Artero, E.G., Lee, I.M., Church, T.S., McAuley, P.A., Stanford, F.C., Kohl, H.W. and Blair, S.N. (2011). Long-term effects of changes in cardiorespiratory fitness and body mass index on all-cause and cardiovascular disease mortality in men: The aerobics centre longitudinal study. *Circulation*, 124: 2483-2490.

Leighton, F. and Urquiaga, I. (2007). Changes in cardiovascular risk factors associated with wine consumption in intervention studies in humans. *Annals of Epidemiology*, 17: S32-S36.

Lippi, G., Franchini, M., Favaloro, E.J. and Targher, G. (2010). Moderate red wine consumption and cardiovascular disease risk: Beyond the 'French Paradox'. *Seminars in Thrombosis and Hemostasis*, 36: 59-70.

Liu, M., Wu, K., Mao, X., Wu, Y. and Ouyang, J. (2010). Astragalus polysaccharide improves insulin sensitivity in KKAy mice: Regulation of PKB/GLUT4 signalling in skeletal muscle. *Journal of Ethnopharmacology*, 127: 32-37.

Luna-Guevara, M.L., Luna Guevara, J.J., Hernandez-Carranza, P., Ruíz-Espinosa, H. and Ochoa-Velasco, C.E. (2019). Phenolic compounds: A good choice against chronic degenerative diseases. pp. 71-108. *In*: Atta-ur-Rhahman, F.R.S. (Ed.). *Studies in Natural Products Chemistry*, Karachi, Pakistan.

Mani, A. (2018). Food preservation by fermentation and fermented food products. *International Journal of Academic Research and Development*, 1: 51-57.

Mankowski, R.T., You, L., Buford, T.W., Leeuwenburgh, C., Manini, T.M., Schneider, S., Qiu, P. and Anton, S.D. (2020). Higher dose of resveratrol elevated cardiovascular disease risk biomarker levels in overweight older adults – A pilot study. *Experimental Gerontology*, 131: 110821.

Marques, D.R., Cassis, M.A., Quelhas, J.O.F., Bertozzi, J., Visentainer, J.V., Oliveira, C.C. and Monteiro, A.R.G. (2017). Characterisation of craft beers and their bioactive compounds. *Chemical Engineering Transactions*, 57: 1747-1752.

Mayer Jr., O., Simon, J. and Rosolová, H. (2001). A population study of the influence of beer consumption on folate and homocysteine concentrations. *European Journal of Clinical Nutrition*, 55: 605-609.

Mayne, S.T. (1996). Beta-carotene, carotenoids, and disease prevention in humans. *The FASEB Journal*, 10: 690-701.

McKay, M., Buglass, A.J. and Lee, C.G. (2011). Fermented beverages: Beers, ciders, wines and related drinks. pp. 63-263. *In*: A.J. Buglass (Ed.). *Handbook of Alcoholic Beverages: Technical, Analytical and Nutritional Aspects*. John Wiley & Sons.

Mendes-Ferreira, A., Cosme, F., Barbosa, C., Falco, V., Inês, A. and Mendes-Faia, A. (2010). Optimization of honey-must preparation and alcoholic fermentation by *Saccharomyces cerevisiae* for mead production. *International Journal of Food Microbiology*, 144: 193-198.

Mgbonyebi, O.P., Russo, J. and Russo, I.H. (1998). Antiproliferative effect of synthetic resveratrol on human breast epithelial cells. *International Journal of Oncology*, 12: 865-869.

Miranda, C.L., Stevens, J.F., Helmrich, A., Henderson, M.C., Rodriguez, R.J., Yang, Y.H., Deinzer, M.L., Barnes, D.W. and Buhler, D.R. (1999). Antiproliferative and cytotoxic effects of prenylated flavonoids from hops (*Humulus lupulus*) in human cancer cell lines. *Food and Chemical Toxicology*, 37: 271-285.

Mokni, M., Hamlaoui, S., Karkouch, I., Amri, M., Marzouki, L., Limam, F. and Aouani, E. (2013). Resveratrol provides cardioprotection after ischemia/reperfusion injury via modulation of antioxidant enzyme activities. *Iranian Journal of Pharmaceutical Research*, 12: 867-875.

Mokni, M., Limam, F., Elkahoui, S., Amri, M. and Aouani, E. (2007). Strong cardioprotective effect of resveratrol, a red wine polyphenol, on isolated rat hearts after ischemia/reperfusion injury. *Archives of Biochemistry and Biophysics*, 457: 1-6.

Müller-Schwarze, D. (2009). Effect of tannins on insect feeding behaviour. *Hands-on Chemical Ecology*, 101-104.

Naczk, M. and Shahidi, F. (2006). Phenolics in cereals, fruits and vegetables: Occurrence, extraction and analysis. *Journal of Pharmeutical and Biomedical Analysis*, 41: 1523-1542.

Nagendra, P.M.N., Sanjay, K.R., Khatokar, M.S., Vismaya, M.N. and Swamy, S.N. (2011). Health benefits of rice bran – A review. *Journal of Nutrition & Food Sciences*, 1-7.

Napoli, C., Balestrieri, M.L., Sica, V., Lerman, L.O., Crimi, E., De Rosa, G., Schiano, C., Servillo, L. and D'Armiento, F.P. (2008). Beneficial effects of low doses of red wine consumption on perturbed shear stress-induced atherogenesis. *Heart Vessels*, 23: 124-133.

Navarro-Cruz, A.R., Ramírez, Y., Ayala, R., Ochoa-Velasco, C., Brambila, E., Avila-Sosa, R., Pérez-Fernández, S., Morales-Medina, J.C. and Aguilar-Alonso, P. (2017). Effect of chronic administration of resveratrol on cognitive performance during aging process in rats. *Oxidative Medicine and Cellular Longevity*, 2017: 8510761.

Nishino, H., Tokuda, H., Satomi, Y., Masuda, M., Osaka, Y., Yogosawa, S., Wada, S., Mou, X.Y., Takayasu, J., Murakoshi, M., Jinnno, K. and Yano, M. (2004). Cancer prevention by antioxidants. *BioFactors*, 22: 57-61.

Nogueira, A., Guyot, S., Marnet, N., Lequéré, J.N., Drilleau, J.F. and Wosiacki, G. (2008). Effect of alcoholic fermentation in the content of phenolic compounds in cider processing. *Brazilian Archives of Biology and Technology*, 51: 1025-1032.

Nozawa, H., Tazumi, K., Sato, K., Yoshida, A., Takata, J., Arimoto-Kobayashi, S. and Kondo, K. (2004). Inhibitory effects of beer on heterocyclic amine-induced mutagenesis and PhIP-induced aberrant crypt foci in rat colon. *Mutation Research*, 559: 177-187.

Ohta, T., Ikeda, Y., Nakamura, R. and Matsuda, I. (1994). Inverse relation of plasma concentrations of apo A-I-containing lipoprotein with no apo A-II (LpA-I) to plasma cholesterol esterification rates and cellular cholesterol reducing capacity of LpA-I. *Atherosclerosis*, 109(1-2): 226.

Oh, W.K., Lee C.H., Lee, M.S., Bae, E.Y., Shon, C.B., Oh, H., Kim, B.Y. and Ahh, J.S. (2005). Antidiabetic effects of extracts from *Psidium guajava*. *Journal of Enthnopharmacology*, 96: 411-415.

Okabe, M., Ikawa, M., Kominami, K., Nakanishi, T. and Nishimune, Y. (1997). 'Green mice' as a source of ubiquitous green cells. *FEBS Letters*, 5: 313-319.

Ozcan, T., Akpinar-Bayizit, A., Yilmaz-Ersan, L. and Delikanli, B. (2014). Phenolics in human health. *International Journal of Chemical Engineering and Applications*, 5: 393-396.

Padro, T., Muñoz-García, N., Vilahur, G., Chagas, P., Deyà, A., Antonijoan, R.M. and Badimon, L. (2018). Moderate beer intake and cardiovascular health in overweight individuals. *Nutrients*, 10: 37.

Panaro, M.A., Carofiglio, V., Acquafredda, A., Cavallo, P. and Cianciulli, A. (2012). Anti-inflammatory effects of resveratrol occur via inhibition of lipopolysaccharide-induced NF-κB activation in Caco-2 and SW480 human colon cancer cells. *British Journal of Nutrition*, 108: 1623-1632.

Pandey, K.B. and Rizvi, S.I. (2009). Plant polyphenols as dietary antioxidants in human health and disease. *Oxidative Medicine and Cellular Longevity,* 2: 270-278.

Panesar, P.S., Joshi, V.K., Bali, V. and Panesar, R. (2017). Technology for production of fortified and sparkling fruit wines. pp. 487-530. *In*: M. Kosseva, V.K. Joshi and P.S. Panesar (Eds.). *Science and Technology of Fruit Wine Production*. Academic Press, London.

Parsons, R.J. and Darling, J.A. (2000). Maguey (Agave spp.) utilisation in mesoamerican civilisation: A case for precolumbian 'pastoralism'. *Boletín de la Sociedad Botánica de México*, 66: 81-91.

Patterson, T. (2010). *Home Winemaking for Dummies*. John Wiley & Sons Ltd., Canada.

Peng, L. and Jang, D. (2018). Resveratrol eliminates cancer stem cells of osteosarcoma by STAT3 pathway inhibition. *PLoS ONE*, 1-13.

Picazo, B., Flores, G.A.C. and Aguilar, C.N. (2019). Aguamiel: Traditional Mexican product as a nutritional alternative. pp. 189-216. *In*: M. Chávez-González, J.J. Buenrostro-Figueroa and C.N. Aguilar (Eds.). *Handbook of Research on Food Science and Technology*, CRC Press.

Poelmans, E. and Swinnen, J.F.M. (2011). A brief economic history of beer. pp. 3-28. *In:* J.F.M. Swinnen (Ed.). *The Economics of Beer*. Oxford: Oxford University Press.

Priyadarshini, A., Rajauria, G., O'Donnell, C.P. and Tiwari, B.K. (2019). Emerging food processing technologies and factors impacting their industrial adoption. *Critical Reviews in Food Science and Nutrition*, 59: 3082-3101.

Quintieri, A.M., Baldino, N., Filice, E., Seta, L., Vitetti, A., Tota, B., De Cindio, B., Cerra, M.C. and Angelone, T. (2013). Malvidin, a red wine polyphenol, modulates mammalian myocardial and coronary performance and protects the heart against ischemia/reperfusion injury. *Journal of Nutritional Biochemistry*, 24: 1221-1231.

Rai, A.K., Pandey, A. and Sahoo, D. (2019). Biotechnological potential of yeasts in functional food industry. *Trends in Food Science and Technology*, 83: 129-137.

Rai, A.K., Prakash, M. and Appaiah, A.K. (2010). Production of *Garcinia* wine: Changes in biochemical parameters, organic acids and free sugars during fermentation of *Garcinia* must. *International Journal of Food Science and Technology*, 45: 1339-1336.

Rai, A.K. and Appaiah, A.K. (2014). Application of native yeast from Garcinia (*Garcinia xanthochumus*) for the preparation of fermented beverage: Changes in biochemical and antioxidant properties. *Food Bioscience*, 5: 101-107.

Ramadori, G., Gautron, L., Fujikawa, T., Vianna, C.R., Elmquist, J.K. and Coppari, R. (2009). Central administration of resveratrol improves diet-induced diabetes. *Endrocrinology*, 150: 5326-5333.

Raman, A. and Lau, C. (1996). Anti-diabetic properties and phytochemistry of *Momordica charantia* L. (Cucurbitaceae), *Phytomedicine*, 2: 349-362.

Ramírez-Rancaño, M. (2000). El rey del pulque: Ignacio Torres Adalid y la industria pulquera, México: Plaza y Valdés, S.A. de C.V. UNAM, Instituto de Investigaciones Sociales.

Riccioni, G. (2009). Carotenoids and cardiovascular disease. *Current Atherosclerosis Reports*, 11: 434-439.

Rodríguez-Ramiro, I., Ramos, S., Bravo, L., Goya, L. and Martín, M.Á. (2012). Procyanidin B2 induces Nrf2 translocation and glutathione S-transferase P1 expression via ERKs and p38-MAPK pathways and protect human colonic cells against oxidative stress. *European Journal of Nutrition*, 51: 881-892.

Rotondo, S., Di Castelnuovo, A. and de Gaetano, G. (2001). The relationship between wine consumption and cardiovascular risk: From epidemiological evidence to biological plausibility. *Italian Heart Journal*, 2: 1-8.

Ruitenberg, A., van Swieten, J.C., Witteman, J.C.M., Metha, K.M., van Duijn, C.M., Hofman, A. and Breteler, M.M.B. (2002). Alcohol consumption and risk of dementia: The Rotterdam study. *Lancet*, 359: 281-286.

Ruíz, G., Valencia-González, H.A., León-Galicia, I., García-Villa, E., García-Carrancá, A. and Gariglio, P. (2018). Inhibition of RAD51 by siRNA and resveratrol sensitises cancer stem cells derived from HeLa cell cultures to apoptosis. *Stem Cells International*, 1-11.

Sánchez-Marroquín, A. and Hope, P.H. (1953). Agave juice, fermentation and chemical composition studies of some species. *Journal of Agricultural and Food Chemistry*, 1: 246-249.

Şanlier, N., Gökcen, B.B. and Sezgin, A.C. (2019). Health benefits of fermented foods. *Critical Reviews in Food Science and Nutrition*, 59: 506-527.

Septembre-Malaterre, A., Remize, F. and Poucheret, P. (2018). Fruits and vegetables, as a source of nutritional compounds and phytochemicals: Changes in bioactive compounds during lactic fermentation. *Food Research International*, 104: 86-99.

Shashirekha, M.N., Mallikarjuna, S.E. and Rajarathnam, S. (2016). Status of bioactive compounds in foods, with focus on fruits and vegetables. *Critical Reviews in Food Science and Nutrition*, 55: 1324-1339.

Shimamura, M., Hazato, T., Ashino, H., Yamamoto, Y., Iwasaki, E., Tobe, H., Yamamoto, K. and Yamamoto, S. (2001). Inhibition of angiogenesis by humulone, a bitter acid from beer hop. *Biochemical and Biophysical Research Communications*, 289: 220-224.

Shukla, M., Gupta, K., Rasheed, Z., Khan, K.A. and Haqqi, T.M. (2008). Consumption of hydrolyzable tannins-rich pomegranate extract suppresses inflammation and joint damage in rheumatoid arthritis. *Nutrition*, 24(7-8): 733-743.

Silva, P. and Vauzour, D. (2018). Wine polyphenols and neurodegenerative diseases: An update on the molecular mechanisms underpinning their protective effects. *Beverages*, 4: 96.

Singh, J., Cumming, E., Manoharan, G., Kalasz, H. and Adeghate, E. (2011). Medicinal chemistry of the anti-diabetic effects of *Momordica charantia*: Active constituents and modes of actions. *The Open Medicinal Chemistry Journal*, 5: 70-77.

Snopek, L., Mlcek, J., Sochorova, L., Baron, M., Hlavacova, I., Jurikova, T., Kizek, R., Sedlackova, E. and Sochor, J. (2018). Contribution of red wine consumption to human health protection. *Molecules*, 23: 1684.

Sohrabvandi, S., Mortazavian, A.M. and Rezaei, K. (2012). Health-related aspects of beer: A review. *International Journal of Food Properties*, 15: 350-373.

Sousa, G.M.J. and Mahajan, P.V. (2011). Stability and shelf-life of fruit and vegetables. pp. 641-656. *In:* Kilcast and Subramaniam (Eds.). *Food and Beverage Stability and Shelf-Life*. Woodhead Publishing Ltd, Cambridge UK.

Srinivasan, M., Sudheer, A.R. and Menon, V.P. (2007). Ferulic acid: Therapeutic potential through its antioxidant property. *Journal of Clinical Biochemistry and Nutrition*, 40: 92-100.

Stahl, W. and Sies, H. (2004). Antioxidant activity of carotenoids. *Molecular Aspects of Medicine*, 24: 345-351.

Svärdsudd, K. (1998). Moderate alcohol consumption and cardiovascular disease: Is there evidence for a preventive effect? *Alcoholism: Clinical and Experimental Research*, 22: 307S-314S.

Švecová, B., Bordovská, M., Kalvachová, D. and Hájek, T. (2015). Analysis of Czech meads: Sugar content, organic acids content and selected phenolic compounds content. *Journal of Food Composition and Analysis*, 38: 80-88.

Swiegers, J.H., Bartowsky, E.J., Henschke, P.A. and Pretorius, I.S. (2005). Yeast and bacterial modulation of wine aroma and flavour. *Australian Journal of Grape and Wine Research*, 11: 139-173.

Szkudelska, K. and Szkudelski, T. (2010). Resveratrol, obesity and diabetes. *European Journal of Pharmacology*, 10: 1-3.

Szkudelski, T. and Szkudelska, K. (2015). Resveratrol and diabetes: From animal to human studies. *Biochimica et Biophysica Acta*, 1852: 1145-1154.

Tapsell, L.C., Neale, E.P., Satija, A. and Hu, F.B. (2016). Foods, nutrients, and dietary patterns: Interconnections and implications for dietary guidelines. *Advances in Nutrition*, 7: 445-454.

Tenta, R., Fragopoulou, E., Tsolukala, M., Xanthopoulou, M., Skyrianou, M., Pratsinis, H. and Kletsas, D. (2017). Antiproliferative effects of red and white wine extracts in PC-3 prostate cancer cells. *Nutrition and Cancer*, 69: 952-961.

Thakkar, K.N., Mhatre, S.S. and Parikh, R.Y. (2010). Biological synthesis of metallic nanoparticles. *Nanomedicine: Nanotechnology Biology Medicine*, 6: 257-262.

Tobe, H., Muraki, Y., Kitamura, K., Komiyama, O., Sato, Y., Sugioka, T., Maruyama, H.B., Matsuda, E. and Nagai, M. (1997). Bone resorption inhibitors from hop extract. *Bioscience, Biotechnology and Biochemistry*, 61: 158-159.

Torres-Maravilla, E., Lenoir, M., Mayorga-Reyes, L., Allain, T., Sokol, H., Langella, P., Sánchez-Pardo, M.E. and Bermúdez-Humarán, L.G. (2016). Identification of novel anti-inflammatory probiotic strains isolated from pulque. *Applied Microbiology and Biotechnology*, 100: 385-396.

Toth, A., Sandor, B., Papp, J., Rabai, M., Botor, D., Horvath, Zs., Kenyeres, P., Juricskay, I., Toth, K. and Czopf, L. (2014). Moderate red wine consumption improves haemorheological parameters in healthy volunteers. *Clinical Haemorheology and Microcirculation*, 56: 13-23.

Tourrel, C., Bailbé, D., Meile, M.J., Kergoat, M. and Portha, B. (2001). Glucagon-like peptide-1 and exendin-4 stimulate beta-cell neogenesis in streptozotocin-treated newborn rats resulting in persistently improved glucose homeostasis at adult age. *Diabetes*, 50: 1562-1570.

Ugliano, M., Bartowsky, E.J., McCarthy, J., Moio, L. and Henschke, P.A. (2006). Hydrolysis and transformation of grape glycosidically bound volatile compounds during fermentation with three *Saccharomyces* yeast strains. *Journal of Agricultural and Food Chemistry*, 54: 6322-6331.

USDA. (2015). National Nutrient Database for Standard Reference. Release 27. U.S. Department of Agriculture. Agricultural Research Service 2015. http://ndb.nal.usda.gov/ndb/foods (accessed March 2020)

Valadez-Blanco, R., Bravo-Villa, G., Santos-Sánchez, N.F., Velasco-Almendarez, S.I. and Montville, T.J. (2012). The artisanal production of pulque, a traditional beverage of the Mexican highlands. *Probiotics and Antimicrobial Proteins*, 4: 140-144.

Verardo, V., Gómez-Caravaca, A.M. and Tabanelli, G. (2020). Bioactive components in fermented foods and food by-products. *Foods*, 9: 153.

Verdu, C., Gatto, J., Freuze, I., Richomme, P., Laurens, F. and Guilet, D. (2013). Comparison of two methods, UHPLC-UV and UHPLC-MS/MS, for the quantification of polyphenols in cider apple juices. *Molecules*, 18: 10213-10227.

Verni, M., Verardo, V. and Giuseppe, R.C. (2019). How fermentation affects the antioxidant properties of cereals and legumes. *Foods*, 8: 362.

Vidavalur, R., Otani, H., Singal, P.K. and Maulik, N. (2006). Significance of wine and resveratrol in cardiovascular disease: French paradox revisited. *Experimental and Clinical Cardiology*, 11: 217-225.

Vieira, H.B., Silva, S.K., Mendonça, B.A., Sawata, M., da Costa, M.M. and Ferreira, P.F. (2019). Beer molecules and its sensory and biological properties: A review. *Molecules*, 24: 1568.

Vinson, J.A., Mandarano, M., Hirst, M., Trevithick, J.R. and Bose, P. (2003). Phenol antioxidant quantity and quality in foods: Beers and the effect of two types of beer on an animal model of atherosclerosis. *Journal of Agricultural and Food Chemistry*, 51: 5528-5533.

Vogel, R.A. (2002). Alcohol, heart disease, and mortality: A review. *Reviews in Cardiovascular Medicine*, 3: 7-13.

Waffo-Téguo, P., Hawthorne, M.E., Cuendet, M., Mérillon, J.-M., Kinghorn, A.D., Pezzuto, J.M. and Mehta, R.G. (2001). Potential cancer-chemopreventive activities of wine stilbenoids and flavans extracted from grape (*Vitis vinifera*) cell cultures. *Nutrition and Cancer*, 40: 173-179.

Wang, P. and Sang, S. (2018). Metabolism and pharmacokinetics of resveratrol and pterostilbene. *BioFactors*, 44: 16-25.

Wang, Z., Huang, Y., Zoul, J., Cao, K., Xu, Y. and Wu, J.M. (2002). Effects of red wine and wine polyphenol resveratrol on platelet aggregation *in vivo* and *in vitro*. *International Journal of Molecular Medicine*, 9: 77-79.

Wannenmacher, J., Gastl, M. and Becker, T. (2018). Phenolic substances in beer: Structural diversity, reactive potential and relevance for brewing process and beer quality. *Comprehensive Reviews in Food Science and Food Safety*, 17: 953-988.

Wei, J., Zhang, Y., Wang, Y., Ju, H., Niu, C., Song, Z., Yuan, Y. and Yue, T. (2020). Assessment of chemical composition and sensorial properties of ciders fermented with different non-Saccharomyces yeasts in pure and mixed fermentations. *International Journal of Food Microbiology*, 318: 108471.

Winkler, C., Wirleitner, B., Schroecksnadel, K., Schennach, H. and Fuchs, D. (2006). Beer down-regulates activated peripheral blood mononuclear cells *in vitro*. *International Immunopharmacology*, 6: 390-395.

Wu, H., Wang, Y., Wu, C., Yang, P., Li, H. and Li, Z. (2016). Resveratrol induces cancer cell apoptosis through MiR-326/PKM2-mediated ER stress and mitochondrial fission. *Journal of Agricultural Food Chemistry*, 64: 9356-9367.

Xu, J.-W., Ikeda, K. and Yamori, Y. (2004). Upregulation of endothelial nitric oxide synthase by cyanidin-3-glucoside, a typical anthocyanin pigment. *Hypertension*, 217-222.

Zhao, F., Nozawa, H., Daikonnya, A., Kondo, K. and Kitanaka, S. (2003). Inhibitors of nitric oxide production from hops (*Humulus lupulus* L.), *Biological & Pharmaceutical Bulletin*, 26: 61-65.

Zou, J.G., Wang, Z.R., Huang, Y.Z., Cao, K.J. and Wu, J.M. (2003). Effect of red wine and wine polyphenol resveratrol on endothelial function in hypercholesterolemic rabbits. *International Journal of Molecular Medicine*, 11: 317-320.

Zuo, W., Zhang, T., Xu, H., Wang, C., Lu, M. and Chen, X. (2019). Effect of fermentation time on nutritional components of red-fleshed apple cider. *Food and Bioproducts Processing*, 114: 276-285.

Fermented Milk Products and Their Potential Health Benefits

S.V.N. Vijayendra

Department of Microbiology and Fermentation Technology, CSIR - Central Food
Technological Research Institute, Mysuru - 570020, India
Present address: CSIR - Central Food Technological Research Institute, Resource Centre,
Hyderabad - 500097, India

1. Introduction

Preserving foods by fermentation, which is a relatively low-cost food-processing method, is an age-old technique and being practised around the world. Fermentation of various food commodities, like milk, vegetables, cereals, etc., is mediated by microorganisms either naturally present in it or intentionally added to convert the sugars present in the commodities into acids and or alcohols (Paulová *et al.*, 2013). Fermentation not only extends the shelf-life of perishable foods, but also improves the sensory quality of the foods as also the nutritional quality of the foods. It provides health benefits to the consumers as reviewed by several researchers (Farnworth, 2005; Parvez *et al.*, 2006; Marsh *et al.*, 2014; Vijayendra and Halami, 2015; Marco *et al.*, 2017; Agyei *et al.*, 2019; Şanlier *et al.*, 2019). Several enzymes present in fermented foods help in digestion of food and provide better nutrition by way of improved absorption by the intestines and reduce the toxic components present in the foods (Reddy and Pierson, 1994). Several lactic acid bacteria (LAB) and bifidobacteria are reported to provide beneficial effects to the consumers (Prasanna *et al.*, 2014; Linares *et al.*, 2017; Wong *et al.*, 2019). In addition to these bacteria, several yeasts present in indigenous fermented foods are also found to have probiotic potential (Syal and Vohra, 2013). Details on preparation of various fermented foods in Indian subcontinent are consolidated earlier (Rati Rao *et al.*, 2006).

Similar to other food commodities, milk, which is a highly perishable food, is extensively being used to prepare various fermented milk products. The commonly known milk-based fermented products are *dahi* (curd), yoghurt, cheese, *kefir*, *kumis*, *villi*, *skyer*, etc. Metchnikoff reported the beneficial properties of fermented milk

Email: svnvijayendra@cftri.res.in, svnvijayendra@yahoo.com

products for the first time in 2008 (Parvez *et al.*, 2006). The role of various lactic cultures in fermented milk products has been reviewed (Panesar, 2011). In many countries, fermented milk products occupy part of the diet and in countries like India, these are consumed regularly. Compared to milk, consumption of fermented milk products imparts therapeutic and nutritional properties to the consumers. The nutritional, therapeutic properties and technological innovations of fermented milk products have been reviewed in the past (Chourasia *et al.*, 2020; Itsaranuwat *et al.*, 2003; Sarkar, 2008; Khetra *et al.*, 2011; Nagpal *et al.*, 2012; Rai *et al.*, 2016; Shiby and Mishra, 2013; Vijayendra and Varadaraj, 2015). Several researchers attributed the potential health benefits of consuming fermented milk products having LAB due to modification of the gastrointestinal microecology (Fernandez *et al.*, 1987; Gilliland, 1990; Gill and Guarner, 2004). With the consumption of fermented milk products, like yoghurt and cheese, significant reduction in colorectal and bladder cancer risk was observed in a meta analysis that covered 61 studies with more than 38,000 cases and participation of 1.9 million people (Zhang *et al.*, 2019). Staying healthy by consuming traditional fermented milk products became a driving force for the milk industry to commercialise several traditional indigenous milk products across the globe. Various health benefits of fermented milk products are shown in Fig. 1.

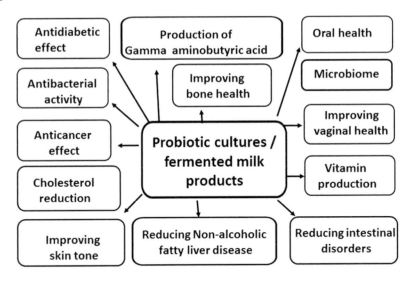

Fig. 1: Health benefits of fermented milk products

2. Role of Probiotic Cultures in Fermented Milk Products

Probiotic cultures are defined as the microorganisms which, when consumed live, provide certain health benefits to the consumers. The beneficial effects include reduction in lactose intolerance, enhancement in immunity levels, reduction in cancer risk, improvement of the health of intestines, reduction in cholesterol, improvement in oral and vaginal health, production of vitamins, etc. Probiotic bacteria are reported to have several health-benefiting attributes for the consumers (Parvez *et al.*, 2006; Roopashri and Varadaraj, 2009). Production of bacteriocins (Mitra *et al.*, 2010; Venkateshwari *et al.*, 2010; Vijayendra *et al.*, 2010), modulating the immune system, blocking proinflammatory molecules and increasing mucin production (Tien *et al.*, 2006), reduction of serum cholesterol content (Lin *et al.*, 1989; Vijayendra and Gupta, 2012), hydrolysis of flatulence causing oligosaccharides (Roopashri and Varadaraj, 2014), are some of the examples to health attributes. Competitive exclusion, stimulation of immune response and competition for nutrients are proposed as mechanisms for action of probiotics (Stanton *et al.*, 2005). Probiotic bacteria are being used along with starter cultures while preparing the fermented milk products, like *dahi*, yoghurt and other fermented dairy foods to increase the therapeutic value of these products (Vijayendra and Gupta, 2010) and non-dairy food products (Vijaya Kumar *et al.*, 2015). Several health benefits of the probiotic cultures, such as suppressing Gram-negative bacteria and pro-inflamatory cytokines of intestines, blocking adhesion cites on intestinal epithelial cells for pathogens, repairing permeability of intestines, increasing immune response in intestines, regulating lipid metabolism, etc. have been reviewed (Sen, 2019). Besides this, several reviews are available that highlight the health and nutritional benefits of probiotics (Gilliland, 1990; Panesar, 2011; Nagpal *et al.*, 2012; Ritchie and Romanuk, 2012; Kechagia *et al.*, 2013; Sharma and Devi, 2014). Some of the specific health benefits of probiotic cultures are presented in this chapter.

2.1   Antibacterial Activity

Several LAB are known to produce antibacterial compounds referred to as bacteriocins, which are small-sized peptides and bactericidal in nature (Klaenhammer, 1988). These include bifidin produced by bifidobacteria (Anand *et al.*, 1984; Cheikhyoussef *et al.*, 2010), acidophilin or acidocin of *Lactobacillus acidophilus* (Shahani *et al.*, 1977; Abo-Amer, 2011), pediocin produced by *Pediococcus* sp. (Devi and Halami, 2014), nisin synthesised by *Lactococcus lactis* (Dussault *et al.*, 2016), bulgarin of *Lb. bulgaricus* (Simova *et al.*, 2008), plantaricin by *Lb. plantarum* (Deegan *et al.*, 2006), etc. These bacteriocins have antibacterial activity against strains closely related to the bacteriocin-producing organisms and/or several food-borne pathogens, like *Escherichia coli*, *Staphylococcus aureus*, *Salmonella*, *Shigella*, *Listeria* spp., etc., and some are known to be heat stable (Zacharof and Lovitt, 2012). Antibacterial activity of *Lb. kefir* against several Gram-positive and Gram-negative bacteria has been reported (Carasi *et al.*, 2014). Detailed information on various aspects of bacteriocins of LAB, such as mechanism of action, production methods, preservative effects, etc. have been provided in several reviews available on this topic (Cintas *et al.*, 2001; Mokoena, 2017; Field *et al.*, 2018; Sidooski *et al.*, 2019; Gao *et al.*, 2019; Venegas-Ortega *et al.*, 2019).

2.2   Anticancer Effect

Cancer is responsible for the death of millions of people all over the world. Among all types of cancers, colorectal cancer (CC) is the most common cancer in men and second most common cancer in women and is linked to Western lifestyle. The number of CC cases has been increasing over a period of time and in 2018, more than 1.8 million new cases of CC were detected (Anon, 2019). Probiotic cultures have great potential to reduce CCs. Several probiotic LAB and bifidobacteria are known to play a role in reducing CC (De Moreno and Perdigon, 2005). Probiotic cultures can inhibit ulcerative cancer cells and suppress the microorganisms which produce enzymes (β-glucuronidase, β-glucosidase, and azoreductase) that can convert procarcinogens into carcinogens, leading to development of cancer (Prasanna *et al.*, 2014) and inactivate carcinogens, like nitroreductase and nitrosamines (Bomba *et al.*, 2002). Human trials involving some probiotic strains of LAB have revealed that probiotic cultures when used as adjuvant cultures can exert anti-carcinogenic effects while treating for cancer (Yu and Li, 2016). Reducing the risk of CC by probiotic cultures of *Lb. acidophilus* and *Bifidobacterium longum* by suppressing the carcinogen-induced genetic material (DNA) was noticed (Oberreuther *et al.*, 2004). Recently the impact of probiotics and prebiotics in preventing CC has been reviewed (Arora *et al.*, 2019). Not only probiotic LAB, but also exopolysaccharides produced by some of the LAB strains have shown anticancer properties (Saadat *et al.*, 2019). Exopolysaccharides of several LAB (*Lb. acidophilus*, *Lb. plantarum* and *Lb. casei*) modulating the natural defence mechanism of the host, were able to protect them from many cancers and displayed anti-tumor properties in several cell lines (Deepak *et al.*, 2016) by down-regulating the genes responsible for proliferation of cancer cells and improving apoptosis (Saber *et al.*, 2017). Górska and co-workers (2019) have very recently reviewed various mechanisms of cancer prevention by LAB and bifidobacteria, which have shown antitumor activities, decrease in cancer cell proliferation and increase in cancer-cell death by producing short-chain fatty acids to help the prevention of carcinogenesis. Probiotic cultures can even eliminate early stage cancer cells by activating phagocytes, indicating that probiotic cultures have a potential in treating certain type of cancers (Górska *et al.*, 2019).

2.3 Antidiabetic Effect

Diabetes is one of the lifestyle diseases and expanding its jaws across the globe – nearly 312 million people were diagnosed to have diabetes in 2015. Unhealthy eating habits, lack of physical exercise and sedentary lifestyle are the major causes of diabetes. Nowadays diabetes has become more common in children too. Probiotics are showing positive effects in controlling diabetes, especially Type-2 diabetes mellitus. Very recently, the potential mechanisms and hypothesis of probiotics intervention in diabetes control have been reviewed (Sun *et al.*, 2019). Probiotics when consumed as dietary supplements can prevent and treat metabolic disorders, like diabetes (Tonucci *et al.*, 2017a,b). Several strategies including feeding probiotics as non-pharmacological therapy to meddle with glucose homeostatis and manipulating gut microbiota for Type 2 diabetic patients have been reviewed recently (Caesar, 2019).

2.4 Improving Oral Health

In the absence of oral hygiene, prevalence of periodontal diseases like caries, plaque or biofilm formation, gingivitis (gum infection), etc. are increasing in the population, especially in children. Use of antibiotics may have side effects or lead to development of resistant bacteria responsible for diseases. Hence, researchers are looking towards probiotics to reduce or replace antibiotic therapy for dental diseases, especially in the last one decade (Twetman and Stecksén-Blicks, 2008; Meurman and Stamatova, 2018; Nadelman *et al.*, 2018; Tester and Al-Ghazzewi, 2018; Mahmoudi *et al.*, 2019). LAB can play a significant role in maintaining the microcosm in oral cavity and reducing the count of *Streptococcus mutans*, the key bacteria responsible for dental diseases (Laleman *et al.*, 2014). There is increased interest in probiotics for oral care products. Halitosis (bad or unpleasant odour in the mouth) is another major issue among the population across the globe. As physical or chemical treatments alone are not solving this condition, researchers are looking towards probiotics, which can modulate the intestinal or oral flora and help eliminate or reduce bad odour in the mouth by inhibiting the growth of other bacteria through its bacteriocins, hydrogen peroxide, etc. (Lalitha, 2011). Consuming probiotic cultures, like *Lactobacillus* and bifidobacteria, on long-term basis has been suggested to prevent caries in children (Twetman and Stecksén-Blicks, 2008). Role of probiotics in curing *Candida* infections in oral cavity was also reported (Jiang *et al.*, 2015). Based on a meta-analysis, it is concluded that dairy products containing probiotic cultures can reduce diseases related to oral cavity, especially dental caries (Nadelman *et al.*, 2018).

3. Microbiome

Microbiome is the term used to refer to the microbial population present in our body, mainly in the oral cavity, small intestine and big intestine. It is a dynamic and complex community of microbial population. Every human being carries nearly 1-2 kg of microbiota, which consists of more than one thousand different species of microbial flora and the cell numbers may be in the range of tens of trillions (Lallés, 2016). They play a major role in maintaining the health and some cause diseases. Fermented foods can modulate the intestinal microbiota (Kato-Kataoka *et al.*, 2016; Rettedal *et al.*, 2019). Probiotic cultures play a major role in influencing the type and number of microorganisms present in the microbiome and thus can improve the health of consumers; it is relatively a recent subject for research (Marco *et al.*, 2017; Volokh *et al.*, 2019; González *et al.*, 2019; Rettedal *et al.*, 2019). The composition of microbiomes can change from person to person and it depends on the diet consumed. Feeding of volunteers with fermented foods containing probiotics (bifidobacteria) has increased the number of beneficial bacteria in the gut microbiome and improved lactose and amino acid metabolism, besides reducing the synthesis of lipopolysaccharides (Volokh *et al.*, 2019). Consumption of fermented milk products has reduced the population of bacteroids and increased fecal levels of *Akkermansia* in the gut microbiome and reversed the intestinal microbiota dysbiosis (González *et al.*, 2019). However, more detailed studies are required to understand the role of probiotics on the gut microbiome.

3.1 Improved Vaginal Health

Vagina of women harbours microorganisms of around 50 different species. They include mainly bacteria (both lactic and non-lactic groups) and yeasts. The composition of vaginal microbiome keeps changing and varies between healthy and unhealthy conditions. The microbiome of the vagina in a healthy woman helps in maintaining the health of the vagina and protects from infecting microbes. Bacterial or yeast vaginosis or urinary tract infection affect the women across the globe and every year more than a billion women are suffering from it. The beneficial roles of probiotics are being explored not only in the gut microbiome, but also in the female reproductive organs. The role of probiotic cultures and Lactobacilli in maintaining the health of the vagina has been reviewed (Falagas *et al*., 2007; Cribby *et al*., 2008; Borges *et al*., 2014; Rostok *et al*., 2018). Several species of Lactobacilli have been identified to be in a healthy vagina (Fredricks *et al*., 2005). During infections, the population of Lactobacilli decreases heavily and vaginal population is dominated by pathogens, Gram-positive and Gram-negative bacteria, and in some cases yeasts, like *Candida* spp. Although medication can reduce vaginosis, repeated usage may lead to development of resistance to antibiotics in microbes, making them ineffective in subsequent usage. Hence, researchers are looking at probiotics as a green alternative to antibiotics or in supplementation to the medication to address vaginosis. Insertion of capsules containing probiotics has given better result (88 per cent cure rate) in curing the bacterial vaginosis than metronidazole (40 per cent cure rate) (Anukam *et al*., 2006). Several clinical trials to mitigate bacterial vaginosis by using probiotics have also taken place in the past and the results are found to be encouraging (Mastromarino *et al*., 2013). Hence, the health of vagina can be improved by probiotic intervention. However, such interventions need to be approved by concerned professional medical associations and followed through international guidelines for this purpose (Buggio *et al*., 2019).

3.2 Improved Bone Health

Osteoporosis, caused by weakening of the bones due to changes in the bone density, is a major public health issue in women and elderly people and it is affecting millions of people across the globe. The bone mineral density modulated by the immune system is mediated by the gut microbiota (Stotzer *et al*., 2003). As probiotic cultures are known to play a role in influencing the gut microbiota, it is also assumed to have an impact on maintaining the health of bones. Probiotics help in bone growth, maintaining its density and structure even under conditions of inflammation, intestinal permeability and dysbiosis. Yousf *et al*. (2015) have reviewed the beneficial effects of probiotics on bone health by proving through several *in vivo* animal experiments. Abboud and Papandreou (2019) very recently reviewed various mechanisms of improved bone health by using probiotics as a supplement. Role of probiotics in regulating the bone and mineral metabolism through multiple mechanisms and on skeletal health has been reviewed very recently (MaCabe and Parameswaran, 2018). Although hormonal therapy is used to prevent osteoporosis in women undergoing menopause, it may have side effects like causing cancer in breast, ovaries and endometrium. As an alternative, people are looking towards dietary supplements to control osteoporosis. Different strains of bifidobacteria and *Lactobacillus* spp. have restored estrogen deficiency-related bone loss (Parvaneh *et al*., 2015). Multi-species probiotic supplementation containing seven deferent probiotic strains of bifidobacteria and *Lactobacillus* have reduced bone resorption in postmenopausal osteopenic women

(McCabe *et al.*, 2013). The role of both probiotics and prebiotics in regulating the bone health by osteoclast and osteoblast in association with intestinal microbiome was reviewed (MaCabe *et al.*, 2015). The role of gut microbiome and probiotics in making the bone stronger with higher bone mineralisation was proved in various studies (Villa *et al.*, 2017a). Besides the presence of calcium and vitamin-D in milk, other vitamins like Vitamin B and Vitamin K synthesised by probiotic cultures was shown to play a critical role in the regulation of bone health (Villa *et al.*, 2017b).

3.3 Improving Immunity Level

Probiotic cultures are known to play a role in modulating both specific and non-specific immunity in the consumers. Consumption of probiotic cultures, especially by elders is helpful as with age the immunity levels decline (Gill and Rutherfurd, 2001). The role of probiotic cultures in stimulating the mucosal system and increasing the antibodies, especially IgA, which plays a major defensive role against microbial infections, is well known (Galdeano *et al.*, 2010). Supplementation of *Lactobacillus casei* 01 reduces inflammation and decreases pro-inflammatory cytokines IL-6 in patients suffering from rheumatoid arthritis disease (Vaghef-Mehrabany *et al.*, 2014). Prior to this, increase in phagocytic activity of the peripheral blood leukocytes in human subjects fed with fermented milk containing *Lb. acidophilus* La1 or *B. bifidum* Bb12 for three weeks was reported (Schiffrin *et al.*, 1997).

3.4 Reducing Intestinal Disorders

Significant reduction in mean duration of symptoms of diarrhoea by 25 hours in patients with acute infectious diarrhoea was noticed upon feeding them with live probiotic cultures; for best results, it was advised to continue for 10-15 days after the withdrawal of symptoms of acute infectious diarrhoea (Abou El-Soud *et al.*, 2015; Wilkins and Sequoia, 2017). Similarly, a meta-analysis of 12 different studies with 51,717 people has indicated that 15 per cent reduction in the risk of traveller's diarrhoea is observed if probiotic cultures are consumed two days before travel and continued till the end of the travel (Wilkins and Sequoia, 2017). Probiotics are also shown to be beneficial in preventing antibiotic-associated diarrhoea in case of *Clostridium difficile* infection and it is recommended to take probiotics for seven to 14 days and probiotics to be consumed parallel to antibiotic treatment, as antibiotics have no effect on these probiotic cultures (Shan *et al.*, 2013; Song *et al.*, 2010). Probiotic cultures even controlled the diarrhoea induced by radiotherapy (Liu *et al.*, 2017). Use of *Bifidobacterium longum* and *B. breve* for prevention and treatment of acute diarrhoea in newborns and infants has gained interest (Di Gioia *et al.*, 2014). The consumption of probiotic *Lactobacillus* sp. as an adjuvant culture along with antibiotic therapy could effectively eradicate *Helicobacter pylori* infection in people suffering with peptic ulcers than in control (Wilkins and Sequoia, 2017).

3.5 Vitamin Production

Although in small quantities, vitamins are essential for our body to carry out various functions, like healing wounds, improving immune system, shoring up of bones, repairing the cellular damage, etc. Vitamin deficiency leads to several disorders in the skin and liver and may affect the brain-glucose metabolism (Russo *et al.*, 2014). As our body cannot synthesise vitamins, we depend on external sources for vitamins. Malnutrition, taking insufficient food or unbalanced diets leads to deficiency in vitamins (LeBlanc *et al.*, 2011). Across the globe, malnutrition is mainly affecting women and children below five years of age. Fermented milk products with their rich vitamin content can address this issue to a certain extent. Several LAB (*Lb. plantarum, Lb. amylovorus, Lb. casei, Streptococcus thermophiles, Propionibacterium freudenreichii, Lb. bulgaricus*), bifidobacteria (*B. infantis, B. animalis*) and yeasts are known to produce vitamins like riboflavin, thiamine, cobalamin, biotin, folate, etc. in the milk during fermentation (Laiño *et al.*, 2013) and it has been exhaustively reviewed in the recent past (Linares *et al.*, 2017; Rai *et al.*, 2019).

3.6 Cholesterol Reduction

Presence of high quantities of cholesterol in the blood leads to cardiovascular diseases. Several researchers have proved the ability of probiotic LAB to reduce the cholesterol levels significantly in the blood (Liu *et al.*, 2006; Wang *et al.*, 2009; Wang *et al.*, 2018). Hassan *et al.* (2019) have exhaustively reviewed the hypocholesterolemia effect of probiotics in reducing atherosclerosis, which causes cardiovascular diseases. Significant reduction in low-density cholesterol, total cholesterol and triglycerides and no change in high-density lipoprotein in serum are noticed in mice fed with cholesterol-rich diet containing *Lb. plantarum* (Wang *et al.*, 2009). Similar observations in serum triglyceride and low-density cholesterol have been observed (Liu *et al.*, 2006). The hypocholesterolemic effect of LAB cultures is presumed to be due to the prevention of exogenous cholesterol absorption by the small intestine, deconjugation of bile salts by enzymes, avoiding reabsorption of bile acids, production of bile salt hydrolyse, incorporation of cholesterol in microbial cells, etc. (Wang *et al.*, 2009). Although reduction in 3-hydroxy-3-methylglutaryl-CoA (HMG CoA) reductase enzyme with feeding of lactic cultures is reported (DeBose-Boyd, 2008), its exact role in hypocholesterolemic effect is yet to be explored (Hassan *et al.*, 2019).

3.7 Other Health Benefits

3.7.1 Reducing Non-alcoholic Fatty Liver Disease

Non-alcoholic fatty acid liver (NAFAL) is a serious disease attributed to insulin resistance and obesity and it is affecting many people including children and increasing across the globe. It leads to excessive fat accumulation in the liver due to unhealthy food habits and sedentary lifestyle. The role of gut-liver axis is suspected in causing this disease. If left untreated, it may lead to liver fibrosis, liver cancer and increase the chances of developing diabetes, chronic renal impairment and cardiovascular disease (Wendy, 2019). Overgrowth of microbiota in the small intestine in 20-75 per cent of the patients suffering from chronic liver disease was observed (Nagata *et al.*, 2007). The importance and potential therapeutic properties of probiotics in NAFAL disease have been reproved in several studies involving animals and human beings (Abenavoli *et al.*, 2013; Wendy, 2018). Besides prebiotics and synbiotics, probiotics cultures, like bifidobacteria and LAB, are known to suppress the inflammatory cascades and reduce the pathogenesis by excluding microorganisms involved in NAFAL-disease condition by improving gut barrier function (Nobili *et al.*, 2018; Elshaghabee *et al.* 2019; Koopman *et al.*, 2019).

3.7.2 Improving Skin Tone

Skin is the largest organ of the body and is directly exposed to the environment. Though skin is the primary defence organ for protecting the body from diseases, it can get affected by various environmental conditions, like low and high temperature, sunrays, which contain ultraviolet rays that can cause skin cancer. Probiotic cultures are not only beneficial in diseases associated with gut and immune system, they also play a role in improving the skin. Probiotic culture of *Lb. rhamnosus* in combination with selenium has proved effective in reducing sunburn complications (Kaur and Rath, 2019). A fermented product (name) containing *Lb. paracasei* producing equol was found effective in preventing skin disorder in experimental mice (Kwon *et al.*, 2019).

3.7.3 Improving Functioning of the Brain and Neurons

γ-aminobutyric acid, popularly called GABA, is a neurotransmitter of the central nervous system and mediates in communicating with the brain and nervous system through chemical signals. It either reduces or inhibits the activity of the neurons. It is considered as a bioactive molecule, which can promote health (Li and Cao, 2010). Many LAB strains, especially those belonging to the genus Lactobacilli, are known to produce GABA (Li and Cao, 2010); some bifidobacteria also produce GABA in relatively small amounts (Park *et al.*, 2005). Probiotic *Lb. plantarum* DM5 has the ability to produce γ-aminobutyric acid (GABA) and can protect against oxidative damage caused due to reactive oxygen species (Das and Goyal, 2015). Dhakal *et al.* (2012) have reviewed production of GABA by microorganisms isolated from fermented milk products. Fermented milk products, like cheese, yoghurt and fermented milk produced bt using GABA-producing LAB, like *Lb. casei Shirota*, *Lb. plantarum*, *Streptococcus salivarius*, and *S. thermophilus*, are found to be enriched with GABA (Inoue *et al.*, 2003; Pouliot *et al.*, 2013; Chen *et al.*, 2016; Linares *et al.*, 2016).

4. Health Benefits of Asian Fermented Milk Products

Several Asians countries are home to traditional fermented milk products. Health benefits of fermented milk products due to bioactive peptides are presented in Table 1. Health benefits of some of the popular fermented milk products are summarised below.

4.1 Dahi

Dahi is a popular fermented milk product consumed almost every day in the Indian sub-continent. There is a reference in *Charaka Samhitā* about the therapeutic

Table 1: Bioactive Peptides in Fermented Milk Products

Milk Product	Starter Cultures	Bioactive Peptides	Bioactivity	References
Fermented milk	*Enterococcus faecalis* CECT 5727	LHLPLP and LVYPFPGPIPNSLPQNIPP SKVYPFPGPI	ACE inhibitory	Quiros *et al.* (2007)
Fermented milk	*Lb. helveticus* LBK-16H	VPP IPP	Antihypertensive	Seppo *et al.* (2002), Seppo *et al.* (2003)
Dahi	*Lb. delbrueckii* ssp. *bulgaricus*	SLVTP	ACE inhibitory	Ashar and Chand (2004)
Gouda Cheese	*Lb. helveticus*	VPP IPP	ACE inhibitory	Butikofer *et al.* (2007), Meyer *et al.* (2009)
Fermented milk	*Lactobacillus rhamnosus* MTCC 5945 (NS4)	MQTDIMIFTIGPA	Anti-hypertensive	Solanki and Hati (2018)
Fermented milk	*Lactobacillus casei* PRA205	VPP, IPP	Anti-hypertensive	Solieri *et al.* (2015)
Yoghurt	*S. salivarius* subsp. *thermophilus* CH 1, *Lb. delbrueckii* subsp. *bulgaricus* LB-12-DRI-VAC, *Lb. helveticus* CH 5 and *Lactobacillus acidophilus* 20552 ATCC	LYQEPVLGPVRGPFPIIV, VLPVPQK, LQDKIHP, YPFPGPIPK,	Anti-hypertensive, anti-oxidative, anti-microbial	Taha *et al.* (2017)
Fermented milk	*Bifidobacterium bifidium* MF 20/5	LVYPFP, LPLP, VLPVPQK	Anti-hypertensive, anti-oxidative	Gonzalez-Gonzalez *et al.* (2013)

importance of *dahi* in the diet (Prajapati and Nair, 2003). Presence of native probiotic lactic cultures, such as *Lb. plantarum*, *Lb. fermentum*, and *Lb. casei* subsp. *tolerans* in homemade *dahi* was noticed (Roopashri and Varadaraj, 2009). The nutritional and therapeutic value of *dahi* was enhanced by incorporating *Lb. acidophilus* and *B. bifidum* along with regular starter cultures (Vijayendra and Gupta, 2012). Preparation of coconut milk *dahi* was reported recently (Shana *et al.*, 2015) with many health-benefitting fatty acids, like capric and lauric acids (Jayadas and Nair, 2006). Reduction in the risk of diabetes by a probiotic *dahi* containing *Lb. acidophilus* and *Lb. casei* in diabetes-induced rats was reported (Yadav *et al.*, 2007). These cultures also delayed the onset of oxidative stress, hyperinsulinemia, dyslipidemia, hyperglycemia. Probiotic *dahi* containing *B. bifidum and Lb. acidophilus* alone or in presence of an adjunct piroxicam may have antiproliferative and antineoplastic effect (Mohania *et al.*, 2014).

Probiotic *dahi* prepared by using *Lb. casei* stimulated non-specific immune response markers and increased phagocytic activity in mice (Jain *et al.*, 2008). Contrary to plain *dahi* and milk, it also reduced the colonisation of *Salmonella enteritidis* remarkably in the spleen and liver, and elevated adaptive and innate immunity (Jain *et al.*, 2009). Feeding of probiotic dahi for 30 days to Wister rats improved the body weight and led to a gain of 129.33g weight (Vijayendra and Gupta, 2012). By reducing the serum cholesterol level by 4.1 mg/dL, the probiotic *dahi* has shown hypocholesterolaemic effect in rats. Presence of several *Lactobacillus* spp. having antimicrobial activity against major food-borne pathogens in commercial *dahi* samples was reported earlier and this action was due to the production of hydrogen peroxide and proteinaceous substance (Balasubramanyam and Varadaraj, 1994, 1995). *Dahi* prepared with *Lb. acidophilus* and *B. bifidum* cultures along with regular starter cultures has shown higher levels of antibacterial activity than *dahi* made with start cultures alone and the highest inhibition was noticed against *E. coli* than with *Salmonella typhimurium* and *Shigella dysenteriae* (Vijayendra and Gupta, 2012). Similarly, probiotic *dahi* prepared with *Lb. acidophilus* and *Lb. casei* increased production of interferon-γ and interleukin (IL)-2, which are the cytokines specific to T helper 1 (Th1) cells and decreased IL-4 and IL-6 cytokenes specific to Th2 cells in experimental mice (Jain *et al.*, 2010). It also suppressed proliferation of ovalbumin-stimulated lymphocytes. The antioxidant status of experimental rats fed with probiotic *dahi* containing *B. bifidum* and *Lb. acidophilus* was enhanced by increased activity of catalase and superoxide dismutase in the red blood cells (Rajpal and Kansal, 2009).

4.2   Shrikhand

Shrikhand is another fermented milk product which is made by using *dahi* as a base material. *Shrikhand*, prepared by using *Lb. acidophilus* and *Lb. Sporogenes*, has better sensory scores than that made with *Lb. rhamnosus* in terms of colour, appearance, aroma, texture, taste and overall acceptability of the product (Swapna *et al.*, 2011). Induced production of IgG was noticed by feeding albino mice with probiotic *shrikhand* and its improvement in the immune system was reported (Subramanian *et al.*, 2005).

4.3   Kefir

Kefir is a fermented milk product, popular in Tibet, Mongolia and Russia and other countries of northwest Asian region. Cauliflower-shaped *kefir* grains, which consist of several strains of LAB and yeast cultures, are used as a starter. *Kefir* grains are reused from one batch to another batch. *Kefir* provides several nutritional and health benefits (hypocholesterolaemic effect, anti-inflammatory effect, control of plasma glucose, antioxidant activity, antibacterial effect, antihypertensive effect, antiallergenic activity, anticarcinogenic activity, antitumor activity, reduction in lactose intolerance) to consumers and these are exhaustively reviewed and the microbial composition of *kefir* grains was also defined (Leite *et al.*, 2013; Rosa *et al.*, 2017; Bengoa *et al.*, 2018). *Kefir* with its varied prophylactic, therapeutic and physiological properties has become a very popular health drink. *Kefir* grains produce exopolysaccharide '*kefirn*' which has antimicrobial activity against several bacteria and yeast, like *Candida* spp., and it has shown healing effect on skin lesions caused by *Staphylococcus aureus* in rat model (Rodrigues *et al.*, 2005). *Kefiran* significantly inhibited eosinophils and other inflammatory-cell release into bronchoalveolar lavage fluid and brought down the interleukins to the normal level in lung inflammation induced by ovalbumin in a murine model of asthma (Kwon *et al.*, 2008).

4.4   Koumiss

Koumiss is an acid-alcohol-based traditional fermented milk product more commonly consumed in Central and West Asian countries, including Russia (Abdel-Salam *et al.*, 2010). *Koumiss* is made using mare's milk and is known to have many health benefits (Zhang and Zhang, 2012; Yao *et al.*, 2017). It is rich in many vitamins, like A, B_1, B_2, B_{12}, C, D and E (Dönmez *et al.*, 2014). Traditionally it has been used to treat several diseases like hepatitis, ulcers and tuberculosis (Wu *et al.*, 2009) and is known to have beneficial effect on various organs of the body, such as endocrine glands, liver, kidneys and healing disorders like intestinal diseases, avitaminosis, anaemia, besides showing its positive impact on cardiovascular, nervous and digestive systems (Mu *et al.*, 2012; Sari *et al.*, 2014; Rong *et al.*, 2015).

5.   Health Benefits of Fermented Milk Products of Europe

5.1   Yoghurt

Yoghurt is a well known and the most commonly consumed fermented milk product in the European region. Yoghurt has higher quantities of several vitamins, including vitamin B_{12}, Vitamin B_2, folic acid, besides minerals like calcium, magnesium, potassium, zinc and protein than milk (Adolfsson *et al.*, 2004; Wang *et al.*, 2013). Yoghurt is also known to have bioactive peptides (Ivey *et al.*, 2015). It had been reported earlier about the modulation of both humoral (Meyer *et al.*, 2007) and cellular (Chaves *et al.*, 2011) immunity with the consumption of yoghurt. Higher immune-protection in probiotic yoghurt than in probiotic whey beverage was noticed (Lollo *et al.*, 2013). In a meta-analysis Zhang *et al.* (2019) noticed a significant reduction in cancer risk with the consumption of yoghurt. Probiotic yoghurt improved the cholesterol metabolism in Type 2 diabetic patients and decreased the risk of cardiovascular disease (Ejtahed *et al.*, 2011). Sarkar (2019) has very recently reviewed various health and nutritional benefits of yoghurt and probiotic yoghurt.

5.2   Cheese

Cheese is an energy-rich, popular, fermented milk product of Western countries. Complex biochemical changes take place in it during initial fermentation and curing stages. It is rich in vitamin B, calcium, conjugated linoleic acids (CLA), sphingolipids, etc. (Ansorena and Astiasarán, 2016; Hur *et al.*, 2016). The LAB present in cheese produce several bioactive peptides, which can provide a range of health benefits by way of its anti-proliferative, antimicrobial, antihypertensive activities due to which it can prevent or treat diseases (Hur *et al.*, 2016; López-Expósito *et al.*, 2017; Rai *et al.*, 2017). CLA present in cheeses impart anti-cancer property to it (Walther *et al.*, 2008). CLA modulate lipid metabolism and act as anti-obesity diet (Kim *et al.*, 2016) and also have a positive impact on biomarkers of atherosclerosis (Sofi *et al.*, 2010).

6.   Health Benefits of African Fermented Milk Products

Similar to Asian countries, African countries are also rich in indigenous, traditional, fermented milk products that have been widely consumed over centuries. Preparation of fermented milk products in the African region, diversity of microflora in these products and several health benefits associated with them have been reviewed very recently (Agyei *et al.*, 2019). Though milk from buffalo, goat, ewe and camel are used, a majority of African fermented milk products are made in large quantities with cow's milk (Jans *et al.*, 2017). Most of these fermented milk products are prepared through spontaneous fermentation. Presence of several bioactive peptides, that can enhance the functional property of African fermented milk products, by LAB present in these products has been reported (Pessione and Cirrincione, 2016) and various bioactive peptides produced were reviewed recently (Agyei *et al.*, 2019). Several probiotic lactic cultures have been isolated from fermented milk products in the African region (Bereda *et al.*, 2014; Digo *et al.*, 2017).

7.   Conclusion

Fermented foods in general and fermented milk in particular are storehouses of health-beneficial microbes, nutrients and bioactive metabolites. Though a majority of studies have highlighted the efficacy of the fermented milk products, very few studies have been carried out involving human subjects in any nation or region as the microbiome is different in people from different regions (reference). Hence, focus is needed to determine the health-benefitting properties in human beings so as to increase the dependency on fermented milk products and reduce consumption of medicines, which may have some side effects when consumed on a long-term basis and also to achieve wellness through food. For this, evidence-based investigations in human subjects are needed. This will help to find the impact of fermented milk products in determining the public health impact and to recommend its use to overcome the deficiency of essential nutrients and reduce the occurrence of infectious diseases. Long-term feeding studies are necessary to prove some beneficial attributes of fermented milk products in improving bone health, etc. Health benefits of less popular traditional and indigenous fermented milk products of all the continents need to be explored and commercialised so that the overall health of the population can be improved.

Abbreviations

CC – Colorectal cancer
CLA – Conjugated linoleic acids
GABA – γ-aminobutyric acid
HMG CoA – 3-hydroxy-3-methylglutaryl-CoA
LAB – Lactic Acid Bacteria
NAFAL – Non-Alcoholic Fatty Acid Liver

References

Abenavoli, L., Scarpellini, E., Rouabhia, S., Balsano, C. and Luzza, F. (2013). Probiotics in non-alcoholic fatty liver disease: Which and when. *Annals of Hepatology*, 12: 357-363.

Abo-Amer, A.E. (2011). Optimisation of bacteriocin production by *Lactobacillus acidophilus* AA11, a strain isolated from Egyptian cheese. *Annals of Microbiology*, 61: 445-452.

Abou El-Soud, N., Said, R., Mosalam, D., Barakat, N. and Sabry, M. (2015). *Bifidobacterium lactis* in treatment of children with acute diarrhoea. A randomised double blind controlled trial., *Open Access Macedonian Journal of Medical Sciences*, 3(3): 403. doi: 10.3889/oamjms.2015.088

Abboud, M. and Papandreou, D. (2019). Gut microbiome, probiotics and bone: An updated mini-review. *Open Access Macedonian Journal of Medical Sciences*, 7(3): 478-481.

Abdel-Salam, A.M., Al-Dekheil, A., Babkr, A., Farahna, M. and Mousa, H.M. (2010). High fibre probiotic fermented mare's milk reduces the toxic effects of mercury in rats. *North American Journal of Medical Sciences*, 2(12): 569. doi:10.4297/najms.2010.2569

Adolfsson, O., Meydani, S.N. and Russell, R.M. (2004). Yogurt and gut function., *The American Journal of Clinical Nutrition*, 80(2): 245-256.

Agyei, D., Owusu-Kwarteng, J., Akabanda, F. and Akomea-Frempong, S. (2019). Indigenous African fermented dairy products: Processing technology, microbiology and health Benefits., *Critical Reviews in Food Science and Nutrition*. doi:10.1080/10408398.2018.1555133

Anand, S.K., Srinivasan, R.A. and Rao, L.K. (1984) Antibacterial activity associated with *Bifidobacterium bifidum* I., *Cultured Dairy Products Journal*, 19(9): 6-8.

Anon (2019). *Colorectal Cancer Statistics – World Cancer Research Fund*. www.wcrf.org. (accessed on 26 September 2019)

Ansorena, D. and Astiasarán, I. (2016). Fermented foods: Composition and health effects. *Encyclopaedia of Food and Health*, Academic Press, Oxford.

Anukam, K.C., Osazuwa, E., Osemene, G.I., Ehigiagbe, F., Bruce, A.W. and Reid, G. (2006). Clinical study comparing probiotic *Lactobacillus* GR-1 and RC-14 with metronidazole vaginal gel to treat symptomatic bacterial vaginosis. *Microbes and Infection*, 8: 2772-2776.

Arora, M., Baldi, A., Kapila, N., Bhandari, S. and Jeet, K. (2019). Impact of probiotics and prebiotics on colon cancer: Mechanistic insights and future approach., *Current Cancer Therapy Reviews*, 15(1): 27-36.

Ashar, M.N. and Chand, R. (2004). Antihypertensive peptides purified from milks fermented with *Lactobacillus delbrueckii* spp. *Bulgaricus*. *Milchwissenschaft*, 59(1): 14-17.

Balasubramanyam, B.V. and Varadaraj, M.C. (1994). *Dahi* as a potential source of lactic acid bacteria active against foodborne pathogenic and spoilage bacteria. *Journal of Food Science and Technology*, 31: 241-243.

Balasubramanyam, B.V. and Varadaraj, M.C. (1995). Antibacterial effect of *Lactobacillus* spp. on foodborne pathogenic bacteria in an Indian milk-based fermented culinary food item. *Cultured Dairy Products Journal*, 30(1): 22-27.

Bengoa, A.A., Iraporda, C., Garrote, G.L. and Abraham, A.G. (2018). *Kefir* micro-organisms: Their role in grain assembly and health properties of fermented milk. *Journal of Applied Microbiology*, 126: 686-670.

Bereda, A., Eshetu, M. and Yilma, Z. (2014). Microbial properties of Ethiopian dairy products: A review. *African Journal of Microbiology Research*, 8(23): 264-271.

Bomba, A., Nemcov, R., Gancarčíková, S., Herich, R., Guba, P. and Mudronov, D. (2002). Improvement of the probiotic effect of microorganisms by their combination with maltodextrins, fructo-oligosaccharides and polyunsaturated fatty acids. *British Journal of Nutrition*, 88(Supplement S1): S95-S99.

Borges, S., Silv, J. and Teixeira, P. (2014). The role of Lactobacilli and probiotics in maintaining vaginal health. *Archives of Gynaecology and Obstetrics*, 289: 479-489.

Buggio, L., Somigliana, E., Borghi, A. and Vercellini, P. (2019). Probiotics and vaginal microecology: Fact or fancy? *BMC Women's Health*, 19, Article number: 25

Butikofer, U., Meyer, J., Sieber, R. and Wechsler, D. (2007). Quantification of the angiotensin converting enzyme-inhibiting tripeptides Val Pro-Pro and Ile-Pro-Pro in hard, semi-hard and soft cheeses. *International Dairy Journal*, 17(8): 968-975.

Caesar, R. (2019). Pharmacologic and nonpharmacologic therapies for the gut microbiota in Type 2 Diabetes. *Canadian Journal of Diabetes*, 43: 224-231.

Carasi, P., Díaz, M., Racedo, S.M., De Antoni, G.D., Urdaci, M.C. and Serradell, M.A. (2014). Safety characterisation and antimicrobial properties of *kefir*-isolated *Lactobacillus kefiri*. *BioMed Research International*, Article ID 208974. http://dx.doi.org/10.1155/2014/208974

Chaves, S., Perdigon, G., De Moreno, A. and De Leblanc, A. (2011). Yoghurt consumption regulates the immune cells implicated in acute intestinal inflammation and prevents the recurrence of the inflammatory process in a mouse model. *Journal of Food Protection*, 74: 801-811.

Cheikhyoussef, A., Cheikhyoussef, N., Chen, H., Zhao, J., Tang, J., Zhang, H. and Chen, W. (2010). Bifidin I – A new bacteriocin produced by *Bifidobacterium infantis* BCRC 14602: Purification and partial amino acid sequence. *Food Control*, 21: 746-753.

Chen, L., Zhao, H., Zhang, C., Lu, Y., Zhu, X. and Lu, Z. (2016). g-Aminobutyric acid-rich yogurt fermented by *Streptococcus salivarius* subsp. *thermophiles* fmb5 appears to have anti-diabetic effect on streptozotocin-induced diabetic mice. *Journal of Functional Foods*, 20: 267-275.

Cintas, L.M., Casaus, M.P., Herranz, C., Nes, I.F. and Hernández, P.E. (2001). Review: Bacteriocins of lactic acid bacteria. *Food Science and Technology International*, 7(4): 281-305.

Chourasia, R., Phukon, L.C., Abedin, M., Sahoo, D., Singh, S.P. and Rai, A.K. (2020). Biotechnological approaches for the production of designer cheese with improved functionality. *Comprehensive Reviews in Food Science and Food Safety.*, doi: 10.1111/1541-4337.12680

Cribby, S., Taylor, M. and Reid, G. (2008). Vaginal microbiota and the use of probiotics. *Interdisciplinary Perspectives on Infectious Diseases*, vol. 2008, Article ID 256490, doi:10.1155/2008/256490

Das, D. and Goyal, A. (2015). Antioxidant activity and γ-aminobutyric acid (GABA) producing ability of probiotic *Lactobacillus plantarum* DM5 isolated from Marcha of Sikkim. *LWT –Food Science and Technology*, 61: 263-268.

DeBose-Boyd, R.A. (2008). Feedback regulation of cholesterol synthesis: Sterol-accelerated ubiquitination and degradation of HMG CoA reductase. *Cell Research*, 18: 609-621.

Deegan, L.H., Cotter, P.D., Colin, H. and Ross, P. (2006). Bacteriocins: Biological tools for bio-preservation and shelf-life extension. *International Dairy Journal*, 16: 1058-1071.

Deepak, V., Pandian, S.R.K., Sivasubramaniam, S.D., Nellaiah, H. and Sundar, K. (2016). Optimization of anticancer exopolysaccharide production from probiotic *Lactobacillus acidophilus* by response surface methodology. *Preparative Biochemistry and Biotechnology*, 46(3): 288-297.

De Moreno, D.L.A. and Perdigon, P. (2005). Antitumour activity of yoghurt. pp. 97-123. *In*: D. Martinez (Ed.). *Focus on Colorectal Cancer Research*. Nova Science Publishers, New York.

Devi, S.M. and Halami, P.M. (2014). Detection and characterisation of pediocin PA-1/AcH like bacteriocin producing lactic acid bacteria. *Current Microbiology*, 63: 181-185.

Dhakal, R., Bajpai, V.K. and Baek, K.H. (2012). Production of gaba (g-aminobutyric acid) by microorganisms: A review. *Brazilian Journal of Microbiology*, 43: 1230-1241.

Di Gioia, D., Aloisio, I., Mazzola, G. and Biavati, B. (2014). Bifidobacteria: Their impact on gut microbiota composition and their applications as probiotics in infants. *Applied Microbiology Biotechnology*, 98: 563-577.

Digo, C.A., Kamau-Mbuthia, E., Matofari, J.W. and. Nġetich, W.K. (2017). Potential probiotics from traditional fermented milk, *mursik* of Kenya. *International Journal of Nutrition and Metabolism*, 10(9): 75-81.

Dönmez, N., Kısadere, I., Balaban, C. and Kadiralieva, N. (2014). Effects of traditional homemade *koumiss* on some hematological and biochemical characteristics in sedentary men exposed to exercise. *Biotechnic and Histochemistry*, 89(8): 558-563.

Dussault, D., Vu, K.D. and Lacroix, M. (2016). Enhancement of Nisin Production by *Lactococcus lactis* subsp. *Lactis*. *Probiotics and Antimicrobial Proteins*, 8(3): 170-175.

Ejtahed, H., Mohtadi-Nia, J., Homayouni-Rad, A., Niafar, M., Asghari-Jafarabadi, M., Mofid, V. and Akbarian-Moghari, A. (2011). Effect of probiotic yogurt containing *Lactobacillus acidophilus* and *Bifidobacterium lactis* on lipid profile in individuals with Type 2 diabetes mellitus. *Journal of Dairy Science*, 94(7): 3288-3294.

Elshaghabee, F.M.F., Rokana, N., Panwar, H., Heller, K.J. and Schrezenmeir, J. (2019). Probiotics as a dietary intervention for reducing the risk of nonalcoholic fatty liver disease. pp. 207-283. *In*: D. Arora, C. Sharma, S. Jaglan and E. Lichtfouse (Eds.). *Pharmaceuticals from Microbes, Environmental Chemistry for a Sustainable World*. Springer, New York.

Falagas, M.E., Betsi, G.I. and Athanasiou, S. (2007). Probiotics for the treatment of women with bacterial vaginosis. *Clinical Microbiology and Infection*, 13: 657-664.

Farnworth, E.R. (2005). The beneficial health effects of fermented foods – Potential probiotics around the world. *Journal of Nutraceuticals, Functional & Medical Foods*, 4: 93-117.

Fernandes, C.F. and Shahani, K.M. (1990). Anticarcinogenic and immunological properties of dietary Lactobacilli. *Journal of Food Protection*, 53: 704-710.

Field, D., Ross, R.P. and Hill, C. (2018). Developing bacteriocins of lactic acid bacteria into next generation biopreservatives. *Current Opinion in Food Science*, 20: 1-6.

Fredricks, D.N., Fiedler, T.L. and Marrazzo, J.M. (2005). Molecular identification of bacteria associated with bacterial vaginosis. *The New England Journal of Medicine*, 353(18): 1899-1911.

Galdeano, C.M., De LeBlanc, A.M., Dogi, C. and Perdigón, G. (2010). Lactic acid bacteria as immunomodulators of the gut-associated immune system. pp. 125-141. *In*: F. Mozzi, R.R. Raya and G.M. Vignolo (Eds.). *Biotechnology of Lactic Acid Bacteria: Novel Applications*. Wiley-Blackwell, Iowa.

Gao, Z, Daliri, E.B., Wang, J., Liu, D., Sian, C., Ye, X. and Ding, A. (2019). Inhibitory effect of lactic acid bacteria on foodborne pathogens: A review. *Journal of Food Protection*, 82(3): 441-453.

Gill, H.S. and Rutherfurd, K.J. (2001). Viability and dose-response studies on the effects of the immuno-enhancing lactic acid bacterium *Lactobacillus rhamnosus* in mice. *British Journal of Nutrition*, 86: 285-289.

Gill, H.S. and Guarner, F. (2004). Probiotics and human health: A clinical perspective. *Postgraduate Medical Journal*, 80: 516-526.

Gilliland, S.E. (1990). Health and Nutritional benefits from lactic acid bacteria. *FEMS Microbiology Reviews*, 87: 175-188.

Górska, A., Przystupski, D., Niemczura, M.J. and Kulbacka, J. (2019). Probiotic bacteria: A promising tool in cancer prevention and therapy. *Current Microbiology*, 76: 939-949.

Gonzalez-Gonzalez, C., Gibson, T. and Jauregi, P. (2013). Novel probiotic-fermented milk with angiotensin I-converting enzyme inhibitory peptides produced by Bifidobacterium bifidum MF 20/5. *International Journal Food Microbiology*, 167: 131-137.

González, S., Fernández-Navarro, T., Arboleya, S., de los Reyes-Gavilán, C.G., Salazar, N. and Gueimonde, M. (2019). Fermented dairy foods: Impact on intestinal microbiota and health-linked biomarkers. *Frontiers in Microbiology*, 10: 1046.

Hassan, A., Din, A.U., Zhu, Y., Zhang, K., Li, T., Wang, Y., Luo, Y. and Wang, G. (2019). Updates in understanding the hypocholesterolemia effect of probiotics on atherosclerosis. *Applied Microbiology and Biotechnology*, 103: 5993-6006.

Hur, S.J., Kim, H.S., Bahk, Y.Y. and Park, Y. (2016). Overview of conjugated linoleic acid formation and accumulation in animal products. *Livestock Science*, 195: 105-111.

Inoue, K., Shirai, T., Ochiai, H., Kasao, M., Hayakawa, K., Kimura, M. and Sansawa, H. (2003). Blood- pressure-lowering effect of a novel fermented milk containing γ-aminobutyric acid (GABA) in mild hypertensives. *European Journal of Clinical Nutrition*, 57: 490-495.

Itsaranuwat, P., Al-Haddad, K.S.H. and Robinson, R.K. (2003). The potential therapeutic benefits of consuming 'health promoting' fermented dairy products: A brief update. *International Journal of Dairy Technology*, 56(4): 203-210.

Ivey, K., Hodgson, J., Kerr, D., Thompson, P., Stojceski, B. and Prince, R. (2015). The effect of yoghurt and its probiotics on blood pressure and serum lipid profile: A randomised controlled trial. *Nutrition, Metabolism and Cardiovascular Diseases*, 25(1): 46-51.

Jain, S., Yadav, H. and Sinha, P.R. (2008). Stimulation of innate immunity by oral administration of *dahi* containing probiotic *Lactobacillus casei* in mice. *Journal of Medicinal Food*, 11: 652-656.

Jain, S., Yadav, H. and Sinha, P.R. (2009). Probiotic *dahi* containing *Lactobacillus casei* protects against *Salmonella enteritidis* infection and modulates immune response in mice. *Journal of Medicinal Food*, 12: 576-583.

Jain, S., Yadav, H., Sinha, P.R., Kapila, S., Naito, Y. and Marotta, F. (2010). Anti-allergic effects of probiotic *dahi* through modulation of the gut immune system. *Turkish Journal of Gastroenterology*, 21: 244-250.

Jans, C., Meile, L., Kaindi, D.W.M., Kogi-Makau, W., Lamuka, P., Renault, P., Kreikemeyer, B., Lacroix, C., Hattendorf, J. and Zinsstag, J. (2017). African fermented dairy products – Overview of predominant technologically important microorganisms focusing on African *Streptococcus infantarius* variants and potential future applications for enhanced food safety and security. *International Journal of Food Microbiology*, 250: 27-36.

Jayadas, N.H. and Nair, K.P. (2006). Coconut oil as base oil for industrial lubricants – Evaluation and modification of thermal, oxidative and low temperature properties. *Tribology International*, 39: 873-878.

Jiang, Q., Stamatova, I., Kari, K. and Meurman, J.H. (2015). Inhibitory activity *in vitro* of probiotic Lactobacilli against oral Candida under different fermentation conditions. *Beneficial Microbes*, 6(3): 361-368.

Kato-Kataoka, A., Nishida, K., Takada, M., Kawai, M., Kikuchi-Hayakawa, H., Suda, K., Ishikawa, H., Gondo, Y., Shimizu, K., Matsuki, T., Kushiro, A., Hoshi, R., Watanabe, O., Igarashi, T., Miyazaki, K., Kuwano, Y. and Rokutan, K. (2016). Fermented milk containing *Lactobacillus casei* strain Shirota preserves the diversity of the gut microbiota and relieves abdominal dysfunction in healthy medical students exposed to academic stress. *Applied and Environmental Microbiology*, 82: 3649-3658.

Kaur, K. and Rath, G. (2019). Formulation and evaluation of UV protective synbiotic skin care topical formulation. *Journal of Cosmetic and Laser Therapy*, 21(6): 332-342.

Kechagia, M., Basoulis, D., Konstantopoulou, S., Dimitriadi, D., Gyftopoulou, K., Skarmoutsou, N. and Fakiri, E.M. (2013). Health Benefits of Probiotics – A Review. *ISRN Nutrition 2013*, Article ID 481651, http://dx.doi.org/10.5402/2013/481651

Khetra, Y., Raju, P.N., Hati, S. and Kanawjia, S.K. (2011). Health benefits of traditional fermented milk products. *Indian Dairyman*, 63(9): 54-60.

Kim, E.K., Ha, A.W., Choi, E.O. and Ju, S.Y. (2016). Analysis of *kimchi*, vegetable and fruit consumption trends among Korean adults: Data from the Korea National Health and Nutrition Examination Survey (1998-2012). *Nutrition Research and Practice*, 10(2): 188-197.

Klaenhammer, T.R. (1988). Bacteriocins of lactic acid bacteria. *Biochimie*, 70(3): 337-349.

Koopman, N., Molinaro, A., Nieuwdorp, M. and Holleboom, A.G. (2019). Review article: Can bugs be drugs? The potential of probiotics and prebiotics as treatment for non-alcoholic fatty liver disease. *Aliment Pharmacology and Therapeutics*, 50: 628- 639.

Kwon, J.E., Lim, J., Bang, I., Kim, I., Kimb, D. and Kang, S.C. (2019). Fermentation product with new equol-producing *Lactobacillus paracasei* as a probiotic-like product candidate for prevention of skin and intestinal disorder. *Journal of the Science of Food and Agriculture*, 99: 4200-4210.

Kwon, O.K., Ahn, K.S., Lee, M.Y., Kim, S.Y., Park, B.Y., Kim, M.K., Lee, I.Y, Oh, S.R. and Lee, H.K. (2008). Inhibitory effect of *kefiran* on ovalbumin-induced lung inflammation in a murine model of asthma. *Archives of Pharmacal Research*, 31: 1590-1596.

Laiño, J.E., Juarez del Valle, M., Savoy de Giori, G. and LeBlanc, J.G. (2013). Development of a high folate concentration yoghurt naturally bio-enriched using selected lactic acid bacteria. *LWT–Food Science and Technology*, 54: 1-5.

Laleman, I., Detailleur, V., Slot, D.E., Slomka, V., Quirynen, M. and Teughels, W. (2014). Probiotics reduce mutans streptococci counts in humans: A systematic review and metaanalysis. *Clinical Oral Investigations*, 18: 1539-1552.

Lalitha, T. (2011). Probiotics and oral health. *Journal of Oral Research and Review*, 3(1): 20-26.

Lallés, J.P. (2016). Microbiota-host interplay at the gut epithelial level, health and nutrition. *Journal of Animal Science and Biotechnology*, 7: 66, doi.10.1186/s40104-016-0123-7

LeBlanc, J.G., Laiño, J.E., del Valle, M.J., Vannini, V., van Sinderen, D., Taranto, M.P., de Valdez, G.F., de Giori, G.S. and Sesma, F. (2011). B-group vitamin production by lactic acid bacteria–current knowledge and potential applications. *Journal of Applied Microbiology*, 111: 1297-1309.

Leite, A.M.O., Miguel, M.A.L., Peixoto, R.S., Rosado, A.S., Silva, J.T. and Paschoalin, V.M.F. (2013). Microbiological, technological and therapeutic properties of *kefir*: A natural probiotic beverage. *Brazilian Journal of Microbiology*, 44: 341-349.

Li, H. and Cao, Y. (2010). Lactic acid bacterial cell factories for γ-aminobutyric acid. *Amino Acids*, 39: 1107-1116.

Lin, S.Y., Ayres, J.W., Winkler, W. and Sandine, W.E. (1989). *Lactobacillus* effects on cholesterol: *In vitro* and *in vivo* results. *Journal of Dairy Research*, 72: 2885-2889.

Linares, D.M., O'Callaghan, T.F., O'Connor, P.M., Ross, R.P. and Stanton, C. (2016). *Streptococcus thermophilus* APC151 strain is suitable for the manufacture of naturally GABA-enriched bioactive yoghurt. *Frontiers in Microbiology*, 7: 1876.

Linares, D.M., Gómez, C., Renes, E., Fresno, J.M, Tornadijo, M.E., Ross, R.P. and Stanton, C. (2017). Lactic acid bacteria and bifidobacteria with potential to design natural biofunctional health-promoting dairy foods. *Frontiers in Microbiology*, 8: 846.

Liu, J.R., Wang, S.Y., Chen, M.J., Chen, H.L., Yueh, P.Y. and Lin, C.W. (2006). Hypocholesterolaemic effects of milk-*kefir* and soya milk *kefir* in cholesterol-fed hamsters. *British Journal of Nutrition*, 95: 939-946.

Liu, M.M., Li, S.T., Shu, Y. and Zhan, H.Q. (2017). Probiotics for prevention of radiation-induced diarrhoea: A meta-analysis of randomised controlled trials. *PLoS ONE*, 12(6): e0178870. doi:10.1371/journal.pone.0178870

Lollo, P.C.B., de Moura, C.S., Morato, P.N., Cruz, A.G., de Freitas Castro, W., Betim, C.B., Nisishima, L., Faria, J.A.F., Junior, M.M., Fernandes, C.O. and Amaya-Farfan, J. (2013). Probiotic yoghurt offers higher immune-protection than probiotic whey beverage. *Food Research International*, 54: 118-124.

Lopez-Exposito, I., Miralles, B., Amigo, L. and Hernandez-Ledesma, B. (2017). Health effects of cheese components with a focus on bioactive peptides A2 – Frias, Juana. pp. 239-273. *In*: C. Martinez-Villaluenga and E. Peñas (Eds.). *Fermented Foods in Health and Disease Prevention*. Academic Press, Boston.

Mahmoudi, R., Moosazad, S. and Aghaei, K. (2019). Oral health by using probiotic products. *In*: R. Mahmoudi (Ed.). *Microbial Health – Role of Prebiotics and Probiotics*. Intech Open Publisher, doi: 10.5772/intechopen.86714

Marco, M.L., Heeney, D., Binda, S., Cifelli, C.J., Cotter, P.D., Foligné, B., Gänzle, M., Kort, R., Pasin, G., Pihlanto, A., Smid, E.J. and Hutkins, R. (2017). Health benefits of fermented foods: Microbiota and beyond. *Current Opinion in Biotechnology*, 44: 94- 102.

Marsh, A.J., Hill, C., Ross, P.R. and Cotter, P.D. (2014). Fermented beverages with health-promoting potential: Past and future perspectives. *Trends in Food Science and Technology*, 38: 113-124.

Mastromarino, P., Vitali, B. and Mosca, L. (2013). Bacterial vaginosis: A review on clinical trials with probiotics. *New Microbiologica*, 36: 229-238.

McCabe, L.R. and Parameswaran, N. (2018). Advances in probiotic regulation of bone and mineral metabolism. *Calcified Tissue International*, 102: 480-488.

McCabe, L.R., Irwin, R., Schaefer, L. and Britton, R.A. (2013). Probiotic use decreases intestinal inflammation and increases bone density in healthy male but not female mice. *Journal of Cell Physiology*, 228: 1793-1798.

McCabe, L.R., Britton, R.A. and Parameswaran, N. (2015). Prebiotic and probiotic regulation of bone health: Role of the intestine and its microbiome. *Current Osteoporosis Reports*, 13: 363-371.

Meurman, J.H. and Stamatova, I.V. (2018). Probiotics: Evidence of oral health implications. *Folia Med (Plovdiv)*, 60(1): 21-29.

Meyer, A.L., Elmadfa, I., Herbacek, I. and Micksche, M. (2007). Probiotic, as well as conventional yogurt, can enhance the stimulated production of proinflammatory cytokines. *Journal of Human Nutrition and Dietetics*, 20: 590-598.

Meyer, J., Butikofer, U., Walther, B., Wechsler, D. and Sieber, R. (2009). Hot topic: Changes in angiotensin-converting enzyme inhibition and concentration of the tripeptides Val-Pro-Pro and Ile-Pro-Pro during ripening of different Swiss cheese varieties. *Journal of Dairy Science*, 92(3): 826-836.

Mitra, S., Chakrabartty, P.K. and Biswas, S.R. (2010). Potential production and preservation of dahi by *Lactococcus lactis* W8, a nisin-producing strain. *LWT–Food Science and Technology*, 43: 337-342.

Mokoena, M.P. (2017). Lactic acid bacteria and their bacteriocins: Classification, biosynthesis and applications against uropathogens: A mini-review. *Molecules*, 22: 1255., doi:10.3390/molecules22081255

Mohania, D., Kansal, V.K., Kruzliak, P. and Kumari, A. (2014). Probiotic *dahi* containing *Lactobacillus acidophilus* and *Bifidobacterium bifidum* modulates the formation of aberrant crypt foci, mucin depleted foci and cell proliferation on 1,2-dimethylhydrazine induced colorectal carcinogenesis in Wistar rats. *Rejuvenation Research*, 17: 325-333.

Mu, Z., Yang, X. and Yuan, H. (2012). Detection and identification of wild yeast in *koumiss*. *Food Microbiology*, 31(2): 301-308.

Nadelman, P., Magno, M.B., Masterson, D., da Cruz, A.G. and Maia, L.C. (2018). Are dairy products containing probiotics beneficial for oral health? A systematic review and meta-analysis. *Clinical Oral Investigations*, 22: 2763-2785.

Nagata, K., Suzuki, H. and Sakaguchi, S. (2007). Common pathogenic mechanism in development progression of liver injury caused by non-alcoholic or alcoholic steatohepatitis. *Journal of Toxicological Science*, 32: 453-468.

Nagpal, R., Behare, P.V., Kumar, M., Mohania, D., Yadav, M., Jain, S., Menon, S., Parkash, O., Marotta, F., Minelli, E., Henry, C.J. and Yadav, H. (2012). Milk, milk products and disease free health: An updated overview. *Critical Reviews in Food Science and Nutrition*, 52: 321-333.

Nobili, V., Putignani, L., Mosca, A., Chierico, F.D., Vernocchi, P., Alisi, A., Stronati, L., Cucchiara, S., Toscano, M. and Drago, L. (2018). Bifidobacteria and Lactobacilli in the gut microbiome of children with non-alcoholic fatty liver disease: Which strains act as health players? *Archives of Medical Science*, 14(1): 81-87.

Oberreuther, M.D.L., Jahreis, G., Rechkemmer, G. and Pool-Zobel, B.L. (2004). Dietary Intervention with the probiotics *Lactobacillus acidophilus* 145 and *Bifidobacterium Longum* 913 modulates the potential of human faecal water to induce damage in HT29 clone19A cells. *British Journal of Nutrition*, 91: 925-932.

Panesar, P.S. (2011). Fermented dairy products: Starter cultures and potential nutritional Benefits. *Food and Nutrition Sciences*, 2: 47-51.

Park, K.B., Ji, G.E., Park, M.S. and Oh, S.H. (2005). Expression of rice glutamate decarboxylase in *Bifidobacterium longum* enhances gamma-aminobutyric acid production. *Biotechnology Letters*, 27: 1681-1684.

Parvaneh, K., Ebrahimi, M., Sabran, M.R., Karimi, G., Hwei, A.N, Abdul-Majeed, S., Ahmad, Z., Ibrahim, Z. and Jamaluddin, R. (2015). Probiotics (*Bifidobacterium longum*) increase bone mass density and upregulate Sparc and Bmp-2 genes in rats with bone loss resulting from ovariectomy. *BioMed Research International*, 2015: 897639. doi:10.1155/2015/897639

Parvez, S., Malik, K.A., Ah Kang, S. and Kim, H.Y. (2006). Probiotics and their fermented food products are beneficial for health. *Journal of Applied Microbiology*, 100: 1171-1185.

Paulová, L., Patáková, P. and Brányik, T. (2013). Advanced fermentation processes. pp. 89-110. *In*: J.A. Teixeira and A.A. Vicente (Eds.). *Engineering Aspects of Food Biotechnology*. CRC Press, New York.

Pessione, E. and Cirrincione, S. (2016). Bioactive molecules released in food by lactic acid bacteria: Encrypted peptides and biogenic amines. *Frontiers in Microbiology*, 7: 876. doi:10.3389/fmicb.2016.00876

Pouliot-Mathieu, K., Gardner-Fortier, C., Lemieux, S., St-Gelais, D., Champagne, C.P. and Vuillemard, J.C. (2013). Effect of cheese containing gamma amino butyric acid- producing lactic acid bacteria on blood pressure in men. *Pharma Nutrition*, 1: 141- 148.

Prajapati, J.B. and Nair, B.M. (2003). The history of fermented foods. pp. 1-25. *In*: E.R. Farnworth (Ed.). *Handbook of Fermented Functional Foods*. CRC Press, Boca Raton.

Prasanna, P.H.P., Grandison, A.S. and Charalampopoulos, D. (2014). Bifidobacteria in milk products: An overview of physiological and biochemical properties, exopolysaccharide production, selection criteria of milk products and health benefits. *Food Research International*, 55: 247-262.

Quiros, A., Ramos, M., Muguerza, B., Delgado, M., Miguel, M., Alexaindre, A. and Recio, I. (2007). Identification of novel antihypertensive peptides in milk fermented with *Enterococcus faecalis*. *International Diary Journal*, 17(1): 33-41.

Rai, A.K., Kumari, R., Sanjukta, S. and Sahoo, D. (2016). Production of bioactive protein hydrolysate using the yeasts isolated from soft *chhurpi*. *Bioresource Technology*, 219: 239-245.

Rai, A.K., Pandey, A. and Sahoo, D. (2019). Biotechnological potential of yeasts in functional food industry. *Trends in Food Science and Technology*, 83: 129-137.

Rai, A.K., Sanjukta, S. and Jeyaram, K. (2017). Production of Angiotensin I converting enzyme inhibitory (ACE-I) peptides during milk fermentation and its role in treatment of hypertension. *Critical Reviews in Food Science and Nutrition*, 57: 2789-2800.

Rajpal, S. and Kansal, V.K. (2009). Probiotic *dahi* containing *Lactobacillus acidophilus* and *Bifidobacterium bifidum* stimulates antioxidant enzyme pathways in rats. *Milchwissenschaft*, 64: 287-290.

Rati Rao, E., Vijayendra, S.V.N. and Varadaraj, M.C. (2006). Fermentation biotechnology of traditional foods of the Indian subcontinent. pp. 1759-1794. *In*: K. Shetty, G. Paliyath, A. Pometto and R.E. Levin (Eds.). *Food Biotechnology*, 2nd ed., CRC Press, Taylor and Francis, Boca Raton, Florida, USA.

Reddy, N.R. and Pierson, M.D. (1994). Reduction in antinutritional and toxic components in plant foods by fermentation. *Food Research International*, 27: 281-290.

Rettedal, E.A., Altermann, E., Roy, N.C. and Dalziel, J.E. (2019). The effects of unfermented and fermented cow and sheep milk on the gut microbiota. *Frontiers in Microbiology*, 10: 458, doi: 10.3389/fmicb.2019.00458

Ritchie, M.L. and Romanuk, T.N. (2012). A meta-analysis of probiotic efficacy for gastrointestinal diseases. *PLoS ONE*, **7**(4): e34938.

Rodrigues, K.L., Caputo, L.R.G., Carvalho, J.C.T., Evangelista, J. and Schneedorf, J.M. (2005). Antimicrobial and healing activity of *kefir* and *kefiran* extract. *International Journal of Antimicrobial Agents*, 25: 404-408.

Rong, J., Zheng, H., Liu, M., Hu, X., Wang, T., Zhang, X., Jin, F. and Wang, L. (2015). Probiotic and anti-inflammatory attributes of an isolate *Lactobacillus helveticus* NS8 from Mongolian fermented *koumiss*. *BMC Microbiology*, 15(1): 196. doi: 10.1186/s12866-015-0525-2

Roopashri, A.N. and Varadaraj, M.C. (2009). Molecular characterisation of native isolates of lactic acid bacteria, bifidobacteria and yeasts for beneficial attributes. *Applied Microbiology and Biotechnology*, 83: 1115-1126.

Roopashri, A.N. and Varadaraj, M.C. (2014). Hydrolysis of flatulence causing oligosaccharides by α-D-Galactosidase of a probiotic *Lactobacillus plantarum* MTCC 5422 in selected legume flours and elaboration of probiotic attributes in soy-based fermented product. *Eur. Food Res. Technol.*, 239: 99-115.

Rosa, D.D., Dias, M.M.S., Grześkowiak, L,M., Reis, S.A, Conceição, L.L. and Peluzio, M.C.G. (2017). Milk *kefir*: Nutritional, microbiological and health benefits. *Nutrition Research Reviews*., doi: 10.1017/S0954422416000275

Rostok, M., Hütt, P., Rööp, T., Smidt, I., Štšepetova, J., Salumets, A. and Mändar, R. (2018). Potential vaginal probiotics: Safety, tolerability and preliminary effectiveness. *Beneficial Microbes*, 10(4): 385-393.

Russo, P., Capozzi, V., Arena, M.P., Spadaccino, G., Dueñas, M.T., López, P., Fiocco, D. and Spano, G. (2014). Riboflavin-overproducing strains of *Lactobacillus fermentum* for riboflavin-enriched bread. *Applied Microbiology and Biotechnology*, 98: 3691-3700.

Saadat, Y.R., Khosroushahi, A.Y. and Gargari, B.P. (2019). A comprehensive review of anticancer, immunomodulatory and health beneficial effects of the lactic acid bacteria exopolysaccharides. *Carbohydrate Polymers*, 217: 79-89.

Saber, A., Alipour, B., Faghfoori, Z. and Khosroushahi, A.Y. (2017). Cellular and molecular effects of yeast probiotics on cancer. *Critical Reviews in Microbiology*, 43: 96-115.

Şanlier, N., Gökcen, B.B. and Sezgin, A.C. (2019). Health benefits of fermented foods. *Critical Reviews in Food Science and Nutrition*, 59: 506-527.

Sari, E., Bakir, B., Aydin, B.D. and Sozmen, M. (2014). The effects of *kefir*, *koumiss*, yoghurt and commercial probiotic formulations on PPARa and PPAR-b/d expressions in mouse kidney. *Biotechnic and Histochemistry*, 89(4): 287-295.

Sarkar, S. (2019). Potentiality of probiotic yoghurt as a functional food – A review. *Nutrition and Food Science*, 49(2): 182-202.

Sarkar, S., Sur, A., Pal, R., Sarkar, K., Majhi, R., Biswas, T. and Banerjee, S. (2011). Potential of *dahi* as a functional food. *Indian Food Industry*, 30(1): 27-32.

Schiffrin, E.J., Brassart, D., Servin, A.L., Rochat, F. and Donnet-Hughes, A. (1997). Immune modulation of blood leukocytes in humans by lactic acid bacteria: Criteria for strain selection. *The American Journal of Clinical Nutrition*, 66(2): 515S-520S.

Sen, M. (2019). Role of probiotics in health and disease – a review. *International Journal of Advanced Life Science Research*, 2(2): 1-11.

Seppo, L., Kerojoki, O., Suomalainen, T. and Korpela, R. (2002). The effect of a *Lactobacillus helveticus* LBK-16 H fermented milk on hypertension: A pilot study on humans. *Milchwissenschaft*, 57(3): 124-127.

Seppo, L., Jauhiainen, T., Poussa, T. and Korpela, R. (2003). A fermented milk high in bioactive peptides has a blood pressure-lowering effect in hypertensive subjects. *American Journal of Clinical Nutrition*, 77(2): 326-330.

Shahani, K.M., Vakil, J.R. and Kilara, A. (1977). Natural antibiotic activity of *Lactobacillus acidophilus* and bulgaricus. II. Isolation of acidophilin from *Lactobacillus acidophilus*. *Cultured Dairy Products Journal*, 12: 8-11.

Shan, L., Hou, P., Wang, Z., Liu, F., Chen, N., Shu, L.H., Zhang, H., Han, X.H., Han, X.X., Cai, X.X., Shang, Y.X. and Vandenplas, Y. (2013). Prevention and treatment of diarrhoea with *Saccharomyces boulardii* in children with acute lower respiratory tract infections. *Beneficial Microbes*, 4(4): 329-334.

Shana, Sridhar, R., Roopa, B.S., Varadaraj, M.C. and Vijayendra, S.V.N. (2015). Optimisation of a novel coconut milk-supplemented *dahi* – A fermented milk product of Indian subcontinent. *Journal of Food Science and Technology*, 52: 7486-7492.

Sharma, M. and Devi, M. (2014). Probiotics: A comprehensive approach toward health foods. *Critical Reviews in Food Science and Nutrition*, 54: 537-552.

Shiby, V.K. and Mishra, H.N. (2013). Fermented milks and milk products as functional foods – A review. *Critical Reviews in Food Science and Nutrition*, 53: 482-496.

Sidooski, T., Brandelli, A., Bertoli, S.L., deSouza, C.K. and de Carvalho, L.F. (2019). Physical and nutritional conditions for optimized production of bacteriocins by lactic acid bacteria – A review. *Critical Reviews in Food Science and Nutrition.* 59(17): 2839-2849 doi: 10.1080/10408398.2018.1474852

Simova, E.D., Beshkova, D.M., Angelov, M.P. and Dimitrov, Zh. P. (2008). Bacteriocin production by strain *Lactobacillus delbrueckii* ssp. *bulgaricus* BB18 during continuous prefermentation of yogurt starter culture and subsequent batch coagulation of milk. *Journal of Industrial Microbiology and Biotechnology*, 35: 559-567.

Sofi, F., Buccioni, A., Cesari, F., Gori, A.M., Minieri, S., Mannini, L., Casini, A., Gensini, G.F., Abbate, R. and Antongiovanni, M. (2010). Effects of a dairy product (pecorino cheese) naturally rich in cis-9, trans-11 conjugated linoleic acid on lipid, inflammatory and haemorheological variables: A dietary intervention study. *Nutrition, Metabolism and Cardiovascular Diseases*, 20: 117-124.

Solanki, D. and Hati, S. (2018). Considering the potential of *Lactobacillus rhamnosus* for producing angiotensin I-converting enzyme (ACE) inhibitory peptides in fermented camel milk (Indian breed). *Food Bioscience*, 23: 16-22.

Solieri, L., Rutella, G.S. and Tagliazucchi, D. (2015). Impact of non-starter Lactobacilli on release of peptides with angiotensin-converting enzyme inhibitory and antioxidant activities during bovine milk fermentation. *Food Microbiology*, 51: 108-116.

Song, H., Kim, J., Jung, S., Kim, S., Park, H., Jeong, Y., Hong, S.P., Cheon, J.H., Kim, W.H., Kim, H.J., Ye, B.D., Yang, S.K., Kim, S.W., Shin, S.J., Kim, H.S., Sung, J.K. and Kim, E.Y. (2010). Effect of probiotic *Lactobacillus* (Lacidofil® Cap) for the prevention of antibiotic-associated diarrhoea: A prospective, randomised, double-blind, multicentre study. *Journal of Korean Medical Science*, 25(12): 1784-1794.

Stanton, C., Ross, R.P., Fitzgerald, G.F. and Sinderen, D.V. (2005). Fermented functional oods based on probiotics and their biogenic metabolites. *Current Opinion in Biotechnology*, 16(2): 198-203.

Stotzer, P.O., Johansson, C., Mellström, D., Lindstedt, G. and Kilander, A.F. (2003). Bone mineral density in patients with small intestinal bacterial overgrowth. *Hepatogastroenterology*, 50(53): 1415-1418.

Subramanian, B.S., Kumar, C.N. and Venkateshaiah, B.V. (2005). Therapeutic properties of dietetic-*shrikhand* prepared using LAB. *Mysore Journal of Agricultural Science*, 39: 399-403.

Sun, Z., Sun, X., Li, J., Li, Z., Hu, Q., Li, L., Hao, X., Song, M. and Li, C. (2019). Using probiotics for type 2 diabetes mellitus intervention: Advances, questions, and potential. *Critical Reviews in Food Science and Nutrition.*, doi: 10.1080/10408398.2018.1547268

Syal, P. and Vohra, A. (2013). Probiotic potential of yeasts isolated from traditional Indian fermented foods. *International Journal of Microbiology Research*, 5: 390-398.

Taha, S., El Abd, M., De Gobba, C., Abdel-Hamid, M., Khalil, E. and Hassan, D. (2017). Antioxidant and antibacterial activities of bioactive peptides in buffalo's yoghurt fermented with different starter cultures. *Food Science and Biotechnology*, 26(5): 1325-1332.

Tester, R. and Al-Ghazzewi, F. (2018), Role of prebiotics and probiotics in oral health. *Nutrition and Food Science*, 48: 16-29.

Tonucci, L.B., dos Santos, K.M.O., Ferreira, C.L.D.L.F., Ribeiro, S.M.R., de Oliveira, L.L. and Martino, H.S.D. (2017a). Gut microbiota and probiotics: Focus on diabetes mellitus. *Critical Reviews in Food Science and Nutrition*, 57: 2296-2309.

Tonucci, L.B., dos Santos, K.M.O., de Oliveira, L.L., Ribeiro, S.M.R. and Martino, H.S.D. (2017b). Clinical application of probiotics in Type 2 diabetes mellitus: A randomised, double-blind, placebo controlled study. *Clinical Nutrition*, 36: 85-92.

Twetman, S. and Stecksén-Blicks, C. (2008). Probiotics and oral health effects in children. *International Journal of Paediatric Dentistry*, 18: 3-10.

Vaghef-Mehrabany, E., Alipour, B., Homayouni-Rad, A., Sharif, S., Asghari-Jafarabadi, M. and Zavvari, S. (2014). Probiotic supplementation improves inflammatory status in patients with rheumatoid arthritis. *Nutrition*, 30(4): 430-435.

Venegas-Ortega, M.G., Flores-Gallegos, A.C., Martínez-Hernández, J.L., Aguilar, C.N. and Nevárez-Moorillón, G.V. (2019). Production of bioactive peptides from lactic acid bacteria: A sustainable approach for healthier foods. *Comprehensive Reviews in Food Science and Food Safety*, 18: 1039-1051.

Venkateshwari, S., Halami, P.M. and Vijayendra, S.V.N. (2010). Characterisation of the heat stable bacteriocin producing and vancomycin-sensitive *Pediococcus pentosaceus* CFR B19 isolated from beans. *Beneficial Microbes*, 1: 159-164.

Vijaya Kumar, B., Vijayendra, S.V.N. and Reddy, O.V.S. (2015). Trends in dairy and non-dairy probiotic products: A review. *Journal of Food Science and Technology*, 52(10): 6112-6124.

Vijayendra, S.V.N., Rajashree, K. and Halami, P.M. (2010). Characterisation of a heat stable anti-listerial bacteriocin produced by vancomycin sensitive *Enterococcus faecium* isolated from *idli* batter. *Indian Journal of Microbiology*, 50(2): 243-246.

Vijayendra, S.V.N. and Gupta, R.C. (2012) Assessment of probiotic and sensory properties of *dahi* and yoghurt prepared using bulk freeze dried cultures in buffalo milk. *Annals of Microbiology*, 62: 939-947.

Vijayendra, S.V.N. and Halami, P.M. (2015). Fermented vegetables – Health benefits. pp. 325-342. *In*: J.P. Thamang (Ed.). *Health Benefits of Fermented Foods and Beverages*. CRC Press, Taylor & Francis, Boca Raton, Florida, USA.

Vijayendra, S.V.N. and Varadaraj, M.C. (2015). *Dahi*, *lassi* and *shrikhand*. pp. 355-376. *In*: A.K. Puniya (Ed.). *Fermented Milk and Dairy Products*. CRC Press, Taylor and Francis, Boca Raton, Florida, USA.

Villa, C.R., Ward, W.E. and Comelli, E.M. (2017). Gut microbiota-bone axis. *Critical Reviews in Food Science and Nutrition*, 57: 1664-1672.

Villa, J.K., Diaz, M.A., Pizziolo, V.R. and Martino, H.S. (2017). Effect of vitamin K in bone metabolism and vascular calcification: A review of mechanisms of action and evidences. *Critical Reviews in Food Science and Nutrition*, 57: 3959-3970.

Volokh, O., Klimenko, N., Berezhnaya, Y., Tyakht, A., Nesterova, P., Popenko, A. and Alexeev, D. (2019). Human gut microbiome response induced by fermented dairy product intake in healthy volunteers. *Nutrients*, 11: 547. doi:10.3390/nu11030547

Walther, B., Schmid, A., Sieber, R. and Wehrmuller, K. (2008). Cheese in nutrition and health. *Dairy Science and Technology*, 88: 389-405.

Wang, Y., Xu, N., Xi, A., Ahmed, Z., Zhang, B. and Bai, X. (2009). Effects of *Lactobacillus plantarum* MA2 isolated from Tibet *kefir* on lipid metabolism and intestinal microflora of rats fed on high-cholesterol diet. *Applied Microbiology and Biotechnology*, 84: 341-347.

Wang, H., Livingston, K.A., Fox, C.S., Meigs, J.B. and Jacques, P.F. (2013). Yoghurt consumption is associated with better diet quality and metabolic profile in American men and women. *Nutrition Research*, 33(1): 18-26.

Wang, L., Guo, M.J., Gao, Q., Yang, J.F., Yang, L., Pang, X.L. and Jiang, X.J. (2018). The effects of probiotics on total cholesterol: A meta-analysis of randomised controlled Trials. *Medicine*, 97(5): e9679, doi:10.1097/MD.0000000000009679

Wendy, M. (2019). The gut-liver axis: New therapeutic targets for non-alcoholic fatty liver disease. *Journal of the Australian Traditional-Medicine Society*, 25: 50-51.

Wilkins, T. and Sequoia, J. (2017). Probiotics for gastrointestinal conditions: A summary of the evidence. *American Family Physician*, 96: 170-178.

Wong, C.B., Odamaki, T. and Xiao, J. (2019). Beneficial effects of *Bifidobacterium longum* subsp. *longum* BB536 on human health: Modulation of gut microbiome as the principal action. *Journal of Functional Foods*, 54: 506-519.

Wu, R., Wang, L., Wang, J., Li, H., Menghe, B., Wu, J., Guo, M. and Zhang, H. (2009). Isolation and preliminary probiotic selection of Lactobacilli from koumiss in Inner Mongolia. *Journal of Basic Microbiology*, 49: 318-326.

Yadav, H., Jain, S. and Sinha, P.R. (2007). Antidiabetic effect of probiotic *dahi* containing *Lactobacillus acidophilus* and *Lactobacillus casei* in high fructose-fed rats. *Nutrition*, 23: 62-68.

Yao, G., Yu, J., Hou, Q., Hui, W., Liu, W., Kwok, L.Y., Menghe, B., Sun, T., Zhang, H. and Zhang, W. (2017). A perspective study of koumiss microbiome by metagenomics analysis based on single-cell amplification technique. *Frontiers in Microbiology*, 8, doi: 10.3389/fmicb.2017.00165

Yousf, H., Tomar, G.B. and Srivastava, R.K. (2015). Probiotics and bone health: It takes guts to improve bone density. *International Journal of Immunotheraphy and Cancer Research*, 1: 18-22.

Yu, A. and Li, L. (2016). The potential role of probiotics in cancer prevention and treatment. *Nutrition and Cancer*, 68: 535-544.

Zacharof, M.P. and Lovitt, R.W. (2012). Bacteriocins produced by lactic acid bacteria – A review article. *APCBEE Procedia*, 2: 50-56.

Zhang, K., Dai, H., Liang, W., Zhang, L. and Deng, Z. (2019). Fermented dairy foods intake and risk of cancer. *International Journal of Cancer*, 144: 2099-2108.

Zhang, W. and Zhang, H. (2012). Fermentation and *koumiss*. pp. 165-172. *In*: Y.H. Hui, E.Ö. Evranuz, R.C. Chandan, L. Cocolin, E.H. Drosinos, L. Goddik, A. Rodríguez and F. Toldrá (Eds.). *Handbook of Animal-based Fermented Food and Beverage Technology*, CRC Press, Tylor & Francis Group, Boca Raton, Florida.

Bioactive Compounds in Fermented Fish and Meat Products: Health Aspects

Sri Charan Bindu Bavisetty[1]*, Soottawat Benjakul[2], Oladipupo Odunayo Olatunde[2] and Ali Muhammed Moula Ali[1]

[1] Faculty of Agro-Industry, King Mongkut's Institute of Technology Ladkrabang, Bangkok-10520, Thailand
[2] Department of Food Technology, Faculty of Agro-Industry, Prince of Songkla University, Hat Yai, Songkhla, 90112, Thailand

1. Introduction

Meat and fish are classified as a protein-rich food group in the food/diet pyramid (Lall and Anderson, 2005; Stadnik and Kęska, 2015), providing an excellent source for well-balanced essential amino acids, especially sulphur-containing amino acids, which have high biological value (Lall and Anderson, 2005; Stadnik and Kęska, 2015). Nevertheless, the perishable nature of fresh meat or meat products, attributed to their biological pH and nutrition value, have brought about the evolution of ancient technologies, such as curing and fermentation to prolong their shelf-life, conserve nutritional value and enhance organoleptic properties (Toldra, 2011). Various fermented foods are prepared via indigenous solid substrate fermentation process, which is either by native microflora or by addition of starter cultures, including bacteria, such as *Lactobacillus, Streptococcus, Lactococcus, Bifidobacterium*, etc. (Pennacchia *et al.,* 2004), along with yeast or mold (Giraffa, 2004). Amongst these microbes, selective viable Lactic Acid Bacteria (LAB) are known to exhibit probiotic properties with the production of several beneficial metabolites during fermentation. These include bacteriocins, lactic acid, acetic acid, alcohols, aldehydes, ketones and many others, which provide a means of producing well-preserved and safe fermented meat products (Kittisakulnam *et al.,* 2017).

In addition to microbial synthesised compounds, during fermentation, many indigenous and microbial enzymes including proteases and lipases show a remarkable breakdown of macromolecules from meat, thus generating more simpler compounds

---

*Corresponding author: sricharanbindu.ba@kmitl.ac.th

with enhanced functional properties and are easy to digest when consumed (Frece *et al.*, 2014; Martínez-Álvarez *et al.*, 2017). Hydrolysis process is associated with several complex biochemical reactions and physical changes, which trigger the formation of flavour and enhance nutritional and organoleptic properties of the final product (Toldrá, 2011). Amongst all the compounds produced, a great number are known to deliver various bio-functionalities, known as bioactive compounds. These may include bioactive peptide, amines, free fatty acids, amino acids, alcohols such as phenyl ethyl alcohol as well as hydroxy acids, such as indole acetic acid, hydroxyl phenyl lactic acid, and phenyl lactic acid and others (Jiménez-Colmenero *et al.*, 2001; Kittisakulnam *et al.*, 2017; Stadnik and Kęska, 2015).

Bioactive compounds are associated with beneficial properties, such as antioxidative, anticancer, antiaging, and antiobesity and with protection against cardiovascular diseases, Alzheimer's disease, ultraviolet radiation (UVR) and migraine prevention (Hayes *et al.*, 2007a; Hayes *et al.*, 2007b; Martín-Cabrejas *et al.*, 2017). Much attention has been drawn towards meat and fish, particularly fermented products based on their bioactive compounds, such as anserine, carnosine, conjugated linoleic acid, L-carnitine, taurine, glutathione and other peptides. Their biological activities involve the preventive effect against cancer and other chronic diseases, such as cardiovascular and diabetes (Rajauria *et al.*, 2016; Stadnik and Kęska, 2015). In addition to bio-preservation against spoilage microbes, some bioactive compounds enhance the safety by inhibiting the growth of various pathogens, such as *Enterococcus faecalis, Staphylococcus aureus* and *Listeria monocytogenes* (Hayes and García-Vaquero, 2016).

This chapter emphasises on all the aspects related to bioactive compounds from fermented fish and meat with potential health benefits, including nutritional factors, probiotic and antioxidant properties. Methods for preparing fermented fish and meat are also included.

2.   Fermentation Process and Products

Fermentation is a process that leads to desirable biochemical changes in food via the activities of enzymes or microorganisms (Campbell-Platt, 1987).

2.1   Fermentation Process

Depending on the food material, fermentation may involve the breakdown of polysaccharides into smaller molecules, such as di- or mono-saccharides and also transformation of proteins into smaller peptides, which are easily digested or absorbed into the body. Fermentation is associated with acceptable sensory qualities of some particular products with unique quality attributes. Fermentation in food could either be spontaneous (uncontrolled) or non-spontaneous (controlled) (Martínez-Álvarez *et al.*, 2017). During fermentation, endogenous and exogenous proteases from microorganisms (bacteria, yeasts or molds) as well as indigenous meat and fish digestive proteolytic enzymes induce hydrolysis of muscle proteins. These enzymes cleave macromolecules, such as proteins into peptides, amino acids, or ammonia and fat/lipids to free fatty acids, which contribute to the characteristic taste (Ghaly *et al.*, 2013). Furthermore, metabolites, such as alcohols, aldehydes, organic acids, ketones, esters and other compounds produced by the fermenting microorganisms also provide the distinctive odour and aroma of fermented fish (Salampessy *et al.*, 2010).

2.1.1 Spontaneous Fermentation

Spontaneous fermentation, also known as natural fermentation, is the oldest method of fermentation. It is usually carried out by the resident microorganisms or microflora naturally present in food (Rajauria *et al.*, 2016). Due to the traditional process employed or properties of the raw material, the sensory qualities of spontaneous fermented food can be more varied than the controlled process (Ojha *et al.*, 2015; Rajauria *et al.*, 2016). Spontaneous fermentation involves salting, which favours the growth of halophilic and halotolerant microorganisms, ripening and occasionally smoking and drying (Skåra *et al.*, 2015). In fish fermentation, it is classified, depending on the amount of salt added, as low-salt (6-8 per cent of total weight), high-salt (more than 20 per cent), and no-salt products (Martínez-Álvarez *et al.*, 2017). The growth of halotolerant and halophilic microorganisms is favoured by the salting process, which reduces the water activity and prevents the growth of both spoilage and pathogen microorganisms, particularly at high-salt concentration (Lopetcharat *et al.*, 2001).

The major microorganisms contributing to fermentation of salted fish depend on the type of fish, salt concentration and temperature. Gassem (2017) reported that bacteria species, such as *Bacillus* (i.e. *B. licheniformis, B. subtilus, B. pumilus, B. mycoides*) and *Staphylococcus* (i.e. *St. epidermidis, St. cahnii* subsp. *cahnii. St. xylosus, St. aureus, St. saprophticus* and *St. hominis*) were the major microorganisms contributing to fermentation in salted fish (Hout-Kasef) having 15 per cent salt. Numerous varieties of microorganisms with diversity have been isolated during the traditional and commercial fermentation of fish, in which proteolytic bacteria, such as *Pseudomonas* and *Bacillus* (Saadi *et al.*, 2015) as well as lactic acid bacteria (LAB) (Martínez-Álvarez *et al.*, 2017) were the most dominant.

LAB, including *Lactobacillus pentosus* and *Lb. plantarum*, were the predominant bacteria in *som-fak*, which is a fermented fish prepared with a blend of fish, sucrose, rice, garlic and salt (Paludan-Müller *et al.*, 2002). Dai *et al.* (2013) identified genera *Vagococcus, Staphylococcus, Lactococcus, Macrococcuss, Enterococcus* and other species of LAB in traditional Chinese fermented-fish product (*chouguiyu*) prepared with fish, brine salt and a mixture of condiments including aniseed, Chinese anise, Chinese red pepper, ginger, cumin, scallion and clover, in which *Lb. sakei* was the dominant species during the fermentation process. *Pediococcus* and *Lactobacillus* and to a lesser extent *Leuconostoc, Tetragenococcus,* and *Lactococcus* were identified as the major bacteria associated with the fermentation of *narezushi* (modern Japanese sushi) (Koyanagi *et al.*, 2011). *Lb. plantarum* and *Pediococcus pentosaceus* were isolated from traditional fermented whole carp fish (Zeng *et al.*, 2015). Asiedu and Sanni (2002) isolated *Pediococcus, Lactobacillus, Staphylococcus, Pseudomonas, Debaromyces, Bacillus, Klebsiella* and *Candida* from traditional fermentation of *Enam Ne–Setaakye*, a West African fermented fish-carbohydrate product made from minced fish flesh (75 per cent), yam (20.5 per cent), onions (1 per cent), ginger (1 per cent) and salt (2.5 per cent). *Lactococcus garvieae, Weissella cibaria, Streptococcus bovis, Pd. pentosaceus, Lb. fermentum*, and *Lb. plantarum* were identified as the major fermenting microorganisms in *plaa-som* (Thai fermented fish with salt, garlic and rice) (Kopermsub and Yunchalard, 2010). *Carnobacterium divergens, Enterococcus faecalis, Lb. delbrueckii* subsp. *delbrueckii* and *Tetragenococcus muriaticus* were identified in *Burong Bangus* (Philippine fermented milk fish *chanos chanos*-rice mixture) (Arcales and Alolod, 2018).

For meat, the most important microorganisms responsible for transformation and maturation of the product are LAB. These belong to genera *Lactobacillus, Staphylococcus, Pediococcus, Enterococcus, Staphylococcus, Micrococcus* and *Leuconostoc* (De Vuyst *et al.*, 2008). However, some fermented sausages are influenced by the development of some surface flora, generally yeasts and molds, belonging to the genus *Hansenii* and *Penicillium*, respectively (Núñez *et al.*, 2015).

2.1.2   Non-spontaneous Fermentation

Non-spontaneous fermentation, also known as controlled fermentation, is implemented to assure the uniformity of the final fermented product, since the microflora of raw materials may vary due to several factors (Martínez-Álvarez *et al.*, 2017). Non-spontaneous fermentation involves the use of a starter culture or enzymes to jump-start the fermentation of food with the purpose of accelerated production of the products with consistent quality and superior sensory properties (Ojha *et al.*, 2015). Starter-cultures used for fermentation are those isolated from the native microflora of spontaneous fermented products, which contribute greatly to the quality of the selected products. Hu *et al.* (2008) demonstrated the improvement of functionality and characteristics of fermented minced silver carp muscle in terms of sensory attributes, nutritional value and digestibility by fermenting the fish with mixed starter cultures including *Pediococcus pentosaceus* ATCC33316, *St. xylosus*-12, *Lb. casei* subsp. casei-1001, and *Lb. plantarum*-15. The quality of *guedj* (a traditional Senegalese fermented-fish product) improved after fermentation with a nisin-producing starter culture (*Lb. lactis* subsp. *lactis* CWBI B1410). Enteric bacteria in the final product were also reduced (Diop *et al.*, 2009).

Some yeasts have also been used for the fermentation of fish, in which undesirable compounds or metabolites can be reduced. Biogenic amine content can be greatly reduced in silver carp sausages after fermentation with mixed starters, including *Lb. plantarum* ZY40 and *Saccharomyces cerevisiae* JM19. In addition, the growth of Enterobacteriaceae and *Pseudomonas*, associated with the accumulation of cadaverine and putrescine, was inhibited in a mixed-culture fermented fish as compared to traditional fermentation (Nie *et al.*, 2014). Barley *koji* (barley steamed and molded with *Aspergillus oryzae*) and halo-tolerant starter microorganisms, such as *Candida versatilis, Zygosaccharomyces rouxii* and *Tetragenococcus halophilus* were used as starters for the fermentation of chum salmon (*Oncorhynchus keta*) sauce, yielding the increased free fatty acid, ethanol contents along with soy-sauce-like flavour (Yoshikawa *et al.*, 2010).

2.2   Products

2.2.1   Fermented Fish

As a general rule, fermented fish products are used as a umami condiment or dishes (Hayes and García-Vaquero, 2016). Methods for preparation of fermented fish between different regions of the world vary. For example, fish is fermented from several months up to one year in the southeast Asian sub-region. The resulting final product can be classified into three forms including: (1) liquid form with different viscosity, which is known as fish sauce, for instance *nam-pla* or *nuoc-mam* from Thailand and Vietnam, respectively; (2) solid-liquid form, also known as fish paste, which could contain tiny pieces of meat, such as *ngapi* (Burma), *prahoc* (Kampuchea), or *bagoong* (Philippines), ground (comminuted by either pounding or grinding) or unground; and (3) solid form (lactofermented fish) in which fish is fermented with a source of sugar (carbohydrates), such as steamed rice. These carbohydrate sources generally accelerate the fermentation process, provide a substrate for beneficial microorganisms, such as LAB, bifidobacteria and yeast and can impact distinctive flavour in the final product (Rajauria *et al.*, 2016). The fermentation of fish in Europe lasts from several days to two years. Occasionally, spices and sugars are added to the brine as additives, while in Africa, the fish are fermented for a shorter time (usually for a few days to several weeks) as compared to Asia and Europe, where the final product is not transformed into a sauce or paste (Anihouvi *et al.*, 2012). In Thailand, for example, *Pla-som* is produced from the mixture of fish, salt and sugar after being fermented for eight days at room temperature by using spontaneous, back-slopping and starter culture fermentation methods (Hwanhlem *et al.*, 2011; Onsurathum *et al.*, 2016). Also, *Som-fug* is produced from a mixture of minced fish, minced garlic, ground steamed rice and salt and fermented for five days at room temperature (Kongkiattikajorn, 2015; Riebroy *et al.*, 2008).

2.2.2   Fermented Meat

A variety of fermented meat products have been produced, depending on the process of preparation and the region where they are produced. Based on the preparation process, fermented meat products are divided into two types – fermented sausages (prepared from comminuted meat mostly using starter cultures) and salted/cured meat products (Bou *et al.*, 2017; Menegas *et al.*, 2013). Fermented sausages are prepared from comminuted meat and fermented with the help of starter culture and are categorised according to the moisture/protein ratio as dry, semi-dry and spreadable (unripened and ripened). The dry sausages are ripened for some period of time and are eaten uncooked, whereas semi-dry sausages are ripened for a short time and need to be cooked before being consumed. Nevertheless, the characteristics of all fermented sausages differ, depending on the variables, such as the type of meat, starter culture, spices, ripening process and other additives (Lücke, 1994). Salted/cured meat products are prepared by ripening the entire muscles or cuts through indigenous enzymes, mostly proteases. Again, the final product may vary, depending on the texture of meat, fat content, ripening parameters, region, nutritional value and microbial load (Toldrá, 2011). Hence, there are a number of fermented meat products in different regions or even within the same region. *Nham* is one of the most popular fermented meat products (sausage) of Thailand. *Nham* is mainly made from minced pork and cooked rice is used as a carbohydrate source for LAB fermentation. The final fermented product usually has a pH of 4.4 to 4.8 (Visessanguan *et al.*, 2006).

3. Microorganisms Associated with Health Aspects

3.1   Probiotic Microorganisms in Fermented Meat

Food safety is a main priority for both consumers as well as meat industry. Over the years, importance of safety assessment related to public health has gained great attention. Different strategies have been designed and implemented to ensure the safety of meat-based products. Among all the strategies, fermentation has gained an increasing interest. In addition to biopreservation, it also favours the development of unique and desirable properties of the final product, such as palatability, flavour, colour and preservation of nutrients (Lücke, 1994). The most common fermented meat products include different types of sausages, salami and ham, and their fermentation can be either spontaneous or by addition of selective starter microorganisms (single species or a mixture of several strains).

LAB have been used as a probiotic and when taken up in adequate amounts, they deliver some health benefits (Parvez *et al.*, 2006). According to WHO/FAO, the main criterion for a probiotic strain is to confer a health benefit on the host (FAO/WHO, 2001; Parvez *et al.*, 2006). Several approaches have been followed to evaluate the probiotic effect of fermented meat on human health. For instance, numerous probiotic strains, especially LAB, have been screened among the naturally present bacteria in fermented meat or those used as starter cultures from meat sources, to evaluate their *in vitro* pathogen-inhibition effect. To be considered safe when consumed, the probiotic microorganism should be of genera or species typically present in the human gut.

In general, to get potential health benefits, several criteria for selection of probiotics have been suggested (Arihara, 2006; De Vuyst *et al.*, 2008; Parvez *et al.*, 2006; Pennacchia *et al.*, 2004). Those are shown as follows:

(a) Antimicrobial activity against pathogenic bacteria.
(b) Tolerant to the action of gastric juices (bile salts, gastric acids and proteolytic enzymes).
(c) Adhere to epithelial and mucosal surfaces.
(d) Interact with resident microbiota and prevent adhesion and colonisation of pathogens (probiotic host interaction, reduce the risk of disease).
(e) Resistant to technological processing (effect of food ingredients, mechanical and heat treatments) and storage conditions.

Another crucial factor is the dose levels of probiotic – a total of 10^8 to 10^9 probiotic microorganisms should be consumed per day to ensure the probiotic effect for the consumer (Kechagia *et al.*, 2013). In general, probiotic microorganisms have to pass through the barriers of stomach as well as the small intestine and survive in the large intestine to confer the health benefits to the host (FAO/WHO, 2001).

In this context, strains of LAB and *Bifidobacteria* are known to be significant. Among them, the genera *Lactobacillus* and *Bifidobacterium* species are most commonly found in fermented meat products. The bacteria belonging to these genera are designated as generally-recognised-as-safe (GRAS) (De Vuyst *et al.*, 2008). Other bacteria and some yeasts found in fermented meat also show probiotic properties, but there are some concerns related to their safety (Table 1). The most frequently isolated LAB strains from fermented meat products are *Lb. plantarum, Lb. curvatus* and *Lb. sakei*. They play an important role in inhibiting the growth of undesired microbiota and in controlling the ripening process (Ranadheera *et al.*, 2017). Several commercially-available probiotic strains have been employed as starter cultures for fermented meat products. *Pd. acidilactici* PA-2, *Lb. sakei* Lb3, *Lb. rhamnosus* GG and LC-705, etc. may be of interest because of their ability to survive under simulated gastrointestinal conditions and deliver various health benefits (Arihara, 2006; Erkkilä *et al.*, 2001; Ranadheera *et al.*, 2017). In addition to health benefits, LAB can help to assure the safety (Table 1).

Probiotic microorganisms exert various processes of action to deliver the beneficial properties, although the exact mechanism of some properties is not fully elucidated. The major bioactive compounds produced by LAB during fermentation include organic acids (mainly lactic and acetic acid), bacteriocins (pediocin, nisin, lacticin, reuterin, reutericyclin, enterocin etc.), peptides or bacteriocin-like substances, which inhibit the growth of undesired microbiota. Other substances, such as hydrogen peroxide, short chain fatty acids, diacetyl, acetoin, alcohols, antifungal compounds (phenyl-lactate, cyclic dipeptides, propionate, 3-hydroxy fatty acids, etc.) and some exopolysaccharides also exhibit antimicrobial property (De Vuyst *et al.*, 2008; Kechagia *et al.*, 2013; Noonpakdee *et al.*, 2003; Parvez *et al.*, 2006). LAB strains like *Brevibacterium longum* ATCC 15708, *Lb. reuteri* ATCC 55730, *Pd. pentosaceus* Mees 1934 and *Lb. rhamnosus* FERM P-15120 play a role in inhibition of deadly pathogens, including *Escherichia coli* O157:H7, *L. monocytogenes* and *Staphylococcus aureus*, which are important for both promoting health and providing food safety (Anastasiadou *et al.*, 2008; Muthukumarasamy and Holley, 2007; Sameshima *et al.*, 1998).

A wide variety of health benefits have been attributed to probiotics and are shown to be strain-specific (Table 1). Some of the well-established claims include improvement of intestinal health, improved digestion, supply and bioavailability of nutrients and growth factors, balancing gut microbiota, modulation/stimulation of immune system, beneficial effects on inflammatory bowel diseases, reduction of serum cholesterol and cancer prevention effects (Caballero *et al.*, 2016; FAO/WHO, 2001; Górska *et al.*, 2014; Valeur *et al.*, 2004). Additional studies are needed to establish and approve the claims, including improvement of lactose metabolism, reduction of atopic diseases and risk of allergies and prevention of antibiotic-associated diarrhoea (Kechagia *et al.*, 2013; Parvez *et al.*, 2006).

3.2   Probiotic Microorganisms in Fermented Fish

Unlike fermented meat, where the starter cultures with various health benefits have been preferred for fermentation, fermentation of fish is mostly a spontaneous process, where the natural microflora are responsible. However, the dominant

Table 1: Starter Cultures Used in Some Fermented Meat Products and Their Health Aspects

Strain	Fermented Product	Reported Health Benefits	References
Lactobacillus			
Lb. curvatus 54M16	fermented sausages, Italy	Antibacterial against various L. monocytogenes stains, Bacillus cereus and the meat spoilage Brochotrix thermosphacta,	Giello et al. (2018)
Lb. johnsonii BS15	Isolated from homemade yogurt, Starter culture for fermented Chinese sausage	Promotes growth and performance, improves lipid metabolism and reduce fat deposition, modulates gut microflora, reduces risk of attenuate inflammation and mitochondrial injury, preventive effect against non-alcoholic fatty liver disease	Wang et al. (2017)
Lb. rhamnosus GG, LC-705 and LOCK900 USE	Dry fermented sausage, Finland	Adheres to human intestinal cells, stimulates intestinal microbial ecosystem, prevents diarrhoea, modulates immune response, lowers faecal enzyme activities, lowers allergies caused by food	Erkkilä et al. (2001), Wójciak et al. (2017)
Lb. plantarum WCFS1 and NCDC No. 020	Raw fermented chicken sausage, India	Modulates immune response to third-party antigens, therapy against inflammatory bowel disease and reverting antigen-dependent immune response in allergies	Geeta and Yadav (2017), Górska et al. (2014)
Lb. casei LOCK 0900	Raw fermented sausage, Poland	Stimulates bone marrow-dendritic cells and embryonal kidney cells	Górska et al. (2014), Trząskowska et al. (2014)
Other lactic acid bacteria			
Lactococcus lactis subsp. lactis strain CWBI B1410,	Fermented lean fish (Podamasys jubelini) and fat fish (Arius heudelotii)	Improves safety	Diop et al. (2009)

Organism	Product	Function	Reference
Enterococcus faecalis EF37	Dry fermented sausage, Italy	limits urinary tract infections, hepatobiliary sepsis, endocarditis, surgical wound infection, bacteraemia and neonatal sepsis	Gardini *et al.* (2008)
Pediococcus pentosaceus Mees 1934	Fermented Greek sausage	Modulates intestinal microflora, not active against LAB strains, Highly active against *L. monocytogenes* and *L. inocua, C. sporogenes* and *C. thyaminolyticum*	Anastasiadou *et al.* (2008)
Lactococcus lactis WNC 20	Thai fermented sausage	Inhibits food borne pathogens *Staphylococcus aureus, Listeria monocytogenes, Clostridium perfringens* and *Bacillus cereus.*	Noonpakdee *et al.* (2003)
Leuconostoc mesenteroides	Fermented minced herring	Improves safety, quality, and nutritive value	Gelman *et al.* (2000)
Pediococcus pentosaceus strains L and S	Fermented mackerel sausage.	Improves safety by decreasing the accumulation of volatile basic nitrogen and undetectable residual nitrite	Yin and Jiang (2001)
Bifidobacterium species			
Bifidobacterium longum KACC 91563	Fermented Sausage, Korea	Prevents and treats allergic disease through mast cell suppression	Song *et al.* (2018)
Bifidobacterium animalis BB-12	Dry-Fermented Pork Neck and Sausage, Poland	Reduces the risk of acute infectious in infancy	Taipale *et al.* (2010), Wójciak *et al.* (2017)
Yeast			
Debaryomyces hansenii	Spanish fermented sausage	Adheres to the intestinal tract, modulates innate immunity	Encinas *et al.* (2000)
Saccharomyces boulardii	Spiced fermented sausage, Denmark	Biotherapeutic agent (diarrhea, clostridium difficile) and treatment of acute childhood diarrhoea	Kurugöl and Koturoğlu (2005), Olesen and Stahnke (2000)

(Contd.)

Table 1: (*Contd.*)

Strain	Fermented Product	Reported Health Benefits	References
Mixed cultures			
Lb. plantarum CCRC10069, *Lc. lactis* subsp. *lactis* CCRC 12315, *Lb. helveticus* CCRC 14092, and their combination	Fermented minced mackerel	Suppresses blood pressure and decreases blood cholesterol and glucose.	Yin *et al.* (2002)
Mixed culture including group one – *Lb. plantarum*-15, *Staphylococcus xylosus*-12 and *Pd. pentosaceus*-ATCC33316 group two: *Lb. plantarum*-15, *Staphylococcus xylosus*-12 and *Lb. casei subs casei*-1.001 group three: *St. xylosus*-12, *Lb. casei subsp. casei*-1.001 and *Pd. pentosaceus*-ATCC33316	Fermented silver carp sausages	Decrease accumulation of by-product from initial microflora Improves safety	Hu *et al.* (2008)
Lb.casei 6002, *Streptococcus lactis* 6018, *Saccharomyces cerevisiae Hansen* 1049 and *Monascus anka* 5037	Fermented grass carp fish	Improves safety by decreasing the accumulation of biogenic amine and also inhibiting the growth of pathogenic bacteria	Liu *et al.* (2010)

microbiota in the final product are LAB. Nevertheless, very limited studies/trails have been conducted on their health benefits to humans, either *in vivo* or *in vitro*. Fermented foods, particularly LAB- and yeast-fermented foods have been known to have probiotic potential, such as the colonisation of the Gastro Intestinal Tract (GIT) with beneficial bacteria or antimicrobial activities of bioactive compounds, including bioactive peptides from food and other secondary metabolites from the fermenting organisms (bacteriocins and antibiotics) against pathogenic bacteria (Linares *et al.*, 2017; Olatunde *et al.*, 2018). Fermented fish could be a source of probiotic either by the activities of the bacteria responsible for the fermentation or by the bioactive peptides produced during the fermentation process (Hayes and García-Vaquero, 2016).

LAB, isolated from fermented fish, are reported to possess antimicrobial properties against pathogenic bacteria. LAB could inhibit the growth and proliferation of pathogenic bacteria by lowering the pH of the system as well as producing bioactive peptides, such as bacteriocins (Olatunde *et al.*, 2018). Liasi *et al.* (2009) isolated three species of *Lactobacillus,* which were identified as *Lb. paracasei* LA02, *Lb. casei* LA17, and *Lb. plantarum* LA22 from *budu* (traditionally fermented fish product). The identified isolates inhibited the growth of both Gram-positive (*Bacillus cereus, L. monocytogenes*, and *St. aureus*) and Gram-negative bacteria (*Salmonella enterica* and *E. coli*), mostly attributed to the ability of the isolate in producing bacteriocin (peptidic antibacterial agent). LAB isolated from *bekasam* (Indonesian fermented fish product) inhibited the growth of tested pathogenic microorganisms, including *L. monocytogenes, Salmonella typhimurium* ATCC14028, *B. cereus, E. coli,* and *St. aureus* (Desniar, 2013). The growth of *Pseudomonas aeruginosa* ATCC 27853 and *St. aureus* ATCC 6538 was inhibited by 27 LAB isolates out of 57 LAB identified from *peda* (Indonesian traditional fermented whole fish) (Putra *et al.*, 2018). Fall *et al.* (2018) reported that 16 strains out of 19 isolated from fermented Machoiron (*Arius latisculatus*), Capitain (*Pseudotolithus brachygnatus*) and Sompatte (*Pomadasys jubelini*) showed promising antimicrobial activity against *E. coli* and *L. monocytogenes*. However, after organic acids and catalase treatment, only nine and four strains, respectively, retained their inhibitory activity against the tested indicators. Probiotic potential of LAB isolated from *bekasam* (Indonesian fermented tilapia fish) was investigated by Nurnaafi and Setyaningsih (2015). Only one isolated LAB identified as *Lb. plantarum* was able to survive for testing at low bile salt concentration (0.5 per cent), pH (2 and 7.2). It could inhibit the growth of *E. coli* and *S. iyphimurium* ATCC 14028.

4. Bioactive Compounds Generated during Fermentation

Several compounds have been considered as beneficial bioactive components for improving health status. A bioactive compound should beneficially affect health without causing or triggering potential damaging effects, such as mutagenicity, allergenicity and toxicity (Sánchez and Vázquez, 2017). Several peptides with bioactive properties have been documented in different fermented fish products.

4.1   Peptide-based Bioactivity

Muscle protein-based peptides are one of the promising bioactive compounds from fermented fish and meat. In fermented meat products, proteolysis is one of the significant biochemical changes, during processing and ripening, in which some sarcoplasmic and myofibrillar proteins undergo degradation. Dry curing of ham is a lengthy process, which takes from several months up to few years. During the process, water activity reduces and proteolysis mediated by indigenous enzymes gradually increases. The major proteases involved in curing of ham are endoproteases (calpains I and II and cathepsins B, D, H and L), peptidases (tripeptidylpeptidases I and II, dipeptidylpeptidases I, II, III and IV) and aminopeptidases (arginyl, alanyl, methyl aminopeptidases and leucyl) (Toldra, 2006; Toldra *et al.*, 1993). On the other hand, fermented sausages are particularised with a shorter process using microorganisms, such as LAB. In addition to indigenous enzymes, proteolytic enzymes from microbial origin are involved in hydrolysis of meat proteins (Anastasiadou *et al.*, 2008). The microbial enzymes from some strains of Lactobacilli and yeasts are known to show a higher degree of protein hydrolysis, especially the myofibrillar and sarcoplasmic proteins (Flores and Toldrá, 2011). The bacterial strains isolated from fermented meat have been reported to have a considerable quantity of aminopeptidases and peptidases including tripeptidase, dipeptidase, arginine aminopeptidase, X-prolyl-dipeptidylpeptidase and aminopeptidase (Flores and Toldrá, 2011; Sanz and Toldrá, 1997; Sanz and Toldrá, 2002). Moreover, LAB also contribute to reduction of pH, which enhances the activity of some muscle proteolytic enzymes (Molly *et al.*, 1997). In some variety of fermented sausages, hydrolysis of muscle protein is also caused by the yeast *Debaryomyces hansenii*, which produces a major class of endoproteases, such as proteases A, B and D and also exopeptidases, including arginyl and prolyl aminopeptidase (Bolumar *et al.*, 2008; Flores and Toldra, 2011).

Considering the diversity of fermented fish products, in some products where the entire fish is used for fermentation, most of the proteolysis is mainly mediated by digestive enzymes in conjunction with proteolysis by indigenous and microbial enzymes (Huss, 1988). Proteinases, such as chymotrypsin, pepsin, trypsin, thermolysin, alcalase and pancreatin are the digestive enzymes commonly found in fish and these are mainly involved in fermentation of fish (Bhat *et al.*, 2015). The proteolytic activity varies, mainly depending on fish species, activity of digestive enzymes and difference in microbiota in environment. In addition to the natural process of fermentation, many fish products, such as fermented fish sauces are produced by adding enzymes from external sources for accelerated production (Salampessy *et al.*, 2010). Trypsin and chymotrypsin are the most common external enzymes used for fermented fish products (Owens and Mendoza, 1985).

The proteolysis by cathepsins, calpains and peptidases during meat fermentation generates a large quantity of peptides and free amino acids. Proteins contain a certain sequence of amino acids, which are inactive within the native protein molecules (Acquah *et al.*, 2019). Once the proteins are hydrolysed, peptides are liberated and can generate functional and regulatory mechanism (Bhat *et al.*, 2015) In view of their promising activity, peptides from meat proteins have possible bioactivities beyond high nutritional value. Bioactive peptides are believed to play a role like a hormone or a drug by eventually binding specifically to target cell receptor and modulating the physiological function in the body (Fitzgerald and Murray, 2006). In most of the cases, bioactive peptides are of two to 20 amino acid residues, which are in an array of specific sequence. The activity mainly depends on the composition and sequence of amino acids, the type of amino acid at N- and C-terminal, length and molecular weight of peptide chain, charge on amino acid, hydrophilic/hydrophobic nature, etc. Several peptides derived from hydrolysis of meat proteins during fermentation have been fractionated and characterised for their bioactivity (Stadnik and Kęska, 2015).

Although there is limited information on generation of bioactive peptides from fermented fish and meat products, an approach using commercial proteases is an efficient way to generate bioactive peptides from fish and meat, which mimic the actual hydrolysis during excessive fermentation. Peptides with antioxidative, antihypertensive (Angiotensin Converting Enzyme (ACE) inhibitory), antimicrobial, hypocholesterolemic and immunomodulatory activities, etc. have been reported in protein hydrolysate (Desniar, 2013; Fitzgerald and Murray, 2006; Geeta and Yadav, 2017; Jiménez-Colmenero *et al.*, 2001). Among the various bioactivities inferred by peptides, antioxidant and ACE-inhibitory are most extensively studied (Fernández *et al.*, 2016; Ohata *et al.*, 2016). Therefore, proteins and peptides from fermented fish and meat have possible bioactivity beyond their nutritional facts.

4.1.1 Proteolysis and Bioactive Peptides in Fermented Fish and Meat

4.1.1.1 Antihypertensive Properties

ACE inhibitory peptides have gained much attention due to their ability to prevent hypertension. ACE mainly converts octapeptide to bradykinin; further inactivation of bradykinin results in depressed action (Donoghue *et al.*, 2000). ACE inhibitory peptide can suppress elevated blood pressure by inhibiting the action of ACE. Various fermented fish products, especially those fermented for long term, such as fish sauces, though they have high sodium chloride (20-25) concentration, they are still useful as a source of antihypertensive peptides (Okamoto *et al.*, 1995). Sardine fermented fish sauce from Japan had substantial ACE inhibitory activity, mainly from peptides containing proline in the sequence. These peptides, when administered orally, have a tendency of lowering blood pressure of spontaneously hypertensive rats (Ichimura *et al.*, 2003). Peptides extracted in ethanol fraction from salmon fish sauce consist of tyrosine as the common amino acid in the sequence exhibiting the strongest ACE inhibitory activity with lowest IC_{50} value. They showed anti-hypotensive effect in SHR when administered intra-peritoneally at a level of 1.8 µM (Okamoto *et al.*, 1995). In addition to fish sauce, different types of fermented fish were reported to possess ACE inhibition effect like *heshiko* and *narezushi,* fermented mackerel products from Japan (Itou *et al.*, 2007), *rakfisk*, a Norwegian fermented trout muscle (Sørensen *et al.,* 2004), and *ngari* a traditional salt-free fermented fish product in the northeastern region of India (Phadke *et al.*, 2014). A strong ACE inhibitory activity was also reported for peptides of *kapi* with MW less than 500 Da (Kleekayai *et al.*, 2015).

ACE inhibitory peptides derived from meat have been studied extensively. These peptides have gained much attention due to their ability in preventing hypertension and cardiovascular diseases (Kęska and Stadnik, 2018). These peptides can serve as indigenous functional compounds present in fermented meat products or as an additive (Table 2). Choe *et al.* (2018) characterised ACE inhibitory peptide from beef loins (whole muscle) stored at refrigeration (5°C) for three days after treating with enzyme thermolysin. Three peptides having ACE inhibitory activity (V-15, V-m1 and V-m2) were fractioned with IC_{50} value of 0.89, 2.69, and 3.09 mM, respectively. Their activities were far superior to other reported bioactive peptides. ACE inhibitory peptides have also been found in fermented meat products. Kęska and Stadnik (2018) reported ACE inhibiting peptides from dry cured pork loins inoculated with probiotic LAB and cured over 360 days. Curing for 180 days was sufficient to generate these ACE inhibitory peptides. In whole products, a total of 48 bioactive peptides were generated. These consisted of 6-22 amino acids, where Pro, Ala, Lys, and Glu amino acids constituted as major amino acids. Most of the peptides showed the potential ACE inhibitory activity, mainly di- and tripeptides, contained hydrophobic amino acids (particularly with aliphatic chains, i.e. Leu, Ile, Gly and Val) at the N-terminal and aromatic residues (Trp, Phe and Tyr) at the C-terminal (Acquah *et al.*, 2019). Fernández *et al.* (2016) studied the liberation of ACE inhibitory peptides in Iberian dry-fermented sausage, *salchichón*, using various starter cultures ripened for 90 days and with another batch having protease added (EPg222). Highest activity and proteolysis were observed with EPg222 at 63 days of ripening (starter culture was around 10^7 CFU g^{-1}). The ACE inhibitory activity of 63-day ripened sample remained stable even after *in vitro* simulated gastrointestinal digestion.

4.1.1.2 Antioxidant Properties

Antioxidative peptides have gained attention since they may exert the antioxidant effect through scavenging of radicals and reactive oxygen species (ROS), chelating of transition metal ions and inhibition of lipid peroxidation (Saiga *et al.*, 2003; Vaštag *et al.*, 2010). Their action can be linked to various factors, such as protein used as substrate, enzyme used, degree of hydrolysis, concentration of peptide, amino acid sequence as well as hydrophobicity. Table 2 shows antioxidant activity of peptides from various fermented fish and meat products.

A considerable antioxidant activities, including radical scavenging, reducing power and metal chelating activities have been found in several traditionally-fermented fish products. Two peptides with higher antioxidant activity were predominantly extracted in the water phase of highly fermented fish product 'pekasam' with ABTS radical scavenging activity and DPPH radical scavenging activity having IC_{50} value of 0.636 and 2.24 mg/ml, respectively. The peptides were identified as Ala-Ile-Pro-Pro-His-Pro-Tyr-Pro and Ile-Ala-Glu-Val-Phe-Leu-Ile-Tre-Asp-Pro-Lys. Their high activity was attributed to the presence of hydrophobic amino acids (Ile, Ala and Pro) and basic amino acids, (Lys) in the peptide sequence (Najafian and Babji, 2018). Similarly, antioxidant peptides from fermented fish sauce *budu* showed high antioxidant activity when peptides contained hydrophobic and basic amino acids in their sequences (Najafian and Babji, 2019). Peralta *et al.* (2008) reported enhancement of antioxidant property of salt-fermented shrimp paste when fermentation proceeded till 360 days. The ethanolic extract of fermented shrimp paste showed higher hydrogen peroxide scavenging activity and DPPH radical scavenging activity as well as greater inhibition of lipid peroxidation. Furthermore, prolonged fermentation could also result in interaction of peptides and amino acids with reducing sugars, especially in fermented seafood products, thus forming desirable Maillard Reaction Products (MRPs) (Faithong *et al.*, 2010). A close correlation between browning intensity and strong antioxidant activity was shown in *kapi* during the fermentation process (Faithong and Benjakul, 2014). The presence of MRPs by increased browning intensity was also noticed in other traditional fish products, such as *som-fug,* a Thai fermented fishery product (Riebroy *et al.*, 2004). Unlike thermally developed MRPs, the MRPs generated during fermentation are presumably less toxic (Faithong and Benjakul, 2014). Additionally, the presence of antioxidant peptides and MRPs could retard lipid oxidation (Pongsetkul *et al.*, 2017a). Pongsetkul *et al.* (2018) demonstrated that use of starter culture with higher extent of proteolysis and lipolysis, such as *Bacillus* spp. K-C3, increased the fermentation rate of *kapi* than the traditional fermentation process. Furthermore, the final fermented product had higher browner colour with enhanced antioxidative properties.

In addition to fermented fish, various fermented meat products have been documented for their antioxidant activities. Sun *et al.* (2009) isolated two peptide fractions (P1 and P2) from Cantonese sausages at different drying periods. P1 fraction showed the highest radical scavenging activity after 18 hours of drying. Further drying process (P2) rendered weaker activity. Histidine was the major amino acid in both the fractions. Glu, Gly and Arg were dominant amino acids and the molecular weight of fractions was less than 2 kDa. Vaštag *et al.* (2010) demonstrated that crude protein hydrolysate from *Petrovac* sausage (Petrovská kolbása) showed higher DPPH radical scavenging activity and reducing power at day 90 of ripening. Broncano *et al.* (2012) determined low-molecular-weight compounds from fermented chorizo sausages, in which DPPH radical scavenging activity was detected in five fractions (1-5). Fraction 1 and 2 showed substantial antioxidant activity (55.3 and 39.8 per cent), while other fractions did not have any activity. Fraction 1 contained natural dipeptides, carnosine and anserine and the tripeptide, Phe-Gly-Gly. The major components in fraction 1 were carnosine and tripeptides. Additionally, some free amino acids were present as minor components. Among them, both basic and acidic amino acids were present. The fraction 2 which contained mainly free amino acids (Val and GABA), L-carnitine and creatinine, rendered slightly lower antioxidant activity. The other fractions, which did not exert antioxidant activity, contained trace amounts of L-carnitine and valine. Therefore, antioxidant activity in fermented chorizo sausages was mainly due to naturally occurring β-alanylpeptides. Many free amino acids, other metabolites and compounds were mostly hydrophilic and contained mainly dipeptide carnosine together with Phe-Gly-Gly and L-carnitine. In another study, Broncano *et al.* (2011) used three different proteases including neutral bacterial protease, fungal protease and fungal protease concentrate, for manufacturing fermented sausages from Iberian pigs. After fermentation, hydrolysate of crude extract (<3 kDa) was assessed for DPPH radical scavenging activity, metal chelating assay, reducing power and inhibition of linoleic acid autoxidation. The hydrolysate from fermented sausages added with enzymes had two-three-fold higher antioxidant activities, compared to the sausages without enzyme addition. Furthermore, the sausage with natural fermentation showed higher value for thiobarbituric acid-reactive substances

Table 2: Bioactivity of Peptides from Fermented Fish and Meat

Bioactive Peptide Source	Peptide	Inhibition Activity	References
Fermented fish			
Fermented fish sauce "*Budu*"	Lue-Asp-Asp-ProVal-Phe-Ile-His and Val-Ala-Ala-Gly-Arg-Thr-Asp-Ala-Gly-Val-His	DPPH radical scavenging activity, reducing power and ABTS radical-scavenging activity assay	Najafian and Babji (2019)
Fermented fish "*Pekasam*"	Ala-Ile-Pro-Pro-His-Pro-Tyr-Pro and Ile-Ala-Glu-Val-Phe-Leu-Ile-Tre-Asp-Pro-Lys	ABTS, DPPH radical scavenging activity	Najafian and Babji (2018)
Fermented shrimp past "*Kapi*"	Aqueous extract	DPPH, ABTS, H_2O_2 radical and singlet oxygen scavenging activities	Pongsetkul *et al.* (2017b)
Fermented fish "*Ngari*"	Aqueous extract	DPPH radical scavenging activity, Ferric reducing antioxidant power (FRAP) and ACE-inhibitory activity	Phadke *et al.* (2014)
Fermented shrimp past "*Kapi*"	Trp-Pro and Ser-Val	ACE-inhibitory activity	Kleekayai *et al.* (2015)
Fermented fish "*heshiko*" and "*narezushi*"	Crude extract	Antihypertensive effect in SHR	Itou and Akahane (2004), Itou *et al.* (2007)
Fermented fish sauce from sardine	Ala-Pro, Lys-Pro, and Arg-Pro	ACE-inhibitory activity, lowering blood pressure in SHR	Ichimura *et al.* (2003)
Fermented fish sauce from salmon	Glu-Tyr, Ile-Tyr, Val-Tyr	Antihypertensive effect in SHR	Okamoto *et al.* (1995)
Fermented meat			
Beef loins (whole muscle)	Leu-Ser-Trp, Phe-Gly-Tyr, and Tyr-Arg-Gln	ACE-inhibitory activity	Choe *et al.* (2018)
Dry cured pork loins inoculated with probiotic LAB	46 peptides composed of 6–22 amino acids	ACE-inhibitory activity	Kęska and Stadnik (2018)

Product	Compound	Activity	Reference
Fermented pork meat sauce	Gln-Tyr-Pro	DPPH radical and OH-radical	Ohata et al. (2016)
Iberian dry-fermented sausage "salchichón"	EPg222 and P200S34	ACE-inhibitory and antioxidant	Fernández et al. (2016)
Harbin dry sausage	Crude fractions	Radical scavenging activity, reducing power, and inhibition of lipid peroxidation	Chen et al. (2015)
Fermented sausages "chorizo"	Activity of five fraction of lower molecular weight peptides di and tripeptides	DPPH radical scavenging assay	Broncano et al. (2012)
Sausages from Iberian pigs	Crude fractions	DPPH radical scavenging activity, metal chelating assay, reducing power and Inhibition of linoleic acid autoxidation	Broncano et al. (2011)
Petrovac sausage "Petrovská Kolbása"	Phe-Gly-Gly	DPPH radical scavenging activity and reducing power	Vaštag et al. (2010)
Cantonese sausage	P1 and P2	DPPH radical scavenging activity	Sun et al. (2009)

(TBARs) and hexanal content. In addition to increasing the bioactivity, the peptides generated also contributed to enhancement of the storage stability of the product (Broncano *et al.*, 2012).

4.2 Lipid-based Bioactivity

4.2.1 Conjugated Linoleic Acid (CLA)

CLA refers to the positional isomer of linoleic acid with conjugated bonds by ruminal microorganisms and animal tissue. They are of great interest because of their potential health-promoting activities – ranging from the ability to reduce body fat and improve bone mass, antiatherogenic, anti-inflammatory, anticarcinogenic and antidiabetic activities (Benjamin and Spener, 2009). The ruminant origin meat has markedly higher CLA concentration. The highest amount was found in lamb and beef (19 and 13 per cent, respectively), CLA content in other meat sources is around 1 per cent. In common marine foods, the level of CLA varied from 0.1 to 0.9 mg/g, which was lower than that of meat (Csápo and Varga-Visi, 2015). In fermented meat products, such as cooked ham and salami, CLA content was similar to that of the raw material. Thus, fermentation and processing had no effect on content of CLA. Since CLA content in unprocessed meat and some fermented meat is low, they hardly impart any health benefits. However, a few fermented meat products contain higher CLA content, especially some types of dry fermented sausages (Bloukas *et al.*, 1997). CLA in these products is produced by microorganisms. Several strains of *Lactobacillus* (*Lb. pentosus, Lb. plantarum,* and *Lb. sakei*) associated with fermented meat are able to produce CLA (Zeng *et al.*, 2009). Gorissen *et al.* (2011) reported that *Lb. sakei* in fermented meat is able to produce CLA in MRS-medium supplemented with linoleic acid. Thus, to exert the health benefit of these meat products, probiotic bacteria with linoleate isomerase activity can be used as an adjunct or starter culture to develop functional fermented meat products with increased levels of CLA. Several studies have been conducted to produce CLA-enriched fermented meat production (Zhang *et al.*, 2010). Production of CLA-enriched meat by dietary supplementation with synthetic CLA has been carried out (Csápo and Varga-Visi, 2015). On the other hand, a few fermented meat products were enriched with synthetic CLA during processing (Arihara, 2006). Nevertheless, synthetic CLA is more expensive than that naturally produced by microbial fermentation. CLA production by microorganisms during fish and meat fermentation needs to be further studied and is a challenge for the future.

4.3 Other Compounds Generated during Fermentation

4.3.1 Gamma Aminobutyric Acid (GABA)

GABA is a non-protein amino acid with four carbon atoms, available mostly as a zwitterion. It is present in both prokaryotes and eukaryotes. In mammals, its principal function is as a neurotransmitter inhibitor of the nervous system and it reduces neuronal excitability and manages stress. It is also responsible for the regulation of muscle tone (Dhakal *et al.*, 2012). GABA naturally enhances immunity under stressed condition and reduces the elevated blood pressure (Pouliot-Mathieu *et al.*, 2013). It can protect against chronic kidney disease, ameliorate oxidative stress induced by nephrectomy, and activate kidney and liver function (Sasaki *et al.*, 2006). In addition, GABA increases the concentration of growth hormone in plasma and modulates protein synthesis in the brain (Tujioka *et al.*, 2007). It protects against various types of cancer, delays/inhibits cancer cell proliferation, stimulatory action on cancer cells apoptosis and suppresses tumor (Schuller *et al.*, 2008). It also controls asthma and breathing problems (Kazemi and Hoop, 1991). In addition, there are many other physiological positive effects of GABA with potential novel application in many therapies. A wide range of traditionally fermented fish and meat products are rich sources of GABA. Production of GABA is catalysed by glutamate decarboxylase and the fermentation conditions can influence the enzyme activity. A number of bacteria and fungi have been reported to produce GABA, using glutamate from fish and meat. The most interesting bacteria for GABA production are LAB, which produce high levels of GABA. GABA-producing LAB include several strains of *Lactobacillus, Enterococcus, Leuconostoc* and *Lactococcus* isolated from various fermented fish and meat products, including Japanese traditional fermented fish, European fermented sausages, Thai fermented shrimp and other meat products. *Lb. delbrueckii.* subsp. *bulgaricus, Lb. plantarum* and *Lb. paracasei* were isolated from Japanese fermented fish (Komatsuzaki *et al.*, 2005). *Lb. futsaii* CS3 was obtained from Thai fermented shrimp (Sanchart *et al.*, 2017). *Lb. plantarum, Lb. farciminis, Lb. futsaii, Lb. reuteri*, and *Pd. pentosaceus* were isolated from small fermented fish in Myanmar (Moe *et al.*, 2015). Due to the numerous health benefits, many studies have been conducted to enhance the production of GABA in fermented meat products, either by optimising the fermentation condition or by using a potential GABA-producing starter culture (Dhakal *et al.*, 2012).

5. Indigenous and Microbial Synthesised Bioactive Compounds

5.1 Minerals and Vitamins

Fermented fish and meat are valuable source of some dietary macro and micronutrients. In the Western diet, fermented meat products, especially fermented sausages are recognised as the main source of essential amino acids, B vitamins, bioavailable iron, zinc, selenium and other minerals (García-Íñiguez de Ciriano *et al.*, 2013). These nutrients are either absent or present in small amounts in other foods. Generally, dietary meat products are known to contain substantial amounts of micronutrients, like 14 per cent iron, 30 per cent zinc, 19 per cent vitamin, 37 per cent niacin, 14 per cent vitamin B_2, 21 per cent vitamin B_6, 22 per cent and vitamin B_{12} (Arihara, 2006). Lack of these nutrients in the human diet could cause adverse health effects since the dietary minerals are essential for muscle, nerve function, bone health, regulation of blood sugar levels, etc. and thus are important in managing diseases, such as cardiovascular, hypertension, diabetes and osteoporosis (Caballero *et al.*, 2016; FAO/WHO, 2001). For instance, selenium is an essential micro nutrient for humans. As an integral part of selenoproteins, it regulates various physiological functions. It plays a key role in redox regulation by removal/decomposition of lipid hydroperoxides and hydrogen peroxide (Ursini *et al.*, 1997). Díaz-Alarcón *et al.*

(1996) reported that dry cure ham or sausages, including salami, are a good source of organic selenium, which ranges from 1 to 3 µg/g of wet weight of the product (Zanardi *et al.*, 2010). The other important essential micronutrients are iron and zinc. Red meat contains an abundance of iron as a part of haem protein in myoglobin, which is in an highly absorbable form of iron. In addition, zinc has also been reported to be highly bioavailable in red meat. Most of the microbes synthesise vitamins exclusively. Fermented meat can also be a source of vitamins, such as vitamin D, E, K, which are not present in raw meat (Marco *et al.*, 2017). Regardless of these beneficial properties, the major concern of these fermented fish and meat products is sodium chloride, which is most frequently used as an ingredient to enhance preservation, taste and textural properties. Excessive amount of sodium intake than the nutritionally recommended level has been linked to harmful effects, such as hypertension, high blood pressure and coronary heart disease (Arihara, 2006). Due to the health risk awareness among the consumer and producers, many strategies have been implemented to reduce/replace sodium chloride with KCl, $CaCl_2$ and $MgCl_2$ to reduce health risk without compromising the quality of final product (Ruusunen and Puolanne, 2005; Zanardi *et al.*, 2010).

The fermented fish, such as Nile tigher fish is a good source of minerals, including barium (Ba), iron (Fe), zinc (Zn), aluminium (Al) and boron (B). These minerals tend to increase through fermentation while phosphorus (P) gets slightly reduced in the traditional fermenting process in comparison with unfermented counterparts (Mohamed and Hamadani, 2016). Udomthawee *et al.* (2012) documented an increase in calcium (Ca) in traditionally salt-fermented shrimp as compared with the fresh counterpart, owing to liquefaction of Ca from the shell into paste during fermentation. Nonetheless, fermented shrimp is not a good source of Fe and sodium (Na). The chemical composition of Atlantic bumper was not adversely affected by fermentation process, in which slight decrease in Ca, P, Na and potassium (K) were recorded for traditionally-fermented Ivorian fish condiment (*Adjuesan*) as compared to fresh fish (Koffi-Nevry *et al.*, 2011). *Pla-ra* (Thailand: traditionally fermented mud carp, fish: salt ratio 4:1.5) and *pa-dag* (Laos: traditionally-fermented mud carp, fish: salt ratio 3:1) had higher Ca and P contents when compared to fresh fish. *Pra-hoc* (Cambodia: traditionally-fermented *gourami*, fish: salt ratio of 3:1) had lower contents than fresh fish, more likely attributed to the differences in the raw materials used (Udomthawee *et al.*, 2012). Traditional *Bruneian belacan* shrimp paste, Korean fermented and dried *saewoojeot* shrimp paste, and Korean dried shrimp paste had high contents of Ca, Fe, magnesium (Mg), Na and P, which could serve as an excellent source of these minerals for humans (Kim *et al.*, 2014).

5.2 Fat and Fatty Acids

Fermented fish and meat products are an important source of fat, apart from proteins. The nutritional and health characteristics are mainly due to the triacylglycerides (TGA) and phospholipids (PLs) which are rich polyunsaturated fatty acids (PUFA), including eicosapentaenoic acid (20:5 EPA), docosapentaenoic acid (22:5 DPA) and docosahexaenoic acid (22:6 DHA) (Muhammed *et al.*, 2015). Generally, fish lipids are known for their well-balanced dietary lipids including some essential fatty acids. In general, fish lipids have a low ratio of n-6/n-3 PUFA (Muhammed *et al.,* 2015). The higher content of n-3 fatty acids are mainly present in PLs than TAG. Furthermore, n-3 fatty acids are highly bioavailable in the form of PLs as they provide more efficient absorption in most of the tissues (Rossmeisl *et al.,* 2012). Consumption of fishery products containing PUFA, in particular n-3 fatty acids, is observed to prevent or modulate certain diseases, including mild hypertension, coronary heart disease, inflammatory disease, autoimmune disorder, impaired brain development in infants, visual acuity and neurological disorders (Valfré *et al.,* 2003; Rai *et al.,* 2013, 2015). Due to the higher content of PUFA, the major concern related to fermented fish product is lipolysis, which could generate free fatty acids (FFA) and oxidation producing hydroperoxides and volatile compounds. Although, fermentation contributes to the development of flavour and safety characteristics, the dynamic changes related with bioactive lipids are not well understood during the process of fermentation, especially in fish. However, changes in lipid profile of traditional fermented fish products have been documented. Pongsetkul *et al.* (2017a) reported that there is 49 per cent increment in FFA; however, phospholipids, including phosphatidyl ethanolamine (4.80 g/100 g) and phosphatidyl choline (5.67 g/100 g), remained unchanged in fermented shrimp *(kapi)*. PUFA, including EPA and DHA, were found as the major (19.03 per cent) fatty acids in *kapi*. Similarly, other studies on traditionally fermented fish and shrimp (using salt/LAB) indicated no significant difference in the content of PUFA between the raw and fermented product (Montaño *et al.,* 2001; Zang *et al.,* 2018), hypothesising the preservative effect of salt and LAB towards PUFA oxidation during the fermentation process.

Nevertheless, fermented meat products are often associated with a negative health image due to the high content of fat, saturated fatty acids (SAF) as well as cholesterol. These compounds are commonly associated with obesity, cardiovascular disease and some types of cancer (Caballero *et al.,* 2016). As per the WHO guidelines for a balance diet, the total dietary fat should range between 15-30 per cent which should contain SFA (6-10 per cent), monounsaturated fatty acids (MUFA) and PUFA (10-15 per cent). Moreover, trans fat should not be more than 1 per cent and intake of cholesterol should be less than 300 mg/day (WHO, 2003). Fat content of fermented meat products varies between 30-50 per cent, depending on the products (Bou *et al.,* 2017). However, from the health viewpoint, excessive fat intake means potential health risks. As per the regulation, many industries are following different strategies to reduce the fat content by at least 30 per cent. Low-fat fermented sausage was produced by reduction of fat content by adding non-lipid components, such as inulin, or cereal, fruit fibre without any effect on the textural properties (Menegas *et al.,* 2013). Other strategies have been implemented by replacing the meat fat by using different lipid materials (mainly oils) of vegetable (olive, linseed, sunflower, soy, canola) and marine origin (fish, algae, etc.) as partial or total substitutes for animal fat (Table 3). For liquid oils, pre-emulsification, mainly using soy protein isolate, was performed. In a total, 25 per cent of fat substitution was acceptable in fermented sausages with respect to sensory and textural properties. However, when a higher percentage was substituted, a drip fat was observed (Arihara, 2006). In addition, an improvement in the lipid profile was achieved by reformulation, depending on the type and amount of oil incorporated. In general, the addition of oil raises MUFA and PUFA content as well as PUFA/SFA ratios while reducing the cholesterol level and n-6/n-3 PUFA ratios (Caballero *et al*., 2016).

Table 3: Some Fermented Meat Products Enriched with Fibre as Fat Replacers

Product	Fibre Ingredient	Fibre Content	References
Hanwoo beef and pork sausage	Rice bran fiber and wheat fiber	1.5%	Jung *et al*. (2018)
Low fat sausage	κ-carrageenan, konjac, and tragacanth	1.5%	Atashkar *et al*. (2018)
Dry-fermented chicken sausage	Inulin	7%	Menegas *et al*. (2013)
Spanish dry-fermented sausage (Salchichón)	Orange fiber	1%	Fernández-López *et al*. (2008)
Dry fermented sausage (Sobrassada)	Carrot dietary fiber	3%	Eim *et al*. (2008)
Thai fermented sausage (Yor)	Carboxymethyl cellulose, locust bean gum and xanthan gum	1%	Chattong *et al*. (2007)

5.3 Squalene

Squalene is a triterpenoid lipid with six conjugated double bonds. In humans, squalene is present at its highest concentration in sebum (~13 per cent). It protects skin from UV-associated skin damage and plays a key factor in cholesterol metabolism. Squalene has been known to have significant dietary benefits, inertness, biocompatibility and other advantageous properties. It is widely used as an excipient in pharmaceutical formations for managing various diseases and therapy. Squalene has several beneficial properties, such as natural antioxidant, skin hydrater and effective quencher of singlet oxygen in UV-associated skin damage. It also has preventive effect against various types of cancer, protective effects against many heart-related diseases and helps in reduction of serum cholesterol, enhancement of immune response to various antigens and is used for drug delivery application (Desai *et al*., 1996; Kelly, 1999; Reddy and Couvreur, 2009; Smith, 2000). In addition, squalene acts as an ideal agent for many therapies. Squalene alone exhibits chemopreventive activity, showing decreased chemotherapy-induced side effects (Reddy and Couvreur, 2009). In addition to this, squalene is reported to be a potential antimicrobial agent towards deadly food-borne pathogens and fungi (Bindu *et al*., 2015). As a triterpene, squalene is well absorbed orally and is used to improve the oral delivery of many therapeutic molecules. For a single person, intakes of 30 mg squalene per day from various sources is recommended (Smith, 2000). Nevertheless, not all dietary components are a good source of squalene, except for some plant oil, such as olive oil, amaranth seed oil and rice bran oil, which are known to be the richest sources and contain 10-560 mg of squalene per 100 g of oil (Popa *et al.*, 2015). Various fresh-water and marine fish species are known to contain substantial amounts of squalene, ranging between 70-1803 mg/kg based on wet weight (Bavisetty and Narayan, 2015; KopiCoVá and VaVreiNoVá, 2007). Thus, various fermented fish products can be an ideal source of dietary squalene. However, shrimp products contain a very low content of squalene (Montaño *et al.*, 2001; Zang *et al.*, 2018). In addition, red meat has a low content of squalene and fermented meat products, such as cured ham and fermented sausages, may not be considered as dietary sources of squalene. In a few European countries, vegetable oil, including olive oil, is commonly used as a fat replacer in various fermented sausages and which can contribute as an enriched dietary source of squalene (Bloukas *et al.*, 1997; Bou *et al.*, 2017).

5.4   Histidyl Dipeptides

Antioxidant-rich food consumption has been known to show preventive effect against oxidative stress in a biological system attributing to detoxification of reactive intermediates or by repairing the damage (Kittisakulnam *et al.*, 2017). Apart from plants, meat and meat products are a good source of various endogenous antioxidants, including carotenoids, ubiquinone, tocopherols, uric acid, ascorbic acid, lipoic acid, spermine, anserine and carnosine, mainly found in the skeletal muscle (Decker *et al.*, 2000). Carnosine (β-alanyl-1-histidine) and anserine (N-βalanyl-1-methyl-1-histidine) are the most abundant antioxidants in meat sources. The concentration of carnosine is much higher than anserine in pork (2700 mg/kg in a pork shoulder). In chicken meat, anserine is more abundant (980 mg/kg) than carnosine (500 mg/kg). The variation in contents depends on the animal type, species and type of muscle (Lücke, 1994). The antioxidant activity of both anserine and carnosine is mainly from their ability to chelate transition metals (Brown, 1981). These peptides have been reported to be beneficial against prevention of oxidative stress-related diseases, wound healing, recovery from fatigue, neuroprotective effect, glucose-lowering effect and energy production (Derave *et al.*, 2019). In addition to naturally available antioxidants in meat, some antioxidants are produced by proteolysis during curing of fermented meat, in which degradation of major proteins takes place to generate polypeptides, small peptides and free amino acids (Ohata *et al.*, 2016). Meat products, especially dry-cured products with longer ripening time, contain more small peptides and free amino acids. For some fermented sausages, where starter cultures are used for controlled fermentation, the formation of smaller peptides and free amino acids occurs within a short period of time. This was attributed to bacterial aminopeptidase. Considering the starter culture for fermented sausages, *Lb. sakei* has been described to exert high exopeptidase activities, including aminopeptidase, x-prolyl-dipeptidylpeptidase, arginine aminopeptidase, tripeptidase and dipeptidase. These enzymes release dipeptides and free amino acids, mainly Leu and Ala (Flores and Toldrá, 2011). For products with longer ripening time, indigenous enzymes play a major role (Tang *et al.*, 2018). Besides fermented meat, traditionally-fermented fish products, such as fish sauce is a good source of antioxidants including carnosine, anserine and some free amino acids. They also play an important role in controlling oxidation during the fermentation process (Ohata *et al.*, 2016).

5.5   L-carnitine

L-carnitine (β-hydroxy-γ-trimethyl amino butyric acid) is quaternary ammonium compound biosynthesised in the human body. It is also found in the skeletal muscle of most animals and is abundant in beef (1300 mg/kg of thigh) (Shimada *et al.,* 2004). The important role of L-carnitine is to transport long chain fatty acids into the mitochondria for oxidation and energy production. Additionally, it is involved in some biological activities in the human body (Adeva-Andany *et al.*, 2017). L-carnitine also helps energy production in muscles while performing hard exercise or physical work and helps the body to absorb calcium to improve the skeletal muscles as well as contributes to lowering of cholesterol level (Adeva-Andany *et al.*, 2017). Oral administration of L-carnitine has many therapeutic benefits; it reduces fatigue and hypertension as well as plays a crucial role in male infertility. It can be used for treatment of primary and secondary carnitine-deficiency syndromes. It modulates immune and inflammatory responses and improves nerve conductions, such as Alzheimer's disease, hepatic encephalopathy and other painful neuropathies (Flanagan *et al.,* 2010; Goa and Brogden, 1987; Lenzi *et al.*, 2003). Fermented meat products specially prepared from red meat are abundant sources of L-carnitine. Demarquoy *et al.* (2004) reported that fermented meat products commonly eaten in Western countries are abundant in L-carnitine (80-580 µmol/100 g). In dry cured ham, the type and concentration of salts does not have any influence on L-carnitine content, but the ripening time and post-salting time affects the L-carnitine content. Chemical changes during fermentation process of cured ham has no significant effect on L-carnitine (Mora *et al.*, 2010). Seafood and fish are relatively low in L-carnitine content (4-36 µmol/100 g) and in which, salmon is the best source (Demarquoy *et al.*, 2004). However, the role of probiotic in L-carnitine metabolism during fermentation process has not been well studied, but synergistic effect of probiotics and L-carnitine has been demonstrated for some potential health benefits, such as modulation of the immune system, control/reduction of inflammatory cytokines, stimulation of antioxidant enzymes, inhibition of IkB kinase and decrease in oxidative stress, etc. (Moeinian *et al.,* 2013). In addition, Demarquoy *et al.* (2004) reported that few fermented sausages from European region, especially '*chorizo*' contain L-carnitine as high as 7-66 mg/100 g. The concentration of L-carnitine increases to substantial amounts during fermentation and drying.

5.6   Taurine

Taurine (2-aminoethanesulfonic acid), a sulphur-containing amino acid, is widely distributed in animal tissues and is a conditionally essential amino acid in humans (Schuller-Levis and Park, 2003). It is important for numerous biological and physiological functions. Since humans can synthesise a limited amount of taurine, dietary intake is necessary. In addition to metabolic activity, taurine has important clinical effects, such as cholestasis prevention, neuromodulation of central nervous system, retinal development and function, endocrine/metabolic effects, antioxidant and anti-inflammatory effect, hypocholesterolemic effect, hypolipedemic effect as well as cardiovascular effects (Laidlaw *et al.*, 1990; Lourenco and Camilo, 2002; Schuller-Levis and Park, 2003; Spencer *et al.,* 2005). As most of the fermented meat products are from pork and beef, the taurine content ranges between 300-500 μmole/100 of meat. In seafood, shrimp contains lower taurine content (80 μmole/100 g of meat).

Higher taurine content is in fish (300-6700 μmole/100 of meat), depending on the fish species (Laidlaw *et al.*, 1990). Peralta *et al.* (2008) reported that the traditional Philippine salt-fermented shrimp paste through prolonged fermentation process had an increased accumulation of taurine along with other FAAs. However, at the very later stages, a significant decrease may occur due to its degradation.

5.7   Coenzyme Q

Coenzyme Q (CoQ) is a natural lipophilic compound which plays a role as cofactor of endogenous enzymes. Its occurrence is ubiquitous in each and every living cell and is also termed as ubiquinone (Ernster and Dallner, 1995). In its reduced state, CoQ10 is known to be an effective antioxidant, protecting against protein and DNA oxidation, lipid peroxidation, and is also capable of functioning synergistically in the presence of other antioxidants (Bentinger *et al.*, 2007). CoQ10 has been reported to have an important role in various clinical aspects. It is known to exert beneficial effect under mitochondrial conditions and neurodegenerative, diabetes, cardiovascular, male infertility and a few other diseases (De Blasio *et al.*, 2015; Maladkar *et al.*, 2016; Yang *et al.*, 2016). Pyo and Oh (2011) determined ubiquinone from various Korean fermented foods. *Jeotgal*, a fermented fish, contained the highest content of CoQs (315.9 μg/g of fermented fish) which was significantly higher than raw fish. It was suggested that increase in CoQs was from diverse microflora, including *Bacillus subtilis, Leuconostoc mesenteroids, Pd. halophilus, Torulopsis* sp., *Sarcina* sp., *Saccharomyces* sp., which were found in Jeotgal. Thai fermented fish (*pla-ra*) was reported to contain ubiquinone nine isoprene (CoQ9) as the major component, which was produced by a species of genus *Virgibacillus*. Among them, one was *Bacillus* and the others were characterised as Gram negative rod-shaped bacterium (Tanasupawat *et al.*, 2010). Tanasupawat *et al.*, (2009) also isolated three halophilic bacteria from Thai fermented fish sauce (*nam-pla*), producing CoQ9 and the bacteria were closely related to *Chromohalobacter salexigens, Halobacteirum salinarum* and *Halococcus saccharolyticus.* Jiang *et al.* (2014) reported *Halomonas shantousis* sp. *Nov*as, the novel bacterial species commonly found in Chinese fermented sauce, which predominantly produces CoQ9. Additionally, they can degrade biogenic amine during the fermentation process. Nevertheless, no literature is available on ubiquinone production in fermented meat.

6.   Formulated Bioactive Compounds

Dietary fibre (DF) plays a major role as a functional ingredient and enhances the desirable physiological properties for various food products, especially fermented meat. The health benefits of dietary fibre have long been realised. Considerable intake of dietary fibre is linked to various health benefits, such as maintenance of gut health (by facilitating survival of probiotics), reducing the risk of coronary heart disease (hypocholesterolaemic effect), preventing carcinogenesis and diabetes Type 2 (ability to reduce glycaemic response), reducing obesity (by providing the sense of satiety) and mainly providing laxative effect (Kaczmarczyk *et al.,* 2012; Slavin, 2013). Fibre-rich fermented meat products have been developed since the past few decades. Very few meat-based products fortified with fibre use fibre in fermented meat (Table 3). Among all, carbohydrates, chiefly oligosaccharides, are the well-known prebiotics and are used as a selective ingredient in fermented meat. Generally, oligosaccharides are resistant to the human digestive process and reach the colon where they act as a food source for gut microflora, mostly LAB and *Bifidobacterium* (Mudgil and Barak, 2013). In fermented meat, fibres (inulin, carrageenan, carrot, peach, lemon, orange, etc.) are used as fat replacers or as functional ingredients (Arihara, 2006; Jung *et al.,* 2018; Menegas *et al.,* 2013) Carrot fibre was added as a fat replacer, which also contributes to increased fibre content. The final product is reported to have good physiochemical and sensory properties (Eim *et al.,* 2008).

7. Conclusion

Functional bioactive compounds in meat and meat products have gained increasing interest as they are known to exert beneficial effects on maintaining human health. This demand provides a great opportunity to the meat industry. Both fermented meat and fish contain bioactives in the form of peptides with potential antioxidant and antihypertensive properties. Fermentation causes changes in certain fatty acids, especially CLA, which is high in fermented meat sausages and is believed to be synthesised by several strains of Lactobacilli. Similarly, a number of bacteria and fungi have been reported to produce GABA in fermented meat and fish. Other bioactive compounds, e.g. histidyl dipeptides, L-carnitine, taurine, coenzyme Q, glutathione, lipoic acid, squalene and astaxanthin are indigenously present in raw meat and fish and are produced by certain microorganisms during fermentation. Moreover, meat, fish and their products are excellent sources of essential and non-essential micronutrients, including various minerals and vitamins. Apart from these, several fatty acids, particularly PUFA from traditionally-fermented meat and fish have a role in nutrition and health characteristics. Overall, fermented meat and fish are known to preserve the essential micronutrients and fatty acids required for nourishing the body and maintaining good health. Regardless of these beneficial properties, the major concern of these fermented fish and meat products is sodium chloride, which is most frequently used to enhance preservation, taste and textural properties. In addition to the compounds generated during processing through fermentation, there are numerous indigenous bioactive compounds in raw meat, where the concentration varies, depending on the sources and type of meat. A limited number of studies have been conducted on possible health benefits of bioactive compounds, either from meat or meat product, directly in humans as most of the conclusions are drawn from the fact that functional compounds may be beneficial to humans. Therefore, future studies to draw strong evidence on the health benefits of meat and meat products need to be conducted. Furthermore, meat scientists and industries are directing their efforts in developing functional meat and meat products with improved health image. These products have been designed by reducing the unhealthy portion of meat or/ and adding functional ingredients, which have been reported to be most safe and efficient.

Abbreviations

ACE – Angiotensin Converting Enzyme
CLA – Conjugated linoleic acids
CoQ – Coenzyme Q
DF – Dietary fibre
FFA – Free fatty acid
GABA – γ-aminobutyric acid
GIT – Gastro Intestinal Tract
GRAS – Generally-recognized-as-safe
LAB – Lactic Acid Bacteria
MUFA – Monounsaturated fatty acids
PLs – Phospholipids
PUFA – Polyunsaturated fatty acids
SAF – Saturated fatty acids
TGA – Triacylglycerides
UVR – Ultraviolet radiation

References

Acquah, C., Chan, Y.W., Pan, S., Agyei, D. and Udenigwe, C.C. (2019). Structure-informed separation of bioactive peptides. *Journal of Food Biochemistry*, 43: e12765.

Adeva-Andany, M.M., Calvo-Castro, I., Fernández-Fernández, C., Donapetry-García, C. and Pedre-Piñeiro, A.M. (2017). Significance of l-carnitine for human health. *IUBMB Life*, 69: 578-594.

Anastasiadou, S., Papagianni, M., Filiousis, G., Ambrosiadis, I. and Koidis, P. (2008). Growth and metabolism of a meat isolated strain of *Pediococcus pentosaceus* in submerged fermentation: Purification, characterisation and properties of the produced pediocin SM-1. *Enzyme and Microbial Technology*, 43: 448-454.

Anihouvi, V., Kindossi, J. and Hounhouigan, J. (2012). Processing and quality characteristics of some major fermented fish products from Africa: A critical review. *International Research Journal of Biological Science*, 1: 72-84.

Arcales, J.A.A. and Alolod, G.A.L. (2018). Isolation and characterisation of lactic acid bacteria in Philippine fermented milkfish chanos chanos-rice mixture (*Burong Bangus*). *Current Research in Nutrition and Food Science Journal*, 6: 500-508.

Arihara, K. (2006). Strategies for designing novel functional meat products. *Meat Science*, 74: 219-229.

Asiedu, M. and Sanni, A.I. (2002). Chemical composition and microbiological changes during spontaneous and starter culture fermentation of Enam Ne–Setaakye, a West African fermented fish-carbohydrate product. *European Food Research and Technology*, 215: 8-12.

Atashkar, M., Hojjatoleslamy, M. and Sedaghat Boroujeni, L. (2018). The influence of fat substitution with κ-carrageenan, konjac and tragacanth on the textural properties of low-fat sausage. *Food Science and Nutrition*, 6: 1015-1022.

Bavisetty, S.C.B. and Narayan, B. (2015). An improved RP-HPLC method for simultaneous analyses of squalene and cholesterol especially in aquatic foods. *Journal of Food Science and Technology*, 52: 6083-6089.

Benjamin, S. and Spener, F. (2009). Conjugated linoleic acids as functional food: An insight into their health benefits., *Nutrition and Metabolism*, 6: 36.

Bentinger, M., Brismar, K. and Dallner, G. (2007). The antioxidant role of coenzyme Q, *Mitochondrion.* 7: S41-S50.

Bhat, Z.F., Kumar, S. and Bhat, H.F. (2015). Bioactive peptides of animal origin: A review. *Journal of Food Science and Technology*, 52: 5377-5392.

Bindu, B.S.C., Mishra, D.P. and Narayan, B. (2015). Inhibition of virulence of *Staphylococcus aureus* – A food borne pathogen–by squalene, a functional lipid. *Journal of Functional Foods*, 18: 224-234.

Bloukas, J., Paneras, E. and Fournitzis, G. (1997). Effect of replacing pork backfat with olive oil on processing and quality characteristics of fermented sausages. *Meat Science*, 45: 133-144.

Bolumar, T., Sanz, Y., Aristoy, M.-C. and Toldrá, F. (2008). Purification and characterisation of proteases A and D from *Debaryomyces hansenii*. *International Journal of Food Microbiology*, 124: 135-141.

Bou, R., Cofrades, S. and Jiménez-Colmenero, F. (2017). Fermented meat sausages. pp. 203-235. *In*: Frias, J. *et al.* (Eds.). *Fermented Foods in Health and Disease Prevention.* (Academic Press, Boston.

Broncano, J.M., Timón, M.L., Parra, V., Andrés, A.I. and Petrón, M.J. (2011). Use of proteases to improve oxidative stability of fermented sausages by increasing low molecular weight compounds with antioxidant activity. *Food Research International*, 44: 2655-2659.

Broncano, J.M., Otte, J., Petrón, M., Parra, V. and Timón, M. (2012). Isolation and identification of low molecular weight antioxidant compounds from fermented "chorizo" sausages. *Meat Science*, 90: 494-501.

Brown, C.E. (1981). Interactions among carnosine, anserine, ophidine and copper in biochemical adaptation. *Journal of Theoretical Biology*, 88: 245-256.

Caballero, B., Finglas, P.M. and Toldrá, F. (2016). Nutrition and health. *In*: Smithers, G. (Ed.). *Reference Module in Food Science.* Elsevier, Amsterdam, Netherlands.

Campbell-Platt, G. (1987). *Fermented Foods of the World. A Dictionary and Guide.* Butterworths, London. Boston: Butterworths.

Chattong, U., Apichartsrangkoon, A. and Bell, A.E. (2007). Effects of hydrocolloid addition and high pressure processing on the rheological properties and microstructure of a commercial ostrich meat product "Yor" (Thai sausage). *Meat Science*, 76: 548-554.

Chen, Q., Kong, B., Sun, Q., Dong, F. and Liu, Q. (2015). Antioxidant potential of a unique LAB culture isolated from Harbin dry sausage: *In vitro* and in a sausage model. *Meat Science*, 110: 180-188.

Choe, J., Seol, K.-H., Kim, H.-J., Hwang, J.-T., Lee, M. and Jo, C. (2018). Isolation and identification of angiotensin I-converting enzyme inhibitory peptides derived from thermolysin-injected beef *M. longissimus*. *Asian-Australasian Journal of Animal Sciences*, 32: 430-436.

Csápo, J. and Varga-Visi, É. (2015). Conjugated linoleic acid production in fermented foods. pp. 75-105. *In*: Holzapfel, W. (Ed.). *Advances in Fermented Foods and Beverages.* Elsevier, Sawston, Cambridge.

Dai, Z., Li, Y., Wu, J. and Zhao, Q. (2013). Diversity of lactic acid bacteria during fermentation of a traditional Chinese fish product, Chouguiyu (*stinky mandarinfish*). *Journal of Food Science*, 78: M1778-M1783.

De Blasio, M.J., Huynh, K., Qin, C., Rosli, S., Kiriazis, H., Ayer, A., Cemerlang, N., Stocker, R., Du, X.-J. and McMullen, J.R. (2015). Therapeutic targeting of oxidative stress with coenzyme Q10 counteracts exaggerated diabetic cardiomyopathy in a mouse model of diabetes with diminished PI3K (p110α) signalling. *Free Radical Biology and Medicine*, 87: 137-147.

De Vuyst, L., Falony, G. and Leroy, F. (2008). Probiotics in fermented sausages. *Meat Science*, 80: 75-78.

Decker, E.A., Livisay, S.A. and Zhou, S. (2000). Mechanisms of endogenous skeletal muscle antioxidants: Chemical and physical aspects. pp. 25-60.. *In*: Decker, E.F.C., Lopez-Bote, C.J. (Eds.). *Antioxidants in Muscle Foods*. John Wiley and Sons, Massachusetts, USA.

Demarquoy, J., Georges, B., Rigault, C., Royer, M.-C., Clairet, A., Soty, M., Lekounoungou, S. and Le Borgne, F. (2004). Radioisotopic determination of l-carnitine content in foods commonly eaten in Western countries. *Food Chemistry*, 86: 137-142.

Derave, W., De Courten, B. and Baba, S.P. (2019). An update on carnosine and anserine research. *Amino Acids*, 51: 1-4.

Desai, K., Wei, H. and Lamartiniere, C. (1996). The preventive and therapeutic potential of the squalene-containing compound, Roidex, on tumor promotion and regression. *Cancer Letters*, 101: 93-96.

Desniar, M. (2013). Characterisation of lactic acid bacteria isolated from an Indonesian fermented fish (*bekasam*) and their antimicrobial activity against pathogenic bacteria. *Emirates Journal of Food and Agriculture*, 25: 489-494.

Dhakal, R., Bajpai, V.K. and Baek, K.-H. (2012). Production of GABA (γ-aminobutyric acid) by microorganisms: A review. *Brazilian Journal of Microbiology*, 43: 1230-1241.

Díaz-Alarcón, J.P., Navarro-Alarcón, M., López-García de la Serrana, H. and López-Martínez, M.C. (1996). Determination of selenium in meat products by hydride generation atomic absorption spectrometry selenium levels in meat, organ meats, and sausages in Spain. *Journal of Agricultural and Food Chemistry*, 44: 1494-1497.

Diop, M.B., Dubois-Dauphin, R., Destain, J., Tine, E. and Thonart, P. (2009). Use of a nisin-producing starter culture of *Lactococcus lactis* subsp. *lactis* to improve traditional fish fermentation in Senegal. *Journal of Food Protection*, 72: 1930-1934.

Donoghue, M., Hsieh, F., Baronas, E., Godbout, K., Gosselin, M., Stagliano, N., Donovan, M., Woolf, B., Robison, K. and Jeyaseelan, R. (2000). A novel angiotensin-converting enzyme–related carboxypeptidase (ACE2) converts angiotensin I to angiotensin 1-9. *Circulation Research*, 87: e1-e9.

Eim, V.S., Simal, S., Rosselló, C. and Femenia, A. (2008). Effects of addition of carrot dietary fibre on the ripening process of a dry fermented sausage (*sobrassada*). *Meat Science*, 80: 173-182.

Encinas, J.-P., López-Díaz, T.-M., García-López, M.-L., Otero, A. and Moreno, B. (2000). Yeast populations on Spanish fermented sausages, *Meat Science*, 54: 203-208.

Erkkilä, S., Suihko, M.-L., Eerola, S., Petäjä, E. and Mattila-Sandholm, T. (2001). Dry sausage fermented by *Lactobacillus rhamnosus* strains. *International Journal of Food Microbiology*, 64: 205-210.

Ernster, L. and Dallner, G. (1995). Biochemical, physiological and medical aspects of ubiquinone function. *Biochimica et Biophysica Acta (BBA)-Molecular Basis of Disease*, 1271: 195-204.

Faithong, N. and Benjakul, S. (2014). Changes in antioxidant activities and physicochemical properties of *kapi*, a fermented shrimp paste, during fermentation. *Journal of Food Science and Technology*, 51: 2463-2471.

Faithong, N., Benjakul, S., Phatcharat, S. and Binsan, W. (2010). Chemical composition and antioxidative activity of Thai traditional fermented shrimp and krill products. *Food Chemistry*, 119: 133-140.

Fall, N.G., Diop, M.B., Tounkara, L.S., Thiaw, O.T. and Thonart, E.P. (2018). Characterisation of bacteriocin-like producing lactic acid bacteria (LAB) isolated from the crude and traditionally fermented fish meat (*guedj*) in Senegal. *International Journal of Fisheries and Aquaculture Research*, 4: 26-33.

FAO/WHO. (2001). Health and nutritional properties of probiotics in food including powder milk with live lactic acid bacteria – Joint Food and Agricultural Organization of the United Nations and World Health Organization Expert Consultation Report, Córdoba, Argentina.

Fernández-López, J., Sendra, E., Sayas-Barberá, E., Navarro, C. and Pérez-Alvarez, J.A. (2008). Physico-chemical and microbiological profiles of *"salchichón"* (Spanish dry-fermented sausage) enriched with orange fibre. *Meat Science*, 80: 410-417.

Fernández, M., Benito, M.J., Martín, A., Casquete, R., Córdoba, J.J. and Córdoba, M.G. (2016). Influence of starter culture and a protease on the generation of ACE-inhibitory and antioxidant bioactive nitrogen compounds in Iberian dry-fermented sausage *"salchichón"*. *Heliyon*, 2: e00093.

Fitzgerald, R.J. and Murray, B.A. (2006). Bioactive peptides and lactic fermentations. *International Journal of Dairy Technology*, 59: 118-125.

Flanagan, J.L., Simmons, P.A., Vehige, J., Willcox, M.D.P. and Garrett, Q. (2010). Role of carnitine in disease. *Nutrition and Metabolism*, 7: 30.

Flores, M. and Toldrá, F. (2011). Microbial enzymatic activities for improved fermented meats. *Trends in Food Science and Technology*, 22: 81-90.

Frece, J., Kovačević, D., Kazazić, S., Mrvčić, J., Vahčić, N., Ježek, D., Hruškar, M., Babić, I. and Markov, K. (2014). Comparison of sensory properties, shelf-life and microbiological safety of industrial sausages produced with autochthonous and commercial starter cultures. *Food Technology and Biotechnology*, 52: 307-316.

García-Íñiguez de Ciriano, M., Berasategi, I., Navarro-Blasco, Í., Astiasarán, I. and Ansorena, D. (2013). Reduction of sodium and increment of calcium and ω-3 polyunsaturated fatty acids in dry fermented sausages: Effects on the mineral content, lipid profile and sensory quality. *Journal of the Science of Food and Agriculture*, 93: 876-881.

Gardini, F., Bover-Cid, S., Tofalo, R., Belletti, N., Gatto, V., Suzzi, G. and Torriani, S. (2008). Modeling the aminogenic potential of *Enterococcus faecalis* EF37 in dry fermented sausages through chemical and molecular approaches. *Applied and Environmental Microbiology*, 74: 2740-2750.

Gassem, M.A. (2017). Microbiological and chemical quality of a traditional salted-fermented fish (*Hout-Kasef*) product of Jazan region, Saudi Arabia. *Saudi Journal of Biological Sciences*, 26: 137-140.

Geeta and Yadav, A.S. (2017). Antioxidant and antimicrobial profile of chicken sausages prepared after fermentation of minced chicken meat with *Lactobacillus plantarum* and with additional dextrose and starch. *LWT*, 77: 249-258.

Gelman, A., Drabkin, V. and Glatman, L. (2000). Evaluation of lactic acid bacteria, isolated from lightly preserved fish products, as starter cultures for new fish-based food products. *Innovative Food Science and Emerging Technologies*, 1: 219-226.

Ghaly, A., Ramakrishnan, V., Brooks, M., Budge, S. and Dave, D. (2013). Fish processing wastes as a potential source of proteins, amino acids and oils: A critical review. *J. Microb. Biochem. Technol.*, 5: 107-129.

Giello, M., La Storia, A., De Filippis, F., Ercolini, D. and Villani, F. (2018). Impact of *Lactobacillus curvatus* 54M16 on microbiota composition and growth of *Listeria monocytogenes* in fermented sausages. *Food Microbiology*, 72: 1-15.

Giraffa, G. (2004). Studying the dynamics of microbial populations during food fermentation. *FEMS Microbiology Reviews*, 28: 251-260.

Goa, K.L. and Brogden, R.N. (1987). L-carnitine. *Drugs*, 34: 1-24.

Gorissen, L., Weckx, S., Vlaeminck, B., Raes, K., De Vuyst, L., De Smet, S. and Leroy, F. (2011). Linoleate isomerase activity occurs in lactic acid bacteria strains and is affected by pH and temperature. *Journal of Applied Microbiology*, 111: 593-606.

Górska, S., Schwarzer, M., Jachymek, W., Srutkova, D., Brzozowska, E., Kozakova, H. and Gamian, A. (2014). Distinct immunomodulation of bone marrow-derived dendritic cell responses to *Lactobacillus plantarum* WCFS1 by two different polysaccharides isolated from *Lactobacillus rhamnosus* LOCK 0900. *Applied and Environmental Microbiology*, 80: 6506.

Hayes, M. and García-Vaquero, M. (2016). Bioactive compounds from fermented food products. pp. 293-310. *In*: Ojha, K.S. and Tiwari, Brijesh K. (Ed.). *Novel Food Fermentation Technologies*. Springer International Publishing, Switzerland.

Hayes, M., Stanton, C., Fitzgerald, G.F. and Ross, R.P. (2007a). Putting microbes to work: Dairy fermentation, cell factories and bioactive peptides. Part II: Bioactive peptide functions. *Biotechnology Journal: Healthcare Nutrition Technology*, 2: 435-449.

Hayes, M., Stanton, C., Slattery, H., O'sullivan, O., Hill, C., Fitzgerald, G. and Ross, R. (2007b). Casein fermentate of *Lactobacillus animalis* DPC6134 contains a range of novel propeptide angiotensin-converting enzyme inhibitors. *Applied and Environmental Microbiology*, 73: 4658-4667.

Hu, Y., Xia, W. and Ge, C. (2008). Characterisation of fermented silver carp sausages inoculated with mixed starter culture. *LWT –Food Science and Technology*, 41: 730-738.

Huss, H.H. (1988). Fresh Fish Quality and Quality Changes: A Training Manual Prepared for the FAO/DANIDA Training Programme on Fish Technology and Quality Control. Food & Agriculture Org., United Nations.

Hwanhlem, N., Buradaleng, S., Wattanachant, S., Benjakul, S., Tani, A. and Maneerat, S. (2011). Isolation and screening of lactic acid bacteria from Thai traditional fermented fish (*Plasom*) and production of Plasom from selected strains. *Food Control*, 22: 401-407.

Ichimura, T., Hu, J., Aita, D.Q. and Maruyama, S. (2003). Angiotensin I-converting enzyme inhibitory activity and insulin secretion stimulative activity of fermented fish sauce. *Journal of Bioscience and Bioengineering*, 96: 496-499.

Itou, K. and Akahane, Y. (2004). Antihypertensive effect of *heshiko*, a fermented mackerel product, on spontaneously hypertensive rats. *Fisheries Science*, 70: 1121-1129.

Itou, K., Nagahashi, R., Saitou, M. and Akahane, Y. (2007). Antihypertensive effect of *narezushi*, a fermented mackerel product, on spontaneously hypertensive rats. *Fisheries Science*, 73: 1344.

Jiang, W., Li, C., Xu, B., Dong, X., Ma, N., Yu, J., Wang, D. and Xu, Y. (2014). Halomonas shantousis sp. nov., a novel biogenic amines degrading bacterium isolated from Chinese fermented fish sauce. *Antonie van Leeuwenhoek*, 106: 1073-1080.

Jiménez-Colmenero, F., Carballo, J. and Cofrades, S. (2001). Healthier meat and meat products: Their role as functional foods. *Meat Science*, 59: 5-13.

Jung, J.-T., Lee, J.-K., Choi, Y.-S., Lee, J.-H., Choi, J.-S., Choi, Y.-I. and Chung, Y.-K. (2018). Effect of rice bran and wheat fibres on microbiological and physicochemical properties of fermented sausages during ripening and storage. *Korean Journal for Food Science of Animal Resources*, 38: 302-314.

Kaczmarczyk, M.M., Miller, M.J. and Freund, G.G. (2012). The health benefits of dietary fibre: Beyond the usual suspects of Type 2 diabetes mellitus, cardiovascular disease and colon cancer. *Metabolism*, 61: 1058-1066.

Kazemi, H. and Hoop, B. (1991). Glutamic acid and gamma-aminobutyric acid neurotransmitters in central control of breathing. *Journal of Applied Physiology*, 70: 1-7.

Kechagia, M., Basoulis, D., Konstantopoulou, S., Dimitriadi, D., Gyftopoulou, K., Skarmoutsou, N. and Fakiri, E.M. (2013). Health benefits of probiotics: A review. *ISRN Nutrition*, 2013: ID 481651.

Kelly, G.S. (1999). Squalene and its potential clinical uses. *Alternative Medicine Review: A Journal of Clinical Therapeutics*, 4: 29-36.

Kęska, P. and Stadnik, J. (2018). Ageing-time dependent changes of angiotensin i-converting enzyme-inhibiting activity of protein hydrolysates obtained from dry-cured pork loins inoculated with probiotic lactic acid bacteria. *International Journal of Peptide Research and Therapeutics*. 25: 1173-1185.

Kim, Y.-B., Choi, Y.-S., Ku, S.-K., Jang, D.-J., Ibrahim, H.H.B. and Moon, K.B. (2014). Comparison of quality characteristics between belacan from *Brunei Darussalam* and Korean shrimp paste. *Journal of Ethnic Foods*, 1: 19-23.

Kittisakulnam, S., Saetae, D. and Suntornsuk, W. (2017). Antioxidant and antibacterial activities of spices traditionally used in fermented meat products. *Journal of Food Processing and Preservation*, 41: e13004.

Kleekayai, T., Harnedy, P.A., O'Keeffe, M.B., Poyarkov, A.A., CunhaNeves, A., Suntornsuk, W. and FitzGerald, R.J. (2015). Extraction of antioxidant and ACE inhibitory peptides from Thai traditional fermented shrimp pastes. *Food Chemistry*, 176: 441-447.

Koffi-Nevry, R., Ouina, T., Koussemon, M. and Brou, K. (2011). Chemical composition and lactic microflora of adjuevan, a traditional Ivorian fermented fish condiment. *Pakistan Journal of Nutrition*, 10: 332-337.

Komatsuzaki, N., Shima, J., Kawamoto, S., Momose, H. and Kimura, T. (2005). Production of γ-aminobutyric acid (GABA) by *Lactobacillus paracasei* isolated from traditional fermented foods. *Food Microbiology*, 22: 497-504.

Kongkiattikajorn, J. (2015). Potential of starter culture to reduce biogenic amines accumulation in som-fug, a Thai traditional fermented fish sausage. *Journal of Ethnic Foods*, 2: 186-194.

Kopermsub, P. and Yunchalard, S. (2010). Identification of lactic acid bacteria associated with the production of plaa-som, a traditional fermented fish product of Thailand. *International Journal of Food Microbiology*, 138: 200-204.

KopiCoVá, Z. and VaVreiNoVá, S. (2007). Occurrence of squalene and cholesterol in various species of Czech freshwater fish. *Czech Journal of Food Sciences*, 25: 195-201.

Koyanagi, T., Kiyohara, M., Matsui, H., Yamamoto, K., Kondo, T., Katayama, T. and Kumagai, H. (2011). Pyrosequencing survey of the microbial diversity of 'narezushi', an archetype of modern Japanese sushi. *Letters in Applied Microbiology*, 53: 635-640.

Kurugöl, Z. and Koturoğlu, G. (2005). Effects of *Saccharomyces boulardii* in children with acute diarrhoea. *Acta Paediatrica*, 94: 44-47.

Laidlaw, S.A., Grosvenor, M. and Kopple, J. (1990). The taurine content of common foodstuffs. *Journal of Parenteral and Enteral Nutrition*, 14: 183-188.

Lall, S. and Anderson, S. (2005). Amino acid nutrition of salmonids: Dietary requirements and bioavailability. *Cahiers Options Méditerranéennes*, 63: 73-90.

Lenzi, A., Lombardo, F., Sgrò, P., Salacone, P., Caponecchia, L., Dondero, F. and Gandini, L. (2003). Use of carnitine therapy in selected cases of male factor infertility: A double-blind crossover trial. *Fertility and Sterility*, 79: 292-300.

Liasi, S., Azmi, T., Hassan, M., Shuhaimi, M., Rosfarizan, M. and Ariff, A. (2009). Antimicrobial activity and antibiotic sensitivity of three isolates of lactic acid bacteria from fermented fish product, *budu*. *Malaysian Journal of Microbiology*, 5: 33-37.

Linares, D.M., Gomez, C., Renes, E., Fresno, J.M., Tornadijo, M.E., Ross, R.P. and Stanton, C. (2017). Lactic acid bacteria and bifidobacteria with potential to design natural biofunctional health-promoting dairy foods. *Frontiers in Microbiology*, 8: 846.

Liu, Z.Y., Li, Z.H., Zhong, P.P., Zhang, P., Zeng, M.Q. and Zhu, C.F. (2010). Improvement of the quality and abatement of the biogenic amines of grass carp muscles by fermentation using mixed cultures. *Journal of the Science of Food and Agriculture*, 90: 586-592.

Lopetcharat, K., Choi, Y.J., Park, J.W. and Daeschel, M.A. (2001). Fish sauce products and manufacturing: A review. *Food Reviews International*, 17: 65-88.

Lourenco, R. and Camilo, M. (2002). Taurine: A conditionally essential amino acid in humans? An overview in health and disease. *Nutricion Hospitalaria*, 17: 262-270.

Lücke, F.-K. (1994). Fermented meat products. *Food Research International*, 27: 299-307.

Maladkar, M., Patil, S. and Sood, S. (2016). Coenzyme Q10: The cardiac bio-energizer in cardiovascular diseases. *Cardiovascular Therapeutics*, 1: 555560.

Marco, M.L., Heeney, D., Binda, S., Cifelli, C.J., Cotter, P.D., Foligné, B., Gänzle, M., Kort, R., Pasin, G. and Pihlanto, A. (2017). Health benefits of fermented foods: Microbiota and beyond. *Current Opinion in Biotechnology*, 44: 94-102.

Martín-Cabrejas, M., Aguilera, Y., Benítez, V. and Reiter, R. (2017). Melatonin synthesis in fermented foods. pp. 105-129. *In*: Juana Frias, C.M.-V.A.E.P. (Ed.). *Fermented Foods in Health and Disease Prevention*. Elsevier, Amsterdam.

Martínez-Álvarez, O., López-Caballero, M., Gómez-Guillén, M. and Montero, P. (2017). Fermented seafood products and health. pp. 177-202. *In*: Peñas, J.F.C.M.-V.E. (Ed.). *Fermented Foods in Health and Disease Prevention*. Elsevier, Amsterdam.

Menegas, L.Z., Pimentel, T.C., Garcia, S. and Prudencio, S.H. (2013). Dry-fermented chicken sausage produced with inulin and corn oil: Physicochemical, microbiological, and textural characteristics and acceptability during storage. *Meat Science*, 93: 501-506.

Moe, N.K.T., Thwe, S.M., Shirai, T., Terahara, T., Imada, C. and Kobayashi, T. (2015). Characterisation of lactic acid bacteria distributed in small fish fermented with boiled rice in Myanmar. *Fisheries Science*, 81: 373-381.

Moeinian, M., Farnaz Ghasemi-Niri, S., Mozaffari, S. and Abdollahi, M. (2013). Synergistic effect of probiotics, butyrate and l-Carnitine in treatment of IBD. *Journal of Medical Hypotheses and Ideas*, 7: 50-53.

Mohamed, E. and Hamadani, L. (2016). Changes in minerals content of traditionally fermented Nile-fish product in Sudan. *Chemistry*, 75: 155-158.

Molly, K., Demeyer, D., Johansson, G., Raemaekers, M., Ghistelinck, M. and Geenen, I. (1997). The importance of meat enzymes in ripening and flavour generation in dry fermented sausages. First results of a European project. *Food Chemistry*, 59: 539-545.

Montaño, N., Gavino, G. and Gavino, V.C. (2001). Polyunsaturated fatty acid contents of some traditional fish and shrimp paste condiments of the Philippines. *Food Chemistry*, 75: 155-158.

Mora, L., Hernández-Cázares, A.S., Sentandreu, M.A. and Toldrá, F. (2010). Creatine and creatinine evolution during the processing of dry-cured ham. *Meat Science*, 84: 384-389.

Mudgil, D. and Barak, S. (2013). Composition, properties and health benefits of indigestible carbohydrate polymers as dietary fibre: A review. *International Journal of Biological Macromolecules*, 61: 1-6.

Muhammed, M., Domendra, D., Muthukumar, S., Sakhare, P. and Bhaskar, N. (2015). Effects of fermentatively recovered fish waste lipids on the growth and composition of broiler meat. *British Poultry Science*, 56: 79-87.

Muthukumarasamy, P. and Holley, R.A. (2007). Survival of *Escherichia coli* O157: H7 in dry fermented sausages containing micro-encapsulated probiotic lactic acid bacteria. *Food Microbiology*, 24: 82-88.

Najafian, L. and Babji, A.S. (2018). Fractionation and identification of novel antioxidant peptides from fermented fish (pekasam). *Journal of Food Measurement and Characterisation*, 12: 2174-2183.

Najafian, L. and Babji, A.S. (2019). Purification and identification of antioxidant peptides from fermented fish sauce (*budu*). *Journal of Aquatic Food Product Technology*, 28: 14-24.

Nie, X., Zhang, Q. and Lin, S. (2014). Biogenic amine accumulation in silver carp sausage inoculated with *Lactobacillus plantarum* plus *Saccharomyces cerevisiae*. *Food Chemistry*, 153: 432-436.

Noonpakdee, W., Santivarangkna, C., Jumriangrit, P., Sonomoto, K. and Panyim, S. (2003). Isolation of nisin-producing *Lactococcus lactis* WNC 20 strain from *nham*, a traditional Thai fermented sausage. *International Journal of Food Microbiology*, 81: 137-145.

Núñez, F., Lara, M.S., Peromingo, B., Delgado, J., Sánchez-Montero, L. and Andrade, M.J. (2015). Selection and evaluation of *Debaryomyces hansenii* isolates as potential bioprotective agents against toxigenic penicillia in dry-fermented sausages. *Food Microbiology*, 46: 114-120.

Nurnaafi, A. and Setyaningsih, I. (2015). Probiotic potential of bekasam lactic acid bacteria of tilapia fish. *Jurnal Teknologi dan Industri Pangan*, 26: 109-114.

Ohata, M., Uchida, S., Zhou, L. and Arihara, K. (2016). Antioxidant activity of fermented meat sauce and isolation of an associated antioxidant peptide. *Food Chemistry*, 194: 1034-1039.

Ojha, K.S., Kerry, J.P., Duffy, G., Beresford, T. and Tiwari, B.K. (2015). Technological advances for enhancing quality and safety of fermented meat products. *Trends in Food Science and Technology*, 44: 105-116.

Okamoto, A., Matsumoto, E., Iwashita, A., Yasuhara, T., Kawamura, Y., Koizumi, Y. and Yanagida, F. (1995). Angiotensin I-converting enzyme inhibitory action of fish sauce. *Food Science and Technology International*, Tokyo 1: 101-106.

Olatunde, O.O., Obadina, A.O., Omemu, A.M., Oyewole, O.B., Olugbile, A. and Olukomaiya, O.O. (2018). Screening and molecular identification of potential probiotic lactic acid bacteria in effluents generated during *ogi* production. *Annals of Microbiology*, 68: 433-443.

Olesen, P.T. and Stahnke, L.H. (2000). The influence of *Debaryomyces hansenii* and *Candida utilis* on the aroma formation in garlic spiced fermented sausages and model minces. *Meat Science*, 56: 357-368.

Onsurathum, S., Pinlaor, P., Haonon, O., Chaidee, A., Charoensuk, L., Intuyod, K., Boonmars, T., Laummaunwai, P. and Pinlaor, S. (2016). Effects of fermentation time and low temperature during the production process of Thai pickled fish (pla-som) on the viability and infectivity of *Opisthorchis viverrini metacercariae*. *International Journal of Food Microbiology*, 218: 1-5.

Owens, J. and Mendoza, L. (1985). Enzymically-hydrolysed and bacterially fermented fishery products. *International Journal of Food Science and Technology*, 20: 273-293.

Paludan-Müller, C., Valyasevi, R., Huss, H.H. and Gram, L. (2002). Genotypic and phenotypic characterisation of garlic-fermenting lactic acid bacteria isolated from som-fak, a Thai low-salt fermented fish product. *Journal of Applied Microbiology*, 92: 307-314.

Parvez, S., Malik, K.A., Ah Kang, S. and Kim, H.Y. (2006). Probiotics and their fermented food products are beneficial for health. *Journal of Applied Microbiology*, 100: 1171-1185.

Pennacchia, C., Ercolini, D., Blaiotta, G., Pepe, O., Mauriello, G. and Villani, F. (2004). Selection of *Lactobacillus* strains from fermented sausages for their potential use as probiotics. *Meat Science*, 67: 309-317.

Peralta, E.M., Hatate, H., Kawabe, D., Kuwahara, R., Wakamatsu, S., Yuki, T. and Murata, H. (2008). Improving antioxidant activity and nutritional components of Philippine salt-fermented shrimp paste through prolonged fermentation. *Food Chemistry*, 111: 72-77.

Phadke, G., Elavarasan, K. and Shamasundar, B.A. (2014). Angiotensin-I converting enzyme (ACE) inhibitory activity and antioxidant activity of fermented fish product ngari as influenced by fermentation period. *International Journal of Pharma and Biosciences*, 5: 134-142.

Pongsetkul, J., Benjakul, S., Sumpavapol, P., Vongkamjan, K. and Osako, K. (2018). Quality of kapi, salted shrimp paste of Thailand, inoculated with *Bacillus* spp. K-C3. *Journal of Aquatic Food Product Technology*, 27: 830-843.

Pongsetkul, J., Benjakul, S., Vongkamjan, K., Sumpavapol, P. and Osako, K. (2017a). Changes in lipids of shrimp (*Acetes vulgaris*) during salting and fermentation. *European Journal of Lipid Science and Technology*, 119: 1700253.

Pongsetkul, J., Benjakul, S., Vongkamjan, K., Sumpavapol, P., Osako, K. and Faithong, N. (2017b). Changes in volatile compounds, ATP-related compounds and antioxidative properties of *kapi*, produced from *Acetes vulgaris*, during processing and fermentation. *Food Bioscience*, 19: 49-56.

Popa, O., Băbeanu, N.E., Popa, I., Niță, S. and Dinu-Pârvu, C.E. (2015). Methods for obtaining and determination of squalene from natural sources. *BioMed Research International*, 2015: ID 367202.

Pouliot-Mathieu, K., Gardner-Fortier, C., Lemieux, S., St-Gelais, D., Champagne, C.P. and Vuillemard, J.-C. (2013). Effect of cheese containing gamma-aminobutyric acid-producing lactic acid bacteria on blood pressure in men. *Pharma Nutrition*, 1: 141-148.

Putra, T., Suprapto, H., Tjahjaningsih, W. and Pramono, H. (2018). The antagonistic activity of lactic acid bacteria isolated from *peda*, an Indonesian traditional fermented fish. *IOP Conference Series: Earth and Environmental Science*, vol. 137, 012060.

Pyo, Y.-H. and Oh, H.-J. (2011). Ubiquinone contents in Korean fermented foods and average daily intakes. *Journal of Food Composition and Analysis*, 24: 1123-1129.

Rai, A.K., Bhaskar, N. and Baskaran V. (2013). Bioefficacy of EPA-DHA from lipids recovered from fish processing wastes through biotechnological approaches. *Food Chemistry*, 136: 80-86.

Rai, A.K., Bhaskar, N. and Baskaran, V. (2015). Effect of feeding lipids recovered from fish processing waste by lactic acid fermentation and enzymatic hydrolysis on antioxidant and membrane bound enzymes in rats. *Journal of Food Science and Technology*, 52: 3701-3710.

Rajauria, G., Sharma, S., Emerald, M. and Jaiswal, A.K. (2016). Novel fermented marine-based products. pp. 235-262. *In*: Ojha, K.S., Tiwari, Brijesh K. (Eds.). *Novel Food Fermentation Technologies*. Springer International Publishing, Switzerland.

Ranadheera, C., Vidanarachchi, J., Rocha, R., Cruz, A. and Ajlouni, S. (2017). Probiotic delivery through fermentation: Dairy vs. non-dairy beverages. *Fermentation*, 3: 67.

Reddy, L.H. and Couvreur, P. (2009). Squalene: A natural triterpene for use in disease management and therapy. *Advanced Drug Delivery Reviews*, 61: 1412-1426.

Riebroy, S., Benjakul, S. and Visessanguan, W. (2008). Properties and acceptability of *som-fug*, a Thai fermented fish mince, inoculated with lactic acid bacteria starters. *LWT–Food Science and Technology*, 41: 569-580.

Riebroy, S., Benjakul, S., Visessanguan, W., Kijrongrojana, K. and Tanaka, M. (2004). Some characteristics of commercial *som-fug* produced in Thailand. *Food Chemistry*, 88: 527-535.

Rossmeisl, M., Jilkova, Z.M., Kuda, O., Jelenik, T., Medrikova, D., Stankova, B., Kristinsson, B., Haraldsson, G.G., Svensen, H. and Stoknes, I. (2012). Metabolic effects of n-3 PUFA as phospholipids are superior to triglycerides in mice fed a high-fat diet: Possible role of endocannabinoids. *PloS ONE*, 7: e38834.

Ruusunen, M. and Puolanne, E. (2005). Reducing sodium intake from meat products. *Meat Science*, 70: 531-541.

Saadi, S., Saari, N., Anwar, F., Hamid, A.A. and Ghazali, H.M. (2015). Recent advances in food biopeptides: Production, biological functionalities and therapeutic applications. *Biotechnology Advances*, 33: 80-116.

Saiga, A., Tanabe, S. and Nishimura, T. (2003). Antioxidant activity of peptides obtained from porcine myofibrillar proteins by protease treatment. *Journal of Agricultural and Food Chemistry*, 51: 3661-3667.

Salampessy, J., Kailasapathy, K. and Thapa, N. (2010). Fermented fish products. pp. 289-307. *In*: Jyoti Prakash Tamang, K.K. (Ed.). *Fermented Foods and Beverages of the World*. CRC Press, London, New York.

Sameshima, T., Magome, C., Takeshita, K., Arihara, K., Itoh, M. and Kondo, Y. (1998). Effect of intestinal *Lactobacillus* starter cultures on the behaviour of *Staphylococcus aureus* in fermented sausage. *International Journal of Food Microbiology*, 41: 1-7.

Sanchart, C., Rattanaporn, O., Haltrich, D., Phukpattaranont, P. and Maneerat, S. (2017). Enhancement of gamma-aminobutyric acid (GABA) levels using an autochthonous *Lactobacillus futsaii* CS3 as starter culture in Thai fermented shrimp (*kung-som*). *World Journal of Microbiology and Biotechnology*, 33: 152.

Sánchez, A. and Vázquez, A. (2017). Bioactive peptides: A review. *Food Quality and Safety*, 1: 29-46.

Sanz, Y. and Toldrá, F. (1997). Purification and characterisation of an aminopeptidase from *Lactobacillus* sake. *Journal of Agricultural and Food Chemistry*, 45: 1552-1558.

Sanz, Y. and Toldrá, F. (2002). Purification and characterisation of an arginine aminopeptidase from *Lactobacillus sakei*. *Applied and Environmental Microbiology*, 68: 1980-1987.

Sasaki, S., Yokozawa, T., Cho, E.J., Oowada, S. and Kim, M. (2006). Protective role of γ-aminobutyric acid against chronic renal failure in rats. *Journal of Pharmacy and Pharmacology*, 58: 1515-1525.

Schuller-Levis, G.B. and Park, E. (2003). Taurine: New implications for an old amino acid. *FEMS Microbiology Letters*, 226: 195-202.

Schuller, H.M., Al-Wadei, H.A. and Majidi, M. (2008). Gamma-aminobutyric acid, a potential tumor suppressor for small airway-derived lung adenocarcinoma. *Carcinogenesis*, 29: 1979-1985.

Shimada, K., Sakuma, Y., Wakamatsu, J., Fukushima, M., Sekikawa, M., Kuchida, K. and Mikami, M. (2004). Species and muscle differences in L-carnitine levels in skeletal muscles based on a new simple assay. *Meat Science*, 68: 357-362.

Skåra, T., Axelsson, L., Stefánsson, G., Ekstrand, B. and Hagen, H. (2015). Fermented and ripened fish products in the northern European countries. *Journal of Ethnic Foods*, 2: 18-24.

Slavin, J. (2013). Fibre and prebiotics: Mechanisms and health benefits. *Nutrients*, 5: 1417-1435.

Smith, T.J. (2000). Squalene: Potential chemopreventive agent. *Expert Opinion on Investigational Drugs*, 9: 1841-1848.

Song, M.-Y., Van-Ba, H., Park, W.-S., Yoo, J.-Y., Kang, H.-B., Kim, J.-H., Kang, S.-M., Kim, B.-M., Oh, M.-H. and Ham, J.-S. (2018). Quality characteristics of functional fermented sausages added with encapsulated probiotic *Bifidobacterium longum* KACC 91563. *Korean Journal for Food Science of Animal Resources*, 38: 981-994.

Sørensen, R., Kildal, E., Stepaniak, L., Pripp, A.H. and Sørhaug, T. (2004). Screening for peptides from fish and cheese inhibitory to prolyl endopeptidase. *Food/Nahrung*, 48: 53-56.

Spencer, A.U., Yu, S., Tracy, T.F., Aouthmany, M.M., Llanos, A., Brown, M.B., Brown, M., Shulman, R.J., Hirschl, R.B. and DeRusso, P.A. (2005). Parenteral nutrition-associated cholestasis in neonates: Multivariate analysis of the potential protective effect of taurine. *Journal of Parenteral and Enteral Nutrition*, 29: 337-344.

Stadnik, J. and Kęska, P. (2015). Meat and fermented meat products as a source of bioactive peptides. *Acta Scientiarum Polonorum, Technologia Alimentaria*, 14: 181-190.

Sun, W., Zhao, H., Zhao, Q., Zhao, M., Yang, B., Wu, N. and Qian, Y. (2009). Structural characteristics of peptides extracted from Cantonese sausage during drying and their antioxidant activities. *Innovative Food Science and Emerging Technologies*, 10: 558-563.

Taipale, T., Pienihäkkinen, K., Isolauri, E., Larsen, C., Brockmann, E., Alanen, P., Jokela, J. and Söderling, E. (2010). *Bifidobacterium animalis* subsp. lactis BB-12 in reducing the risk of infections in infancy. *British Journal of Nutrition*, 105: 409-416.

Tanasupawat, S., Chamroensaksri, N., Kudo, T. and Itoh, T. (2010). Identification of moderately halophilic bacteria from Thai-fermented fish (pla-ra) and proposal of *Virgibacillus siamensis* sp. nov., *The Journal of General and Applied Microbiology*, 56: 369-379.

Tanasupawat, S., Namwong, S., Kudo, T. and Itoh, T. (2009). Identification of halophilic bacteria from fish sauce (*nam-pla*) in Thailand. *Journal of Culture Collections*, 6: 69-75.

Tang, K.X., Shi, T. and Gänzle, M. (2018). Effect of starter cultures on taste-active amino acids and survival of pathogenic *Escherichia coli* in dry fermented beef sausages. *European Food Research and Technology*, 244: 2203-2212.

Toldrá, F. (2006). The role of muscle enzymes in dry-cured meat products with different drying conditions. *Trends in Food Science and Technology*, 17: 164-168.

Toldrá, F. (2011). Improving the sensory quality of cured and fermented meat products. pp. 508-526. *In*: Kerry, J.P. and Kerry, J.F. (Eds.). *Processed Meats*, 1st ed., Woodhead Publishing, Sawston, Cambridge.

Toldra, F., Rico, E. and Flores, J. (1993). Cathepsin B, D, H and L activities in the processing of dry-cured ham. *Journal of the Science of Food and Agriculture*, 62: 157-161.

Trząskowska, M., Kołożyn-Krajewska, D., Wójciak, K. and Dolatowski, Z. (2014). Microbiological quality of raw-fermented sausages with *Lactobacillus casei* LOCK 0900 probiotic strain. *Food Control*, 35: 184-191.

Tujioka, K., Okuyama, S., Yokogoshi, H., Fukaya, Y., Hayase, K., Horie, K. and Kim, M. (2007). Dietary γ-aminobutyric acid affects the brain protein synthesis rate in young rats. *Amino Acids*, 32: 255.

Udomthawee, K., Chunkao, K., Phanurat, A. and Nakhonchom, K. (2012). Protein, calcium and phosphorus composition of fermented fish in the lower Mekong basin. *Chiang Mai J. Sci.*, 39: 327-335.

Ursini, F., Maiorino, M. and Roveri, A. (1997). Phospholipid hydroperoxide glutathione peroxidase (PHGPx): More than an antioxidant enzyme. *Biomedical and Environmental Sciences, BES*, 10: 327-332.

Valeur, N., Engel, P., Carbajal, N., Connolly, E. and Ladefoged, K. (2004). Colonisation and immunomodulation by *Lactobacillus reuteri* ATCC 55730 in the human gastrointestinal tract. *Applied and Environmental Microbiology*, 70: 1176-1181.

Valfré, F., Caprino, F. and Turchini, G. (2003). The health benefit of seafood. *Veterinary Research Communications*, 27: 507-512.

Vaštag, Ž., Popović, L., Popović, S., Petrović, L. and Peričin, D. (2010). Antioxidant and angiotensin-I converting enzyme inhibitory activity in the water-soluble protein extract from Petrovac Sausage (*petrovská kolbása*). *Food Control*, 21: 1298-1302.

Visessanguan, W., Benjakul, S., Riebroy, S., Yarchai, M. and Tapingkae, W. (2006). Changes in lipid composition and fatty acid profile of *nham*, a Thai fermented pork sausage, during fermentation. *Food Chemistry*, 94: 580-588.

Wang, H., Ni, X., Qing, X., Zeng, D., Luo, M., Liu, L., Li, G., Pan, K. and Jing, B. (2017). Live probiotic *Lactobacillus johnsonii* BS15 promotes growth performance and sowers fat deposition by improving lipid metabolism, intestinal development, and gut microflora in broilers. *Frontiers in Microbiology*, 8: 1073.

Wójciak, K.M., Libera, J., Stasiak, D.M. and Kołożyn-Krajewska, D. (2017). Technological aspect of *Lactobacillus acidophilus* Bauer, *Bifidobacterium animalis* BB-12 and *Lactobacillus rhamnosus* LOCK900 USE in dry-fermented pork neck and sausage. *Journal of Food Processing and Preservation*, 41: e12965.

Yang, X., Zhang, Y., Xu, H., Luo, X., Yu, J., Liu, J. and Chang, R.C.-C. (2016). Neuroprotection of coenzyme Q10 in neurodegenerative diseases. *Current Topics in Medicinal Chemistry*, 16: 858-866.

Yin, L.J. and Jiang, S.T. (2001). *Pediococcus pentosaceus* L and S utilisation in fermentation and storage of mackerel sausage. *Journal of Food Science*, 66: 742-746.

Yin, L.J., Pan, C.L. and Jiang, S.T. (2002). Effect of lactic acid bacterial fermentation on the characteristics of minced mackerel. *Journal of Food Science*, 67: 786-792.

Yoshikawa, S., Kurihara, H., Kawai, Y., Yamazaki, K., Tanaka, A., Nishikiori, T. and Ohta, T. (2010). Effect of halotolerant starter microorganisms on chemical characteristics of fermented chum salmon (*Oncorhynchus keta*) sauce. *Journal of Agricultural and Food Chemistry*, 58: 6410-6417.

Zanardi, E., Ghidini, S., Conter, M. and Ianieri, A. (2010). Mineral composition of Italian salami and effect of NaCl partial replacement on compositional, physico-chemical and sensory parameters. *Meat Science*, 86: 742-747.

Zang, J., Xu, Y., Xia, W., Jiang, Q., Yang, F. and Wang, B. (2018). Phospholipid molecular species composition of Chinese traditional low-salt fermented fish inoculated with different starter cultures. *Food Research International*, 111: 87-96.

Zeng, X., Xia, W., Jiang, Q. and Guan, L. (2015). Biochemical and sensory characteristics of whole carp inoculated with autochthonous starter cultures. *Journal of Aquatic Food Product Technology*, 24: 52-67.

Zeng, Z., Lin, J. and Gong, D. (2009). Identification of lactic acid bacterial strains with high conjugated linoleic acid-producing ability from natural sauerkraut fermentations. *Journal of Food Science*, 74: M154-M158.

Zhang, W., Xiao, S., Samaraweera, H., Lee, E.J. and Ahn, D.U. (2010). Improving functional value of meat products. *Meat Science*, 86: 15-31.

Part IV
Recent Advances in Food Fermentation

Microbial Transformation during Gut Fermentation

Rounak Chourasia[1], Loreni Chiring Phukon[1], Md Minhajul Abedin[1],
Dinabandhu Sahoo[1,2] and Amit Kumar Rai[1]*

[1] Institute of Bioresources and Sustainable Development, Sikkim Centre, Tadong, India
[2] Department of Botany, University of Delhi, Delhi, India

1. Introduction

Human nutrition and energy harvest are affected by microorganisms from fermented foods and the gastrointestinal (GI) tract. With colon bacterial mass of approximately 1.5 kg and a density of 10^{12} cells per gram of intestinal content, the gut microbiota is recognised as an essential metabolic organ (Et *et al.*, 2011). Disturbance of the gut microbial ecosystem by antibiotics, diet change and pathogens lead to dysbiosis (gut microbial imbalance) which results in a negative impact on health (Shi *et al.*, 2018). Diseases related to dysbiosis of the gut microbial population include colitis, Crohn's disease, obesity, cardiovascular diseases and even mental health diseases, thus highlighting the importance of the gut-brain axis (Carding *et al.*, 2015). Diet plays an important role in defining and maintenance of the microbial content in the gut of an individual with fermented foods being the source of live microorganisms called probiotics, that enhance gastrointestinal health (Marco *et al.*, 2017).

Gut metagenome sequencing studies have revealed the presence of a vast pool of genes dedicated to breakdown and transformation of varied dietary components, most important of which are the genes involved in catabolism of biogenic metabolites ingested upon consumption of fermented foods (Rajakovich and Balskus, 2019; Stanton *et al.*, 2005). The biogenic metabolites include dietary fibres, such as exopolysaccharides (EPS) and galactooligosaccharides (GOS), a vast array of polyphenols, such as flavonoids (isoflavones, flavanols, flavones, anthocyanins, flavonols), lignans and phenolic acids (Magnusdottir and Thiele, 2018). EPS and GOS produced by starter or adjunct cultures in fermented foods have a prebiotic potential, serving as fermentable substrates for gut microbiota. Fermentation of EPS by CAZymes (Carbohydrate-Active Enzymes)-secreting gut microbiota result in

Correspondence author: amitraikvs@gmail.com

generation of molecules that serve as substrate for the growth of health-stimulating gut microbial strains and also exert bioactivity (Liu *et al.*, 2018).

Polyphenol aglycones produced by starter strains during food fermentation pass through the small intestine, avoiding absorption and enter the colon where gut microbial strains metabolise these aglycones into bioactive metabolites that are absorbed by the enterocytes via passive diffusion (Pathak *et al.*, 2018). Polyphenol aglycones undergo enterohepatic circulation and thus circulate throughout the body in conjugated and deconjugated bioactive forms (Marín *et al.*, 2015). Gut fermentation of polyphenol aglycones results in production of metabolites demonstrating antioxidant, immunostimulatory, antihypertensive, anticancer and antidiabetic properties. In addition, prevention of cardiovascular diseases and maintenance of healthy gut-microbial population has also been observed (Actis-Goretta *et al.*, 2003; Galleano *et al.*, 2010; Soobrattee *et al.*, 2006). This chapter focuses on the biotransformation of fermented food-derived biogenic metabolites by gut microbial strains into bioactive metabolites and their impact on the health of the consumer.

2.   Dietary Fibres

Undigested dietary fibres which the human digestive system is unable to breakdown in the small intestine are the main source of nutrition and energy on which bacterial species residing on the gut are dependent. The gut microbiota breaks down these fibres by saccharolytic bacterial fermentation and releases health-beneficial metabolites. Decrease in carbohydrate source compels gut microbiota to turn to alternate energy sources which lead to generation of metabolites with detrimental effect to human health (Han *et al.*, 2017). Short-chain fatty acids (SCFAs) are the major fermentation products of undigested dietary fibres by gut microbial strains. Dietary fibres are categorised as soluble and insoluble, based on their relative solubility in water. Both kinds of fibres are found in food and their concentrations depend upon the food source and the form of consumption. For example, vegetables and fruits contain higher concentrations of soluble fibres as compared to insoluble fibres, which are present in high amounts in cereals (Prasad and Bondy, 2018). β-glycan, inulin-resistant starch, pectin and guar fibres are soluble fibres whereas insoluble fibres contain cellulose, hemicellulose and lignin (Lattimer and Haub, 2010; Ötles and Ozgoz, 2014). Soluble fibres form gel in the intestine after absorbing water and are easily fermented by the gut microbiota (McRorie and McKeown, 2017).

The bioactive properties of insoluble fibres include their ability to bind carcinogens, mutagens and other toxic chemicals during food digestion and the excretion of these bound forms of insoluble fibres form the faeces (Prasad and Bondy, 2018). Bioactivity of soluble fibres is dependent on the fermentation of these fibres in the colon by gut microbiota, resulting in generation of SCFA, including acetate, butyrate and propionate (Galvez *et al.*, 2005; Papandreou *et al.*, 2015). The concentrations of SCFA generated from soluble fibres depend on the amount of fibres present in the food during consumption, bacterial strains present in the gut microbiota and the gut transition time of the fibres (Prasad and Bondy, 2018).

2.1   Exopolysaccharides

Microbial strains involved as starter cultures and/or adjunct cultures in fermentation of food substrates synthesise complex polysaccharides, among which EPSs are well known to impart various physiological and health benefits to the consumers (Feng *et al.*, 2012; Joshi, 2016; Khalil *et al.*, 2018). EPSs are high molecular weight carbohydrate polymers, present in long chains consisting of branched repeating sugar units, substituted sugars or sugar derivatives, including phosphate and acetyl substituting groups (Welman, 2015). Health benefits provided by EPS of fermented foods include competitive exclusion of pathogenic bacteria by adhering and promoting colonisation of probiotic bacteria and by acting as a physical barrier to pathogens, blood and serum cholesterol reduction, immunomodulation, immunostimulant, antioxidant, antihypertensive, antidiabetic, antitumor, antiviral, antiproliferative effects and generation of SCFAs in the gut upon degradation by gut microbiota (Kheni and Vyas, 2017; Liu *et al.*, 2019; Paynich *et al.*, 2017; Zhang *et al.*, 2017).

Fermented dairy products, including yoghurt, cheese, *kefir*, sour milk and buttermilk have been observed to contain EPS-producing lactic acid bacteria (LAB) (Welman, 2015). Other sources include fermented soybeans: *kinema, natto, hawaizaar* (where the ropy texture of the EPS produced by *Bacillus* spp. is observed in abundance), *kimchi,* and *ganjang* (fermented soy sauce) (Marvasi *et al.*, 2010; Palaniyandi *et al.*, 2018; Song *et al.*, 2011; Tamang *et al.*, 2016). Many studies have reported the production of bioactive EPS by microorganisms in fermented foods. The production of antioxidant EPSs in Sichuan pickle prepared upon fermentation of vegetables by *Bacillus* sp. S-1 was reported (Hu *et al.*, 2019). EPS produced by *Bacillus* spp. demonstrate antiviral and immunomodulating properties in addition to heavy metal chelation, bio-emulsification and bio-flocculation (Alizadeh-Sani *et al.*, 2018; Arena *et al.*, 2006; Gupta and Diwan, 2017). LAB produce exopolysaccharides during milk fermentation and ripening of fermented dairy products. LAB generate heteropolysaccharides that are secreted outside the cell and result in the formation of either EPS capsules attached to the cell or are released into the substrate, forming ropy texture (Hassan, 2008). These EPSs have a GRAS (Generally Recognised as Safe) status and consist mainly of rhamnose, glucose and galactose which are present in varying concentrations.

The gut microbiota consists of a diversity of microbial species where the large numbers of genes are shared within the wide population in order to attain metabolic co-operation, resulting in benefit to the host (Liu *et al.*, 2018). Breakdown of complex polysaccharides by a large pool of CAZymes (carbohydrate active enzymes) shared by the resident strains of the gut microbiota is an example of such metabolic co-operation. These CAZymes collectively break down complex polysaccharides that cannot be catabolised by the human digestive system (Kaoutari *et al.*, 2013). The products generated upon catabolism of complex polysaccharides include vitamins, monosaccharides and organic acids which can be utilised by the human body or are cross-fed to other gut microbial strains, thus aiding in maintenance of beneficial gut microbial population (Turnbaugh *et al.*, 2009; Welman, 2015).

EPS derived from fermented foods demonstrate prebiotic potential, a prebiotic being defined as 'a non-digestible selectively fermented ingredient that allows specific changes, both in the composition and/or activity in the gastrointestinal microbiota that confers benefits upon host well-being and health' (Carlson *et al.*, 2018). EPS produced by the yoghurt starter strains, including *Streptococcus thermophilus* SFi12 and SFi39, are degraded by gut microorganisms that have been isolated from faecal slurry (Ruijssenaars *et al.*, 2000). Interestingly, EPS from *St. thermophilus* SFi20 and *Lactobacillus helveticus* Lh59, and EPS in the Finnish ropy milk fermented by *Lactococcus lactis* subsp. *cremoris* B40 could not be broken down by the same gut microbial community. This difference in EPS degradation was later found to be due to the presence of an extra β-galactosyl side-chain in the EPS produced by *Lc. lactis* subsp. *cremoris* B40. EPSs produced by the faecal bacteria *Bifidobacterium pseudocatenulatum* promote the growth of other faecal strains, including *Faecalibacterium prausnitzii,* and *Desulfovibrio* spp. (Milani *et al.*, 2017). Furthermore, EPS produced by *Bifidobacterium longum* promotes growth of *Oscillospira* spp., *Anaerostipes* spp. and *Prevotella* spp. (Salazar *et al.*, 2008). Therefore, it is evident that EPS produced by bifidobacteria in fermented dairy products and gastrointestinal in origin serve as substrates for gut metabolism. *Bifidobacterium bifidum* EPS exert growth-stimulatory effects on total anaerobic bacteria and Lactobacilli of the gut, while inhibiting the growth of *Bacteroides fragilis,* enterococci and enterobacteria, thus reducing the diversity of unwanted enterobacterial strains (Li *et al.*, 2014).

EPS fermentation in gut is associated with production of high levels of SCFAs which apart from being metabolised by the liver as an energy source demonstrates various bioactivities, including anti-colon cancer activity, immunomodulation and antimicrobial activity (Tan *et al.*, 2014). High levels of SCFAs are generated upon fermentation of bifidobacterial EPSs by other gut bacterial strains (Salazar *et al.*, 2008). The major SCFAs produced during gut fermentation of EPS include propionate, butyrate and acetate. These SCFAs demonstrate diverse bioactive properties and have been studied for their potential in disease prevention. Butyrate is the major SCFA involved in maintaining gut homeostasis and preventing various diseases (Canani *et al.*, 2011). Sodium butyrate slows down cancer initiation by inhibiting growth of neuroblastoma cells, enhances the cell-killing effect of X-radiation, cyclic AMP stimulating agents and chemotherapeutic agents (Chen *et al.*, 2019). In addition, sodium butyrate accelerates apoptosis in cancer cells by inhibiting histone deacetylase, which leads to hyperacetylation of the basic proteins and loosening of their attachment to DNA and resulting in increased apoptotic gene transcription (Jiang *et al.*, 2012; Prasad, 1980; Prasad and Hsie, 1971). Other effects of sodium butyrate in cancer inhibition include down regulation of microRNA in colon cancer cells, resulting in apoptosis, prevention of colony formation and cell invasion by tumor cells (Ma *et al.*, 2020). Inversely, up regulation of the microRNA miR-31 by sodium butyrate leads to the senescence of breast cancer cells (Londin *et al.*, 2015; MacFarlane and R. Murphy, 2010).

Apart from inhibition and prevention of cancer, butyrate is found to be effective against inflammatory disorders, neurodegenerative diseases, diabetes and cardiovascular diseases (Aguilar *et al.*, 2016; Bourassa *et al.*, 2016; Endesfelder *et al.*, 2016;, Furusawa *et al.*, 2013; Khan and Jena, 2016; Paiva *et al.*, 2017; Zhang *et al.*, 2016; Zimmerman *et al.*, 2012). The other two major SCFAs produced during fermentation of dietary fibres in the gut – propionate and acetate – also show bioactivity. Propionate has been observed to lower serum cholesterol levels, lipids and reducing the incidence of diseases like cancer (Byrne *et al.*, 2016; Chambers *et al.*, 2015; Hosseini *et al.*, 2011). Similarly, induction of apoptosis in colorectal cancer cells by sodium acetate has been reported (Oliveira *et al.*, 2015). Sodium acetate and propionate were reported to reduce appetite, thus preventing obesity which has been nearing epidemic proportions worldwide and leads to the causality of other diseases (Frost *et al.*, 2014).

2.2 Prebiotic Oligosaccharides

The most profoundly available prebiotic oligosaccharides include GOS, fructo-oligosaccharides (FOS) and inulin which have been found to enhance gut population of bifidobacterial and LAB strains (Welman, 2015). GOS are available in fermented dairy products, including fermented milks, yoghurts and in some cheese (Fischer and Kleinschmidt, 2018). Glycosyl transfer of D-galactosyl units on the D-galactose moiety of lactose, catalysed by β-galactosidase results in the generation of GOS (Martínez-Villaluenga *et al.*, 2008). Commensal microbial population, including members of Lactobacilli and bifidobacteria are capable of fermenting and utilising GOS (Davis *et al.*, 2011). LacS permease of these bacteria transport GOS inside the cells which are then broken down by intracellular galactosidases, and finally the bacterial β-galactosidase hydrolyses the terminal lactose (Azcarate-Peril *et al.*, 2017). Administration of GOS leads to a bifidogenic response which results in an increase in the population of β-galactosidase producing bifidobacterial strains. This increase in lactose-fermenting bacteria in the gut has been associated with enhanced digestion of dairy products by lactose-intolerant individuals (Azcarate-Peril *et al.*, 2017). However, the enrichment of bifidobacteria after GOS consumption comes at the expense of certain *Bacteroides* species (Davis *et al.*, 2011). Apart from exerting an increase in the gut barrier function in lactose-intolerant individuals, GOS fermentation by probiotics in the gut leads to the generation of bioactive compounds, such as SCFAs, lactate and vitamins (Ahmad and Khalid, 2018).

FOS and inulin are naturally-occurring oligosaccharides that are added in the food matrix during fermentation to produce symbiotic foods, such as symbiotic milk, cheese, non-dairy fermented beverages and fermented meat (Bis-Souza *et al.*, 2020; Freire *et al.*, 2017; Oliveira *et al.*, 2013). Besides influencing the viability of probiotic strains in fermented foods and promoting their transport to the large intestine, fermentation of FOS and inulin in the gut results in favourable stimulation of growth of *Bifidobacterium* spp. and *Lactobacilli* spp. (Chen *et al.*, 2017). FOS fermentation in the gut leads to a butyrogenic effect similar to the fermentation of EPS and GOS where SCFAs, including butyrate, propionate and acetate, are produced that demonstrate health-beneficial properties (Mariano *et al.*, 2020; Shoaib *et al.*, 2016).

3. Polyphenols

Polyphenols are a large family of bioactive secondary metabolites found in almost all plant foods and are classified into flavonoids, phenolic acids, lignans and stilbenes (Galleano and Verstraeten, 2010). Bioavailability plays an important role in bioactivity of polyphenols and studies indicate that 90 per cent of total dietary intake of polyphenols avoids digestion and absorption in the small intestine, thereby reaching the large intestine (Conlon and Bird, 2014). Lack of bioavailability of polyphenols is attributed to the presence of glycosylated, esterified, or polymeric forms in plants (Pandey and Rizvi, 2009). β-glucosidase activity of food-fermenting cultures has been observed to remove glycosylation from flavonoids, releasing bioactive aglycone forms which can be absorbed by the small intestine (Michlmayr and Kneifel, 2014).

However, further biotransformation reactions are necessary for conversion of aglycone forms to their bioactive metabolites, which are observed to be mediated by microbiota in the gut (Scalbert and Williamson, 2000). Although certain intestinal enzymes hydrolyse polyphenols, the resulting metabolite may not be the desired absorbable product. For example, the hydrolysis of quercetin-3-O-rhamnoglucoside and quercetin-3-O-rhamnoside can be mediated by both human intestinal enzymes and by enzymes of the gut microflora, but the readily available form, quercetin, is released by enzymatic hydrolysis by the gut microbiota (Pathak *et al.*, 2018). The major polyphenols offered by fermented food products include isoflavones, phenolic acids, quercetin, anthocyanins, lignans, catechins and the metabolism of these compounds in the gut results in generation of metabolites with notable bioactivity (Fig. 1).

3.1   Flavonoids

Flavonoids belong to low-molecular-weight polyphenols that are formed by the combination of derivatives of phenylalanine via the shikimic acid pathway and acetic acid (Ghasemzadeh and Ghasemzadeh, 2011). The first step during flavonoid formation is the formation of phenylalanine from phenylpyruvate which is then transformed to trans-cinnamic acid. Hydrolysis of trans-cinnamic acid releases p-coumaric acid (C-9 acids) which condenses with three units of C-2 malonyl-CoA to form a C-15 chalcone (Kuhnau, 1976). Subsequent ring closures and hydration result in formation of compounds, such as 3,4-diolflavonoids (flavonols) and 3-hydroxyflavonoids (flavanols). The flavonoid nucleus consists of three phenolic rings, referred to as the A, B, and C rings, where the A ring (benzene) is condensed with the C ring (a six-membered ring), which carries a phenyl benzene ring (B ring) as a substituent at the 2-position (Yi *et al.*, 2011). The heterocyclic pyran ring C yields anthocyanidins and flavanols that yield flavanones, flavonols and flavones which together are termed as 4-oxo-flavonoids as they carry a carbonyl group on C-4 of the C ring (Aherne and Brien, 2002). The various subgroups of flavonoids are a result of a variety of substitution reactions in the A, B, and C rings. Changes in oxidation level on the flavonoid C ring lead to the formation of anthocyanidins, flavanols, flavones, flavanones, flavonols and isoflavones (Panche *et al.*, 2016). Conversion of

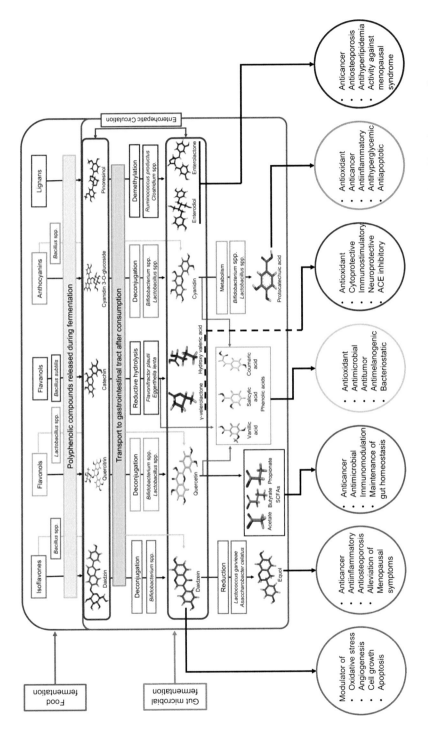

Fig. 1: Gut microbial fermentation of fermented food derived polyphenolic compounds and associated bioactive properties

Table 1: Gut Microbial Enzyme Catalyzed Conversion of Flavonoid Compounds to Bioactive Metabolites and Associated Bioactive Properties

Flavonoid Group	Dietary Compound	Gut Metabolism	Enzyme	Gut microbial Strains	Bioactive Metabolite	Bioactivity	References
Isoflavones	Daidzin Genistin Glycitin	Deconjugation	β-galactosidase	*Bifidobacterium* spp.	Daidzein Genistein Glycitein	Antioxidant Anticancer Chemoprotective Antidiabetic	Poschner *et al.* (2017), Das *et al.* (2018)
	Daidzein	Reduction	L-DZNR L-THDR	*Lactococcus garvieae Adlercreutzia* sp. *Asacharobacter celatus*	S-equol	Anticancer Antiosteoporosis Anti-inflammatory Anticancer	Iino *et al.* (2019), Mayo *et al.* (2019)
	Genistein	Reduction	DZNR DHDR THDR	*Slackia* spp.	5-Hydroxy-equol	Anti-inflammatory Antiosteoporosis Anticancer Chemoprotective	Jin *et al.* (2010), Lee *et al.* (2017)
Flavonols	Quercitrin	Deglycosylation	β-glucosidase	*Lactobacillus* spp. *Bifidobacterium* spp.	Quercetin	Antioxidant Antihyperglycemic Anti-inflammatory Hepatoprotective Cytoprotective	Qin *et al.* (2017), Rodriguez-Castaño *et al.* (2019)
	Rutin	Deglycosylation	β-glucosidase	*Lactobacillus* spp. *Bifidobacterium* spp. *Fusobacterium* spp. *Enterococcus* spp.	Quercetin	Antioxidant Antihyperglycemic Anti-inflammatory Hepatoprotective Cytoprotective	Braune and Blaut (2016), Shin *et al.* (2016), De Souza *et al.* (2017), Yang *et al.* (2019)

	Compound	Transformation	Enzyme	Microorganisms	Metabolites	Bioactivity	References
Flavonols	Quercetin	Dioxygenation, cleavage	Dioxygenases	*Eubacterium ramulus*, *Clostridium perfringens*, *Enterococcus* sp., *Bacteroides fragilis*	Phenolic acids	Antioxidant Antimicrobial Antitumor	Feng et al. (2018), Kawabata et al. (2019)
Flavanols	Catechin	C-ring cleavage	Dioxygenases	*Flavonifractor plautii* *Eggerthella lenta*	Hydroxyvaleric acids	Antioxidant Cytoprotective	Kutschera et al. (2011), Stevens and Maier (2016)
	Catechin	A-ring cleavage	Reverse Claisen reaction: Thiolases	*Lactobacillus* spp. *Flavonifractor plautii* *Eggerthella lenta*	γ-valerolactones	Antioxidant ACE inhibition Antitumor	Stevens and Maier (2016), Mena et al. (2019)
Anthocyanins	Cyanidin 3-O-glucoside	Deconjugation	β-glucosidase	*Bifidobacterium* spp. *Lactobacillus* spp.	Cyanidin	Antioxidant Anticancer Anti-inflammatory	Kalt, (2019), Lila et al. (2016)
	Cyanidin	B-ring cleavage	Decarboxylases	*Bifidobacterium* spp. *Lactobacillus* spp.	Protocatechuic acid	Antihyperglycemic Anti-inflammatory Antioxidant Antimicrobial	Kawabata et al. (2019), Majdoub et al. (2019)

Note: DZNR: Daidzein reductase; DHDR: Dihydrodaidzein reductase; THDR: Tetrahydrodaidzein reductase; ACE: Angiotensin Converting Enzyme.

these flavonoid subgroups to metabolites with increased bioactivity is catalysed by gut microbial enzymes (Table 1).

3.1.1   Isoflavones

Soybean (*Glycine max*) is the most cultivated plant in the world and is known to be rich in isoflavones among other compounds, including proteins, anthocyanins, lipids and oligosaccharides (Barne, 2010). Soy isoflavones are diphenolic compounds called phytoestrogens, that bind to estrogen receptors with higher affinity for ERβ than ERα, eliciting antiproliferative and antioxidative properties (da Silva *et al.*, 2011). In addition, soy isoflavones have inhibitory effects on topoisomerase, tyrosine kinase and angiogenesis which might reduce cancer risk (Zhong *et al.*, 2018). β-glucosides genistin, daidzin and glycitin are present in soybeans which are hydrolysed into their aglycone forms – genistein, daidzein and glycitein, respectively during soybean fermentation (Rai *et al.*, 2017; Zhang and Yu, 2019). Fermented soy foods are consumed intensively in Asia and this is complimented by the fact that lactase-phlorizin hydrolase, a mucosal β-glucosidase enzyme of the small intestine, is absent in most Asian individuals (Yoshiara *et al.*, 2018). This implies that soy isoflavones are consumed in aglycone forms by major Asian population (Barnes, 2010; Rivera-Sagredo *et al.*, 1992). The aglycones have a less complex structure and are absorbed in the gastrointestinal tract by a higher rate than the glycoside forms, thus fermentation of soybeans is known to increase the bioavailability of isoflavones (Choi *et al.*, 2007). Hydrolysis of isoflavone glucosides to aglycones by β-glucosidases in the anaerobic condition of the gut is followed by reductive modification of the aglycone heterocyclic ring to generate metabolites with higher bioactivity (Szeja *et al.*, 2017). S-equol is a reduced metabolite of aglycones – daidzein and genistein – with stronger affinity for estrogen receptor ERβ than the R- (+) -isomer. R-equol is present in the racemic mixture with S- (-) -isomer in chemically synthesised equol as opposed to natural equol synthesised by human gut microbiota where only S- equol is generated (Setchell *et al.*, 2005).

Gut microbiota have an important role in metabolism, bio availability and bioactivity of dietary isoflavones in the human system. Isoflavone glycosides, daidzin and genistin are not easily absorbed by the intestinal absorptive cells due to their large hydrophilic structures and require hydrolysis mediated by intestinal bacterial enzymes that release the unconjugated isoflavone aglycones, daidzein and genistein which are more estrogenic and easily absorbed (Rafii, 2015). *Bifidobacterium* spp., residing in the human gut, express probiotic properties, demonstrate high β-glucosidase activity and are responsible for the initial hydrolysis of isoflavone glucosides into their aglycones and subsequent absorption (Stojanov and Kreft, 2020). A Gram-positive bacterial strain, HGH6 and an *Escherechia coli* strain, HGH21, found from faeces of a healthy individual, show high β-glucosidase activity indicating the role of gut microbiota in deconjugation of isoflavone glucosides (Hur *et al.,*, 2000). Fermented soy foods supply isoflavones in the aglycone forms that can be directly absorbed by the intestinal absorptive cells (Simmons *et al.*, 2012). After the aglycones enter the circulation upon absorption through the intestine, they may be transported to the liver where conjugation by liver enzymes form glycosylated, sulphated and glucuronidated isoflavones that are more water soluble (Teng and Chen, 2019). These liver conjugates of isoflavones are excreted to the bile which are then passed to the intestinal lumen, thus completing the enterohepatic circulation of isoflavones (Barnes, 2010). The liver-conjugated isoflavones along with ingested and deconjugated isoflavones in the small intestine then enter the colon where further transformation occurs, resulting in liberation of bioactive aglycone metabolites. Enzymes, from colonic microflora, deconjugate the conjugated isoflavone forms, which may then be reabsorbed through the enterohepatic circulation or may be subjected to further metabolism to form metabolites that are more estrogenic than parent compounds (Křížová *et al.*, 2019). One such metabolite is equol, which is known to be the most effective isoflavone metabolite in stimulating an estrogenic response (Setchell and Clerici, 2010).

Biotransformation of equol from isoflavone aglycones involves formation of intermediates and association of several strains of colonic bacteria. *Eggerthella* spp. along with *Adlercreutzia equolifaciens*, *Asaccharobacter celatus*, *Slackia isoflavoniconvertens* and *Lactococcus garvieae* are found in the gut and are associated with production of equol (Mayo *et al.*, 2019). *Lc. garvieae*, which is found in some Italian cheeses, was isolated from the human intestinal tract and was observed to efficiently convert daidzein to equol (Iino *et al.*, 2019). This strain is among the very rare bacteria that can completely convert daidzein to equol. In *Lc. garvieae*, the enzyme L-daidzein reductase converts daidzein to (R)-dihydrodaidzein which is rapidly racemised by the novel enzyme dihydrodaidzein racemase to produce (S)-dihydrodaidzein. Dihydrodaidzein reductase belongs to the short-chain dehydrogenase/reductase family and converts (S)-dihydrodaidzein to trans-tetrahydrodaidzein (Tsuji *et al.*, 2012). In the final step, trans-tetrahydrodaidzein is converted to (S)-equol by the enzyme L-tetrahydrodaidzein reductase (THDR) (Setchell and Clerici, 2010). *Bifidobacterium* strains present in the gut, including *Bifidobacterium pseudolongum*, *Bifidobacterium animalis* and *B. Longum*, have been observed to deconjugate malonyl, acetyl, and β-glucoside conjugates of daidzin to produce daidzein, including transformation of daidzein to equol in fermented soymilk (Rafii, 2015). *Sl. isoflavoniconvertens* is another human colonic bacterium that metabolises daidzein to equol through intermediate dihydrodaidzein and is also observed to transform genistein to 5-hydroxyequol (Schröder *et al.*, 2013). Daidzein reductase produced by the bacterium converts genistein and daidzein to dihydrogenistein and dihydrodaidzein, respectively. THDR reduces dihydrodaidzein to equol whereas dihydrogenistein is partially transformed to 5-hydroxyequol. *Sl. equolifaciens* was capable to convert 150 μM of daidzein to equol in 24 hours with an 85 per cent yield and also transformed genistein to 5-hydroxyequol (Jin *et al.*, 2010). *A. equolifaciens,* which is 93 per cent similar to *Eggerthella sinensis*, was isolated from the gut and was observed to convert equol from daidzein (Maruo *et al.*, 2008). *A. equolifaciens* differentiates from *Eggerthella* strains in cell wall peptidoglycan types apart from differences in biochemical reactions. Another bacterial strain with similarity to *Eggerthella* and capable of transforming daidzein to equol with the intermediate dihydrodaidzein was isolated from rat intestine and due to its distinct biochemical characteristics, was named as *A. celatus* (Vázquez *et al.*, 2017).

Among the various other strains present in the gut and observed to metabolise isoflavone aglycones, some strains transform daidzein partially to the intermediate dihydrodaidzein while other strains specifically transform dihydrodaidzein to S-equol (Mayo *et al.*, 2019). These strains work in synergy for the complete transformation of the aglycone to generate the bioactive metabolite. Novel strain, Julong 732, was isolated from human colon with 92.8 per cent similarity to *Eggerthella hongkongensis* HKU10 and the strain could transform equol from dihydrodaidzein, but not daidzein (Rafii, 2015). Further, Julong 732 was combined with *Lactobacillus* sp. Niu-O16 to form a mixed culture of the two anaerobic bacterial strains and production of S-equol from daidzein was observed. *Lb.* sp. Niu-O16 was isolated from bovine rumen and has been known to transform daidzein to dihydrodaidzein, thus providing dihydrodaidzein to Julong 732 as substrate for metabolism to S-equol (Wang *et al.*, 2007). Daidzein was observed to be completely transformed to equol in 24 hours along with increased population of the strain Julong 732 and suppressed growth of *Lb.* sp. Niu-O16. Thus, the action of two metabolic pathways from two biotransforming gut microbial strains working in synergy can lead to transformation of daidzein to equol.

3.1.2 Flavonols and Flavones

Flavonols and flavones are the most common plant-based food flavonoids that enter human digestion upon consumption. The major flavonols found in food are quercetin, myricetin and kaempferol while apigenin and luteolin are the two major dietary flavones (Kozłowska and Szostak-Węgierek, 2018). The health benefits associated with flavonols include antioxidant, hypotensive, anticarcinogenic, antidiabetic, anti-inflammatory, anticholesterol, antiosteoporosis and antifungal properties (Kozłowska and Szostak-Węgierek, 2018; Panche *et al.*, 2016). Quercetin has been observed to be effective against cancer, neurodegenerative diseases, coronary heart disease, arthritis and asthma (David *et al.*, 2016; Li *et al.*, 2016). The food sources of flavonols differ significantly, depending upon the country and culture of origin of the food source. In the Middle East and other Asian countries like Korea, India and China, where fermented foods are staple diets, flavonol and flavone consumption and absorption are known to be higher than in countries with unfermented food sources of the two flavonoids (Ahn-Jarvis *et al.*, 2019; Hostetler *et al.*, 2017; Melini *et al.*, 2019; Swain and Padhy, 2016).

Quercetin is the main flavonol found in food, such as *Tarhana* which is a traditional fermented food in Turkey. It is prepared by mixing yoghurt, Baker's yeast, wheat flour, onion, pepper and tomato followed by lactic acid fermentation for five to seven days. During this period, conversion of quercetin-3-glucoside, quercetin-4'-glucoside, quercetin-3,4'-diglucoside, and isorhamnetin-4'-glucoside to quercetin occurs (Demirci *et al.*, 2019). *Tarhana* is powdered and consumed as soup which results in high quercetin intake upon consumption. Increase of quercetin intake upon consumption of fermented foods is attributed to enzymatic hydrolysis of quercetin conjugates and release of the aglycone form (Degirmencioglu *et al.*, 2016). Legume seeds, such as soybeans contain flavonols and flavones as glycoside, conjugates, which are deconjugated and released in aglycone forms upon fermentation (Egert and Rimbach, 2011). These aglycones present strong antioxidant activity upon consumption as compared to glycoside conjugates of flavonoids. Fermented tea, such as *kombucha*, is a source of deconjugated flavones, which are absorbed efficiently upon consumption and exert stronger antioxidant properties (Bhattacharya *et al.*, 2013).

Absorption of flavonols by the human digestive system varies for different flavonol glycosides obtained from different food sources and which may depend upon the sugar moiety attached to the flavonol aglycone (Kumar *et al.*, 2013). The glycoside moiety imparts hydrophilic nature to the flavonol upon conjugation and thus the flavonol glycosides require active transport in the digestive system for absorption (Thilakarathna and Rupasinghe, 2013). In contrast, flavonol aglycones are hydrophobic in nature and are easily absorbed by passive diffusion, which explains the higher rate of absorption of the aglycone forms of flavonoids. Like other flavonoids, flavonols and flavones undergo entero-hepatic circulation after absorption, which leads to conjugation of the flavonoids with glucuronic acid and/or sulphate (Scalbert and Williamson, 2018). The unabsorbed and conjugated flavonols pass to the colon where gut microbiota metabolises the flavonols to aglycones and phenolic acids. The gut microflora consists of *Bifidobacterium* spp. and *Lactobacillus* spp. which produce flavonoid glycoside-hydrolysing enzymes, including β-glucosidases, β-rhamnosidases and β-galactosidases (Li *et al.*, 2019; Mueller *et al.*, 2018; Wei *et al.*, 2020). These enzymes are responsible for hydrolysis of flavonol glycosides resulting in the generation of flavonol aglycones, which are easily absorbed and demonstrate high levels of bioactivity. Quercitrin and rutin are hydrolysed in the human gut to quercetin, while robinin is hydrolysed by the gut microbiota to generate the flavonol, aglycone kaempferol (Braune and Blaut, 2016; De Souza *et al.*, 2017; Yang *et al.*, 2019). These aglycones are further metabolised by the gut microflora into metabolites with efficient absorption and higher bioactivity. However, some gut bacteria metabolise flavonols to generate metabolites, which show decreased antioxidant properties as compared to the parent aglycone (Kawabata *et al.*, 2019).

Quercetin is metabolised to generate the primary metabolite 2-(3,4-dihydroxyphenyl) acetic acid while 2-(3,5-dihydroxyphenyl) acetic acid is the primary metabolite generated from the metabolism of myricetin (Vissiennon *et al.*, 2012). These metabolites are further subjected to dihydroxylation and methylation in the liver, followed by conversion of hydroxyphenylacetic acids and hydroxyphenylpropionic acids by gut bacteria to produce derivatives of benzoic acid (Zhang *et al.*, 2014). Phenolic acids generated by the gut metabolism of flavonols demonstrate efficient absorption and high bioactivity. Phloroglucinol, 3,4-dihydroxybenzaldehyde, 3,4-dihydroxytoluene, acetate and butyrate are other metabolites produced upon quercetin gut metabolism (Dueñas *et al.*, 2013; Feng *et al.*, 2018). Different gut-bacterial species are responsible for flavonol deglycosylation and ring fission, resulting in generation of diverse metabolites. *Eubacterium ramulus* transforms the flavonol glycoside quercetin-3-O-glucoside in the gut to generate the metabolites 3,4-dihydroxyphenylacetic acid, acetate and butyrate (Braune *et al.*, 2016). Similarly, the gut bacteria *Enterococcus casseliflavus* produces lactate, acetate and formate by metabolising the sugar moiety of quercetin-3-O-glucoside (Baik *et al.*, 2015). The metabolite 3,4-dihydroxyphenylacetic acid produced from fermentation of quercetin by *Clostridium perfringens* and *B. fragilis* demonstrate higher antioxidant activity than that of other strains. Yet a decline in antioxidant activity of 3,4-dihydroxyphenylacetic acid is observed as compared to the activity of quercetin (Zhang *et al.*, 2014). Thus, based on the metabolite produced, antioxidant activity of quercetin can decrease or increase upon gut bacterial fermentation in the human colon. Flavonol and flavone metabolites produced by gut fermentation of glucurono and sulphate conjugates are reabsorbed and enter the enterohepatic cycle where the metabolites are again subjected to conjugation to form flavonol and flavone glucuronides and sulphates (Zeng *et al.*, 2016). These compounds are polar and water soluble and are readily excreted in urine and bile. Flavonol excretions in the bile are passed into the duodenum, where the compounds are metabolised by gut fermentation. Thus, consumption of flavonols results in repeated absorption of the aglycone form and conjugation reactions followed by gut fermentation of conjugates, leading to release of aglycone forms (Williamson *et al.*, 2018). This metabolism strategy ensures restricted flavonol excretion (0.8-1.4 per cent of flavonol intake) and leads to flavonol concentration in the plasma upon repeated intake of flavonol food sources (Rodríguez-García *et al.*, 2019).

3.1.3   Flavanols: Catechins, Epicatechins and Procyanidins

Flavanols are a major class of polyphenol flavonoids that are known to demonstrate strong antioxidant property, efficiently scavenging various free radical scavenging forms (Kumar and Pandey, 2013). The most important flavanols associated with bioactivity include catechin, epicatechin and procyanidins (Mendes *et al.*, 2019). The ability of flavanols to inhibit low density lipoprotein (LDL) oxidation, resulting in lowering of plasma cholesterol levels and prevention of platelet aggregation might impart prevention of cardiovascular diseases (da Silva *et al.*, 2019; Martins *et al.*, 2020). In addition, catechin intake demonstrates protective activity against degenerative diseases and reduces the risk of coronary heart disease (CHD)-related mortality. Epigallocatechin gallate (EGCG) is found to act against various carcinogens and has been reported to inhibit carcinogenesis of the small intestine, bladder, colon, stomach, liver, skin, lungs, prostate, oesophagus and mammary glands (Granja *et al.*, 2016; Sur *et al.*, 2016; Wang *et al.*, 2017). The main sources of flavanols include legumes, such as soybeans, broad beans and green beans with increase in flavanol concentration observed after fermentation (Sies *et al.*, 2012). Catechin, epicatechin and epicatechin gallate are the major polyphenols available in fruit wine (Arts *et al.*, 2000). Increase in catechin aglycone content of soybean upon fermentation by *Bacillus subtilis* as compared to unfermented soybeans implies the importance of fermentation in plant flavanol aglycone release and absorption after consumption (Samruan *et al.*, 2014).

Two-thirds of the ingested flavanols are unabsorbed by the small intestine and pass to the colon where the gut microbiota metabolise catechins to form phenylvaleric acids and phenylvalerolactones (Mena *et al.*, 2019). The gut microbes associated with biotransformation of dietary catechins into hydroxyvaleric acid and valerolactone metabolites are *Flavonifractor plautii* and *Eggerthella lenta* (Kutschera *et al.*, 2011). *F. plautii* transforms catechins to form 4-hydroxy-5-(3,4-hydroxyphenyl) valeric acid

(4-H-3,4-diHPVA) and 5-(3,4-dihydroxyphenyl)-γ-valerolactone (diHPVL) whereas *E. lenta* converts catechins to 1-(3,4-dihydroxyphenyl)-3-(2,4,6-trihydroxyphenyl) propan-2-ol. These metabolites formed after gut biotransformation are absorbed and undergo conjugation reactions, such as glucuronidation and sulfation (Braune and Blaut, 2016). These conjugates include the valerolactone 3'-O-glucuronide conjugate of diHPVL, 4-hydroxyphenylacetic acid and ,4-dihydroxyphenylacetic acid. Superior antioxidant activity of diHPVL as compared to the activity of ascorbic acid highlights the importance of gut transformation of flavanols in increase of polyphenol bioactivity (Sánchez-Patán *et al.*, 2012). Phenylvalerolactones interact with various intracellular pathways and are reported to have cytoprotective effects against oxidative stress in brown adipocytes (Mele *et al.*, 2017). The flavanol metabolite 5-(3',5'-dihydroxyphenyl)-γ-valerolactone exerts immunostimulatory activity by activating CD4(+) T cells and enhancing natural killer (NK) cells cytotoxicity (Kim *et al.*, 2016). In addition, it also shows neuroprotective activity that results in increase of nerve cell proliferation and neurogenesis. Angiotensin-converting enzyme (ACE) inhibitory activity was observed for both 5-(3,4,5-trihydroxyphenyl)-γ-valerolactone and 5-(3',5'-dihydroxyphenyl)-γ-valerolactone along with hydroxyphenyl valeric acids in hypertensive rats (Takagaki and Nanjo, 2015). Futhermore, inhibition of acetylcholinesterase activity by the metabolite γ-valerolactone has been observed (Okello *et al.*, 2012).

3.1.4 Anthocyanins

Anthocyanins are plant secondary metabolites with a planar C_6-C_3-C_6 carbon structure typical of flavonoids. This group of flavonoids includes derivatives of 2-phenylbenzopyrylium, which is responsible for water-soluble pigmentation (violet, purple, blue and black) of plant parts, including fruits, vegetables, seeds and flowers (Colanero *et al.*, 2020). The beneficial effects of anthocyanins in human health include anticancer and antioxidant properties, heart disease prevention, blood cholesterol reduction and UV protection (Blesso, 2019). The antioxidant activity of anthocyanins is found to be greater than both vitamins C and E (Khoo *et al.*, 2017). Fermentation of common beans and soybeans is found to boost anthocyanin levels and associated antioxidant activities (Juan *et al.*, 2010; Tsamo *et al.*, 2019). Increased anthocyanin content and antioxidant activity of fermented soybean preparations as compared to unfermented beans have been observed with increase in antioxidant activity associated with increased anthocyanin content, depending on the starter microorganism used for fermentation (Jiang *et al.*, 2020). This dependency is related to metabolism of conjugated anthocyanin molecules by the fermenting microorganism into their aglycone forms. The aglycone forms of anthocyanins are known as anthocyanidins and include cyanidin, petunidin, delphinidin, peonidin, malvidin and pelargonidin which are of acute dietary importance (Khoo *et al.*, 2017). Unconjugated anthocyanin molecules are rapidly absorbed through the small intestine via involvement of specific enzymes, such as bilitranslocases (Kalt, 2019). However, the conjugated anthocyanins, including glucuronidated and sulphated forms, have very low bioavailability which leads to their passing to the colon either in undigested form from the small intestine or after conjugation through the enterohepatic circulation (Kalt, 2019; Lila *et al.*, 2016). In the colon, anthocyanins undergo deconjugation and ring fission, including deglycosylation and degradation of anthocyanin diglucosides and monoglucosides. The heterocyclic C-ring of anthocyanins is broken down and anthocyanin A-ring and B-ring are degraded into phloroglucinol derivatives and benzoic acids respectively as a result of gut metabolism (Kubow *et al.*, 2017; Rocchetti *et al.*, 2019).

Gut microbiota break down the anthocyanin nucleus and release phenolic acids as products which can be further metabolised by the gut microbiota (Kawabata *et al.*, 2019; Majdoub *et al.*, 2019). Cyanidin is converted to protocatechuic acid, peonidin to vanillic acid (3-methoxy-4-hydroxybenzoic acid), pelargonidin to 4-hydroxybenzoic acid and malvidin to syringic acid (3,4-dimethoxybenzoic acid). Transformation of cyanidin-3-rutinoside to the corresponding glucoside and then its conversion to phenolic acid was observed (Oksuz *et al.*, 2019). Methoxyl derivatives of anthocyanins are metabolised by gut microbiota followed by O-demethylation. The β-glucosidase activity of gut microbiota is responsible for metabolism of the anthocyanidins malvidin-3-glucoside and delphinidin-3-glucoside into their bioactive forms (Li *et al.*, 2018). *Bifidobacterium* spp. and *Lactobacillus* spp. are the main gut microbiota that serve as sources of β-glucosidase for metabolism of anthocyanins in the colon (Ávila *et al.*, 2009). Anthocyanin metabolism in the gut also results in growth modulation of these specific microbiota (*Bifidobacterium* spp. and *Lactobacillus* spp.) responsible for polyphenol metabolism and other prebiotic functions. *Lactobacillus* spp. have been shown to stimulate the host immune system by enhancing the gut barrier function, resulting in prevention of allergies and diarrhoea, modulation of lipid metabolism and activation of provitamins (Li et al. 2019). Enhanced growth of *Bifidobacterium* spp. and *Lactobacillus* spp. upon incubation of the anthocyanidin malvidin-3-glucoside with faecal slurry is observed while no change was observed on the growth of *Bacteroides* spp. (Hidalgo *et al.*, 2012). The anthocyanin gut metabolite, gallic acid, has been observed to reduce the growth of the harmful bacteria *Clostridium histolyticum* and *Bacteroides* spp. without affecting the growth of beneficial gut microbiota strains (Igwe *et al.*, 2019; Sun *et al.*, 2018). A decrease in growth of *C. histolyticum* was observed upon incubation of anthocyanin-rich red wine extract with human faecal bacteria (Sánchez-Patán *et al.*, 2012).

3.2   Phenolic Acids

Phenolic acids are polyphenols containing a carboxylic acid function and are categorised into two distinctive groups: hydrobenzoic acids (salicylic acid, vanillic acid, gallic acid) and hydroxycinnamic acids (p-coumaric acid, caffeic acid, ferulic acid) (Kumar and Goel, 2019). Phenolic acids are formed from the B-ring when flavonoids undergo ring fission, resulting in degradation of the C-ring of the flavonoid nucleus (Saibabu *et al.*, 2015). Flavones, flavanones and flavanols are metabolised to phenylpropionic acid derivatives which undergo further metabolism into benzoic acid derivatives. Hydroxycinnamic acids are bound to plant cell walls and are additionally esterified to organic acids, lipids, or sugars to alter their solubility in cell organelles, such as vacuoles (Marchiosi *et al.*, 2020). De-esterification of the additional groups is required for the release of hydroxycinnamic acids and this is observed in the gut by esterase production of the colonic microflora which is the major site for chlorogenic acid metabolism (Kelly *et al.*, 2018). Apart from isoflavones (daidzein and genistein) being major phenolic compounds present in fermented soy products, studies have reported the presence of ferulic acid, p-coumaric acid, gentisic acid, caffeic acid, syringic acid, vanillic acid and p-hydroxylbenzoic acid in only very low concentrations (Li *et al.*, 2020; Xie *et al.*, 2019). Black soybean contains eleven phenolic acids, including syringic acid, vanillic acid, caffeic acid, ferulic acid, gentisic acid, chlorogenic acid, benzoic acid, p-hydroxylbenzoic acid, p-coumaric acid, trans-cinnamic acid and salicylic acid (Lee *et al.*, 2020). It has been reported that gallic acid is present in soybean fermented food along with the flavanols catechin and epicatechin (Cho *et al.*, 2009). Phenolic acids are the product of gut metabolism of various flavonoid groups and are often included in the diet through fermented foods (Pasinetti *et al.*, 2018). These compounds are extensively studied for their bioactivities, including antioxidant, antitumor and antimicrobial properties among others. Gallic acid has been reported to have antioxidant, antimelanogenic, bacteriostatic and antineoplastic properties and has shown anticancer property in prostate carcinoma cells (Saffari-Chaleshtori *et al.*, 2019). Gallic acid, due to its ability to suppress cell viability, proliferation, invasion and angiogenesis of human glioma cells, has been proposed as a potent compound for the treatment of brain tumours (Lu *et al.*, 2010). Furthermore, it was reported and explained that gallic acid is an active anti-Herpes Simplex Virus (HSV)-2 agent (Kratz *et al.*, 2008). The ability of gallic acid in inducing HeLa cervical cancer cell-death by apoptosis and necrosis was proved (You *et al.*, 2010). Vanillic acid has been reported to express anthelmintic activities and suppress hepatic fibrosis in chronic liver injury (Itoh *et al.*, 2010). Genistic acid contains antirheumatic, analgesic and anti-inflammatory activities and inhibits low density lipoprotein oxidation in human plasma in addition to being a cytostatic agent (Khadem and Marles, 2010). Salicylic acid is known for antiseptic, antifungal, antipyretic and anti-inflammatory properties (Przybylska-Balcerek and Stuper-Szablewska, 2019). Caffeic and ferulic acids have been reported as antimicrobial agents against bacterial and fungal pathogens (Heleno *et al.*, 2015) and have shown high DPPH (2,2-diphenyl1-picrylhydrazyl) radical scavenging activity along with ABTS (2,20-azino-bis (3-thylbenzothiazoline-6-sulphonic acid) radical scavenging activity and oxygen radical absorbance capacity (ORAC), confirming their status as potent antioxidant molecules (Piazzon *et al.*, 2012). Also p-coumaric acid was reported to demonstrate bactericidal activity via dual mechanisms by disrupting bacterial cell membranes, and binding to genomic DNA of bacteria and inhibiting cellular functions that lead to cell death (Lou *et al.*, 2012). Promising antimicrobial activity of p-coumaric acid was observed against several bacterial and fungal pathogens (Heleno *et al.*, 2015). In addition to antimicrobial activity, p-coumaric acid demonstrates antioxidant activity against free radicals and antitumor activity against MCF7 (breast), NCI-H460 and HCT15 carcinoma cell lines.

The absorption and conjugation of phenolic acid follow the same pathways as other polyphenols when they are consumed in free forms (Bento-Silva *et al.*, 2019). Phenolic acid bioavailability study is important for understanding the conjugations and bioactivity of phenolic acid in the consuming organism. Upon consumption of free form of phenolic acid, rapid absorption occurs in the small intestine after which conjugation reactions take place along with conjugation in liver, including glucuronidation, methylation and sulphation (Kumar and Goel, 2019). Conjugation reactions are essential as they inhibit the emergence of potential toxic effects of phenolic acids (detoxification), but an increase in hydrophilicity of conjugated metabolites assures elimination of phenolic acids from the urinary or the biliary route (Lorigooini *et al.*, 2020). This also results in the presence of low aglycone concentrations in the blood after polyphenol consumption. After liver conjugations and circulation, polyphenols are released by the biliary pathway in the duodenum, where they come in contact with gut microbial enzymes, especially β-glucuronidases, after which the free forms are again absorbed, conjugated and recirculated. However, the chemical form of the phenolic acid ingested can influence conjugation and metabolism of these compounds by gut microbiota and their bioactivities (Stevens and Maier, 2016). Chlorogenic acid, a complex of caffeic acid ester linked to quinic acid, cannot be metabolised by the human cells as the cells lack the esterase enzyme capable of releasing caffeic acid from the chlorogenic acid complex (Liang and Kitts, 2015). Thus, esterification of the phenolic acid changes the overall biological property (Kishida and Matsumoto, 2019). Consequently, chlorogenic acid can be de-esterified by gut microbial esterase in the colon (Liu *et al.*, 2020). According to several studies, when caffeic acid and its esters (chlorogenic acid and caftaric acid) are incubated with human gut microbiota, de-esterification occurs, resulting in formation of 3-(3-hydroxyphenyl) propionic acid from caffeic acid by reduction of a double bond and dihydroxylation at the C4 position. In addition, shortening of side chain by β-oxidation leads to formation of benzoic acid to a low degree (Barberan *et al.*, 2014; Marín *et al.*, 2015; Vollmer *et al.*, 2017). Similarly, hydroxycinnamic acids, including ferulic acid, bound to plant cell wall can only be released by esterases and xylanases secreted by gut microbiota and transformed to 3-(3-hydroxyphenyl) propionic acid (Kelly *et al.*, 2018). Interestingly, it has been observed that hydrolysis of esterified phenolic acids by the gut microbiota results in degradation of the aglycones, thus releasing simple aromatic acids, which implies reduction in efficiency of phenolic acid absorption when present in esterified forms rather than free form (Marín *et al.*, 2015).

3.3   Lignans

Lignans are diphenolic compounds categorised as phytoestrogens due to their plant origin as well as their oestrogenic and anti-oestrogenic effects in humans (Yoder *et al.*, 2015). Dietary lignans include secoisolariciresinol diglucoside (SDG), precursor of secoisolariciresinol (SECO), pinoresinol (PIN), syringaresinol (SYR), matairesinol (MAT), sesamin (SES), medioresinol (MED), lariciresinol (LARI), arctigenin (ART), isolariciresinol and 7'-hydroxymatairesinol (Peterson *et al.*, 2010). The sources of lignans include mainly flaxseeds, which upon fermentation result in increase of functional properties (De Silva and Alcorn, 2019). Isolariciresinol is not the major lignan analysed from plant sources; however, it is found to be abundant in fermented soy-based food (21 per cent of the samples) only after the most abundant lignan SYR (50 per cent of the samples) (Durazzo *et al.*, 2018). Enterolactone (ENL) and enterodiol (END) are the bioactive forms of lignan and are produced by biotransformation of dietary lignans by the gastrointestinal microbiota. ENL and END are also known as enterolignans or mammalian lignans and their bioavailability is dependent on the presence and action of the gut microbiota (Yoder *et al.*, 2015). Metabolism of plant lignans by gut microbiota includes demethylation, ring cleavage, deglucosylation, dehydroxylation and oxidation. The intestinal bacteria involved in conversion of plant lignans to enterolignans include *Clostridium* spp., *Bacteroides* spp., *Peptostreptococcus* spp. and *Eggerthella* spp (Woting *et al.*, 2010). END and its oxidation product ENL show bioactivity against cardiovascular diseases, breast cancer, colon cancer, osteoporosis, hyperlipidemia, menopausal syndrome and prostate cancer (De Silva and Alcorn, 2019; Kezimana *et al.*, 2018). The presence of phytoestrogens, like isoflavones, is high in soy fermented foods of Asian origin and provide much higher protection against chronic diseases upon consumption as compared to that of the Western diet. However, the lignan content of Western and Asian diets is similar and this can be attributed to the ubiquity of lignans in plant-based foods. Yet, it should be considered that the major sources of lignan in Asian diet are fermented foods and the bioavailability and concentration of lignan intake upon consumption should be higher than that of unfermented food (Bartkiene *et al.*, 2020; Rodríguez-García *et al.*, 2019).

Enterolignans are the major metabolites produced upon metabolism of lignans by the gut microbiota which are then absorbed in the colon. ENL and END are the end metabolites formed after faecal incubation of the lignans MAT, SECO, LARI (Kähkönen *et al.*, 2001). SYR is partly converted to ENL and END, whereas enterofuran is observed in small amounts for LARI and PIN. Interestingly, isolariciresinol incubation with faecal microbiota result in no enterolignan formation (Gupta *et al.*, 2016). Biotransformation of plant lignans by gut microbiota to enterolignans begins by hydrolysis of the sugar moiety of the lignan (hydrolysis of SDG to release SECO) which is followed by dehydroxylation and demethylation to generate the enterolignan END. Oxidation of END by the gut microflora results in the formation of the final bioactive enterolignan ENL (Gaya *et al.*, 2016). The metabolism of different plant lignan includes additional reduction steps, which depend on the structure of the lignan. For example, metabolism of LARI and PIN to enterolignans requires one or two additional reduction reactions by the gut microbiota (Clavel *et al.*, 2006). Biotransformation of the lignan arctin which consists of a butyrolactone structure is similar to the transformation of SDG where initially the lignan glucosides are hydrolysed to their aglycone which is followed by demethylation of a methoxy group and a subsequent chain of dehydroxylation and demethylation reactions lead to the formation of ENL (Setchell *et al.*, 2014). As opposed to the gut transformation of SDG, which is readily metabolised by gut microbiota to enterolignans, SES is partially metabolised to END and ENL in the gut. It is absorbed and metabolised to hydroxylated metabolites in the liver, followed by excretion in the bile. The excreted lignan metabolites that reach the bile then undergo enterohepatic circulation before reaching the colon where gut microbiota transform the metabolite to enterolignans (Farbood *et al.*, 2019).

Gut microbiota responsible for various transformation reactions of plant lignans to enterolignans include *Bacteroides* spp., *Clostridium* spp., *Bifidobacterium* spp., *Eubacterium* spp., and *Ruminococcus* spp. *B. fragilis, Bacteroides distasonis, Clostridium* sp. SDG-Mt85-3Db, and *Clostridium cocleatum* which are responsible for the deglycosylation of SDG to SECO with *B. pseudocatenulatum* WC 401, show the highest rate of SDG deglycosylation (Clavel *et al.*, 2006). Further demethylation of SECO in the gut can be attributed to the activity of *Peptostreptococcus productus, Eubacterium callanderi, Eubacterium limosum, Butyribacterium methylotrophicum* and *Ruminococcus productus* (Gaya *et al.*, 2017; Landete *et al.*, 2017; Quartieri *et al.*, 2016). In addition to SDG, these gut microbes also metabolise other plant lignans to enterolignans. For example, *R. productus* is involved in demethylation of LARI, PIN, MAT and other methylated aromatic compounds (Corona *et al.*, 2020). *Eubacterium* sp. ARC-2 is associated with the demethylation of SECO to the metabolite 2,3-bis (3,4-dihydroxybenzyl)-1,4-butanediol and demethylation of the arctin aglycone arctigenin (Jin *et al.*, 2010). *E. lenta* SDG-2 biotransforms SDG to END and ENL by phenolic p-dehydroxylation of the plant lignan and its metabolite intermediates. Dehydroxylation of SECO is observed to be catalysed by the gut bacteria, *E. lenta* and *Clostridium scindens* (Cortés-Martín *et al.*, 2020). Enantioselective oxidation of END to ENL is catalysed by *Ruminococcus* sp. END-1 and another intestinal bacterial strain END-2 (Jin *et al.*, 2010). Defined gut microbial community, including *Blautia producta, E. lenta, Lactonifactor longoviformis* and *Clostridium saccharogumia*, can transform the plant lignan SDG to the mammalian lignans, END and ENL, via the intermediate SECO (Woting *et al.*, 2010). In addition, *E. lenta* is also associated with transformation of PIN and LARI to SECO while *Enterococcus faecalis* strain PDG-1 converts the plant lignan PIN to LARI (Yoder *et al.*, 2015). Thus, various gut microbial strains catalyse different lignan biotransformations that ultimately lead to the release of bioactive enterolignans, END and ENL.

4   Conclusion

With the help of a large amount of metabolic functionality provided by gut microbial species, the human host has been able to process and digest a wide range of dietary substrates (Rowland *et al.*, 2018). The catalytic power of gut microbiota to metabolise undigested fibres, such as EPS obtained from fermented foods and generation of functional compounds, such as SCFAs, have resulted in protection and well-being of the human body. The immense capacity of the gut microbial species to metabolise various polyphenol phytochemicals and release bioactive compounds has been elucidated. Differences in metabolism of polyphenols between individuals have been considered as a consequence of differences in gut microbiota composition (Cardona *et al.*, 2013). However, various gut bacterial species have shown the capability of similar metabolic pathways leading to the generation of the same functional metabolite. CAZymes have been found to be expressed in various different species (Soverini *et al.*, 2017). Metabolites derived from gut microbiota metabolism are absorbed from the gut and these compounds enter exogenous and endogenous pathways of the host. The compounds, after expressing their function in the host system, undergo conjugation, modification and along with metabolites produced by the host are secreted into the gut via enterohepatic circulation wherein they serve as substrates for the gut microbiota (Hoyles and Swann, 2019). Thus, a crosstalk between host metabolic processes and gut microbiome lead to expression of complete metabolic phenotype of the host. Future studies involving host-gut microbial interactions and discovery of metabolic pathways that result in generation of novel functional metabolites will increase the importance of gut microbiota on human health and diseases.

Acknowledgements

Authors would like to acknowledge Department of Biotechnology (DBT), Government of India, for financial support.

Abbreviations

EPS – Exopolysaccharides
SCFA – Short chain fatty acid
GI – Gastrointestinal
GOS – Galactooligosaccharides
CAZymes – Carbohydrate-active enzymes
LAB – Lactic Acid Bacteria
GRAS – Generally Recognised As Safe
AMP – Adenosine monophosphate
DNA – Deoxyribonucleic acid
RNA – Ribonucleic acid
FOS – Fructooligosaccharides
DZNR – Daidzein reductase
DHDR – Dihydrodaidzein reductase
THDR – Tetrahydrodaidzein reductase
ACE – Angiotensin converting enzyme
ER – Estrogen receptor
LDL – Low Density Lipoprotein
HDL – High Density Lipoprotein
EGCG – Epigallocatechin gallate
CHD – Coronary Heart Disease
NK – Natural killer cells
DPPH – 2,2-diphenyl1-picrylhydrazyl
ABTS – 2,2'-azino-bis (3-ethylbenzothiazoline-6-sulphonic acid)
ORAC – Oxygen radical absorbance capacity
SDG – Secoisolariciresinoldiglucoside
SECO – Secoisolariciresinol
PIN – Pinoresinol
SYR – Syringaresinol
MAT – Matairesinol

SES – Sesamin
MED – Medioresinol
LARI – Lariciresinol
ART – Arctigenin
ENL – Enterolactone
END – Enterodiol

References

Actis-Goretta, L., Ottaviani, J.I., Keen, C.L. and Fraga, C.G. (2003). Inhibition of angiotensin converting enzyme (ACE) activity by flavan-3-ols and procyanidins. *FEBS Letters*, 555: 597-600. https://doi.org/10.1016/S0014-5793(03)01355-3

Aguilar, E.C., Santos, L.C. dos, Leonel, A.J., de Oliveira, J.S., Santos, E.A., Navia-Pelaez, J.M., da Silva, J.F., Mendes, B.P., Capettini, L.S.A., Teixeira, L.G., Lemos, V.S. and Alvarez-Leite, J.I. (2016). Oral butyrate reduces oxidative stress in atherosclerotic lesion sites by a mechanism involving NADPH oxidase down-regulation in endothelial cells. *The Journal of Nutritional Biochemistry*, 34: 99-105. https://doi.org/10.1016/j.jnutbio.2016.05.002

Aherne, S.A. and Brien, N.M.O. (2002). Dietary flavonols: Chemistry, food content, and metabolism. *Nutrition*, 9007: 75-81.

Ahmad, A. and Khalid, S. (2018). Therapeutic aspects of probiotics and prebiotics. pp. 53-91. *In*: A. Grumezescu and A.M. Holban (Eds.). *Diet, Microbiome and Health*. Elsevier. https://doi.org/10.1016/B978-0-12-811440-7.00003-X

Ahn-Jarvis, J.H., Parihar, A. and Doseff, A.I. (2019). Dietary flavonoids for immunoregulation and cancer: Food design for targeting disease. *Antioxidants*, 8: 202. https://doi.org/10.3390/antiox8070202

Alizadeh-Sani, M., Hamishehkar, H., Khezerlou, A., Azizi-Lalabadi, M., Azadi, Y., Nattagh-Eshtivani, E., Fasihi, M., Ghavami, A., Aynehchi, A. and Ehsani, A. (2018). Bioemulsifiers derived from microorganisms: Applications in the drug and food industry. *Advanced Pharmaceutical Bulletin*, 8: 191-199. https://doi.org/10.15171/apb.2018.023

Arena, A., Maugeri, T.L., Pavone, B., Iannello, D., Gugliandolo, C. and Bisignano, G. (2006). Antiviral and immunoregulatory effect of a novel exopolysaccharide from a marine thermotolerant *Bacillus licheniformis*. *International Immunopharmacology*, 6: 8-13. https://doi.org/10.1016/j.intimp.2005.07.004

Arts, C.W.I., Putte, van de B. and Hollman, P.C.H. (2000). Catechin contents of foods commonly consumed in The Netherlands. 2. Tea, wine, fruit juices, and chocolate milk. *Journal of Agricultural and Food Chemistry*, 48: 1752-1757. https://doi.org/10.1021/JF000026+

Ávila, M., Hidalgo, M., Sánchez-Moreno, C., Pelaez, C., Requena, T. and Pascual-Teresa, S. (2009). Bioconversion of anthocyanin glycosides by bifidobacteria and *Lactobacillus*. *Food Research International*, 42: 1453-1461. https://doi.org/10.1016/j.foodres.2009.07.026

Azcarate-Peril, M.A., Ritter, A.J., Savaiano, D., Monteagudo-Mera, A., Anderson, C., Magness, S.T. and Klaenhammer, T.R. (2017). Impact of short-chain galactooligosaccharides on the gut microbiome of lactose-intolerant individuals. *Proceedings of the National Academy of Sciences of the United States of America*, 114: 367-375. https://doi.org/10.1073/pnas.1606722113

Baik, J.H., Shin, K.-S., Park, Y., Yu, K.-W., Suh, H.J. and Choi, H.-S. (2015). Biotransformation of catechin and extraction of active polysaccharide from green tea leaves via simultaneous treatment with tannase and pectinase. *Journal of the Science of Food and Agriculture*, 95: 2337-2344. https://doi.org/10.1002/jsfa.6955

Barberan, T.-F., García-Villalba, R., Quartieri, A., Raimondi, S., Amaretti, A., Leonardi, A. and Rossi, M. (2014). *In vitro* transformation of chlorogenic acid by human gut microbiota. *Molecular Nutrition & Food Research*, 58: 1122-1131. https://doi.org/10.1002/mnfr.201300441

Barnes, S. (2010). The biochemistry, chemistry and physiology of the isoflavones in soybeans and their food products. *Lymphatic Research and Biology*, 8: 89-98. https://doi.org/10.1089/lrb.2009.0030

Bartkiene, E., Mozuriene, E., Lele, V., Zokaityte, E., Gruzauskas, R., Jakobsone, I., Juodeikiene, G., Ruibys, R. and Bartkevics, V. (2020). Changes of bioactive compounds in barley industry by-products during submerged and solid state fermentation with antimicrobial *Pediococcus acidilactici* strain LUHS29. *Food Science & Nutrition*, 8: 340-350. https://doi.org/10.1002/fsn3.1311

Bento-Silva, A., Koistinen, V.M., Mena, P., Bronze, M.R., Hanhineva, K., Sahlstrøm, S., Kitrytė, V., Moco, S. and Aura, A.M. (2019). Factors affecting intake, metabolism and health benefits of phenolic acids: Do we understand individual variability? *European Journal of Nutrition*, 59: 1275-1293. https://doi.org/10.1007/s00394-019-01987-6

Bhattacharya, S., Gachhui, R. and Sil, P.C. (2013). Effect of Kombucha, a fermented black tea in attenuating oxidative stress mediated tissue damage in alloxan induced diabetic rats. *Food and Chemical Toxicology*, 60: 328-340. https://doi.org/10.1016/j.fct.2013.07.051

Bis-Souza, C.V., Pateiro, M., Domínguez, R., Penna, A.L.B., Lorenzo, J.M. and Silva Barretto, A.C. (2020). Impact of fructooligosaccharides and probiotic strains on the quality parameters of low-fat Spanish Salchichón. *Meat Science*, 159: 107936. https://doi.org/10.1016/j.meatsci.2019.107936

Blesso, C.N. (2019). Dietary anthocyanins and human health. *Nutrients*, 11: 2107. https://doi.org/10.3390/nu11092107

Bourassa, M.W., Alim, I., Bultman, S.J. and Ratan, R.R. (2016). Butyrate, neuroepigenetics and the gut microbiome: Can a high fiber diet improve brain health? *Neuroscience Letters*, 625: 56-63. https://doi.org/10.1016/j.neulet.2016.02.009

Braune, A. and Blaut, M. (2016). Bacterial species involved in the conversion of dietary flavonoids in the human gut. *Gut Microbes*, 9: 1158395. https://doi.org/10.1080/19490976.2016.1158395

Braune, A., Engst, W., Elsinghorst, P.W., Furtmann, N., Bajorath, J., Gütschow, M. and Blaut, M. (2016). Chalcone isomerase from Eubacterium ramulus catalyses the ring contraction of flavanonols. *Journal of Bacteriology*, 198: 2965-2974. https://doi.org/10.1128/JB.00490-16

Byrne, C.S., Chambers, E.S., Alhabeeb, H., Chhina, N., Morrison, D.J., Preston, T., Tedford, C., Fitzpatrick, J., Irani, C., Busza, A., Garcia-Perez, I., Fountana, S., Holmes, E., Goldstone, A.P. and Frost, G.S. (2016). Increased colonic propionate reduces anticipatory reward responses in the human striatum to high-energy foods. *The American Journal of Clinical Nutrition*, 104: 5-14. https://doi.org/10.3945/ajcn.115.126706

Canani, R.B., Costanzo, M. Di, Leone, L., Pedata, M., Meli, R. and Calignano, A. (2011). Potential beneficial effects of butyrate in intestinal and extraintestinal diseases. *World Journal of Gastroenterology*, 17: 1519-1528. https://doi.org/10.3748/wjg.v17.i12.1519

Carding, S., Verbeke, K., Vipond, D.T., Corfe, B.M. and Owen, L.J. (2015). Dysbiosis of the gut microbiota in disease. *Microbial Ecology in Health and Disease*, 26: 26191. https://doi.org/10.3402/mehd.v26.26191

Cardona, F., Andrés-Lacueva, C., Tulipani, S., Tinahones, F.J. and Queipo-Ortuño, M.I. (2013). Benefits of polyphenols on gut microbiota and implications in human health.

The Journal of Nutritional Biochemistry, 24: 1415–1422. https://doi.org/10.1016/j.jnutbio.2013.05.001

Carlson, J.L., Erickson, J.M., Lloyd, B.B. and Slavin, J.L. (2018). Health effects and sources of prebiotic dietary fibre. *Current Developments in Nutrition*, 2: nzy005. https://doi.org/10.1093/cdn/nzy005

Chambers, E.S., Viardot, A., Psichas, A., Morrison, D.J., Murphy, K.G., Zac-Varghese, S.E.K., MacDougall, K., Preston, T., Tedford, C., Finlayson, G.S., Blundell, J.E., Bell, J.D., Thomas, E.L., Mt-Isa, S., Ashby, D., Gibson, G.R., Kolida, S., Dhillo, W.S., Bloom, S.R., Morley, W., Clegg, S. and Frost, G. (2015). Effects of targeted delivery of propionate to the human colon on appetite regulation, body weight maintenance and adiposity in overweight adults. *Gut*, 64: 1744-1754. https://doi.org/10.1136/gutjnl-2014-307913

Chen, D., Yang, X., Yang, J., Lai, G., Yong, T., Tang, X., Shuai, O., Zhou, G., Xie, Y. and Wu, Q. (2017). Prebiotic effect of fructooligosaccharides from Morinda officinalis on Alzheimer's disease in rodent models by targeting the microbiota-gut-brain axis. *Frontiers in Aging Neuroscience*, 9: 403. https://doi.org/10.3389/fnagi.2017.00403

Chen, J., Zhao, K.N. and Vitetta, L. (2019). Effects of intestinal microbial-elaborated butyrate on oncogenic signaling pathways. *Nutrients*, 11: 1026. https://doi.org/10.3390/nu11051026

Cho, K., Hong, S., Math, R., Lee, J., Kambiranda, D., Kim, J., Islam, S., Yun, M., Cho, J. and Lim, W. (2009). Biotransformation of phenolics (isoflavones, flavanols and phenolic acids) during the fermentation of *cheonggukjang* by *Bacillus pumilus* HY1. *Food Chemistry*, 114: 413-419. https://doi.org/10.1016/j.foodchem.2008.09.056

Choi, Y.M., Kim, Y.S., Ra, K.S. and Suh, H.J. (2007). Characteristics of fermentation and bioavailability of isoflavones in Korean soybean paste (*doenjang*) with application of Bacillus sp. KH-15. *International Journal of Food Science & Technology*, 42: 1497-1503. https://doi.org/10.1111/j.1365-2621.2006.01371.x

Clavel, T., Borrmann, D., Braune, A., Doré, J. and Blaut, M. (2006). Occurrence and activity of human intestinal bacteria involved in the conversion of dietary lignans. *Anaerobe*, 12: 140-147. https://doi.org/10.1016/j.anaerobe.2005.11.002

Colanero, S., Tagliani, A., Perata, P. and Gonzali, S. (2020). Alternative splicing in the anthocyanin fruit gene encoding an R2R3 MYB transcription factor affects anthocyanin biosynthesis in tomato fruits. *Plant Communications*, 1: 100006. https://doi.org/10.1016/j.xplc.2019.100006

Conlon, M.A. and Bird, A.R. (2014). The impact of diet and lifestyle on gut microbiota and human health. *Nutrients*, 7: 17-44. https://doi.org/10.3390/nu7010017

Corona, G., Kreimes, A., Barone, M., Turroni, S., Brigidi, P., Keleszade, E. and Costabile, A. (2020). Impact of lignans in oilseed mix on gut microbiome composition and enterolignan production in younger healthy and premenopausal women: An *in vitro* pilot study. *Microbial Cell Factories*, 19: 1-14. https://doi.org/10.1186/s12934-020-01341-0

Cortés-Martín, A., Selma, M.V., Tomás-Barberán, F.A., González-Sarrías, A. and Espín, J.C. (2020). Where to look into the puzzle of polyphenols and health? The postbiotics and gut microbiota associated with human metabotypes. *Molecular Nutrition & Food Research*, 64: 1900952. https://doi.org/10.1002/mnfr.201900952

Das, D., Sarkar, S., Bordoloi, J., Wann, S.B., Kalita, J. and Manna, P. (2018). Daidzein, its effects on impaired glucose and lipid metabolism and vascular inflammation associated with Type 2 diabetes. *BioFactors*, 44: 407-417. https://doi.org/10.1002/biof.1439 not in text

da Silva, L.H., Celeghini, R.M.S. and Chang, Y.K. (2011). Effect of the fermentation of whole soybean flour on the conversion of isoflavones from glycosides to aglycones. *Food Chemistry*, 128: 640–644. https://doi.org/10.1016/J.FOODCHEM.2011.03.079

da Silva, R.R., da Conceição, P.J.P., de Menezes, C.L.A., de Oliveira Nascimento, C.E., Machado Bertelli, M., Pessoa Júnior, A., de Souza, G.M., da Silva, R. and Gomes, E. (2019). Biochemical characteristics and potential application of a novel ethanol and glucose-tolerant β-glucosidase secreted by *Pichia guilliermondii* G1.2. *Journal of Biotechnology*, 294: 73-80. https://doi.org/10.1016/j.jbiotec.2019.02.001

David, A.V.A., Arulmoli, R. and Parasuraman, S. (2016). Overviews of biological importance of quercetin: A bioactive flavonoid. *Pharmacognosy Reviews*, 10: 84. https://doi.org/10.4103/0973-7847.194044

Davis, L.M.G., Martínez, I., Walter, J., Goin, C. and Hutkins, R.W. (2011). Barcoded pyrosequencing reveals that consumption of galactooligosaccharides results in a highly specific bifidogenic response in humans. *PLoS ONE*, 6: e25200. https://doi.org/10.1371/journal.pone.0025200

De Silva, S.F. and Alcorn, J. (2019). Flaxseed lignans as important dietary polyphenols for cancer prevention and treatment: Chemistry, pharmacokinetics, and molecular targets. *Pharmaceuticals*, 12: 68. https://doi.org/10.3390/ph12020068

De Souza, L.A., Tavares, W.M.G., Lopes, A.P.M., Soeiro, M.M. and De Almeida, W.B. (2017). Structural analysis of flavonoids in solution through DFT 1H NMR chemical shift calculations: Epigallocatechin, Kaempferol and Quercetin. *Chemical and Physical Lettets*, 676: 46-52. https://doi.org/10.1016/j.cplett.2017.03.038

Degirmencioglu, N., Gürbüz, O., Nur Herken, E. and Yurdunuseven Yıldız, A. (2016). The impact of drying techniques on phenolic compound, total phenolic content and antioxidant capacity of oat flour tarhana. *Food Chemistry*, 194: 587-594. https://doi.org/10.1016/j.foodchem.2015.08.065

Demirci, A.S., Palabiyik, I., Ozalp, S. and Sivri, G.T. (2019). Effect of using *kefir* in the formulation of traditional tarhana. *Food Science and Technology*, 39: 358-364. https://doi.org/10.1590/fst.29817

Dueñas, M., Dueñas, D., Surco-Laos, F., González-Manzano, S., González-Paramás, A.M., Gómez-Orte, E., Cabello, J. and Santos-Buelga, C. (2013). Deglycosylation is a key step in biotransformation and lifespan effects of quercetin-3-O-glucoside in *Caenorhabditis elegans*. *Pharmacological Research*, 76: 41-48. https://doi.org/10.1016/j.phrs.2013.07.001

Durazzo, A., Lucarini, M., Camilli, E., Marconi, S., Gabrielli, P., Lisciani, S., Gambelli, L., Aguzzi, A., Novellino, E., Santini, A., Turrini, A. and Marletta, L. (2018). Dietary lignans: Definition, description and research trends in databases development. *Molecules*, 23: 3251. https://doi.org/10.3390/molecules23123251

Egert, S. and Rimbach, G. (2011). Which sources of flavonoids: Complex diets or dietary supplements? *Advances in Nutrition*, 2: 8-14. https://doi.org/10.3945/an.110.000026

Endesfelder, D., Engel, M., Davis-Richardson, A.G., Ardissone, A.N., Achenbach, P., Hummel, S., Winkler, C., Atkinson, M., Schatz, D., Triplett, E., Ziegler, A.-G. and Castell, W. (2016). Towards a functional hypothesis relating anti-islet cell autoimmunity to the dietary impact on microbial communities and butyrate production. *Microbiome*, 4: 17. https://doi.org/10.1186/s40168-016-0163-4

Et, J., Vlieg, V.H., Veiga, P., Zhang, C., Derrien, M. and Zhao, L. (2011). Impact of microbial transformation of food on health – From fermented foods to fermentation in the gastro-intestinal tract. *Current Opinion Biotechnology*, 22: 211-219. https://doi.org/10.1016/j.copbio.2010.12.004

Farbood, Y., Ghaderi, S., Rashno, Masome, Khoshnam, S.E., Khorsandi, L., Sarkaki, A. and Rashno, Md. (2019). Sesamin: A promising protective agent against diabetes-associated cognitive decline in rats. *Life Sciences*, 230: 169-177. https://doi.org/10.1016/j.lfs.2019.05.071

Feng, M., Chen, X., Li, C., Nurgul, R. and Dong, M. (2012). Isolation and identification of an exopolysaccharide-producing lactic acid bacterium strain from Chinese *paocai* and biosorption of Pb(II) by its exopolysaccharide. *Journal of Food Science*, 77: 111-117. https://doi.org/10.1111/j.1750-3841.2012.02734.x

Feng, X., Li, Y., Brobbey Oppong, M. and Qiu, F. (2018). Insights into the intestinal bacterial metabolism of flavonoids and the bioactivities of their microbe-derived ring cleavage metabolites. *Drug Metabolism Reviews*, 50: 343-356. https://doi.org/10.1080/03602532 .2018.1485691

Fischer, C. and Kleinschmidt, T. (2018). Synthesis of galactooligosaccharides in milk and whey: A review. *Comprehensive Reviews in Food Science and Food Safety*, 17: 678-697. https://doi.org/10.1111/1541-4337.12344

Freire, A.L., Ramos, C.L. and Schwan, R.F. (2017). Effect of symbiotic interaction between a fructooligosaccharide and probiotic on the kinetic fermentation and chemical profile of maize blended rice beverages. *Food Research International*, 100: 698-707. https://doi. org/10.1016/j.foodres.2017.07.070

Frost, G., Sleeth, M.L., Sahuri-Arisoylu, M., Lizarbe, B., Cerdan, S., Brody, L., Anastasovska, J., Ghourab, S., Hankir, M., Zhang, S., Carling, D., Swann, J.R., Gibson, G., Viardot, A., Morrison, D., Louise Thomas, E. and Bell, J.D. (2014). The short-chain fatty acid acetate reduces appetite *via* a central homeostatic mechanism. *Nature Communications*, 5: 3611. https://doi.org/10.1038/ncomms4611

Furusawa, Y., Obata, Y., Fukuda, S., Endo, T.A., Nakato, G., Takahashi, D., Nakanishi, Y., Uetake, C., Kato, K., Kato, T., Takahashi, M., Fukuda, N.N., Murakami, S., Miyauchi, E., Hino, S., Atarashi, K., Onawa, S., Fujimura, Y., Lockett, T., Clarke, J.M., Topping, D.L., Tomita, M., Hori, S., Ohara, O., Morita, T., Koseki, H., Kikuchi, J., Honda, K., Hase, K. and Ohno, H. (2013). Commensal microbe-derived butyrate induces the differentiation of colonic regulatory T cells. *Nature*, 504: 446-450. https://doi.org/10.1038/nature12721

Galleano, M. and Verstraeten, S.V. (2010). Basic biochemical mechanisms behind the health benefits of polyphenols. *Molecular Aspects of Medicine*, 31: 435-445. https://doi. org/10.1016/J.MAM.2010.09.006

Galleano, M., Verstraeten, S.V., Oteiza, P.I. and Fraga, C.G. (2010). Antioxidant actions of flavonoids: Thermodynamic and kinetic analysis. *Archives of Biochemistry and Biophysics*, 501: 23-30. https://doi.org/10.1016/J.ABB.2010.04.005

Gaya, P., Medina, M., Sánchez-Jiménez, A. and Landete, J.M. (2016). Phytoestrogen metabolism by adult human gut microbiota. *Molecules*, 21: 1034. https://doi.org/10.3390/ molecules21081034

Gaya, P., Peirotén, Á. and Landete, J.M. (2017). Transformation of plant isoflavones into bioactive isoflavones by lactic acid bacteria and bifidobacteria. *Journal of Functional Foods*, 39: 198-205. https://doi.org/10.1016/j.jff.2017.10.029

Ghasemzadeh, A. and Ghasemzadeh, N. (2011). Flavonoids and phenolic acids: Role and biochemical activity in plants and human. *Journal of Medicinal Plants Research*, 5: 6697-6703. https://doi.org/10.5897/JMPR11.1404

Granja, A., Pinheiro, M. and Reis, S. (2016). Epigallocatechin gallate nanodelivery systems for cancer therapy. *Nutrients*, 8: 3078. https://doi.org/10.3390/NU8050307

Gupta, P. and Diwan, B. (2017). Bacterial exopolysaccharide mediated heavy metal removal: A review on biosynthesis, mechanism and remediation strategies. *Biotechnology Reports*, 13: 58-71. https://doi.org/10.1016/J.BTRE.2016.12.006

Gupta, S., Allen-Vercoe, E. and Petrof, E.O. (2016). Fecal microbiota transplantation: In perspective. *Therapeutic Advances in Gastroenterology*, 9: 229-239. https://doi. org/10.1177/1756283X15607414

Han, M., Wang, C., Liu, P., Li, D., Li, Y. and Ma, X. (2017). Dietary fibre gap and host gut microbiota. *Protein & Peptide Letters*, 24: 388-396. https://doi.org/10.2174/0929866524 666170220113312

Hassan, A.N. (2008). Possibilities and challenges of exopolysaccharide-producing lactic cultures in dairy foods. *Journal of Dairy Science*, 91: 1282-1298. https://doi.org/10.3168/JDS.2007-0558

Heleno, S.A., Martins, A., João, M., Queiroz, R.P. and Ferreira, I.C.F.R. (2015). Bioactivity of phenolic acids: Metabolites versus parent compounds: A review. *Food Chemistry*, 173: 501-513. https://doi.org/10.1016/j.foodchem.2014.10.057

Hidalgo, M., Oruna-concha, M.J., Walton, G.E., Kallithraka, S., Spencer, J.P.E., Gibson, G.R. and Pascual-teresa, S.D. (2012). Metabolism of anthocyanins by human gut micro flora and their influence on gut bacterial growth. *Journal of Agricultural and Food Chemistry*, 60: 3882-3890.

Hosseini, E., Grootaert, C., Verstraete, W. and Van de Wiele, T. (2011). Propionate as a health-promoting microbial metabolite in the human gut. *Nutrition Reviews*, 69: 245-258. https://doi.org/10.1111/j.1753-4887.2011.00388.x

Hostetler, G.L., Ralston, R.A. and Schwartz, S.J. (2017). Flavones: Food sources, bioavailability, metabolism, and bioactivity. *Advances in Nutrition: An International Review Journal*, 8: 423-435. https://doi.org/10.3945/an.116.012948

Hoyles, L. and Swann, J. (2019). Influence of the human gut microbiome on the metabolic phenotype. pp. 535-560. *In*: J.C. Lindon, J.K. Nicholson and E. Holmes (Eds.). *The Handbook of Metabolic Phenotyping*. Elsevier. https://doi.org/10.1016/B978-0-12-812293-8.00018-9

Hu, X., Pang, X., Wang, P.G. and Chen, M. (2019). Isolation and characterisation of an antioxidant exopolysaccharide produced by *Bacillus* sp. S-1 from Sichuan pickles. *Carbohydrate Polymers*, 204: 9-16. https://doi.org/10.1016/J.CARBPOL.2018.09.069

Hur, H.G., Lay, J.O., Beger, R.D., Freeman, J.P. and Rafii, F. (2000). Isolation of human intestinal bacteria metabolizing the natural isoflavone glycosides daidzin and genistin. *Archives of Microbiology*, 174: 422-428.

Igwe, E.O., Charlton, K.E., Probst, Y.C., Kent, K. and Netzel, M.E. (2019). A systematic literature review of the effect of anthocyanins on gut microbiota populations. *Journal of Human Nutrition and Dietetics*, 2019: 12582. https://doi.org/10.1111/jhn.12582

Iino, C., Shimoyama, T., Iino, K., Yokoyama, Y., Chinda, D., Sakuraba, H., Fukuda, S. and Nakaji, S. (2019). Daidzein intake is associated with equol producing status through an increase in the intestinal bacteria responsible for equol production. *Nutrients*, 11: 433. https://doi.org/10.3390/nu11020433

Itoh, A., Isoda, K., Kondoh, M., Kawase, M., Watari, A., Kobayashi, M., Tamesada, M. and Yagi, K. (2010). Hepatoprotective effect of syringic acid and vanillic acid on CCl4-induced liver injury. *Biological and Pharmaceutical Bulletin*, 33: 983-987. https://doi.org/10.1248/bpb.33.983

Jiang, W., Guo, Q., Wu, J., Guo, B., Wang, Y., Zhao, S., Lou, H., Yu, X., Mei, X., Wu, C., Qiao, S. and Wu, Y. (2012). Dual effects of sodium butyrate on hepatocellular carcinoma cells. *Molecular Biology Reports*, 39: 6235-6242. https://doi.org/10.1007/s11033-011-1443-5

Jiang, X., Sun, J. and Bai, W. (2020). Anthocyanins in food. pp. 1-52. *In*: S.D. Sarker, J. Xiao and Y. Asakawa (Eds.). *Handbook of Dietary Phytochemicals*. Springer. https://doi.org/10.1007/978-981-13-1745-3_13-1

Jin, J.S., Kitahara, M., Sakamoto, M., Hattori, M. and Benno, Y. (2010). Slackia equolifaciens sp. nov., a human intestinal bacterium capable of producing equol. *International Journal of Systematic and Evolutionary Microbiology*, 60: 1721-1724. https://doi.org/10.1099/ijs.0.016774-0

Joshi, V.K. (2016). *Indigenous Fermented Foods of South Asia*. CRC Press, Boca Raton, Florida.

Juan, M., Wu, C. and Chou, C. (2010). Fermentation with *Bacillus* spp. as a bioprocess to enhance anthocyanin content, the angiotensin converting enzyme inhibitory effect and the reducing activity of black soybeans. *Food Microbiology*, 27: 918-923. https://doi. org/10.1016/j.fm.2010.05.009

Kähkönen, M., Hopia, A. and Heinonen, M. (2001). Berry phenolics and their antioxidant activity. *Journal of Agricultural and Food Chemistry*, 49: 4076-4082. https://doi. org/10.1021/jf010152t

Kalt, W. (2019). Anthocyanins and their C6-C3-C6 metabolites in humans and animals. *Molecules*, 24: 024 https://doi.org/10.3390/molecules24224024

Kaoutari, A. El, Armougom, F., Gordon, J.I., Raoult, D. and Henrissat, B. (2013). The abundance and variety of carbohydrate-active enzymes in the human gut microbiota. *Nature Reviews Microbiology*, 11: 497-504. https://doi.org/10.1038/nrmicro3050

Kawabata, K., Yoshioka, Y. and Terao, J. (2019). Role of intestinal microbiota in the bioavailability and physiological functions of dietary polyphenols. *Molecules*, 24: 370. https://doi.org/10.3390/molecules24020370

Kelly, S.M., O'Callaghan, J., Kinsella, M. and Sinderen, D. (2018). Characterisation of a hydroxycinnamic acid esterase from the *Bifidobacterium longum* subsp. *longum* Taxon. *Frontiers in Microbiology*, 9: 2690. https://doi.org/10.3389/fmicb.2018.02690

Kezimana, P., Dmitriev, A.A., Kudryavtseva, A.V., Romanova, E.V. and Melnikova, N.V. (2018). Secoisolariciresinol diglucoside of flaxseed and its metabolites: Biosynthesis and potential for nutraceuticals. *Frontiers in Genetics*, 9: 641. https://doi.org/10.3389/ fgene.2018.00641

Khadem, S. and Marles, R.J. (2010). Monocyclic phenolic acids; hydroxy- and polyhydroxybenzoic acids: Occurrence and recent bioactivity studies. *Molecules*, 15: 7985-8005. https://doi.org/10.3390/molecules15117985

Khalil, E.S., Manap, M.Y., Mustafa, S., Amid, M., Alhelli, A.M. and Aljoubori, A. (2018). Probiotic characteristics of exopolysaccharides-producing *Lactobacillus* isolated from some traditional Malaysian fermented foods. *CyTA – Journal of Food*, 16: 287-298. https://doi.org/10.1080/19476337.2017.1401007

Khan, S. and Jena, G. (2016). Sodium butyrate reduces insulin-resistance, fat accumulation and dyslipidemia in type-2 diabetic rat: A comparative study with metformin. *Chemico-Biological Interactions*, 254: 124-134. https://doi.org/10.1016/j.cbi.2016.06.007

Kheni, K. and Vyas, T.K. (2017). Characterisation of exopolysaccharide produced by *Ganoderma* sp. TV1 and its potential as antioxidant and anticancer agent. *Journal of Biologically Active Products from Nature*, 7: 72-80. https://doi.org/10.1080/22311866.2 017.1306461

Khoo, H.E., Azlan, A., Tang, S.T. and Lim, S.M. (2017). Anthocyanidins and anthocyanins: Coloured pigments as food, pharmaceutical ingredients, and the potential health benefits. *Food & Nutrition Research*, 61: 1361779. https://doi.org/10.1080/16546628.2017.13617 79

Kim, Y.H., Won, Y.S., Yang, X., Kumazoe, M., Yamashita, S., Hara, A., Takagaki, A., Goto, K., Nanjo, F. and Tachibana, H. (2016). Green tea catechin metabolites exert immunoregulatory effects on CD4+ T cell and natural killer cell activities. *Journal of Agricultural and Food Chemistry*, 64: 3591-3597. https://doi.org/10.1021/acs.jafc.6b01115

Kishida, K. and Matsumoto, H. (2019). Urinary excretion rate and bioavailability of chlorogenic acid, caffeic acid, p-coumaric acid, and ferulic acid in non-fasted rats maintained under physiological conditions. *Heliyon*, 5: e02708. https://doi.org/10.1016/j.heliyon.2019. e02708

Kozłowska, A. and Szostak-Węgierek, D. (2018). Flavonoids – Food sources, health benefits, and mechanisms involved. pp. 1-27. *In*: J.M. Merillon (Ed.). *Bioactive Molecules in Food*, Springer. https://doi.org/10.1007/978-3-319-54528-8_54-1

Kratz, J., Andrighetti-Fröhner, C., Leal, P., Nunes, R., Yunes, R., Trybala, E., Bergström, T., Barardi, C. and Simões, C. (2008). Evaluation of anti-HSV-2 activity of gallic acid and pentyl gallate. *Biological and Pharmaceutical Bulletin*, 31: 903-907. https://doi.org/10.1248/bpb.31.903

Křížová, L., Dadáková, K., Kašparovská, J. and Kašparovský, T. (2019). Isoflavones. *Molecules*, 24: 1076. https://doi.org/10.3390/molecules24061076

Kubow, S., Iskandar, M.M., Melgar-Bermudez, E., Sleno, L., Sabally, K., Azadi, B., How, E., Prakash, S., Burgos, G. and Felde, T. (2017). Effects of simulated human gastrointestinal digestion of two purple-fleshed potato cultivars on anthocyanin composition and cytotoxicity in colonic cancer and non-tumorigenic cells. *Nutrients*, 9: 953. https://doi.org/10.3390/nu9090953

Kumar, N. and Goel, N. (2019). Phenolic acids: Natural versatile molecules with promising therapeutic applications. *Biotechnology Reports*, 24: e00370. https://doi.org/10.1016/j.btre.2019.e00370

Kumar, S. and Pandey, A.K. (2013). Chemistry and biological activities of flavonoids: An overview. *The Scientific World Journal*, 2013: 1-16. https://doi.org/10.1155/2013/162750

Kutschera, M., Engst, W., Blaut, M. and Braune, A. (2011). Isolation of catechin-converting human intestinal bacteria. *Journal of Applied Microbiology*, 111: 165-175. https://doi.org/10.1111/j.1365-2672.2011.05025.x

Kuhnau, J. (1976). The Flavonoids. A class of semi-essential food components: Their role in human nutrition. *World Review of Nutrition and Dietetics*, 24: 117-191. https://doi.org/10.1159/000399407

Landete, J.M., Gaya, P., Rodríguez, E., Langa, S., Peirotén, Á., Medina, M. and Arqués, J.L. (2017). Probiotic bacteria for healthier aging: Immunomodulation and metabolism of phytoestrogens. *BioMed Research International*, 2017: 1-10. https://doi.org/10.1155/2017/5939818

Lee, K.J., Baek, D.-Y., Lee, G.-A., Cho, G.-T., So, Y.-S., Lee, J.-R., Ma, K.-H., Chung, J.-W. and Hyun, D.Y. (2020). Phytochemicals and antioxidant activity of Korean black soybean (*Glycine max* L.) landraces. *Antioxidants*, 9: 213. https://doi.org/10.3390/antiox9030213

Lee, P.G., Joonwon, K,, Eun, J.K., Sang, H.L., Kwon, Y.C., Romas, J.K. and Byung, G.K. (2017). Biosynthesis of (-)-5-hydroxy-equol and 5-hydroxy-dehydroequol from soy isoflavone, genistein using microbial whole cell bioconversion. *ACS Chemical Biology*, 12(11): 2883-2890. https://doi.org/10.1021/acschembio.7b00624

Li, B.C., Zhang, T., Li, Y.Q. and Ding, G.B. (2019). Target discovery of novel α-l-rhamnosidases from human fecal metagenome and application for biotransformation of natural flavonoid glycosides. *Applied Biochemistry and Biotechnology*, 189: 1245-1261. https://doi.org/10.1007/s12010-019-03063-5

Li, J., He, Y.J., Zhou, L., Liu, Y., Jiang, M., Ren, L. and Chen, H. (2018). Transcriptome profiling of genes related to light-induced anthocyanin biosynthesis in eggplant (*Solanum melongena* L.) before purple colour becomes evident. *BMC Genomics*, 19: 201. https://doi.org/10.1186/s12864-018-4587-z

Li, J., Wu, T., Li, N., Wang, X., Chen, G. and Lyu, X. (2019). Bilberry anthocyanin extract promotes intestinal barrier function and inhibits digestive enzyme activity by regulating the gut microbiota in aging rats. *Food and Function*, 10: 333-343. https://doi.org/10.1039/c8fo01962b

Li, S., Chen, T., Xu, F., Dong, S., Xu, H., Xiong, Y. and Wei, H. (2014). The beneficial effect of exopolysaccharides from *Bifidobacterium bifidum* WBIN03 on microbial diversity in mouse intestine. *Journal of the Science of Food and Agriculture*, 94: 256-264. https://doi.org/10.1002/jsfa.6244

Li, S., Jin, Z., Hu, D., Yang, W., Yan, Y., Nie, X., Lin, J., Zhang, Q., Gai, D., Ji, Y. and Chen, X. (2020). Effect of solid-state fermentation with *Lactobacillus casei* on the nutritional value, isoflavones, phenolic acids and antioxidant activity of whole soybean flour. *LWT*, 125: 109264. https://doi.org/10.1016/j.lwt.2020.109264

Li, Y., Yao, J., Han, C., Yang, J., Chaudhry, M.T., Wang, S., Liu, H. and Yin, Y. (2016). Quercetin, inflammation and immunity. *Nutrients*, 8: 167. https://doi.org/10.3390/nu8030167

Liang, N. and Kitts, D.D. (2015). Role of chlorogenic acids in controlling oxidative and inflammatory stress conditions. *Nutrients*, 8: 16. https://doi.org/10.3390/nu8010016

Lila, M.A., Burton-Freeman, B., Grace, M. and Kalt, W. (2016). Unraveling anthocyanin bioavailability for human health. *Annual Review of Food Science and Technology*, 7: 375-393. https://doi.org/10.1146/annurev-food-041715-033346

Liu, L., Li, M., Yu, M., Shen, M., Wang, Q., Yu, Y. and Xie, J. (2019). Natural polysaccharides exhibit anti-tumor activity by targeting gut microbiota. *International Journal of Biological Macromolecules*, 121: 743-751. https://doi.org/10.1016/j.ijbiomac.2018.10.083

Liu, L., Li, M., Yu, M., Shen, M., Wang, Q., Yu, Y. and Xie, J. (2018). Natural polysaccharides exhibit anti-tumor activity by targeting gut microbiota. *International Journal of Biological Macromolecules*, 121: 743-751. https://doi.org/10.1016/j.ijbiomac.2018.10.083

Liu, Y., Xie, M., Wan, P., Chen, G., Chen, C., Chen, D., Yu, S., Zeng, X. and Sun, Y. (2020). Purification, characterisation and molecular cloning of a dicaffeoylquinic acid-hydrolysing esterase from human-derived *Lactobacillus fermentum* LF-12. *Food and Function*, 11: 3235-3244. https://doi.org/10.1039/D0FO00029A

Londin, E., Loher, P., Telonis, A.G., Quann, K., Clark, P., Jing, Y., Hatzimichael, E., Kirino, Y., Honda, S., Lally, M., Ramratnam, B., Comstock, C.E.S., Knudsen, K.E., Gomella, L., Spaeth, G.L., Hark, L., Katz, L.J., Witkiewicz, A., Rostami, A., Jimenez, S.A., Hollingsworth, M.A., Yeh, J.J., Shaw, C.A., McKenzie, S.E., Bray, P., Nelson, P.T., Zupo, S., Van Roosbroeck, K., Keating, M.J., Calin, G.A., Yeo, C., Jimbo, M., Cozzitorto, J., Brody, J.R., Delgrosso, K., Mattick, J.S., Fortina, P. and Rigoutsos, I. (2015). Analysis of 13 cell types reveals evidence for the expression of numerous novel primate- and tissue-specific microRNAs. *Proceedings of the National Academy of Sciences*, 112: 1106-1115. https://doi.org/10.1073/pnas.1420955112

Lorigooini, Z., Jamshidi-kia, F. and Hosseini, Z. (2020). Analysis of aromatic acids (phenolic acids and hydroxycinnamic acids). pp. 199-219. *In*: S.F. Nabavi, S.M. Nabavi, M. Saeedi and A.S. Silva (Eds.). *Recent Advances in Natural Products Analysis*. Elsevier. https://doi.org/10.1016/B978-0-12-816455-6.00004-4

Lou, Z., Wang, H., Rao, S., Sun, J., Ma, C. and Li, J. (2012). p-Coumaric acid kills bacteria through dual damage mechanisms. *Food Control*, 25: 550-554. https://doi.org/10.1016/j.foodcont.2011.11.022 not in text

Lu, Y., Jiang, F., Jiang, H., Wu, K., Zheng, X., Cai, Y., Katakowski, M., Chopp, M. and To, S.S.T. (2010). Gallic acid suppresses cell viability, proliferation, invasion and angiogenesis in human glioma cells. *European Journal of Pharmacology*, 641: 102-107. https://doi.org/10.1016/j.ejphar.2010.05.043

Ma, X., Zhou, Z., Zhang, X., Fan, M., Hong, Y., Feng, Y., Dong, Q., Diao, H. and Wang, G. (2020). Sodium butyrate modulates gut microbiota and immune response in colorectal cancer liver metastatic mice. *Cell Biology and Toxicology*, 36: 1-7. https://doi.org/10.1007/s10565-020-09518-4

MacFarlane, L.A. and Murphy, P. (2010). MicroRNA: Biogenesis, function and role in cancer., *Current Genomics*, 11: 537-561. https://doi.org/10.2174/138920210793175895

Magnusdottir, S. and Thiele, I. (2018). Science Direct Modeling metabolism of the human gut microbiome ttir and Ines Thiele. *Current Opinion in Biotechnology*, 51: 90-96. https://doi.org/10.1016/j.copbio.2017.12.005

Majdoub, Y.O. El, Diouri, M., Arena, P., Arigò, A., Cacciola, F., Rigano, F., Dugo, P. and Mondello, L. (2019). Evaluation of the availability of delphinidin and cyanidin-3-O-sambubioside from *Hibiscus sabdariffa* and 6-gingerol from *Zingiber officinale* in colon using liquid chromatography and mass spectrometry detection. *European Food Research and Technology*, 245: 2425-2433. https://doi.org/10.1007/s00217-019-03358-1

Marchiosi, R., dos Santos, W.D., Constantin, R.P., de Lima, R.B., Soares, A.R., Finger-Teixeira, A., Mota, T.R., de Oliveira, D.M., Foletto-Felipe, M. de P., Abrahão, J. and Ferrarese-Filho, O. (2020). Biosynthesis and metabolic actions of simple phenolic acids in plants. *Phytochemistry Reviews*, 6: 1-42. https://doi.org/10.1007/s11101-020-09689-2

Marco, M.L., Heeney, D., Binda, S., Cifelli, C.J., Cotter, P.D., Foligné, B., Gänzle, M., Kort, R., Pasin, G., Pihlanto, A., Smid, E.J. and Hutkins, R. (2017). Health benefits of fermented foods: Microbiota and beyond. *Current Opinion in Biotechnology*, 44: 94-102. https://doi.org/10.1016/j.copbio.2016.11.010

Mariano, T.B., Higashi, B., Sanches Lopes, S.M., Pedroza Carneiro, J.W., de Almeida, R.T.R., Pilau, E.J., Gonçalves, J.E., Correia Gonçalves, R.A. and de Oliveira, A.J.B. (2020). Prebiotic fructooligosaccharides obtained from escarole (*Cichorium endivia* L.) roots. *Bioactive Carbohydrates and Dietary Fibre*, 24: 100233. https://doi.org/10.1016/j.bcdf.2020.100233

Marín, L., Miguélez, E.M., Villar, C.J. and Lombó, F. (2015). Bioavailability of dietary polyphenols and gut microbiota metabolism: Antimicrobial properties. *BioMed Research International*, 2015: 905215.

Martínez-Villaluenga, C., Cardelle-Cobas, A., Corzo, N. and Olano, A. (2008). Study of galactooligosaccharide composition in commercial fermented milks. *Journal of Food Composition and Analysis*, 21: 540-544. https://doi.org/10.1016/j.jfca.2008.05.008

Martins, T.F., Palomino, O.M., Álvarez-Cilleros, D., Martín, M.A., Ramos, S. and Goya, L. (2020). Cocoa flavanols protect human endothelial cells from oxidative stress. *Plant Foods for Human Nutrition*, 75: 161-168. https://doi.org/10.1007/s11130-020-00807-1

Maruo, T., Sakamoto, M., Ito, C., Toda, T. and Benno, Y. (2008). *Adlercreutzia equolifaciens* gen. nov., sp. nov., an equol-producing bacterium isolated from human faeces, and emended description of the genus *Eggerthella*. *International Journal of Systematic and Evolutionary Microbiology*, 58: 1221-1227. https://doi.org/10.1099/ijs.0.65404-0

Marvasi, M., Visscher, P.T. and Casillas Martinez, L. (2010). Exopolymeric substances (EPS) from *Bacillus subtilis*: Polymers and genes encoding their synthesis. *FEMS Microbiology Letters*, 313: 1-9. https://doi.org/10.1111/j.1574-6968.2010.02085.x

Mayo, B., Vázquez, L. and Flórez, A.B. (2019). Equol: A bacterial metabolite from the daidzein isoflavone and its presumed beneficial health effects. *Nutrients*, 11: 2231. https://doi.org/10.3390/nu11092231

Mele, L., Carobbio, S., Brindani, N., Curti, C., Rodriguez-Cuenca, S., Bidault, G., Mena, P., Zanotti, I., Vacca, M., Vidal-Puig, A. and Del Rio, D. (2017). Phenyl-γ-valerolactones, flavan-3-ol colonic metabolites, protect brown adipocytes from oxidative stress without affecting their differentiation or function. *Molecular Nutrition & Food Research*, 61: 1700074. https://doi.org/10.1002/mnfr.201700074

Melini, F., Melini, V., Luziatelli, F., Ficca, A.G. and Ruzzi, M. (2019). Health-promoting components in fermented foods: An up-to-date systematic review. *Nutrients*, 11: 1189. https://doi.org/10.3390/nu11051189

Mena, P., Bresciani, L., Brindani, N., Ludwig, I.A., Pereira-Caro, G., Angelino, D., Llorach, R., Calani, L., Brighenti, F., Clifford, M.N., Gill, C.I.R., Crozier, A., Curti, C. and Rio, D. Del. (2019). Phenyl-γ-valerolactones and phenylvaleric acids, the main colonic metabolites of flavan-3-ols: Synthesis, analysis, bioavailability, and bioactivity. *Natural Product Reports*, 36: 714-752. https://doi.org/10.1039/C8NP00062J

Mendes, T.M.N., Murayama, Y., Yamaguchi, N., Sampaio, G.R., Fontes, L.C.B., Torres, E.A.F. da S., Tamura, H. and Yonekura, L. (2019). Guaraná (*Paullinia cupana*) catechins and procyanidins: Gastrointestinal/colonic bioaccessibility, Caco-2 cell permeability and the impact of macronutrients. *Journal of Functional Foods*, 55: 352-361. https://doi.org/10.1016/j.jff.2019.02.026

Michlmayr, H. and Kneifel, W. (2014). β-glucosidase activities of lactic acid bacteria: Mechanisms, impact on fermented food and human health. *FEMS Microbiology Letters*, 352: 1-10. https://doi.org/10.1111/1574-6968.12348

Milani, C., Duranti, S., Bottacini, F., Casey, E., Turroni, F., Mahony, J., Belzer, C., Delgado Palacio, S., Arboleya Montes, S., Mancabelli, L., Lugli, G.A., Rodriguez, J.M., Bode, L., de Vos, W., Gueimonde, M., Margolles, A., van Sinderen, D. and Ventura, M. (2017). The first microbial colonisers of the human gut: Composition, activities, and health implications of the infant gut microbiota. *Microbiology and Molecular Biology Reviews*, 81: 1-67. https://doi.org/10.1128/mmbr.00036-17

Mueller, M., Zartl, B., Schleritzko, A., Stenzl, M., Viernstein, H. and Unger, F.M. (2018). Rhamnosidase activity of selected probiotics and their ability to hydrolyse flavonoid rhamnoglucosides. *Bioprocess and Biosystems Engineering*, 41: 221-228. https://doi.org/10.1007/s00449-017-1860-5

Okello, E., Leylabi, R. and McDougall, G. (2012). Inhibition of acetylcholinesterase by green and white tea and their simulated intestinal metabolites. *Food and Function*, 3: 651-661. https://doi.org/10.1039/c2fo10174b

Oksuz, T., Tacer-Caba, Z., Nilufer-Erdil, D. and Boyacioglu, D. (2019). Changes in bioavailability of sour cherry (*Prunus cerasus* L.) phenolics and anthocyanins when consumed with dairy food matrices. *Journal of Food Science and Technology*, 56: 4177-4188. https://doi.org/10.1007/s13197-019-03888-2

Oliveira, C.S.F., Pereira, H., Alves, S., Castro, L., Baltazar, F., Chaves, S.R., Preto, A. and Côrte-Real, M. (2015). Cathepsin D protects colorectal cancer cells from acetate-induced apoptosis through autophagy-independent degradation of damaged mitochondria., *Cell Death & Disease*, 6: 1788-1788. https://doi.org/10.1038/cddis.2015.157

Oliveira, R.P.S., Casazza, A.A., Aliakbarian, B., Perego, P., Converti, A. and Oliveira, M.N. (2013). Influence of fructooligosaccharides on the fermentation profile and viable counts in a symbiotic low fat milk., *Brazilian Journal of Microbiology*, 44: 431-434. https://doi.org/10.1590/S1517-83822013000200014

Paiva, I., Pinho, R., Pavlou, M.A., Hennion, M., Wales, P., Schütz, A.-L., Rajput, A., Szegő, É.M., Kerimoglu, C., Gerhardt, E., Rego, A.C., Fischer, A., Bonn, S. and Outeiro, T.F. (2017). Sodium butyrate rescues dopaminergic cells from alpha-synuclein-induced transcriptional deregulation and DNA damage., *Human Molecular Genetics*, 26: 2231-2246. https://doi.org/10.1093/hmg/ddx114

Palaniyandi, S.A., Damodharan, K., Suh, J.-W. and Yang, S.H. (2018). Functional characterisation of an exopolysaccharide produced by *Bacillus sonorensis* MJM60135 isolated from Ganjang. *Journal of Microbiology and Biotechnology*, 28: 663-670. https://doi.org/10.4014/jmb.1711.11040

Panche, A.N., Diwan, A.D. and Chandra, S.R. (2016). Flavonoids: An overview. *Journal of Nutritional Science*, 5: e47. https://doi.org/10.1017/jns.2016.41

Pandey, K.B. and Rizvi, S.I. (2009). Plant polyphenols as dietary antioxidants in human health and disease. *Oxidative Medicine and Cellular Longevity*, 2: 270-278. https://doi.org/10.4161/oxim.2.5.9498

Pasinetti, G.M., Singh, R., Westfall, S., Herman, F., Faith, J. and Ho, L. (2018). The role of the gut microbiota in the metabolism of polyphenols as characterised by gnotobiotic mice. *Journal of Alzheimer's Disease*, 63: 409. https://doi.org/10.3233/JAD-171151

Pathak, S., Kesavan, P., Banerjee, Anushka, Banerjee, Antara, Celep, G.S., Bissi, L. and Marotta, F. (2018). Metabolism of dietary polyphenols by human gut microbiota and their health benefits. pp. 347-359. *In*: R. Watson, V. Preedy and S. Zibadi (Eds.). *Polyphenols:*

Mechanisms of Action in Human Health and Disease. Elsevier Inc.https://doi.org/10.1016/B978-0-12-813006-3.00025-8

Paynich, M.L., Jones-Burrage, S.E. and Knight, K.L. (2017). Exopolysaccharide from *Bacillus subtilis* induces anti-inflammatory M2 macrophages that prevent T cell-mediated disease. *Journal of Immunology*, 198: 2689-2698. https://doi.org/10.4049/jimmunol.1601641

Peterson, J., Dwyer, J., Adlercreutz, H., Scalbert, A., Jacques, P. and McCullough, M.L. (2010). Dietary lignans: Physiology and potential for cardiovascular disease risk reduction. *Nutrition Reviews*, 68: 571. https://doi.org/10.1111/j.1753-4887.2010.00319.x

Piazzon, A., Vrhovsek, U., Masuero, D., Mattivi, F., Mandoj, F. and Nardini, M. (2012). Antioxidant activity of phenolic acids and their metabolites: Synthesis and antioxidant properties of the sulfate derivatives of ferulic and caffeic acids and of the acyl glucuronide of ferulic acid. *Journal of Agricultural and Food Chemistry*, 60: 12312-12323. https://doi.org/10.1021/jf304076z

Poschner, S., Alexandra, M.S., Martin, Z., Judith, W., Daniel, D., Bettina, P., Konstantin, L.S. and Walter, J. (2017). The impacts of genistein and daidzein on estrogen conjugations in human breast cancer cells: A targeted metabolomics approach. *Frontiers in Pharmacology*, 8: 699. https://doi.org/10.3389/fphar.2017.00699 not in text

Prasad, K.N. (1980). Butyric acid: A small fatty acid with diverse biological functions. *Life Sciences*, 27: 1351-1358. https://doi.org/10.1016/0024-3205(80)90397-5

Prasad, K.N. and Hsie, A.W. (1971). Morphologic differentiation of mouse neuroblastoma cells induced *in vitro* by dibutyryl adenosine 3':5'-cyclic monophosphate. *Nature: New Biology*, 233: 141-142.

Przybylska-Balcerek, A. and Stuper-Szablewska, K. (2019). The effect of phenolic acids on living organisms. *Indian Journal of Medical Research and Pharmaceutical Sciences*, 6: 1-14. https://doi.org/10.5281/zenodo.3446898

Qin, X.L., Hui, Y.S., Wu, Y., Yong, J.L., Lin, Z., Ting, L. and Yong, H. (2017). Analysis of metabolites of quercitrin in rat intestinal flora by using UPLC-ESI-Q-TOF-MS/MS. *Zhongguo Zhongyao Zazhi*, 42: 357-362. https://doi.org/10.19540/j.cnki.cjcmm.20161222.018. not in

Quartieri, A., García-Villalba, R., Amaretti, A., Raimondi, S., Leonardi, A., Rossi, M. and Tomàs-Barberàn, F. (2016). Detection of novel metabolites of flaxseed lignans *in vitro* and *in vivo*. *Molecular Nutrition & Food Research*, 60: 1590-1601. https://doi.org/10.1002/mnfr.201500773

Rafii, F. (2015). The role of colonic bacteria in the metabolism of the natural isoflavone daidzin to equol. *Metabolites*, 5: 56-73. https://doi.org/10.3390/metabo5010056

Rai, A.K., Sanjukta, S., Chourasia, R., Bhat, I., Bhardwaj, P.K. and Sahoo, D. (2017). Production of bioactive hydrolysate using protease, β-glucosidase and α-amylase of *Bacillus* spp. isolated from *kinema*. *Bioresource Technology*, 235: 358-365. https://doi.org/10.1016/j.biortech.2017.03.139

Rajakovich, L.J. and Balskus, E.P. (2019). Metabolic functions of the human gut microbiota: The role of metalloenzymes. *Natural Product Reports*, 36: 593-625. https://doi.org/10.1039/C8NP00074C

Rivera-Sagredo, A., Cañada, F.J., Nieto, O., Jimenez-Barbero, J. and Martín-Lomas, M. (1992). Substrate specificity of small-intestinal lactase. Assessment of the role of the substrate hydroxyl groups. *European Journal of Biochemistry*, 209: 415-422.

Rocchetti, G., Lucini, L., Giuberti, G., Bhumireddy, S.R., Mandal, R., Trevisan, M. and Wishart, D.S. (2019). Transformation of polyphenols found in pigmented gluten-free flours during in vitro large intestinal fermentation. *Food Chemistry*, 298: 125068. https://doi.org/10.1016/j.foodchem.2019.125068

Rodriguez-Castaño, G.P., Matthew, R.D., Xingbo, L., Bradley, W.B., Alejandro, A-G. and

Federico, E.R. (2019). Bacteroides thetaiotaomicron starch utilisation promotes quercetin degradation and butyrate production by *Eubacterium ramulus*. *Frontiers in Microbiology*, 10: 1145. https://doi.org/10.3389/fmicb.2019.01145. not

Rodríguez-García, C., Sánchez-Quesada, C., Gaforio, J.J. and Gaforio, J.J. (2019). Dietary flavonoids as cancer chemopreventive agents: An updated review of human studies. *Antioxidants*, 8: 137. https://doi.org/10.3390/antiox8050137

Rowland, I., Gibson, G., Heinken, A., Scott, K., Swann, J., Thiele, I. and Tuohy, K. (2018). Gut microbiota functions: Metabolism of nutrients and other food components. *European Journal of Nutrition*, 57: 1-24. https://doi.org/10.1007/s00394-017-1445-8

Ruijssenaars, H.J., Stingele, F. and Hartmans, S. (2000). Biodegradability of food-associated extracellular polysaccharides. *Current Microbiology*, 40: 194-199. https://doi.org/10.1007/s002849910039

Saffari-Chaleshtori, J., Shafiee, S.M., Ghatreh-Samani, K. and Jalilian, N. (2019). The study of drug resistance properties of ABCG2 (ATP-binding cassette G2) in contact with thymoquinone, gallic acid, and hesperetin antioxidants. *Journal of HerbMed Pharmacology*, 8: 108-113. https://doi.org/10.15171/jhp.2019.17

Saibabu, V., Fatima, Z., Khan, L.A. and Hameed, S. (2015). Therapeutic potential of dietary phenolic acids. *Advances in Pharmacological Sciences*, 2015: 1-10. https://doi.org/10.1155/2015/823539

Salazar, N., Gueimonde, M., Hernández-Barranco, A.M., Ruas-Madiedo, P. and Gavilán, C.G. (2008). Exopolysaccharides produced by intestinal *Bifidobacterium* strains act as fermentable substrates for human intestinal bacteria. *Applied and Environmental Microbiology*, 74: 4737-4745. https://doi.org/10.1128/AEM.00325-08

Samruan, W., Gasaluck, P. and Oonsivilai, R. (2014). Total phenolic and flavonoid contents of soybean fermentation by *Bacillus subtilis* SB-MYP-1. *Advanced Materials Research*, 931: 1-6. https://doi.org/10.4028/www.scientific.net/AMR.931-932.1587

Sánchez-Patán, F., Cueva, C., Monagas, M., Walton, G.E., Gibson, G.R., Quintanilla-López, J.E., Lebrón-Aguilar, R., Martín-Álvarez, P.J., Moreno-Arribas, M.V. and Bartolomé, B. (2012). *In vitro* fermentation of a red wine extract by human gut microbiota: Changes in microbial groups and formation of phenolic metabolites. *Journal of Agricultural and Food Chemistry*, 60: 2136-2147. https://doi.org/10.1021/jf2040115

Scalbert, A. and Williamson, G. (2018). Chocolate: Modern science investigates an ancient medicine dietary intake and bioavailability of polyphenols 1. *The Journal of Nutrition*, 112: 2073-2085.

Scalbert, A. and Williamson, G. (2000). Dietary intake and bioavailability of polyphenols. *The Journal of Nutrition*, 130: 2073S-2085S. https://doi.org/10.1093/jn/130.8.2073S

Schröder, C., Matthies, A., Engst, W., Blaut, M. and Braune, A. (2013). Identification and expression of genes involved in the conversion of daidzein and genistein by the equol-forming bacterium *Slackia isoflavoniconvertens*. *Applied and Environmental Microbiology*, 79: 3494. https://doi.org/10.1128/AEM.03693-12

Setchell, K., Brown, N., Zimmer-Nechemias, L., Wolfe, B., Jha, P. and Heubi, J. (2014). Metabolism of secoisolariciresinol-diglycoside the dietary precursor to the intestinally derived lignan enterolactone in humans. *Food and Function*, 5: 491-501. https://doi.org/10.1039/C3FO60402K

Setchell, K.D., Clerici, C., Lephart, E.D., Cole, S.J., Heenan, C., Castellani, D., Wolfe, B.E., Nechemias-Zimmer, L., Brown, N.M., Lund, T.D., Handa, R.J. and Heubi, J.E. (2005). S-Equol, a potent ligand for estrogen receptor β, is the exclusive enantiomeric form of the soy isoflavone metabolite produced by human intestinal bacterial flora. *The American Journal of Clinical Nutrition*, 81: 1072-1079. https://doi.org/10.1093/ajcn/81.5.1072

Setchell, K.D.R. and Clerici, C. (2010). Equol: History, chemistry and formation. *The Journal of Nutrition*, 3: 1355-1362. https://doi.org/10.3945/jn.109.119776.1355S

Shi, Y., Kellingray, L., Le Gall, G., Zhao, J., Zhang, H., Narbad, A., Zhai, Q. and Chen, W. (2018). The divergent restoration effects of *Lactobacillus* strains in antibiotic-induced dysbiosis. *Journal of Functional Foods*, 51: 142-152. https://doi.org/10.1016/J.JFF.2018.10.011

Shin, N.R., J.S., Moon, S.Y., Shin, L., Li, Y.B., Lee, T.J.K. and Han, N.S. (2016). Isolation and characterisation of human intestinal enterococcus avium efel009 converting rutin to quercetin. *Letters in Applied Microbiology*, 62: 68-74. https://doi.org/10.1111/lam.12512

Shoaib, M., Shehzad, A., Omar, M., Rakha, A., Raza, H., Sharif, H.R., Shakeel, A., Ansari, A. and Niazi, S. (2016). Inulin: Properties, health benefits and food applications. *Carbohydrate Polymers*, 147: 444-454. https://doi.org/10.1016/j.carbpol.2016.04.020

Sies, H., Hollman, P.C.H., Grune, T., Stahl, W., Biesalski, H.K. and Williamson, G. (2012). Protection by flavanol-rich foods against vascular dysfunction and oxidative damage: 27th Hohenheim Consensus Conference. *Advances in Nutrition*, 3: 217-221. https://doi.org/10.3945/an.111.001578

Silva, L.H., da, R., Celeghini, M.S. and Yoon, K.C. (2011). Effect of the fermentation of whole soybean flour on the conversion of isoflavones from glycosides to aglycones. *Food Chemistry*, 128: 640-644. https://doi.org/10.1016/J.FOODCHEM.2011.03.079.

Silva, R.R., Conceição, P.J.P., Menezes, C.L.A., Nascimento, C.E.O., Bertelli, M.M., Pessoa, A., Souza, G.M., Silva, R. and Eleni, G. (2019). Biochemical characteristics and potential application of a novel ethanol and glucose-tolerant β-glucosidase secreted by *Pichia guilliermondii* G1.2. *Journal of Biotechnology*, 294: 73-80. https://doi.org/10.1016/j.jbiotec.2019.02.001.

Silva, S.F. and Alcorn, J. (2019). Flaxseed lignans as important dietary polyphenols for cancer prevention and treatment: Chemistry, pharmacokinetics, and molecular targets. *Pharmaceuticals*, 12: 68. https://doi.org/10.3390/ph12020068

Simmons, A.L., Chitchumroonchokchai, C., Vodovotz, Y. and Failla, M.L. (2012). Isoflavone retention during processing, bioaccessibility, and transport by caco-2 cells: Effects of source and amount of fat in a soy soft pretzel. *Journal of Agricultural and Food Chemistry*, 60: 12196-12203. https://doi.org/10.1021/jf3037209

Song, Y.-R., Song, N.-E., Kim, J.-H., Nho, Y.-C. and Baik, S.H. (2011). Exopolysaccharide produced by *Bacillus licheniformis* strains isolated from *kimchi*. *The Journal of General and Applied Microbiology*, 57: 169-175.

Soobrattee, M.A., Bahorun, T. and Aruoma, O.I. (2006). Chemopreventive actions of polyphenolic compounds in cancer. *BioFactors*, 27: 19-35. https://doi.org/10.1002/biof.5520270103

Souza, L.A.D., Tavares, W.M.G., Lopes, A.P.M., Soeiro, M.M. and Wagner, A.B.D. (2017). Structural analysis of flavonoids in solution through DFT 1H NMR chemical shift calculations: Epigallocatechin, kaempferol and quercetin. *Chemical Physics Letters*, 676: 46-52. https://doi.org/10.1016/j.cplett.2017.03.038.

Soverini, M., Turroni, S., Biagi, E., Quercia, S., Brigidi, P., Candela, M. and Rampelli, S. (2017). Variation of carbohydrate-active enzyme patterns in the gut microbiota of Italian healthy subjects and Type 2 diabetes patients. *Frontiers in Microbiology*, 8: 2079. https://doi.org/10.3389/fmicb.2017.02079

Stanton, C., Ross, R.P., Fitzgerald, G.F. and Sinderen, D.V. (2005). Fermented functional foods based on probiotics and their biogenic metabolites. *Current Opinion in Biotechnology*, 16: 198-203. https://doi.org/10.1016/j.copbio.2005.02.008

Stevens, J.F. and Maier, C.S. (2016). The chemistry of gut microbial metabolism of polyphenols. *Phytochemistry Reviews: Proceedings of the Phytochemical Society of Europe*, 15: 425. https://doi.org/10.1007/s11101-016-9459-z

Stojanov, S. and Kreft, S. (2020). Gut microbiota and the metabolism of phytoestrogens.

Revista Brasileira de Farmacognosia, 30: 145-154. https://doi.org/10.1007/s43450-020-00049-x

Sun, H., Chen, Y., Cheng, M., Zhang, X., Zheng, X. and Zhang, Z. (2018). The modulatory effect of polyphenols from green tea, oolong tea and black tea on human intestinal microbiota *in vitro*. *Journal of Food Science and Technology*, 55: 399-407. https://doi.org/10.1007/s13197-017-2951-7

Sur, S., Pal, D., Roy, R., Barua, A., Roy, A., Saha, P. and Panda, C.K. (2016). Tea polyphenols EGCG and TF restrict tongue and liver carcinogenesis simultaneously induced by N-nitrosodiethylamine in mice. *Toxicology and Applied Pharmacology*, 300: 34-46. https://doi.org/10.1016/j.taap.2016.03.016

Swain, S.S. and Padhy, R.N. (2016). Isolation of ESBL-producing gram-negative bacteria and in silico inhibition of ESBLs by flavonoids. *Journal of Taibah University Medical Sciences*, 11: 217-229. https://doi.org/10.1016/j.jtumed.2016.03.007

Szeja, W., Grynkiewicz, G. and Rusin, A. (2017). Isoflavones, their glycosides and glycoconjugates, synthesis and biological activity. *Current Organic Chemistry*, 21: 218-235. https://doi.org/10.2174/1385272820666160928120

Takagaki, A. and Nanjo, F. (2015). Biotransformation of (-)-epigallocatechin and (-)-gallocatechin by intestinal bacteria involved in isoflavone metabolism. *Biological and Pharmaceutical Bulletin*, 38: 325-330. https://doi.org/10.1248/bpb.b14-00646

Tamang, J.P., Shin, D.-H., Jung, S.-J. and Chae, S.-W. (2016). Functional properties of microorganisms in fermented foods. *Frontiers in Microbiology*, 7: 578. https://doi.org/10.3389/fmicb.2016.00578

Tan, J., McKenzie, C., Potamitis, M., Thorburn, A.N., Mackay, C.R. and Macia, L. (2014). The role of short-chain fatty acids in health and disease. *Advances in Immunology*, 121: 91-119. https://doi.org/10.1016/B978-0-12-800100-4.00003-9

Teng, H. and Chen, L. (2019). Polyphenols and bioavailability: An update. *Critical Reviews in Food Science and Nutrition*, 59: 2040-2051. https://doi.org/10.1080/10408398.2018.1437023

Thilakarathna, S.H. and Rupasinghe, V.H.P. (2013). Flavonoid bioavailability and attempts for bioavailability enhancement. *Nutrients*, 5: 367. https://doi.org/10.3390/nu5093367

Tsamo, A.T., Mohammed, M., Ndibewu, P.P. and Dakora, F.D. (2019). Identification and quantification of anthocyanins in seeds of Kersting's groundnut [*Macrotyloma geocarpum* (Harms) Marechal & Baudet] landraces of varying seed coat pigmentation. *Journal of Food Measurement and Characterisation*, 13: 2310-2317. https://doi.org/10.1007/s11694-019-00150-3

Tsuji, H., Moriyama, K., Nomoto, K. and Akaza, H. (2012). Identification of an enzyme system for daidzein-to-equol conversion in slackia sp. strain NATTS. *Applied and Environmental Microbiology*, 78: 1228-1236. https://doi.org/10.1128/AEM.06779-11

Turnbaugh, P.J., Hamady, M., Yatsunenko, T., Cantarel, B.L., Duncan, A., Ley, R.E., Sogin, M.L., Jones, W.J., Roe, B.A., Affourtit, J.P., Egholm, M., Henrissat, B., Heath, A.C., Knight, R. and Gordon, J.I. (2009). A core gut microbiome in obese and lean twins. *Nature*, 457: 480-484. https://doi.org/10.1038/nature07540

Vázquez, L., Guadamuro, L., Giganto, F., Mayo, B. and Flórez, A.B. (2017). Development and use of a real-time quantitative PCR method for detecting and quantifying equol-producing bacteria in human faecal samples and slurry cultures. *Frontiers in Microbiology*, 8: 1155. https://doi.org/10.3389/fmicb.2017.01155

Vissiennon, C., Nieber, K., Kelber, O. and Butterweck, V. (2012). Route of administration determines the anxiolytic activity of the flavonols kaempferol, quercetin and myricetin – Are they prodrugs? *Journal of Nutritional Biochemistry*, 23: 733-740. https://doi.org/10.1016/j.jnutbio.2011.03.017

Vollmer, M., Schröter, D., Esders, S., Neugart, S., Farquharson, F.M., Duncan, S.H., Schreiner, M., Louis, P., Maul, R. and Rohn, S. (2017). Chlorogenic acid versus amaranth's caffeoylisocitric acid – Gut microbial degradation of caffeic acid derivatives. *Food Research International*. https://doi.org/10.1016/j.foodres.2017.06.013

Wang, X., Kim, H., Kang, S., Kim, S. and Hur, H. (2007). Production of phytoestrogen s-equol from daidzein in mixed culture of two anaerobic bacteria. *Archives of Microbiology*, 187: 155-160. https://doi.org/10.1007/s00203-006-0183-8

Wang, X., Ye, T., Chen, W.J., Lv, Y., Hao, Z., Chen, J., Zhao, J.Y., Wang, H.P. and Cai, Y.K. (2017). Structural shift of gut microbiota during chemopreventive effects of epigallocatechin gallate on colorectal carcinogenesis in mice. *World Journal of Gastroenterology*, 23: 8128-8139. https://doi.org/10.3748/wjg.v23.i46.8128

Wei, B., Wang, Y.K., Qiu, W.H., Wang, S.J., Wu, Y.H., Xu, X.W. and Wang, H. (2020). Discovery and mechanism of intestinal bacteria in enzymatic cleavage of C–C glycosidic bonds. *Applied Microbiology and Biotechnology*. https://doi.org/10.1007/s00253-019-10333-z

Welman, A.D. (2015). Exopolysaccharides from fermented dairy products and health promotion. pp. 23-38. *In*: W. Holzapfel (Ed.). *Advances in Fermented Foods and Beverages*. Elsevier Ltd. https://doi.org/10.1016/B978-1-78242-015-6.00002-5

Williamson, G., Kay, C.D. and Crozier, A. (2018). The bioavailability, transport and bioactivity of dietary flavonoids: A review from a historical perspective. *Comprehensive Reviews in Food Science and Food Safety*, 17: 1054-1112. https://doi.org/10.1111/1541-4337.12351

Woting, A., Clavel, T., Loh, G. and Blaut, M. (2010). Bacterial transformation of dietary lignans in gnotobiotic rats. *FEMS Microbiology Ecology*, 72: 507-514. https://doi.org/10.1111/j.1574-6941.2010.00863.x

Xie, C., Zeng, H., Li, J. and Qin, L. (2019). Comprehensive explorations of nutritional, functional and potential tasty components of various types of *sufu*, a Chinese fermented soybean appetiser. *Food Science and Technology*, 39: 105-114. https://doi.org/10.1590/fst.37917

Yang, J., Lee, H., Sung, J., Kim, Y., Jeong, H.S. and Lee, J. (2019). Conversion of rutin to quercetin by acid treatment in relation to biological activities. *Preventive Nutrition and Food Science*, 24: 313-320. https://doi.org/10.3746/pnf.2019.24.3.313

Yi, L., Jin, X., Chen, C.Y., Fu, Y.J., Zhang, T., Chang, H., Zhou, Y., Zhu, J.D., Zhang, Q.Y. and Mi, M.T. (2011). Chemical structures of 4-oxo-flavonoids in relation to inhibition of oxidised low-density lipoprotein (LDL)-induced vascular endothelial dysfunction. *International Journal of Molecular Sciences*, 12: 5471-5489. https://doi.org/10.3390/ijms12095471

Yoder, S.C., Lancaster, S.M., Hullar, M.A.J. and Lampe, J.W. (2015). Gut microbial metabolism of plant lignans: Influence on human health. pp. 103-117. *In*: K. Tuohy, D.D. Rio (Eds.). *Diet-Microbe Interactions in the Gut*. Elsevier. https://doi.org/10.1016/B978-0-12-407825-3.00007-1

Yoshiara, L.Y., Madeira, T.B., De Camargo, A.C., Shahidi, F. and Ida, E.I. (2018). Multistep optimisation of β-glucosidase extraction from germinated soybeans (*Glycine max* L. Merril) and recovery of isoflavone aglycones., *Foods*, 7: 110. https://doi.org/10.3390/foods7070110

You, B., Moon, H., Han, Y. and Park, W. (2010). Gallic acid inhibits the growth of HeLa cervical cancer cells via apoptosis and/or necrosis. *Food and Chemical Toxicology*, 48: 1334-1340. https://doi.org/10.1016/j.fct.2010.02.034

Zeng, M., Sun, R., Basu, S., Ma, Y., Ge, S., Yin, T., Gao, S., Zhang, J. and Hu, M. (2016). Disposition of flavonoids via recycling: Direct biliary excretion of enterically or extrahepatically derived flavonoid glucuronides. *Molecular Nutrition and Food Research*, 60: 1006-1019. https://doi.org/10.1002/mnfr.201500692

Zhang, H. and Yu, H. (2019). Enhanced biotransformation of soybean isoflavone from glycosides to aglycones using solid-state fermentation of soybean with effective microorganisms (EM) strains. *Journal of Food Biochemistry*, 43: e12804. https://doi.org/10.1111/jfbc.12804

Zhang, J., Zhao, X., Jiang, Y., Zhao, W., Guo, T., Cao, Y., Teng, J. and Hao, X. (2017). Antioxidant status and gut microbiota change in an aging mouse model as influenced by exopolysaccharide produced by *Lactobacillus plantarum* YW11 isolated from Tibetan *kefir*. *Journal of Dairy Science*, 100: 6025-6041. https://doi.org/10.3168/jds.2016-12480

Zhang, Z., Peng, X., Li, S., Zhang, N., Wang, Y. and Wei, H. (2014). Isolation and identification of quercetin degrading bacteria from human fecal microbes. *PLoS ONE*, 9: e90531. https://doi.org/10.1371/journal.pone.0090531

Zhang, Z., Shi, L., Pang, W., Liu, W., Li, J., Wang, H. and Shi, G. (2016). Dietary fibre intake regulates intestinal microflora and inhibits ovalbumin-induced allergic airway inflammation in a mouse model. *PLoS ONE*, 11: e0147778. https://doi.org/10.1371/journal.pone.0147778

Zhong, X., Ge, J., Chen, S., Xiong, Y., Ma, S. and Chen, Q. (2018). Association between dietary isoflavones in soy and legumes and endometrial cancer: A systematic review and meta-analysis. *Journal of the Academy of Nutrition and Dietetics*, 118: 637-651. https://doi.org/10.1016/j.jand.2016.09.036

Zimmerman, M.A., Singh, N., Martin, P.M., Thangaraju, M., Ganapathy, V., Waller, J.L., Shi, H., Robertson, K.D., Munn, D.H. and Liu, K. (2012). Butyrate suppresses colonic inflammation through HDAC1-dependent Fas upregulation and Fas-mediated apoptosis of T cells. *American Journal of Physiology-Gastrointestinal and Liver Physiology*, 302: 1405-1415. https://doi.org/10.1152/ajpgi.00543.2011

Index